AI and Ecological Change for Sustainable Development

Jabulani Garwi
University of the Free State, South Africa

Reason Masengu
Middle East College, Oman

IGI Global
Scientific Publishing
Publishing Tomorrow's Research Today

Vice President of Editorial	Melissa Wagner
Director of Acquisitions	Mikaela Felty
Director of Book Development	Jocelynn Hessler
Production Manager	Mike Brehm
Cover Design	Jose Rosado

Published in the United States of America by
IGI Global Scientific Publishing
701 East Chocolate Avenue
Hershey, PA, 17033, USA
Tel: 717-533-8845
Fax: 717-533-7115
Website: https://www.igi-global.com E-mail: cust@igi-global.com

Library of Congress Cataloging-in-Publication Data

Names: Garwi, Jabulani, 1984- editor. | Masengu, Reason, 1982- editor.
Title: AI and ecological change for sustainable development / edited by Jabulani Garwi, Reason Masengu.
Description: Hershey, PA : IGI Global Scientific Publishing, [2025] | Includes bibliographical references and index. | Summary: "This book seeks to advance the field of sustainable development by critically examining the transformative potential of AI in addressing ecological challenges in the Global South. It explores how AI technologies, such as machine learning, predictive analytics, and data-driven decision-making, can not only mitigate ecological degradation but also foster resilient socio-economic systems by promoting sustainable practices"-- Provided by publisher.
Identifiers: LCCN 2024060591 (print) | LCCN 2024060592 (ebook) | ISBN 9798337306803 (hardcover) | ISBN 9798337306810 (paperback) | ISBN 9798337306827 (ebook)
Subjects: LCSH: Sustainable development--Technological innovations. | Artificial intelligence.
Classification: LCC HC79.E5 A3683 2025 (print) | LCC HC79.E5 (ebook) | DDC 338.9/27028563--dc23/eng/20250217
LC record available at https://lccn.loc.gov/2024060591
LC ebook record available at https://lccn.loc.gov/2024060592

British Cataloguing in Publication Data
A Cataloguing in Publication record for this book is available from the British Library.

Table of Contents

Detailed Table of Contents

Ecological contexts provide a critical framework for understanding the intricate relationships between living organisms and their environments. This encompasses a variety of foundational principles, from biodiversity and ecosystems to the dynamics of biogeochemical cycles. Understanding these contexts is essential to address pressing environmental challenges such as climate change, habitat loss, and pollution. By applying ecological concepts, researchers can analyze how species interactions, energy flow, and nutrient cycling contribute to ecosystem health and resilience. Moreover, ecological contexts offer valuable insights into sustainable practices and conservation strategies. Integrated approaches combining ecological knowledge with technological advancements can mitigate the effects of anthropogenic pressures. Therefore, an understanding of ecological contexts is vital for addressing contemporary environmental issues, fostering a balanced relationship between human activities and natural systems.

Biodiversity is under threat due to numerous factors, including habitat destruction, climate change, poaching, and invasive species. As human activities continue to put immense pressure on ecosystems, endangered species, and their habitats become more critical than ever. In this context, Artificial Intelligence (AI) has emerged as a transformative tool for biodiversity conservation. By combining advanced predictive analytics, machine learning, and real-time habitat monitoring, AI can provide unprecedented insights into ecosystem health and species survival prospects. This chapter delves into how AI can revolutionize biodiversity conservation, focusing on two major areas: predictive analytics for endangered species management and

AI-driven habitat monitoring. We explore current AI applications, case studies, and future directions in these fields, offering a comprehensive look at how AI can enhance conservation efforts and guide sustainable decision-making.

Artificial Intelligence (AI) is transforming environmental monitoring by providing advanced tools to address ecological challenges. This chapter examines Ai applications in monitoring deforestation and pollution, bridging gaps in traditional methods that lack real-time data, precision and scalability. The study explores how Ai technologies enhance environmental assessment and decision-making, particularly in identifying and managing deforestation and pollution. Using a case study approach, the chapter reviews applications of machine learning, remote sensing, and predictive analytics. Methodologies include analyzing satellite imagery to track forest loss and using Ai models to monitor air and water pollution. Key findings reveal that Ai significantly improves the speed and accuracy of detecting environmental changes, delivering actionable insights for policymakers and conservationists. The findings emphasize Ai's potential to close gaps in environmental monitoring and support sustainable resource management, highlighting challenges such as data access and technological disparities.

The chapter of the book delves into the revolutionary impact of artificial intelligence (AI) on water quality evaluation. This chapter offers a fundamental comprehension of water quality criteria, conventional techniques, and the possibilities of artificial intelligence. The primary components encompassed within this framework consist

time monitoring of soil, weather patterns, and pest control. These technologies address critical challenges like food security and environmental conservation. However, the adoption of AI in agriculture faces obstacles, including high implementation costs, data privacy, and limited access to technology. Future directions involve advancing AI capabilities to support climate-resilient farming, promoting digital inclusion, and developing policy frameworks to address socioeconomic concerns. By leveraging AI, agriculture can transition toward sustainable practices, ensuring food security while minimizing environmental impact. In light of the above perspective, this chapter highlights AI's potential to transform agriculture amidst ongoing challenges.

The present chapter delves into the revolutionary role of AI technologies in promoting environmentally responsible and economically viable farming practices. The chapter highlights case studies where AI-driven tools, like, drones, sensors, etc., have successfully enabled real-time monitoring of crop health, pest control, and yield prediction. Socioeconomic implications of AI adoption, emphasizing AI potential to bridge the gap between smallholder farmers and large-scale agricultural enterprises are also discussed. This chapter anticipates future trends, including AI's role in developing resilient crops, mitigating climate risks, and facilitating circular agricultural systems. The ethical and environmental considerations of AI deployment in agriculture are critically analysed to ensure sustainable and equitable outcomes. This comprehensive analysis provides a roadmap for harnessing AI to achieve the United Nations Sustainable Development Goals (SDGs) related to agriculture, food security, and environmental conservation.

Disaster Management (DM) is preparing for, responding to, and recovering from disasters to reduce the impact on people and property. Landslides are a significant hydrogeological hazard affecting large parts of India. Remote Sensing provides crucial and timely data about natural disasters through spatial and multi-temporal remote sensing inputs. Geographical Information System plays a crucial role in disaster management by providing support for collecting and processing remote sensing data. Advances in AI models enable accurate landslide identification by processing the RS satellite data using GIS. The successful integration of AI, RS, and GIS in landslide detection has paved the way for a standard change in disaster risk reduction and response. This chapter elucidates the significance of integrating RS, GIS, and Artificial Intelligence in landslide disaster prediction. This chapter explains the characteristics and importance of RS data, processes involved in identifying the landslide causative factors, and AI-based object detection principles useful for target landslide prediction.

AI is emerging as a potentially revolutionary tool in addressing climate-related challenges, optimizing resource management, and facilitating sustainable development. This necessitates the creation of innovative methods to enhance resilience and adaptability. This narrative study explores the application of AI in climate resilience strategies, focusing on predictive analytics, disaster risk management, carbon footprint reduction, and sustainable urban planning. It synthesizes findings from existing literature to showcase AI-driven innovations, including machine learning models for climate forecasting, AI-powered decision support systems for policymakers, and automated monitoring of environmental indicators. Additionally, it examines the ethical considerations, challenges, and opportunities associated with using AI for climate resilience. The study underscores the potential of AI to accelerate the transition to sustainability but emphasizes the need for interdisciplinary collaboration, policy integration, and inclusive governance to fully realize its benefits.

Climate change has become an emerging global issue in recent decades. As a result, the 'Climate Neutrality' concept has become a key framework for countries for balancing the carbon emissions generated. The chapter examines how AI technologies can enhance resource efficiency, optimize various human practices, and improve energy management, thereby contributing to climate neutrality. Case studies from developing countries illustrate successful implementations of AI-driven solutions

This chapter discusses rule-based control for energy management optimized with artificial intelligence in AC/DC hybrid systems. A case study at an educational institution is considered with the photovoltaic-battery energy storage system in order to show the significance of optimized demand response method in reducing energy consumption costs. The energy cost is minimized while considering state of charge of battery thresholds and other constraints of battery. The study demonstrates particle swarm optimization's efficiency over genetic algorithm with respect to run time. It is observed that autumn season achieves 39.44% cost savings by prioritizing PV energy and grid injection, while winter achieves 76.47% savings by aligning discharge with peak demand.

The use of artificial intelligence (AI) is becoming more powerful for transformative community engagement and raising environmental awareness, particularly in the Global South. This chapter examines how Ai-driven technologies can promote sustainable practices in resource-limited environments. To enable local communities make knowledgeable decisions about waste management, energy conservation and sustainable agriculture, the study makes use of case studies and illustrations of Ai applications, including chatbots, mobile apps and predictive analytics. Findings demonstrate that Ai systems effectively disseminate environmental knowledge, support participatory decision-making and drive behavioral change towards sustainability. Ai-enabled platforms improved access to localized environmental

Precision agriculture has emerged as a vital approach to enhance productivity, reduce waste, and promote environmental sustainability. Artificial intelligence (AI) is enhancing precision farming with innovative technology. Artificial intelligence facilitates precise and efficient agricultural practices by integrating sensors, drones, and IoT systems for data acquisition with machine learning and AI technologies for predictive analytics, resource optimization, and pest detection. Applications encompass sustainable land management, enhanced crop monitoring, and effective resource utilization, all of which contribute to reducing environmental impact and increasing productivity. Significant concerns include data security, accessibility for small-scale farms, and the necessity of human expertise to complement artificial intelligence solutions. AI revolutionizes agriculture through technologies such as robotic weed management and climate-resilient farming systems offers promising avenues for sustainable food production and global food security.

Artificial Intelligence (AI) has emerged as a disruptive influence in water management, providing instruments to optimize water utilization, improving conservation initiatives, and foster sustainability. Applications for AI in tackling water-related issues. In water management, AI enhances early leak detection, predictive maintenance, and demand forecasting, hence promoting efficient water distribution, cost reduction, and infrastructure robustness. AI-driven smart irrigation and precision agriculture technologies enhance water efficiency, augment crop yields, and minimize waste in sustainable agriculture. In water quality monitoring, AI facilitates real-time pollutant detection and enhances treatment procedures, thereby protecting public health and reducing chemical usage. Moreover, AI augments public awareness by delivering tailored water conservation techniques and educational resources, facilitating behavioral modification and encouraging sustainable practices. Future breakthroughs in machine learning and data analytics possess significant promise to transform water conservation.

Preface

INTRODUCTION

As the world faces mounting environmental crises, including climate change, biodiversity loss, and resource depletion, artificial intelligence (AI) has emerged as a powerful force in advancing sustainable solutions. Once confined to theoretical discussions, AI now plays a pivotal role in ecological management and sustainability efforts. The rapid intersection of technological innovation and environmental urgency highlights the need to examine AI's transformative potential in ecological conservation and resource governance. As environmental disruptions intensify, integrating AI-driven solutions provides a strategic means of strengthening resilience and fostering a more sustainable future.

This book, *AI and Ecological Change for Sustainable Development*, offers an extensive exploration of AI's contributions to ecological sustainability while addressing the complexities of its application. By embedding AI within environmental science, conservation, and resource management, new possibilities emerge for mitigating climate change, enhancing biodiversity protection, and refining sustainable practices. While AI's impact across various industries has been widely studied, its role in environmental sustainability remains a developing field with notable knowledge gaps. Existing research often focuses on either AI's technological capabilities or broader sustainability challenges, with few comprehensive analyses of their intersection. This volume seeks to bridge that gap by providing an interdisciplinary perspective that integrates technological progress with ecological stewardship.

A central focus of this book is AI's transformative role in agriculture, where cutting-edge technologies are reshaping farming practices to enhance efficiency and sustainability. With food security increasingly at risk due to climate change and a rapidly expanding global population, AI-powered precision agriculture presents a viable solution for optimizing yields while minimizing environmental impact.

Additionally, the book examines AI's vital contributions to disaster management, climate risk mitigation, and environmental health, illustrating its broad applicability. By critically assessing both the opportunities and challenges AI presents, this volume offers a balanced discussion on its role in strengthening ecological resilience.

The book is systematically structured into thematic sections that explore AI's contributions to biodiversity conservation, environmental monitoring, agriculture, disaster response, and water resource management. Each chapter presents cutting-edge research, practical case studies, and real-world applications that illustrate AI's potential in ecological sustainability. Through predictive analytics, remote sensing, machine learning, and Internet of Things (IoT) technologies, AI is revolutionizing how environmental challenges are tackled. By leveraging AI-driven insights, policymakers, researchers, and industry leaders can refine decision-making, develop adaptive strategies, and implement proactive measures to safeguard natural ecosystems.

Chapter 1 lays the groundwork by outlining core ecological principles and their relevance to contemporary environmental challenges. It introduces key concepts such as biodiversity, ecosystem dynamics, and natural resource cycles, emphasizing their critical role in maintaining ecological balance. The discussion highlights how scientific frameworks can be applied to address global concerns such as climate change, habitat degradation, and pollution. By integrating ecological knowledge with sustainable methodologies, this chapter underscores the necessity of interdisciplinary approaches to conservation and resource management. Given the complexity of these environmental challenges, technological advancements, particularly in artificial intelligence, are increasingly shaping modern sustainability strategies.

Building on this foundation, Chapter 2 explores AI's profound influence on ecological sustainability, particularly its ability to drive economic development while reinforcing environmental protection. It examines AI's role in optimizing resource allocation, improving energy efficiency, and advancing precision agriculture. Real-world applications demonstrate how AI harmonizes technological progress with ecological responsibility, reinforcing the premise that sustainability and innovation can coexist. As AI's role in environmental stewardship expands, its applications go beyond resource optimization to active conservation efforts.

Subsequent chapters explore into AI-driven strategies for addressing urgent environmental threats. Chapter 3 focuses on biodiversity risks, illustrating how AI technologies enable early threat detection and species conservation. By equipping conservationists with critical data for targeted interventions, AI strengthens conservation strategies, ensuring proactive rather than reactive responses to biodiversity loss. Expanding on this, Chapter 4 examines AI's role in broader environmental monitoring, particularly in tracking deforestation, pollution, and habitat loss. These capabilities allow for timely and effective interventions, demonstrating AI's potential as an indispensable tool in ecological preservation.

Later chapters explore essential themes such as water resource management, disaster preparedness, and the transition to renewable energy. AI's capacity to optimize resource distribution, enhance energy production, and promote urban sustainability is examined in depth. Throughout these discussions, ethical considerations, including data privacy, accessibility, and algorithmic fairness, are interwoven to advocate for the responsible deployment of AI in environmental contexts. Addressing these concerns is essential to ensuring that technological advancements align with principles of inclusivity, transparency, and long-term sustainability.

CHAPTER OVERVIEWS

Chapter 1: Ecological Contexts to Environmental Challenges: An Overview of Foundations and Applications

This chapter lays the groundwork for understanding ecological systems and their relevance to sustainability. It explores core concepts like biodiversity, ecosystem services, and nutrient cycling. The authors argue for integrating ecological knowledge into global environmental strategies. Emphasis is placed on applying ecological literacy to AI-driven solutions.

Chapter 2: AI for Biodiversity Conservation: Harnessing Predictive Analytics and Habitat Monitoring to Protect Endangered Species and Ecosystems

AI is shown to enhance biodiversity conservation through predictive analytics and habitat monitoring. The chapter highlights current applications for tracking endangered species and evaluating ecosystem health. It discusses the integration of big data and machine learning to inform conservation planning. Challenges and future directions are also explored.

Chapter 3: AI Applications in Environmental Monitoring: Case Studies on Deforestation and Pollution Assessment

This chapter presents case studies demonstrating AI's effectiveness in monitoring deforestation and pollution. Techniques like satellite imaging and deep learning are used for environmental assessments. The authors explain how these tools offer timely and accurate insights for sustainable policymaking. Limitations in data and system implementation are also addressed.

Chapter 4: AI-Driven Solutions for Water Quality Assessment

AI is explored as a tool for monitoring water quality using sensor networks and machine learning models. The chapter discusses predictive capabilities, data visualization, and real-time decision-making. It highlights successful applications in urban and rural settings. Ethical and infrastructural challenges are examined.

Chapter 5: AI's Role in Environmental Health and Tackling Ecological Impacts through Multidimensional Views

This chapter addresses how AI supports environmental health by analyzing complex ecological and human health data. It emphasizes early detection of pollution risks and targeted interventions. Multidimensional approaches for data interpretation are introduced. The need for ethical AI frameworks in health-environment applications is also discussed.

Chapter 6: The Transformative Potential of Artificial Intelligence in Building Sustainable Agricultural Systems

AI is positioned as a critical tool for sustainable agriculture through precision farming and resource optimization. The chapter covers yield prediction, pest control, and soil monitoring. It advocates for equitable technology access in rural communities. Future directions include AI's role in food security and climate adaptation.

Chapter 7: Empowering Sustainable Agriculture: Harnessing AI for Enhanced Yields, Resource Efficiency, and Eco-Friendly Farming Practices

Building on Chapter 6, this chapter details how AI optimizes crop health, irrigation, and environmental impact. It includes global case studies demonstrating productivity and ecological benefits. Socioeconomic impacts and digital literacy challenges are discussed. The chapter concludes with recommendations for AI policy in agriculture.

Chapter 8: Artificial Intelligence-Based Landslide Prediction Using Remote Sensing Data

Focusing on geohazards, this chapter outlines how AI enhances landslide prediction through remote sensing and GIS. It details how satellite data is processed to identify risk zones. A case study validates AI model accuracy. The approach is framed as vital for proactive disaster risk management.

Chapter 9: Enhancing Climate Resilience and Sustainable Development through the Integration of Artificial Intelligence

This chapter explores AI's contribution to climate resilience via modeling, early warning systems, and adaptive planning. It synthesizes tools used across energy, agriculture, and disaster response. Interdisciplinary collaboration and ethical deployment are emphasized. The authors advocate for inclusive AI-driven sustainable development frameworks.

Chapter 10: Climate Neutrality-Based Sustainable Practices and AI for Developing Countries

AI's potential to support climate-neutral initiatives in developing countries is explored. The chapter covers localized applications in waste management, transportation, and clean energy. Barriers to adoption—such as funding and infrastructure—are analyzed. The authors highlight AI as a catalyst for equitable climate action.

Chapter 11: Harnessing Solar Energy for a Sustainable Society 5.0: Legal and Regulatory Challenges in India

This chapter evaluates India's solar energy sector in the context of Society 5.0. It critiques regulatory bottlenecks and proposes policy reforms for wider AI and IoT adoption. Legal frameworks are examined for their impact on sustainability goals. Future prospects for solar innovation are outlined.

Chapter 12: Rule-Based Energy Management Optimized with Artificial Intelligence and a Case Study Analysis

AI-enhanced rule-based systems are discussed as tools for managing hybrid energy networks. A case study illustrates how smart systems can reduce energy waste and improve sustainability. Emphasis is placed on adaptive control strategies and grid optimization. The chapter addresses real-world deployment challenges.

Chapter 13: AI-Enabled Urban Planning for Smart Cities

This chapter discusses how AI can streamline urban planning by analyzing traffic, pollution, and infrastructure data. Smart city frameworks are introduced with a focus on sustainability. Use cases highlight improved public services and environmental outcomes. The role of governance in ethical AI use is emphasized.

Chapter 14: Artificial Intelligence in Waste Management: Turning Environmental Burden into Economic Opportunity

AI's role in optimizing waste collection, segregation, and recycling is explored. The chapter details how machine learning models can enhance efficiency and reduce landfill use. Economic implications and environmental benefits are discussed. It also touches on public engagement and smart policy implementation.

Chapter 15: Ethical Considerations in the Use of AI for Environmental Sustainability

This chapter addresses ethical dilemmas in using AI for environmental management. It explores data privacy, algorithmic bias, and decision accountability. Frameworks for ethical AI deployment are proposed. The authors call for inclusive policies that prioritize equity and transparency.

Chapter 16: Future Trends and Research Directions in AI and Ecological Change for Sustainable Development

Concluding the volume, this chapter highlights emerging trends such as explainable AI, federated learning, and AI-ecosystem modeling. It synthesizes interdisciplinary approaches needed for long-term sustainability. Research gaps and future priorities are outlined. The authors advocate for continued innovation guided by ethical and ecological principles.

The final chapter synthesizes key insights and presents a forward-looking analysis of AI's role in achieving sustainable development. It reflects on emerging challenges, such as AI's environmental footprint and regulatory complexities, while proposing policy recommendations to support responsible innovation. By integrating ethical AI frameworks with policy development, this chapter provides a strategic roadmap for future research, governance, and cross-sector collaboration in addressing environmental imperatives.

CONCLUSION

This book serves as an invaluable resource for academics, researchers, and practitioners seeking to understand AI's role in ecological sustainability. It aims to inspire meaningful collaboration between technology developers, environmental scientists, and policymakers, fostering an interdisciplinary dialogue that drives sustainable innovation. By contributing to the evolving discourse on AI and ecological resilience, this volume highlights the profound potential of technology in shaping a more sustainable world.

We extend our deepest gratitude to the contributors whose expertise and dedication have been instrumental in shaping this discourse on AI-driven ecological change. Their insights have greatly enriched this work. We also acknowledge the readers—researchers, students, and professionals—who engage with and contribute to this critical field. It is our hope that this book serves as both a guiding resource and a source of inspiration for those committed to advancing sustainability through AI-driven solutions.

Jabulani Garwi
University of the Free State, South Africa

Reason Masengu
Middle East College, Oman

Chapter 1
Ecological Contexts to Environmental Challenges:
An Overview of Foundations and Applications

Oindrilla Ghosh
https://orcid.org/0000-0001-9609-0781
Christ University, India

Manjeri Subin Sunder Raj
https://orcid.org/0000-0002-2429-4395
Christ University, India

Binod Kumar
https://orcid.org/0009-0002-9721-1367
Radha Govind University, India

ABSTRACT

Ecological contexts provide a critical framework for understanding the intricate relationships between living organisms and their environments. This encompasses a variety of foundational principles, from biodiversity and ecosystems to the dynamics of biogeochemical cycles. Understanding these contexts is essential to address pressing environmental challenges such as climate change, habitat loss, and pollution. By applying ecological concepts, researchers can analyze how species interactions, energy flow, and nutrient cycling contribute to ecosystem health and resilience. Moreover, ecological contexts offer valuable insights into sustainable practices and conservation strategies. Integrated approaches combining ecological

DOI: 10.4018/979-8-3373-0680-3.ch001

knowledge with technological advancements can mitigate the effects of anthropogenic pressures. Therefore, an understanding of ecological contexts is vital for addressing contemporary environmental issues, fostering a balanced relationship between human activities and natural systems.

INTRODUCTION

The interplay between ecological contexts and environmental challenges is a paramount area of focus in contemporary environmental sciences (Das & Hazra, 2020). As human activities increasingly affect the natural world, understanding these challenges through the lens of ecology becomes essential for developing sustainable solutions (Sahana et al., 2021). Ecological contexts refer to the specific biological communities and physical environments in which species interact, and these contexts play a crucial role in shaping the resilience and vulnerabilities of ecosystems faced with pressing environmental issues such as climate change, biodiversity loss, pollution, and habitat destruction (Kanjilal et al., 2010).

The study of ecological contexts provides insights into how different species and communities respond to environmental stressors (Ahsan & Warner, 2014). It encompasses a multifaceted approach that integrates physical, chemical, and biological factors, acknowledging that ecosystems are not static but dynamic entities that evolve over time (Danda et al., 2019). This dynamism is essential for understanding how ecosystems adapt, overlaying a temporal dimension that highlights the importance of historical changes alongside current conditions (Panda et al., 2016).

One of the most significant aspects of ecological contexts is their capacity to influence and mitigate various environmental challenges (Kanjilal et al., 2013). For instance, healthy wetlands can absorb excess nutrients, reducing the impacts of agricultural runoff; resilient forests can store carbon, helping to mitigate climate change; and diverse ecosystems are more likely to withstand and recover from disturbances (Savacool, 2018). Therefore, addressing environmental challenges requires a comprehensive understanding of ecological contexts, which serves as a foundation for conservation efforts, resource management, and policy-making (Mistri, 2019).

Moreover, the growing recognition of the interconnectedness of ecological and human systems highlights the need for interdisciplinary collaborations that unify ecology, sociology, economics, and political science (Alam & Bhardwaj, 2020). This understanding is crucial for developing holistic solutions that address not only the symptoms of environmental challenges but also their root causes (Roy, 2018). By positioning ecological contexts at the forefront of environmental discourse, it is possible to foster a more profound appreciation for nature's inherent values and promote stewardship that transcends disciplinary boundaries (Rashad Al Mahamud et al., 2020).

Historical Evolution

Understanding the historical evolution of ecological contexts in relation to environmental challenges involves tracing the development of ecological thought and how it has shaped our perception and response to environmental issues (Saif et al., 2016). The roots of ecology can be traced back to ancient civilizations, where early observations of nature laid the groundwork for an understanding of ecological relationships (Biswas, 2013). However, it wasn't until the late 19th and early 20th centuries that ecology began to emerge as a formal scientific discipline (Pramanik et al., 2021).

The introduction of concepts such as the food web and ecological succession marked significant milestones in the field (Chowdhury, 2019). Pioneer ecologists such as Ernst Haeckel, who coined the term "ecology" in 1866, laid the foundation for investigating the relationships between organisms and their environments (Adams et al., 2020). As ecologists began to understand the intricate interactions between species and their habitats, the concept of ecosystems was further developed, gaining traction throughout the 20th century (Molinari, 2017). The work of figures like Charles Elton and Arthur Tansley led to the establishment of ecology as a science focused on the interactions within and between communities, populations, and their physical environments (Satterthwaite & Mitlin, 2014).

As the late 20th century progressed, the lenses through which ecological contexts were examined expanded considerably (Satterthwaite, Huq, Reid, Pelling, & Romero Lankao, 2007). The emergence of systems ecology, which focuses on complex interactions within biological systems, allowed for a more holistic understanding of ecosystem dynamics (Soltesova, Brown, Dayal, & Dodman, 2014). Meanwhile, the rise of globalization in the 21st century emphasized the interconnectedness of ecosystems worldwide, complicating local environmental challenges with global consequences such as climate change, invasive species, and loss of biodiversity (Ayers & Forsyth, 2009).

During this time, key conferences and reports, such as the 1972 United Nations Conference on the Human Environment in Stockholm and the 1992 Earth Summit in Rio de Janeiro, highlighted the necessity of integrating ecological understanding into policy-making (Cannon & Muller-Mahn, 2010). These international platforms underscored the importance of sustainable practices and environmental governance that align with ecological realities (Twigg, 2007).

More recently, discussions about ecological contexts have evolved to center around sustainability and resilience, emphasizing how ecosystems can adapt in the face of human-induced changes (Pervin et al., 2013). Concepts such as ecosystem services and biodiversity conservation have become pivotal in addressing environmental challenges, influencing policies that advocate for sustainable development and resource management (Olhoff & Schaer, 2010).

Theoretical Frameworks

Understanding environmental challenges requires a robust theoretical framework that intricately weaves together ecological, social, economic, and political dimensions (Anguelovski & Carmin, 2011). Several theoretical frameworks provide valuable lenses through which these multi-faceted issues can be examined (Leck & Simon, 2012). Among the most prominent frameworks are the Ecosystem Services Framework, the Social-Ecological Systems (SES) Framework, and the Risk and Vulnerability Framework (Roy, Hulme, and Jahan, 2013).

The *Ecosystem Services Framework* emphasizes the benefits humans derive from ecosystems, which include provisioning (food and water), regulating (climate and disease control), cultural (recreational and spiritual), and supporting services (nutrient cycling and soil formation) (Anguelovski & Carmin, 2011). By quantifying ecosystem services, policymakers can better understand the economic value of natural resources and prioritize conservation strategies, making this framework immensely relevant in addressing environmental challenges such as deforestation, pollution, and climate change (Tyler & Moench, 2012).

Conversely, the *Social-Ecological Systems (SES) Framework* provides a more integrative approach by recognizing the interconnectedness of ecological and human systems (Pelling, 2011). It underscores that ecological phenomena cannot be fully understood without accounting for the social contexts in which they occur, including governance structures, community dynamics, and cultural beliefs (Agrawal, Kononen, and Perrin, 2009). This framework is particularly beneficial in assessing environmental challenges that require collective action, such as water scarcity, biodiversity loss, and habitat destruction (Dodman and Mitlin, 2013). By examining feedback loops and interactions between social and ecological systems, the SES Framework empowers local communities to engage in adaptive management practices, fostering resilience in the face of ecological change (Hurlimann & March, 2012).

The *Risk and Vulnerability Framework* further complements these theories by focusing on the uncertainty surrounding environmental challenges (Few, Brown, and Tompkins, 2007). This framework is particularly useful in areas prone to natural disasters, such as floods, hurricanes, and droughts (Carmin et al., 2012). It emphasizes understanding the vulnerabilities of different populations and ecosystems, taking into account socio-economic factors, geographical contexts, and adaptive capacity (Roberts et al., 2012). By integrating risk assessment and vulnerability analysis, this framework aids in developing strategies that enhance community resilience and adaptive capacity, ensuring that interventions are both context-specific and equitable (Leck & Simon, 2012).

In addition to these frameworks, the *Human-Environment Interaction model* serves as a foundational concept for understanding how human activities influence ecological contexts (Almansi et al., 2014). This model demonstrates that human actions ranging from urbanization and industrialization to conservation and restoration efforts directly impact ecological integrity (Syam, 2013). Recognizing this interaction is crucial for devising effective environmental policies and interventions (Brown, Dayal, & Rumbaitis del Rio, 2012).

Governance

Governance in ecological contexts is vital for effectively managing the increasingly complex environmental challenges facing our planet (Corfee-Morlot et al., 2011). As environmental issues extend beyond local boundaries, governance frameworks that embrace ecological contexts can facilitate collaborative and integrative approaches to resource management (Dewulf, 2013). Effective governance not only addresses ecological problems but also considers the social, economic, and political dimensions that influence environmental outcomes (Norris, Stevens, Pfefferbaum, Wyche, & Pfefferbaum, 2008).

At the core of governance in ecological contexts is the concept of adaptive management (O'Brien, 2012). This approach recognizes the dynamic nature of ecosystems and the uncertainties inherent in environmental management (Agrawal, 2010). By employing methods that are flexible and responsive to changing conditions, stakeholders can implement strategies that account for ecological variability (Folke et al., 2005). Such adaptive governance encourages continuous learning and refinement of policies, ensuring that management practices remain relevant in the face of new information and shifting ecological parameters (Smit and Wandel, 2006).

Moreover, the principle of inclusivity is paramount in governance frameworks focused on ecological contexts (Berkes and Ross, 2013). This involves engaging multiple stakeholders, including government agencies, non-governmental organizations, local communities, and Indigenous peoples (Adger et al., 2009). Collaborative efforts harness diverse knowledge systems, ensuring that various perspectives and experiences are taken into account (Berkes, 2009). This participatory approach not only builds trust among stakeholders but also leads to more effective and equitable decision-making (Mertz et al., 2009). For instance, when local communities are involved in resource management, they can impart valuable insights into local ecological conditions and traditional practices that promote sustainability (Eakin and Luers, 2006).

Furthermore, governance systems must incorporate robust legal and institutional frameworks that support ecological sustainability (Gómez-Baggethun et al., 2012). Policies that protect ecosystems and enforce conservation efforts are critical for mitigating environmental challenges (Yohe and Tol, 2002). For example, establishing protected areas or implementing regulations on resource extraction can safeguard vital habitats and promote biodiversity (Cutter et al., 2003). However, governance must also address the socio-economic factors that drive environmental degradation, such as poverty and inequality (Armitage, 2005). By integrating social justice principles into environmental governance, we can create a more equitable system that considers the rights and needs of all stakeholders (Adger et al., 2005).

Technological advancements also play a key role in modern governance for ecological contexts (Biesbroek et al., 2013). Tools such as Geographic Information Systems (GIS), remote sensing, and ecological modeling facilitate data collection and analysis, enhancing the capacity for informed decision-making (Smit and Pilifosova, 2003). These technologies enable real-time monitoring of ecosystems, allowing for timely interventions in response to ecological changes (Berkes, 2009). Moreover, they can aid in predicting the impacts of various governance strategies, promoting evidence-based approaches to environmental management (Midgley, 1995).

USE OF AI AND ML

Artificial Intelligence (AI) and Machine Learning (ML) have emerged as transformative tools in addressing ecological challenges and enhancing our understanding of environmental systems (Elliot, 1993). These technologies harness vast amounts of data generated by environmental monitoring, thus allowing for more informed decision-making (Sovacool, 2011). By analyzing complex ecological interactions and predicting environmental changes, AI and ML play crucial roles in shaping policies and strategies to mitigate the impacts of climate change, habitat destruction, and resource depletion (Moser and Ekstrom, 2010).

One notable application of AI and ML is in biodiversity monitoring (Pelling, 2003). Machine learning techniques can analyze camera trap images and sound recordings to identify species presence, abundance, and distribution (Agrawal, 2008). This automation reduces the labor-intensive nature of traditional wildlife surveys, providing rapid insights into ecological changes (Pelling & Manuel-Navarrete, 2011). Platforms like eBird use machine learning to help ornithologists track bird migrations and populations, allowing conservationists to address critical threats more efficiently (Mosher, 1979).

AI and ML also contribute significantly to climate modeling and risk assessment (Omer, 1979). These technologies can simulate climate scenarios, integrating multiple data sources to predict future conditions under various policy scenarios (Paiva, 1977). For example, AI-powered models can analyze atmospheric data, soil moisture levels, and land-use patterns to forecast droughts, floods, and other extreme weather events (Gray, 2006). This predictive capability aids governments and organizations in developing adaptive strategies to mitigate climate-related risks and allocate resources effectively (Healy, 2008).

In the realm of precision agriculture, AI and ML facilitate sustainable farming practices (Peeters, 2012). By analyzing data from sensors in the field, these technologies provide real-time insights on soil health, water requirements, and crop conditions (Erickson, 2012). Farmers can make data-driven decisions to optimize resource use, reducing waste and environmental degradation (Dominelli, 2010). Predictive analytics can also assist in pest and disease management, enabling timely interventions that minimize damage to crops while lessening the need for agrochemicals (Dubey et al., 2017).

Augmented Reality (AR)

Augmented Reality (AR) enhances the real world with digital overlays, and its applications in ecological contexts are revolutionizing environmental education and awareness (Mishra et al., 2021). By integrating AR technology into ecological studies, researchers and educators can provide immersive learning experiences that promote a better understanding of biodiversity and ecosystems (Biswas & Nautiyal, 2020). For instance, AR applications can allow users to visualize endangered species in their natural habitats, thereby fostering a sense of connection and urgency for conservation efforts (Rudra & Chattopadhyay, 2019).

One notable application of AR is in urban planning and landscape design, where it aids in visualizing the impact of development projects on existing ecosystems (Rudra & Chattopadhyay, 2018). Planners and stakeholders can utilize AR to simulate future environments, assess potential ecological effects, and promote sustainable practices (Satio et al., 2020). For example, an AR toolkit can visualize future tree coverage, water drainage, or wildlife corridors, enabling more informed decision-making that prioritizes ecological health (Sarkar & Sanyal, 2019).

In educational contexts, AR can bridge the gap between theoretical knowledge and real-world applications, particularly in subjects like ecology and environmental science (Hafiz Iqbal, 2020). By transforming complex concepts into interactive experiences, AR can deepen student engagement, promote critical thinking, and inspire the next generation of environmental stewards (Roy, 2019). Overall, AR represents a promising frontier in addressing ecological contexts by making environmental challenges more tangible and encouraging proactive participation (Mozumdar et al., 2018).

Virtual Reality (VR)

Virtual Reality (VR) offers immersive experiences that can profoundly impact ecological education and environmental advocacy (Dasgupta et al., 2021). By fully immersing users in digitally-created environments, VR can simulate ecological systems, allow exploration of remote natural world environments, and raise awareness about urgent environmental challenges (Ahsan & Warner, 2014). For instance, students and adults alike can virtually explore the Amazon rainforest, witnessing both rich biodiversity and the devastating effects of deforestation without leaving their classrooms or homes (Smit and Skinner, 2002).

VR applications can serve as powerful tools for visualizing climate change impacts, such as rising sea levels or changing weather patterns (Gómez-Baggethun et al., 2012). Users can "walk through" scenarios depicting future conditions in their own communities, which fosters a personal sense of responsibility toward mitigating environmental issues (Carmin et al., 2012). By experiencing firsthand the potential consequences of inaction, individuals may be inspired to adopt more sustainable lifestyles or engage in advocacy efforts (Almansi, Hardoy, & Pandiella, 2014).

Furthermore, VR has been employed in training conservationists and environmental scientists (DiGregorio, 2013). Simulations can replicate challenging scenarios, such as wildlife rescue missions or sustainable resource management practices, providing trainees with hands-on experience in a risk-free setting (Sen, 2019). This practical application of VR enhances skill acquisition and prepares participants with essential tools to tackle real-world ecological problems (Anwar et al., 2023).

Mixed Reality (MR)

Mixed Reality (MR) combines elements of both Augmented and Virtual Reality, allowing users to interact with digital content superimposed on the real world while remaining aware of their physical surroundings (Dey and Hazra, 2021). This hybrid environment has significant potential to address ecological contexts and environmental challenges by facilitating deeper interactions with ecological data and raising awareness about environmental issues (Jalais, 2007).

One application of MR is in environmental monitoring and conservation efforts. MR can assist researchers and conservationists in visualizing and understanding complex ecological data in real-time (Unogwu et al., 2024). For example, by wearing MR headsets, users could identify various species and their health metrics through digital overlays as they traverse their habitats (Szabo et al., 2016). This capability can enhance fieldwork efficiency, enabling more accurate data collection and intuitive insights into ecosystem health and changes over time (Rahman et al., 2021).

Moreover, MR holds promise for public engagement in conservation initiatives. Interactive experiences, such as virtual tours of threatened habitats with integrated ecological data, can effectively inform and inspire communities to take action (Dasgupta et al., 2014). By visualizing the impact of human activities on local environments, MR not only educates users about ecological stewardship but also invites them to participate in conservation efforts actively (Danda & Sriskanthan, 2011).

Advantages

Ecological contexts play a critical role in shaping our understanding and response to environmental challenges (Ghosh, 2012). By situating environmental issues within their ecological frameworks, we gain a comprehensive perspective that can drive effective policy-making, conservation efforts, and sustainable practices (Bhowmick and Poddar, 2023).

Holistic Understanding: Ecological contexts allow us to appreciate the interconnectedness of different environmental factors (StC,D. E., 2008). Recognizing that ecosystems are complex webs of interactions where flora, fauna, and abiotic elements like climate and geology interrelate facilitates a more complete understanding of how human activities disrupt these relationships (Mahadevia & Vikas, 2012). This holistic view helps identify the root causes of environmental degradation rather than merely addressing the symptoms (Afroz and Alam, 2013).

Enhanced Policy Formulation: Policymakers equipped with ecological insights can craft regulations and interventions that reflect ecological realities (Pitchaikani, 2020). Instead of implementing generic solutions, they can tailor policies that consider the specific needs of ecosystems (Doshi et al., 2024). For example, conservation measures for endangered species can be designed around the particular habitat and ecological requirements of those species, leading to more successful outcomes (Hiran et al., 2021).

Improved Resource Management: Ecological contexts lead to more sustainable resource management practices (Moore, 2023). By understanding the carrying capacity of ecosystems and their regeneration rates, societies can manage resources like water, forests, and fisheries more sustainably (Mondal, Bowers, & Hossen, 2021). This ensures that human consumption does not exceed the ecosystem's ability to replenish itself, safeguarding future resource availability (Islam, Dey, Sanyal, & Bandyopadhyay, 2022).

Promoting Biodiversity Conservation: Ecological contexts emphasize the importance of biodiversity, functioning as a foundation for a stable and resilient environment (Ray, 2020). Recognizing the multiple benefits that biodiversity provides such as ecosystem services, resilience to climate change, and aesthetic and cultural values brings attention to the need for conservation efforts that go beyond single-species protection (Begum, Lobry De Bruyn, Amirul Islam, & Kristiansen, 2022). This approach fosters the preservation of entire ecosystems, promoting a richer and more diverse natural world (Chanda & Akhand, 2023).

Community Engagement and Education: Bringing ecological contexts into the discussion of environmental issues fosters greater public understanding and engagement (Jamal, Ghosh, Hazarika, & Sen, 2022). Educational initiatives that emphasize ecological relationships can empower individuals and communities to participate in conservation efforts, promoting stewardship and sustainable practices within their own habitats (Nicholls et al., 2020).

Integration with Technology: The consideration of ecological contexts often leads to innovative solutions using technology (Thakur et al., 2021). For example, geographic information systems (GIS) can analyze ecological data to identify areas most vulnerable to environmental changes, guiding targeted interventions (Mukhopadhyay, 2009). By using modern tools and methodologies grounded in ecological understanding, stakeholders can more efficiently allocate resources and develop adaptive strategies (Biswas, 2013).

Challenges

While integrating ecological contexts into environmental challenges provides numerous advantages, various challenges can complicate this approach (Sarwar Hossain et al., 2016). Understanding these challenges is essential for developing effective strategies to address environmental problems and improve ecological health (Pramanik et al., 2021).

Complexity and Uncertainty: One of the primary challenges of focusing on ecological contexts is the inherent complexity and uncertainty of ecosystems (Biswas & Nautiyal, 2020). Ecosystems are dynamic and constantly changing due to natural processes and human influences (Adams et al., 2020). Predicting how multiple variables interact can be exceedingly difficult, leading to uncertainties in outcomes when applying ecological principles to real-world scenarios (Soltesova, Brown, Dayal, & Dodman, 2014). This unpredictability can undermine confidence in management decisions (Few et al., 2007).

Data Gaps: Effective application of ecological contexts requires comprehensive data on ecosystems, species distributions, and environmental conditions (Roberts, 2008). However, there are often significant data gaps, especially in less-studied regions or ecosystems (Roberts, 2013). Inadequate data can lead to misinformed decisions, potentially causing more harm than good. In many cases, researchers may need to rely on outdated or incomplete information, making it challenging to align policies with current ecological realities (Sivaramakrishnan, 2007).

Conflicting Interests: Decision-makers often face conflicting interests among various stakeholders, including industries, community groups, and environmental organizations (Leck & Simon, 2012). Recognizing ecological contexts might not align with all stakeholders' goals, leading to resistance or backlash against initiatives that aim to prioritize ecological integrity (TERI, 2011). Consensus-building and negotiation can be both challenging and time-consuming, often delaying necessary actions (Kernaghan & da Silva, 2013).

Implementation Challenges: Even when policies are formulated with ecological contexts in mind, implementing those policies can be fraught with obstacles (Almansi, 2010). Factors such as resource limitations, lack of political will, and bureaucratic inertia can impede progress (Almansi et al., 2014). Effective implementation requires continuous stakeholder engagement, adequate funding, and a commitment to overcoming institutional barriers (Pelling & Manuel-Navarrete, 2011).

Global and Local Discrepancies: Environmental challenges often manifest differently at local, regional, and global levels (Adger, 2003). A solution that works well in one context may be ineffective or even harmful in another due to variations in ecological conditions, social structures, and economic realities (Sahana et al., 2021). This complexity necessitates a flexible and adaptive approach, but achieving such adaptability can be difficult in practice (Honadle, 1999).

Implementation of Regulatory Frameworks

The implementation of regulatory frameworks in ecological contexts serves as a critical mechanism for addressing environmental challenges (Uddin, 2019). Given the complexity of ecological systems and the multitude of threats they face from human activities, effective regulatory frameworks are vital (Okoye et al., 2024). These frameworks are designed to enforce standards, promote sustainable practices, and safeguard natural resources (Aggarwal et al., 2024). Various stakeholders, including government agencies, non-governmental organizations (NGOs), and the private sector, play pivotal roles in implementing these regulations (Pal and Poddar, 2024).

A successful implementation of regulatory frameworks begins with comprehensive research and understanding of the ecological context (Tyagi et al., 2020). For instance, before establishing regulations, it is crucial to assess the specific environmental challenges prevalent in a region, such as habitat loss, pollution, climate change, or over-extraction of resources (Unnikrishnan and Shankar, 2007). This foundational research informs policymakers about the most pressing issues and guides the development of targeted regulations that address these challenges (Polin & Alam, 2020). Collaborative approaches involving scientists, community members, and industry representatives can further enrich this understanding and ensure that regulations are practically applicable and culturally relevant (Chanda & Akhand, 2023).

Once the regulatory framework has been established, enforcement becomes the next crucial step (Francesco & Rahman, 2020). Effective enforcement mechanisms must be in place to ensure compliance with regulations (Ghosh, 2023). This may include regular monitoring of environmental indicators, reporting requirements for industries, and penalties for non-compliance (Mondal, Bowers, & Hossen, 2021). For instance, the Clean Water Act in the United States imposes strict limits on pollutants discharged into water bodies, and regular inspections by the Environmental Protection Agency (EPA) ensure that industries adhere to these standards (Mitra, 2024). The presence of robust enforcement mechanisms can deter non-compliance and reinforce the importance of regulatory frameworks (Alam & Bhardwaj, 2020).

Additionally, regulatory frameworks must be adaptable to emerging environmental challenges (Adger, 2003). Climate change, for instance, presents evolving threats that require continuous reassessment of existing regulations (Jamal, Ghosh, Hazarika, & Sen, 2022). Policymakers must remain vigilant and willing to update regulations to better suit changing ecological contexts and scientific advancements (Begum, Lobry De Bruyn, Amirul Islam, & Kristiansen, 2022). Incorporating adaptive management practices allows for flexibility and responsiveness in regulatory frameworks, ensuring that they evolve alongside environmental conditions.

CONCLUSION

As we confront pressing issues like climate change, habitat destruction, and biodiversity loss, the integration of ecological considerations into regulatory policies becomes paramount. These frameworks facilitate the development of guidelines that reflect the complexities of ecosystems while promoting sustainable practices that benefit both the environment and society.

The foundations of regulatory frameworks must be built on solid scientific research and ecological understanding. Recognizing the interconnectedness of natural systems enables policymakers to design regulations that not only protect individual species or habitats but also sustain ecosystem services that humans depend on. Furthermore, involving multiple stakeholders from scientists to local communities enhances the credibility and effectiveness of these regulations. Collaborative efforts foster shared ownership, which is crucial for compliance and success.

Ultimately, the foundations and applications of regulatory frameworks in ecological contexts underscore a vital truth: the health of our planet is inextricably linked to human actions. Through thoughtful implementation of regulatory measures that consider the ecological landscape, society can make significant strides towards overcoming environmental challenges. The pathway to sustainability lies in our ability to integrate ecological wisdom with regulatory effectiveness, ensuring that present and future generations inherit a world that is not only livable but flourishing. Thus, as we look to the future, a commitment to these principles is essential for fostering a harmonious relationship between humanity and the environment.

FUTURE RESEARCH DIRECTIONS

As the landscape of environmental challenges continues to evolve, future research must address urgent and emerging areas of concern.

Ecological Restoration Techniques: Investigate innovative restoration techniques that can enhance ecosystem resilience and restore functionality. Research could explore the effectiveness of different methods, such as rewilding or assisted migration, especially as climates change.

Microbial Ecology and Ecosystem Health: Delve deeper into the role of microbial communities in various ecosystems. Understanding their contributions to nutrient cycling, soil health, and ecosystem stability can lead to better land management practices and biodiversity conservation efforts.

Impact of Pollution on Ecological Contexts: Investigate the differential impacts of various types of pollution (e.g., plastic, air, and water pollution) on ecosystems. Understanding how pollutants interact with local ecological contexts can inform cleaner production methods and pollution remediation strategies.

Socio-Ecological Systems Research: Advance the study of socio-ecological systems, focusing on the interdependencies between human and ecological processes. Research can explore how socio-economic changes impact ecological health and vice versa.

Sustainable Practices in Agriculture: Research sustainable agricultural practices that align with ecological contexts, examining their impacts on local biodiversity and ecosystem services. This includes studies on agroecology, permaculture, and regenerative farming techniques.

Climate Refugees and Ecological Disruption: Explore the ecological implications of human migration driven by environmental challenges, including climate change. Understanding how displaced populations interact with new ecosystems and the resulting ecological changes can provide insights for policy and management.

Role of Traditional Ecological Knowledge: Investigate the importance of Traditional Ecological Knowledge (TEK) in shaping sustainable practices and responses to environmental changes. Research can explore how indigenous knowledge systems contribute to biodiversity conservation and ecosystem management.

Predictive Models for Future Challenges: Develop predictive models to simulate future ecological contexts under various climate scenarios. Such models can inform planning and preparedness strategies for potential environmental changes.

Ultimately, as we face increasingly complex environmental challenges, understanding and addressing the ecological contexts surrounding these challenges will be crucial. Increased interdisciplinary collaboration, innovative research methodologies, and active community engagement are key to promoting more effective solutions for a sustainable future.

REFERENCES

Adams, H., Adger, W. N., Ahmad, S., Ahmed, A., Begum, D., Matthews, Z., Rahman, M. M., Nilsen, K., Gurney, G. G., & Streatfield, P. K. (2020). Multi-dimensional well-being associated with economic dependence on ecosystem services in deltaic social-ecological systems of Bangladesh. *Regional Environmental Change, 20*(42), 2–16. DOI: 10.1007/s10113-020-01620-x

Adger, W. N. (2003). *Social aspects of adaptive capacity*. Climate Change, Adaptive Capacity and Development, 29-49. DOI: 10.1142/9781860945816_0003

Adger, W. N., Dessai, S., Goulden, M., Hulme, M., Lorenzoni, I., Nelson, D. R., Naess, L. O., Wolf, J., & Wreford, A. (2009). Are there social limits to adaptation to climate change? *Climatic Change, 93*(3-4), 335–354. DOI: 10.1007/s10584-008-9520-z

Adger, W. N., Hughes, T. P., Folke, C., Carpenter, S. R., & Rockström, J. (2009). Social-ecological resilience to coastal disasters. *Science, 309*(5737), 1036–1039. DOI: 10.1126/science.1112122 PMID: 16099974

Aggarwal, K., Doshi, R., Mijwil, M. M., Hiran, K. K., Gök, M., & Bala, I. (2024). A Simple Way to Comprehend the Difference and the Significance of Artificial Intelligence in Agriculture. *Intelligent Techniques for Predictive Data Analytics*, 189-201.

Agrawal, A. (2008). *The role of local institutions in adaptation to climate change.* Social Dimensions of Climate Change: Equity and Vulnerability in a Warming World, 173-198. https://hdl.handle.net/10986/28274

Ahsan, M. N., & Warner, J. (2014). The socioeconomic vulnerability index: A pragmatic approach for assessing climate change led risks-A case study in the south-western coastal Bangladesh. *International Journal of Disaster Risk Reduction, 8*, 32–49. DOI: 10.1016/j.ijdrr.2013.12.009

Al Mahmud, R., Titumir, M., Afrin, T., Islam, M. S., Titumir, R. A. M., Afrin, T., Islam, M. S., Onneshan, U., & Dhaka, B. (2020). *Traditional Knowledge, Institutions and Human Sociality in Sustainable Use and Conservation of Biodiversity of the Sundarbans of Bangladesh.* Managing Socio-ecological Production Landscapes and Seascapes for Sustainable Communities in Asia. Springer (pp. 67–91).

Alam, A. (2012). Some Indian bryophytes known for their biologically active compounds. *International Journal of Applied Biology and Pharmaceutical Technology, 3*(2), 239–246.

Alam, M., & Bhardwaj, S. (2020). *Temporal variation in glacier's area and identification of glacial lakes in Sikkim*. DOI: 10.1007/978-981-15-2097-6_8

Ali, A. M. (2006). Rice to shrimp: Land use land cover changes and soil degradation in Southwestern Bangladesh. *Land Use Policy, 23*(4), 421–435. DOI: 10.1016/j.landusepol.2005.02.001

Ali, B. (2020). Community participation and quality of life in naturebased tourism: Exploring the antecedents and moderators. *Journal of Hospitality & Tourism Research (Washington, D.C.), 20*(10), 1–32. DOI: 10.1177/1096348020980094

Almansi, F. (2010). *The relationship between disaster risks and urban planning in Argentina,* Case study prepared for ISDR Global Assessment Report 2011.

Almansi, F., & Hardoy, J. (2013). *Climate change impact in coastal areas of the Rio de la Plata River: Actions for prevention and adaptation*. DOI: 10.4324/9781315658735

Almansi, F., Hardoy, J., & Pandiella, G. (2014). Impacto del cambio climático en ciudades costeras del estuario del Rio de la Plata. *Medio Ambiente y Urbanización, 79*, 217–237. DOI: 10.1080/17565529.2014.918868

Anguelovski, I., & Carmin, J. (2011). Something borrowed, everything new – current opinion: Innovation and institutionalisation in urban climate governance. *Current Opinion in Environmental Sustainability, 3*(3), 169–175. DOI: 10.1016/j.cosust.2010.12.017

Anurag Danda, A., Ghosh, N., Bandyopadhyay, J., & Hazra, S. (2019). Managed retreat: Adaptation to climate change in the Sundarbans ecoregion in the Bengal Delta. *Journal of the Indian Ocean Region, 15*(3), 317–335. DOI: 10.1080/19480881.2019.1652974

Anwar, G., Arianto, W., Simarmata, M., Bertham, Y. H., Yuwana, Y., Romeida, A., . . . Ruslan, D. (2023). Floristic Composition and Species Richness of Mangrove Ecosystems in Pantai Panjang–Pulau Baai Nature Park, Bengkulu Province, Indonesia. *International Journal of Advancement in Life Sciences Research*, 6(3), 10-18. 10-18. https://doi.org/DOI: DOI: 10.31632/ijalsr.2023.v06i03

Armitage, D. (2005). Adaptive capacity and community-based natural resource management. *Environmental Management, 35*(6), 703–715. DOI: 10.1007/s00267-004-0076-z PMID: 15940398

Ascoli, F. D. (1921). *A revenue history of the Sundarbans from 1870 to 1920*. Bengal Secretariat Book Depot, Calcutta. https://cslrepository.nvli.in//handle/123456789/7362

Ayeb-Karlsson, S., van der Geest, K., Ahmed, I., Huq, S., & Warner, K. (2016). A people-centred perspective on climate change, environmental stress, and livelihood resilience in Bangladesh. *Sustainability Science, 11*(4), 679–694. DOI: 10.1007/s11625-016-0379-z PMID: 30174739

Ayers, J. (2011). Resolving the adaptation paradox: Exploring the potential for deliberative adaption policy-making in Bangladesh. *Global Environmental Politics, 11*(1), 62–88. DOI: 10.1162/GLEP_a_00043

Ayers, J., & Forsyth, T. (2009). Community-based adaptation to climate change, strengthening resilience through development. *Environment, 51*(4), 23–31. DOI: 10.3200/ENV.51.4.22-31

Aziz, M. A., Tollington, S., Barlow, A., Goodrich, J., Shamsuddoha, M., Islam, M. A., & Groombridge, J. J. (2017). Investigating patterns of tiger and prey poaching in the Bangladesh Sundarbans: Implications for improved management. *Global Ecology and Conservation, 9*, 70–81. DOI: 10.1016/j.gecco.2016.12.001

Begum, F., Lobry De Bruyn, L., Amirul Islam, M., & Kristiansen, P. (2022). *The Role of Women's Participation in Co-Management of the Sundarban Mangrove Forest of Bangladesh*. University of New England. Retrieved from https://hdl.handle.net/1959.11/55652

Berger, R. M., & Kelly, J. J. (1993). Social Work in the Ecological Crisis. *Social Work, 38*, 521–526. https://www.jstor.org/stable/23717148

Berkes, F. (2009). Evolution of co-management: Role of knowledge generation, bridging organizations and social learning. *Journal of Environmental Management, 90*(5), 1692–1702. DOI: 10.1016/j.jenvman.2008.12.001 PMID: 19110363

Berkes, F. (2009). Indigenous ways of knowing and the study of environmental change. *Journal of the Royal Society of New Zealand, 39*(4), 151–156. DOI: 10.1080/03014220909510568

Berkes, F., & Ross, H. (2013). Community resilience: Toward an integrated approach. *Society & Natural Resources, 26*(1), 5–20. DOI: 10.1080/08941920.2012.736605

Besthorn, F. H., & McMillen, D. P. (2002). '*The Oppression of Women and Nature: Ecofeminism as a Framework for Social Justice Oriented Social Work*', Families and Society. *The Journal of Contemporary Human Service, 83*, 221–232. DOI: 10.1606/1044-3894.20

Beveridge, H. (1876). Favorite Share Flag Were the Sundarbans inhabited in ancient times. *Journal of the Asiatic Society of Bengal, 45*, •••. https://archive.org/details/biostor-281250

Bhowmik, D., & Poddar, S. (2023). Impacts of Flood in India from 1953 to 2020. [AMT]. *Advancement in Management and Technology*, *4*(2), 12–33.

Biesbroek, G. R., Klostermann, J. E. M., Termeer, C. J. A. M., & Kabat, P. (2013). On the nature of barriers to climate change adaptation. *Regional Environmental Change*, *13*(5), 1119–1129. DOI: 10.1007/s10113-013-0421-y

Biswas, B. (2013). "The God of Death Takes Half of Our Children": Health of Women and Children in the Sundarbans Islands. *Journal of Health Care for the Poor and Underserved*, 24(2), 730–740. https://doi.org/. 2013.0069DOI: 10.1353/hpu

Biswas, S., & Nautiyal, S. (2020). An assessment of socio-economic vulnerability at the household level: A study on villages of the Indian Sundarbans. *Environment, Development and Sustainability*, *23*(7), 11120–11137. DOI: 10.1007/s10668-020-01085-2

Breil, M., & Johnson, K. (2012). *Conceptualizing urban adaptation to climate change*. Review of Environment, Energy and Economics (Re3). Available at https://ssrn.com/abstract=2125728

Brown, A., Dayal, A., & Rumbaitis del Rio, C. (2012). From practice to theory: Emerging lessons from Asia for building urban climate change resilience. *Environment and Urbanization*, *24*(2), 531–556. DOI: 10.1177/0956247812456490

Bulkeley, H., & Tuts, R. (2013). Understanding urban vulnerability, adaptation and resilience in the context of climate change. *Local Environment*, *18*(6), 646–662. DOI: 10.1080/13549839.2013.788479

Cannon, T., & Mueller-Mahn, D. (2010). Vulnerability, resilience and development discourses in context of climate change. *Natural Hazards*, *55*(3), 621–635. DOI: 10.1007/s11069-010-9499-4

Carmin, J., Anguelovski, I., & Roberts, D. (2012). Urban climate adaptation in the global South: Planning in an emerging policy domain. *Journal of Planning Education and Research*, *32*(1), 18–32. DOI: 10.1177/0739456X11430951

Chakrabarti, R. (2009). *Local People and the Global Tiger: An Environmental History of the Sundarbans*. Global Environment, Vol. 2, (2009),(No.3), pp. 72-95. https://www.jstor.org/stable/43201488

Chanda, A., & Akhand, A. (2023, August). Challenges towards the Sustainability and Enhancement of the Indian Sundarban Mangrove's Blue Carbon Stock. *Life (Basel, Switzerland)*, *13*(8), 1–19. DOI: 10.3390/life13081787 PMID: 37629645

Chattopadhyay, H. (1999). *The Mystery of the Sunderbans* (First ed.). A Mukherjee & Co. https://archive.org/details/the-mystery-of-the-sundarbans/page/18/mode/2up

Chowdhury, A. N. (2019). Environment and wellbeing: Eco-psychiatry in Sundarban Delta, India. *International Journal of Management and Social Sciences*, *37*, 2. DOI: 10.21013/jmss.v14.n2sp.p5

Corfee-Morlot, J., Cochran, I., Hallegatte, S., & Teasdale, P. (2011). Multilevel risk governance and urban adaptation policy. *Climatic Change*, *104*(1), 169–197. DOI: 10.1007/s10584-010-9980-9

Cox, D., & Pawar, M. (2006). *International Social Work: Issues, Strategies and Programs*. SAGE., DOI: 10.4135/9781544308685

Cutter, S. L., Boruff, B. J., & Shirley, W. L. (2003). Social vulnerability to environmental hazards. *Social Science Quarterly*, *84*(2), 242–261. DOI: 10.1111/1540-6237.8402002

Danda, A. A. (2019, November). *Environmental Security in the Sundarban in the Current Climate Change Era Strengthening India-Bangladesh Cooperation*. Observer Research Foundation, ORF Occasional Paper No. 220. https://www.orfonline .org/research/environmental- security- in- the- sundarban- in- the -current-climate-change-era-strengthening-india-bangladesh-cooperation-57191

Das, S., & Hazra, S. (2020). Trapped or resettled: Coastal communities in the Sundarbans Delta. *Climate Crisis and Local Communities*, *64*, 15–17.

Dasgupta, S., Wheeler, D., & Ghosh, S. (2021). Fishing in Salty Waters: Poverty, Occupational Saline Exposure, and Women's Health in the Indian Sundarban. *Journal of Management and Sustainability*, *12*(1), 1. DOI: 10.5539/jms.v12n1p1

Dewulf, A. (2013). Contrasting frames in policy debates on climate change adaptation. *Wiley Interdisciplinary Reviews: Climate Change*, *4*(4), 321–330. DOI: 10.1002/wcc.227

Dey, M. K., & Hazra, A. K. (2021). Abundance of microarthropods population in different sites of Midnapore east coast of West Bengal, India. *International Journal of Advancement in Life Sciences Research*, 13-25. https://doi.org/DOI: DOI: 10.31632/ijalsr.2021.v04i03.003

DiGregorio, M. (2013a). Planning for climate resilience: Learning from grassroots, Presentation at *CBA 7 conference*, Dhaka, April 2013. DOI: 10.4324/9781315658735

Dodman, D., & Mitlin, D. (2013). Challenges for community-based adaptation: Discovering the potential for transformation. *Journal of International Development, 25*(5), 640–659. DOI: 10.1002/jid.1772

Dodman, D., & Satterthwaite, D. (2008). Institutional capacity, climate change adaptation and the urban poor. *IDS Bulletin, 39*(4), 67–74. DOI: 10.1111/j.1759-5436.2008.tb00478.x

Dominelli, L. (2010). '*Globalization, Contemporary Challenges and Social Work Practice*'. *International Social Work, 53*(5), 599–612. DOI: 10.1177/0020872810371201

Dominelli, L. (2011). '*Climate Change: Social Workers' Roles and Contributions to Policy Debates and Interventions*'. *International Journal of Social Welfare, 20*(4), 430–438. DOI: 10.1111/j.1468-2397.2011.00795.x

Doshi, R., Dadhich, M., Poddar, S., & Hiran, K. K. (2024). *Integrating generative AI in education to achieve sustainable development goals.*

Dubey, S., Trivedi, R., Chand, B., Mandal, B., & Rout, S. (2017). Farmers' perception of climate change, impacts on freshwater aquaculture and adaptation strategies in climatic change hotspots: A case of the Indian Sundarban delta. *Environmental Development, 21*, 38–51. DOI: 10.1016/j.envdev.2016.12.002

Eakin, H., & Lemos, M. C. (2006). Adaptation and the state: Latin America and the challenge of capacity-building under globalization. *Global Environmental Change, 16*(1), 7–18. DOI: 10.1016/j.gloenvcha.2005.10.004

Eakin, H., & Luers, A. L. (2006). Assessing the vulnerability of social-environmental systems. *Annual Review of Environment and Resources, 31*(1), 365–394. DOI: 10.1146/annurev.energy.30.050504.144352

Elliot, D. (2012). *Social Development and Social Work.* In Healy, L. M., & Link, R. J. (Eds.), *Handbook of International Social Work* (pp. 102–108). Oxford University Press., DOI: 10.1093/acprof:oso/9780195333619.003.0014

Elliott, D. (1993). Social Work and Social Development: Towards an Integrative Model for Social Work Practice. *International Social Work, 36*(1), 21–36. DOI: 10.1177/002087289303600103

Erickson, C. L. (2011). *Environmental Degradation and Preservation.* In Healy, L. M., & Link, R. J. (Eds.), *Handbook of International Social Work* (pp. 184–189). Oxford University Press., DOI: 10.1093/acprof:oso/9780195333619.003.0027

Eriksen, S. H., Brown, K., & Kelly, P. M. (2011). The dynamics of vulnerability: Locating coping strategies in Kenya and Tanzania. *The Geographical Journal, 177*(4), 287–305. DOI: 10.1111/j.1475-4959.2005.00174.x

Eriksen, S. H., Nightingale, A. J., & Eakin, H. (2015). Reframing adaptation: The political nature of climate change adaptation. *Global Environmental Change, 21*(2), 670–680. DOI: 10.1016/j.gloenvcha.2015.09.014

Few, R., Brown, K., & Tompkins, E. L. (2007). Public participation and climate change adaptation: Avoiding the illusion of inclusion. *Climate Policy*, *7*(1), 45–59. DOI: 10.1080/14693062.2007.9685637

Folke, C., Hahn, T., Olsson, P., & Norberg, J. (2005). Adaptive governance of social-ecological systems. *Annual Review of Environment and Resources, 30*(1), 441–473. DOI: 10.1146/annurev.energy.30.050504.144511

Forsyth, T. (2013). Community-based adaptation: A review of past and future challenges. *Wiley Interdisciplinary Reviews: Climate Change*, *4*(5), 439–446. Advance online publication. DOI: 10.1002/wcc.231

Francesco, E. D., & Rahman, M. (2020, December). *Natural, social and economic regeneration in the Sundarbans of Bangladesh.* Circular Conversations. Retrieved from https://www.circularconversations.com/earth-heroes/bangladesh-environment-and-development-society

Germain, C. B. (1978). General-systems Theory and Ego Psychology: An Ecological Perspective. *The Social Service Review*, *52*(4), 535–550. https://www.jstor.org/stable/30015668. DOI: 10.1086/643676

Ghosh, A. (2012). *Living with changing climate–impact, vulnerability and adaptation challenges in Indian Sundarbans.* Centre for Science and Environment.

Ghosh, A. (2019). *The Tides of Life*. Indira Gandhi National Centre for the Arts.

Ghosh, O. (2023). Environmental hazards and disasters - A response towards mitigating disaster management. *AIP Conference Proceedings*, *2763*(1), 050008. DOI: 10.1063/5.0158892

Gómez-Baggethun, E., Corbera, E., & Reyes-García, V. (2012). Traditional ecological knowledge and global environmental change: Research findings and policy implications. *Ecology and Society*, *18*(4), 72. DOI: 10.5751/ES-06288-180472 PMID: 26097492

Government of West Bengal. (2020). (*rep.*). State Action Plan on Climate Change.

Gray, M. (2006). The Progress of Social Development in South Africa. *International Journal of Social Welfare, 15*(s1, Suppl. 1), S53–S64. DOI: 10.1111/j.1468-2397.2006.00445.x

Gray, M., & Coates, J. R. (2012). Environmental Ethics for Social Work: Social Work's Responsibility to the Non-human World. *International Journal of Social Welfare, 21*(3), 239–247. DOI: 10.1111/j.1468-2397.2011.00852.x

Gray, M., Mazibuko, F., & O'Brien, F. (1996). Social Work Education for Social Development. *Journal of Social Development in Africa, 11*(1), 33–42. https://api.semanticscholar.org/CorpusID:221926574

Grothmann, T., & Patt, A. (2005). Adaptive capacity and human cognition: The process of individual adaptation to climate change. *Global Environmental Change, 15*(3), 199–213. DOI: 10.1016/j.gloenvcha.2005.01.002

Gutierrez, K. S., & LePrevost, C. E. (2016). Climate justice in rural southeastern United States: A review of climate change impacts and effects on human health. *International Journal of Environmental Research and Public Health, 13*(2), 189. Advance online publication. DOI: 10.3390/ijerph13020189 PMID: 26848673

Hafiz Iqbal, M. (2020). Valuing ecosystem services of Sundarbans mangrove forest for improved conservation: Approach of randomized conjoint experiment. *Forestry Economics Review, 2*(1), 2631–3030. DOI: 10.1108/FER-04-2020-0008

Healy, L. M. (2008). *International Social Work: Professional Action in an Interdependent World* (2nd ed.). Oxford University Press., Available at https://global.oup.com/ushe/product/international-social-work-9780190922252?cc=in&lang=

Hiran, K. K., Khazanchi, D., Vyas, A. K., & Padmanaban, S. (Eds.). (2021). *Machine learning for sustainable development* (Vol. 9). Walter de Gruyter GmbH & Co KG., DOI: 10.1515/9783110702514

Honadle, G. (1999). *How Context Matters: Linking Environmental Policy to People and Place*. Kumarian Press., https://api.semanticscholar.org/CorpusID:152578133

Hurlimann, A. C., & March, A. P. (2012). The role of spatial planning in adapting to climate change. *Wiley Interdisciplinary Reviews: Climate Change, 3*(5), 477–488. DOI: 10.1002/wcc.183

Islam, S. T., Dey, A., Sanyal, N., & Bandyopadhyay, S. (2022). *Troubles in The Sundarbans: A study of social and ecological issues in Hingalganj Block, North 24 Parganas West Bengal*. ActionAid International.

Jabeen, H., Johnson, C., & Allen, A. (2010). Built-in resilience: Learning from grassroots coping strategies for climate variability. *Environment and Urbanization*, *22*(2), 415–431. DOI: 10.1177/0956247810379937

Jahan, R. (1995). *The elusive agenda: Mainstreaming women in development*. London: ZED Books. https://www.jstor.org/stable/41260001

Jalais, A. (2007). The Sundarbans: Whose World Heritage Site? *Conservation & Society*, *5*(3), 335–342. https://www.jstor.org/stable/26392892

Jamal, S., Ghosh, A., Hazarika, R., & Sen, A. (2022, July). Livelihood, conflict and tourism: An assessment of livelihood impact in Sundarbans, West Bengal. *International Journal of Geoheritage and Parks*, *10*(1), 383–399. Advance online publication. DOI: 10.1016/j.ijgeop.2022.07.004

Kanjilal, B., Bose, S., Patra, N., Barman, D., Ghosh, U., Mandal, A., Vardrevu, L., & Sengupta, P. (2013). *How Healthy are the Children of Indian Sundarbans?*http://r4d.dfid.gov.uk/pdf/outputs/FutureHealth_RPC/SHWforweb.pdf

Kanjilal, B., Mazumdar, P., Mukherjee, M., Mondal, S., Barman, D., Singh, S., & Mandal, A. (2010). Health care in the Sundarbans (India): Challenges and plan for a better future. *Future Health Systems Research Programme* (Issues 1–91). DOI: 10.6133/apjcn.012016.01

Kanjilal, B., Mazumdar, P. G., Mukherjee, M., Mondal, S., Barman, D., & Singh, S. M. (2010, January). *Health care in the Sundarbans (India): Challenges. Future Health Systems Research Programme*, 21-30. Retrieved from https://assets.publishing.service.gov.uk/media/57a08b24ed915d3cfd000b68/sundarbans.pdf

Kemp, S. P., Whittaker, J. K., & Tracy, E. M. (1997) *Person–Environment Practice: The Social Ecology of Interpersonal Helping*. New York: Aldine De Gruyter. https://api.semanticscholar.org/CorpusID:141687227

Kernaghan, S., & da Silva, J. (2013). Initiating and sustaining action: Experiences building resilience to climate change in Asian cities. *Urban Climate*, *7*, 47–63. DOI: 10.1016/j.uclim.2013.10.008

Khan, S. A. R., Umar, M., Yu, Z., & Nawaz, M. T. (2023). A Recent Digitalization in Recycling Industry Attaining Ecological Sustainability: A Comprehensive Outlook and Future Trend. *Environmental Science and Pollution Research International*, *30*(47), 103760–103775. DOI: 10.1007/s11356-023-29537-y PMID: 37695483

Leck, H., & Simon, D. (2012). Fostering multiscalar collaboration and co-operation for effective governance of climate change adaptation. *Urban Studies (Edinburgh, Scotland), 50*(6), 1221–1238. Advance online publication. DOI: 10.1177/0042098012461675

Leichenko, R. (2011). Climate change and urban resilience. *Current Opinion in Environmental Sustainability, 3*(3), 164–168. DOI: 10.1016/j.cosust.2010.12.014

Mahadevia, K., & Vikas, M. (2012). Climate change—Impact on the Sunderbans: A case study. *International Scientific Journal-Environmental Science, 2*(1), 7–15.

Mertz, O., Mbow, C., Reenberg, A., & Diouf, A. (2009). Farmers' perceptions of climate change and agricultural adaptation strategies in rural Sahel. *Environmental Management, 43*(5), 804–816. DOI: 10.1007/s00267-008-9197-0 PMID: 18810526

Midgley, J. (1994). Defining Social Development: Historical Trends and Conceptual Formulations. *Social Development Issues, 16*(3), 3–19. DOI: 10.4135/9781446221839.n2

Midgley, J. (1995). *Social Development: The Developmental Perspective in Social Welfare*. SAGE., DOI: 10.4135/9781446221839.n2

Miller, S. E., Hayward, R. A., & Shaw, T. V. (2012). Environmental Shifts for Social Work: A Principles Approach. *International Journal of Social Welfare, 21*(3), 270–277. DOI: 10.1111/j.1468-2397.2011.00848.x

Mimi, R., & Haynes, L. (2018). *The role of livelihood activities and income portfolios in shaping women's perceptions of wellbeing and adaptation choices, in the Indian Bengal Delta Statement of Originality.*

Mishra, M., Acharyya, T., Santos, C. A. G., da Silva, R. M., Kar, D., Mustafa Kamal, A. H., & Raulo, S. (2021). Geo-ecological impact assessment of severe cyclonic storm Amphan on Sundarban mangrove forest using geospatial technology. *Estuarine, Coastal and Shelf Science, 260*, 107486. DOI: 10.1016/j.ecss.2021.107486

Mistri, A. (2019). Is the migration from Indian Sundarban an environmental migration? Investigating through Sustainable Livelihood Approach (SLA). *Asian Profile, 47*(3), 195–218.

Mitlin, D., & Satterthwaite, D. (2012). Addressing poverty and inequality new forms of urban governance in Asia. *Environment and Urbanization, 24*(2), 395–401. DOI: 10.1177/0956247812458062

Mitra, R. (2024). *Nature-based Solutions can Help Sundarbans Survive. DownToEarth.* Retrieved from https://www.downtoearth.org.in/news/climate-change/nature-based-solutions-can-help-sundarbans -survive-95230

Mizanur Rahman, M., Rakib Hossain, M., & Nazrul Islam, M. (2021). Climate Change Impact on Sundarbans: Challenges for Mitigation Strategies. *Climate Change in Bangladesh: A Cross-Disciplinary Framework*, 47-64.

Molinari, N. (2017). Intensifying Insecurities: The impact of climate change on vulnerability to human trafficking in the Indian Sundarbans. *Anti Trafficking Review, 363*(8), 1–11. DOI: 10.14197/atr.20121784

Mondal, M. K., Bhuiyan, S. I., & Franco, D. T. (2001). Soil salinity reduction and prediction of salt dynamics in the coastal rice lands of Bangladesh. *Agricultural Water Management, 47*(1), 9–23. DOI: 10.1016/S0378-3774(00)00098-6

Mondal, T., Bowers, W. W., & Hossen, M. S. (2021, March 25). Sustainable Management of Sundarbans: Stakeholder Attitudes Towards Participatory Management and Conservation of Mangrove Forests. *Journal of Sustainable Development, 14*(3), 23–40. DOI: 10.5539/jsd.v14n3p23

Moore, J. (2023, March 6). *PayBitoPro saves Sundarbans' child education.* Retrieved from PayBitoPro: https://www.paybito.com/lost-in -the -mangroves -paybito -steps -up -to -save -child -education -in -sundarbans/ 1

Moser, S. C., & Ekstrom, J. A. (2010). A framework to diagnose barriers to climate change adaptation. *Proceedings of the National Academy of Sciences of the United States of America, 107*(51), 22026–22031. DOI: 10.1073/pnas.1007887107 PMID: 21135232

Mozumdar, M., Shamsuzzaman, M., Un-Nabi, R., & Harun-Al-Rashid, A. (2018). Socio-economic characteristics and fishing operation activities of the artisanal fishers in the Sundarbans Mangrove Forest, Bangladesh. *Turkish Journal of Fisheries and Aquatic Sciences, 18*(6), 788–799. DOI: 10.4194/1303-2712-v18_6_05

Mukhopadhyay, A. (2009). *Cyclone Aila and the Sundarbans: An Enquiry into the Disaster and Politics of Aid and Relief.* In Research Report on CRG Annual Winter Course on Forced Migration.

Nicholls, R. J., Adger, W. N., Hutton, C. W., & Hanson, S. E. (2020). *Deltas in the Anthropocene.* Palgrave Macmillan Springer Nature., DOI: 10.1007/978-3-030-23517-8

Norgaard, K. M. (2017, September). The Sociological Imagination in a Time of Climate Change. *Global and Planetary Change, Elsevier*. DOI: DOI: 10.1016/j.gloplacha.2017.09.018

Norris, F., Stevens, S., Pfefferbaum, B., Wyche, K., & Pfefferbaum, R. (2008). Community resilience as metaphor, theory, set of capacities, and strategy for disaster readiness. *American Journal of Community Psychology*, *41*(1–2), 127–150. DOI: 10.1007/s10464-007-9156-6 PMID: 18157631

Nyong, A., Adesina, F., & Elasha, B. O. (2007). The value of indigenous knowledge in climate change mitigation and adaptation strategies in the African Sahel. *Mitigation and Adaptation Strategies for Global Change*, *12*(5), 787–797. DOI: 10.1007/s11027-007-9099-0

O'Brien, K. (2012). Global environmental change II: From adaptation to deliberate transformation. *Progress in Human Geography*, *36*(5), 667–676. DOI: 10.1177/0309132511425767

O'Brien, K., Eriksen, S., Nygaard, L. P., & Schjolden, A. (2007). Why different interpretations of vulnerability matter in climate change discourses. *Climate Policy*, *7*(1), 73–88. DOI: 10.1080/14693062.2007.9685639

Okoye, K., Nganji, J. T., Hiran, K. K., & Hosseini, S. (2024, May). Impact and implications of AI methods and tools for the future of education. [). Frontiers Media SA.]. *Frontiers in Education*, *9*, 1434052. DOI: 10.3389/feduc.2024.1434052

Olhoff, A., & Schaer, C. (2010). *Screening tools and guidelines to support the mainstreaming of climate change adaptation into development assistance a stocktaking report*. New York: UNDP. http://www.undp.org/climatechange/library.shtml

Omer, S. (1979). *'Social Development'*. *International Social Work*, *22*(3), 11–26. DOI: 10.1177/002087287902200303

Pal, P., & Poddar, S. (2024). Applications of Smart Farming Sensors: A Way Forward. *Data Science for Agricultural Innovation and Productivity*, 175.

Panda, S., Sadhu, C., Pramanik, G., Pahari, S., & Hossain, J. (2016). Concerning public health situation of under-nutrition in children and anemia in women in Indian Sundarbans delta: A community based cross sectional investigation. *BMC Nutrition*, *2*(1), 1–10. DOI: 10.1186/s40795-016-0105-3

Peeters, J. (2012). The Place of Social Work in Sustainable Development: Towards Ecosocial Practice. *International Journal of Social Welfare*, *21*(3), 287–298. DOI: 10.1111/j.1468-2397.2011.00856.x

Pelling, M. (2003). *The vulnerability of cities: Natural disasters and social resilience*. Earthscan., DOI: 10.1080/15693430500111884

Pelling, M. (2011). *Adaptation to climate change: From resilience to transformation*. Routledge., DOI: 10.4324/9780203889046

Pelling, M., & Manuel-Navarrete, D. (2011). From resilience to transformation: The adaptive cycle in two Mexican urban centers. *Ecology and Society*, *16*(2), 11. DOI: 10.5751/ES-04038-160211

Pervin, M., Sultana, S., Phirum, A., Camara, I., Nzau, M., Phonnasane, V., & Anderson, S. (2013). *A framework for mainstreaming climate resilience into development planning*. IIED Working Paper, IIED, London https://www.iied.org/10050iied

Pitchaikani, J. S. (2020, March). Integrated coastal zone management practices for Sundarbans, India. *Indian Journal of Geo-Marine Sciences*, *49*(03), 352–356. https://core.ac.uk/download/pdf/326041666.pdf

Polin, F., & Alam, D. (2020, March 13). *Enhancing Community Participation in Forest Management: Case Study of the Sundarbans, Bangladesh*. South-West Region (Khulna Division). Retrieved from https://www.researchgate.net/publication/339916369_Enhancing_Community_Participation_in_Forest_Management_Case_Study_of_the_Sundarbans_Bangladesh

Pramanik, M., Szabo, S., Pal, I., Udmale, P., Pongsiri, M., & Chilton, S. (2021). Population health risks in multi-hazard environments: Action needed in the Cyclone Amphan and COVID-19 - hit Sundarbans region, India. *Climate and Development*, *14*(2), 99–104. DOI: 10.1080/17565529.2021.1889948

Rahman, M., Giedraitis, V. R., Lieberman, L. S., Akhtar, T., & Taminskienė, V. (2013). Shrimp cultivation with water salinity in Bangladesh: The implications of an ecological model. *Universal Journal of Public Health*, *1*(3), 131–142. DOI: 10.13189/ujph.2013.010313

Ray, P. S. (2020). Socio-Economic History in Sundarbans. *Journal of Critical Reviews*, *7*(2), 1030–1034.

Reid, H., Alam, M., Berger, R., Cannon, T., Huq, S., & Milligan, A. (2009). *Community-based adaptation to climate change: An overview*. Participatory Learning and Action: Community Based Adaptation to Climate Change, 60, 11–33. Available at https://www.researchgate.net/publication/272791313_Community-based_adaptation_to_climate_change_An_overview

Ribot, J. C., Najam, A., & Watson, G. (2009). *Climate Variation, Vulnerability and Sustainable Development in the Semi-arid Tropics*. In Schipper, E. L. F., & Burton, I. (Eds.), *The Earthscan Reader on Adaptation to Climate Change* (pp. 117–160). Earthscan., Available at https://www.researchgate.net/publication/291177311_Climate_variation_vulnerability_and_sustainable_development_in_the_semi-arid_tropics

Roberts, D. (2008). Thinking globally, acting locally institutionalising climate change at the the local government level in Durban, South Africa. *Environment and Urbanization, 20*(2), 521–537. DOI: 10.1177/0956247808096126

Roberts, D. (2013). *Busting the Myths – An African reality check on Community Based Adaptation in Durban, South Africa*. Presentation at *CBA 7 conference*, Dhaka, April 2013. DOI: 10.1177/0956247808096126

Rogers, P. P., Jalal, K. F., & Boyd, J. A. (2007). *An Introduction to Sustainable Development*. Harvard University Press., DOI: 10.4324/9781849770477

Roy, D. (2020). 'On the Horns of a Dilemma'! Climate change, forest conservation and the marginal people in Indian Sundarbans. *Forum for Development Studies*, 47(2), 307–326. https://doi.org/. 1786452DOI: 10.1080/08039410.2020

Roy, M., Hulme, D., & Jahan, F. (2013). Contrasting adaptation responses by squatters and low-income tenants in Khulna, Bangladesh. *Environment and Urbanization, 25*(1), 157–176. DOI: 10.1177/0956247813477362

Roy, S. (2018). Livelihood resilience of the indigenous Munda community in the Bangladesh Sundarbans Forest. *Handbook Of Climate Change Resilience*, 10(1), 1–21. DOI: 10.1007/978-3-319-71025-9_10-1

Roy, S. (2019). *Women's Use of the Sundarbans Forest Resources*. Springer Briefs in Environment, Security. *Development and Peace*, *29*, 59–71. DOI: 10.1007/978-981-13-6776-2_4

Rudra, A., & Chattopadhyay, A. (2018). *Environmental vulnerabilities and risks in population: A comparison of two Sundarban Delta*. Population Association of America, May, 1–14. https://www.researchgate.net/publication/326201083

Rudra, A., & Chattopadhyay, A. (2019). Environmental change of coastal Sundarbans: Impact on livelihood and standard of living status of indigenous people. *Environmental Quality Management, 29*(2), 77–84. DOI: 10.1002/tqem.21667

Sahana, M., Rehman, S., Ahmed, R., & Sajjad, H. (2021). Analyzing climate variability and its effects in Sundarban Biosphere Reserve, India: Reaffirmation from local communities. *Environment, Development and Sustainability, 23*(2), 2465–2492. DOI: 10.1007/s10668-020-00682-5

Saif, S., Russell, A. M., Nodie, S. I., Inskip, C., Lahann, P., Barlow, A., Greenwood Barlow, C., Islam, M. A., & MacMillan, D. C. (2016). Local usage of tiger parts and its role in tiger killing in the Bangladesh Sundarbans. *Human Dimensions of Wildlife, 21*(2), 95–110. DOI: 10.1080/10871209.2015.1107786

Saito, O., Subramanian, S., Hashimoto, S., & Takeuchi, K. (2020). *Managing socio-ecological production landscapes and seascapes for sustainable communities in Asia mapping and navigating stakeholders, policy and action.* Springer https://www.springer.com/series/11884

Sarkar, S., & Sanyal, P. (2019). Community perception towards tourism development in the Indian Sundarbans: Case study of Jharkhali Island. *International Journal of Engineering Research & Technology (Ahmedabad), 8*(12), 526–540. www.ijert.org

Sarwar Hossain, M., Eigenbrod, F., Amoako Johnson, F., & Dearing, J. A. (2016). Unravelling the interrelationships between ecosystem services and human wellbeing in the Bangladesh delta. *International Journal of Sustainable Development and World Ecology, 24*(2), 120–134. DOI: 10.1080/13504509.2016.1182087

Satterthwaite, D., Huq, S., Reid, H., Pelling, M., & Romero Lankao, P. (2007). *Adapting to climate change in urban areas: the possibilities and constraints in low-and middle-income nations.* Human Settlements Working Paper, IIED, London. https://www.iied.org/10549iied

Satterthwaite, D., & Mitlin, D. (2014). *Reducing urban poverty in the global south.* Routledge., DOI: 10.4324/9780203104316

Schipper, E. L., Ayers, J., Reid, H., Huq, S., & Rahman, A. (Eds.), *Community-based adaptation to climate change: scaling it up* (pp. 214–225). Routledge., DOI: 10.4324/9780203105061

Sen, H. S. (2019). *The Sundarbans: A disaster-prone ecoregion.* Berlin: Springer. 10, 978-3 https://link.springer.com/content/pdf/10.1007/978-3-030-00680-8.pdf

Sharma, D., & Tomar, S. (2010). Mainstreaming climate change adaptation in Indian cities. *Environment and Urbanization, 22*(2), 451–465. DOI: 10.1177/0956247810377390

Sharma, R., Webster, P., & Bhattacharyya, S. (2014). Factors affecting the performance of community health workers in India: A multi-stakeholder perspective. *Global Health Action, 7*(1), 25352. DOI: 10.3402/gha.v7.25352 PMID: 25319596

Sivaramakrishnan, K. C. (2007). *Democracy in urban India.* Urban Age. https://urbanage.lsecities.net/essays/democracy-in-urban-india

Smit, B., & Pilifosova, O. (2003). *From adaptation to adaptive capacity and vulnerability reduction.* In Climate Change, Adaptive Capacity and Development (pp. 9-28). Imperial College Press. DOI: 10.1142/9781860945816_0002

Smit, B., & Skinner, M. W. (2002). Adaptation options in agriculture to climate change: A typology. *Mitigation and Adaptation Strategies for Global Change*, *7*(1), 85–114. DOI: 10.1023/A:1015862228270

Smit, B., & Wandel, J. (2006). Adaptation, adaptive capacity and vulnerability. *Global Environmental Change*, *16*(3), 282–292. DOI: 10.1016/j.gloenvcha.2006.03.008

Soltesova, K., Brown, A., Dayal, A., & Dodman, D. (2014). *Community participation in urban adaptation to climate change: Potential and limits for CBA approaches*. Community-Based Adaptation to Climate Change., DOI: 10.4324/9780203105061

Sovacool, B. K. (2011). An international comparison of four polycentric approaches to climate and energy governance. *Energy Policy*, *39*(6), 3832–3844. DOI: 10.1016/j.enpol.2011.04.014

Sovacool, B. K. (2018). Bamboo beating bandits: Conflict, inequality, and vulnerability in the political ecology of climate change adaptation in Bangladesh. *World Development*, *102*, 183–194. DOI: 10.1016/j.worlddev.2017.10.014

StC. (2008). D. E. *Convention on the Conservation of Migratory Species of Wild Animals*.

Sundarbans National Park. (1987). *UNESCO World Heritage Centre*. https://whc.unesco.org/en/list/452

Syam, D. (2013). *National engagement and replication in Indonesia*. Presentation at the CBA7 conference, Dhaka, April 2013. DOI: 10.1177/0956247808096126

Szabo, S., Hossain, M. S., Adger, W. N., Matthews, Z., Ahmed, S., Lázár, A. N., & Ahmad, S. (2016). Soil salinity, household wealth and food insecurity in tropical deltas: Evidence from south-westcoast of Bangladesh. *Sustainability Science*, *11*(3), 411–421. DOI: 10.1007/s11625-015-0337-1 PMID: 30174734

Thakur, S., Maity, D., Mondal, I., Basumatary, G., Ghosh, P. B., Das, P., & De, T. K. (2021). Assessment of changes in land use, land cover, and land surface temperature in the mangrove forest of Sundarbans, northeast coast of India. *Environment, Development and Sustainability*, *23*(2), 1917–1943. DOI: 10.1007/s10668-020-00656-7

Tripodi, T., & Potocky-Tripodi, M. (2007). *International Social Work Research: Issues and Prospects*. Oxford University Press., DOI: 10.1093/acprof:oso/9780195187250.001.0001

Twigg, J. (2007). *Characteristics of a disaster-resilient community, a guidance note*. Available at https://www.researchgate.net/publication/315799865_Characteristics_of_a_Disaster-resilient_Community_A_Guidance_Note_Characteristics_of_a_Disaster-resilient_Community_A_Guidance_Note

Tyagi, S. K. S., Mukherjee, A., Pokhrel, S. R., & Hiran, K. K. (2020). An intelligent and optimal resource allocation approach in sensor networks for smart agri-IoT. *IEEE Sensors Journal*, *21*(16), 17439–17446. DOI: 10.1109/JSEN.2020.3020889

Tyler, S., & Moench, M. (2012). A framework for urban climate resilience. *Climate and Development*, *4*(4), 311–326. DOI: 10.1080/17565529.2012.745389

Uddin, S. M. (2019). Religion, Nature, and Life in the Sundarbans. *Asian Ethnology*, *78*(2), 289–310. https://www.jstor.org/stable/26845329

Unogwu, O. J., Doshi, R., Hiran, K. K., Mijwil, M. M., Catherine, A. T., & Abotaleb, M. (2024). *Exploring the Implications of Emerging Artificial Intelligence Technologies at Edge Computing in Higher Education*. In Integrating Generative AI in Education to Achieve Sustainable Development Goals (pp. 169-182). IGI Global. DOI: DOI: 10.4018/979-8-3693-2440-0.ch009

Yohe, G., & Tol, R. S. J. (2002). Indicators for social and economic coping capacity moving toward a working definition of adaptive capacity. *Global Environmental Change*, *12*(1), 25–40. DOI: 10.1016/S0959-3780(01)00026-7

ADDITIONAL READING

Andtfolk, M., Nyholm, L., Eide, H., Rauhala, A., & Fagerström, L. (2022). Attitudes Toward the Use of Humanoid Robots in Healthcare—A Cross-Sectional Study. *AI & Society*, *37*(4), 1739–1748. DOI: 10.1007/s00146-021-01271-4

Asiedu-Berkoe, F., Chandi, G. M., Bandoh, D. A., Atsu, B. K., Lokossou, V. K., Antara, S. N., Sarkodie, B., & Kenu, E. (2022). State of Public Health Emergency Preparedness and Response Capacity of Ghana. *Journal of Interventional Epidemiology and Public Health*, *5*(4), 1–16. DOI: 10.37432/jieph.supp.2022.5.4.05.4

Chauhan, P., Bali, A., & Kaur, S. (2024). Breaking Barriers for Accessible Health Programs: The Role of Telemedicine in a Global Healthcare Transformation. In *Transformative Approaches to Patient Literacy and Healthcare Innovation*. IGI Global., DOI: 10.4018/979-8-3693-3661-8.ch014

KEY TERMS AND DEFINITIONS

Indigenous Rights: Indigenous rights refer to the collective and individual rights of indigenous peoples to maintain their culture, language, land, and governance systems. These rights also encompass the right to participate in decision-making processes that affect their lives, particularly regarding land use, environmental conservation, and health services.

Traditional Ecological Knowledge (TEK): Traditional Ecological Knowledge encompasses the understanding and practices developed by indigenous communities over generations concerning their local ecosystems. TEK includes insights about biodiversity, resource management, and sustainable practices that can inform contemporary environmental approaches and policies, contributing to both conservation and sustainable development.

Health Disparities: Health disparities are the differences in health outcomes and access to healthcare services between different population groups. In the context of the Sundarbans, indigenous communities may face unique health challenges due to factors like marginalization, environmental degradation, and limited access to healthcare, significantly impacting their overall well-being.

Sustainable Development: Sustainable development refers to a development approach that meets the needs of the present without compromising the ability of future generations to meet their own needs. In the Sundarbans, this concept emphasizes the integration of economic, social, and environmental goals while respecting indigenous rights and promoting community participation in development initiatives.

Resilience: Resilience in this context refers to the capacity of indigenous communities in the Sundarbans to adapt to health challenges and environmental changes while maintaining their cultural identity and practices. Building resilience involves leveraging indigenous wisdom, fostering community cohesion, and developing adaptive strategies to cope with external pressures such as climate change and health crises.

Chapter 2
AI for Biodiversity Conservation:
Harnessing Predictive Analytics and Habitat Monitoring to Protect Endangered Species and Ecosystems

G. R. Manjunatha
https://orcid.org/0009-0002-4476-5710
Bangalore University, India

H. R. Raveesha
Bangalore University, India

ABSTRACT

Biodiversity is under threat due to numerous factors, including habitat destruction, climate change, poaching, and invasive species. As human activities continue to put immense pressure on ecosystems, endangered species, and their habitats become more critical than ever. In this context, Artificial Intelligence (AI) has emerged as a transformative tool for biodiversity conservation. By combining advanced predictive analytics, machine learning, and real-time habitat monitoring, AI can provide unprecedented insights into ecosystem health and species survival prospects. This chapter delves into how AI can revolutionize biodiversity conservation, focusing on two major areas: predictive analytics for endangered species management and AI-driven habitat monitoring. We explore current AI applications, case studies, and future directions in these fields, offering a comprehensive look at how AI can enhance conservation efforts and guide sustainable decision-making.

DOI: 10.4018/979-8-3373-0680-3.ch002

1. INTRODUCTION

Biodiversity is the cornerstone of healthy ecosystems and is essential for the well-being of our planet. It encompasses the immense variety of life forms found on Earth, from minuscule microorganisms to towering trees, which play critical roles in maintaining ecological balance. This diversity provides invaluable medicinal resources and contributes to climate regulation and stability. Unfortunately, human activities pose significant threats to this delicate web of life. The alarming rate of environmental degradation, characterized by deforestation, habitat loss, pollution, and climate change, has resulted in a substantial decline in biodiversity (Muluneh, 2021). Kahrić et al. (2022) note that species are increasingly facing extinction.

Innovative conservation strategies are necessary. Traditional approaches to conservation, while effective to some extent, scope and scale. The intricate nature of ecosystems and the rapid pace of environmental changes highlight the need for new technologies to enhance conservation efforts (Tuia et al., 2022). Metolo and Fovet (2024) highlight that AI has emerged as a powerful ally in the fight for biodiversity conservation. By leveraging AI technologies, researchers and conservationists can analyze vast amounts of data to predict environmental changes, monitor wildlife populations, and refine conservation strategies (Fergus et al., 2024).

AI enables faster and more accurate assessments of ecosystems, making it possible to foresee shifts in species distribution, track endangered species, and intervene more effectively against threats such as habitat fragmentation and poaching (Fergus et al., 2024).

This chapter will explore the transformative potential of AI in biodiversity conservation, focusing on its applications in predictive analytics, habitat monitoring, and species protection. By examining these areas, we will illustrate how AI can help address the current biodiversity crisis and foster more proactive and sustainable conservation practices. From anticipating environmental changes to providing real-time data on ecosystem health, AI offers unprecedented opportunities for safeguarding biodiversity. Ayoola et al. (2024) argue that it ultimately empowers conservationists to make informed decisions and take timely actions, contributing to the natural world for future generations.

1.1 The Role of AI in Biodiversity Conservation

Artificial intelligence (AI) is a fast-evolving area of computer science, machines with human intelligence. It empowers them to learn from data, predict outcomes, and carry out tasks independently (Soori et al., 2023). In biodiversity conservation, AI processes vast amounts of ecological data—such as satellite imagery, sensor data from the environment, and footage from camera traps—to identify patterns and forecast the future state of ecosystems (Olawade et al., 2024B). This capability enables researchers and conservation managers to make well-informed decisions, allocate resources more efficiently, and develop effective strategies to protect the planet's biodiversity.

For example, Olawade et al. (2024B) and Alotaibi and Nassif (2024) explain that AI can analyze satellite images to monitor deforestation or track habitats, providing insights that would be difficult or time-consuming. AI systems can also process data from environmental sensors to track climate variations or pollution levels, predicting potential threats to species and ecosystems, as indicated in Table 1. A practical application of AI in conservation is its use in camera trap networks; AI algorithms automatically identify and count species captured on camera, making monitoring animal populations in remote areas more manageable. AI-driven tools, conservationists are empowered to act more swiftly and accurately to preserve biodiversity. Fergus et al. (2024) emphasize that AI is invaluable in several key areas of biodiversity conservation.

1.2 Predicting Ecological Trends

The ability of AI to process and analyze vast amounts of historical and real-time environmental data has transformed ecological trends. By uncovering patterns in this data, AI systems can identify emerging ecological changes and forecast shifts in species populations and habitat conditions (Olawade et al., 2024B; Bol6n-Canedo et al., 2024). Veiga et al. (2023) explain that by leveraging machine learning algorithms (MLA), AI can integrate diverse data sources, such as weather patterns, habitat characteristics, and behavioral data of species, to create predictive models.

These models are powerful tools for anticipating how ecosystems may change in response to various scenarios, including the effects of climate change or habitat destruction. For example, AI algorithms can analyze temperature trends, precipitation data, and land-use patterns to predict shifts in species distributions. Such insights allow conservationists to identify areas where certain species may face increased extinction risks, providing opportunities for targeted intervention and proactive conservation measures (Llewelyn et al., 2023).

By delivering detailed forecasts, AI supports long-term biodiversity conservation planning. It enables researchers and policymakers to understand the potential impacts of natural processes and human activities on ecosystems. For instance, predictive models can guide habitat restoration efforts, inform species reintroduction programs, or aid in developing policies that mitigate threats to biodiversity. Rawas (2024) suggests that in this way, AI empowers conservationists with a deeper understanding of ecological dynamics, ensuring more effective and sustainable strategies for protecting the natural world.

1.3 Real-Time Habitat Monitoring

Monitoring ecosystems is essential for tracking environmental changes and evaluating the success of conservation efforts. While valuable, traditional methods for habitat monitoring often face challenges such as being time-consuming, labor-intensive, and limited in geographic or temporal coverage. In contrast, AI-powered monitoring systems provide a transformative approach by enabling real-time observation and analysis of ecosystems. Using data from sources such as satellites, drones, and environmental sensors, AI can continuously monitor habitats with unprecedented accuracy and efficiency (Alotaibi & Nassif, 2024).

One of the most significant advantages of AI systems is their ability to detect subtle environmental changes that might go unnoticed with manual methods. For example, AI algorithms can analyze temperature changes. These variations may serve as early warning signs of environmental threats, including pollution, deforestation, or the impacts of climate change (United Nations Environment Programme, 2022). This capability allows conservationists to identify and address potential issues before they escalate into larger problems.

Additionally, AI-driven monitoring combats illegal activities such as poaching or unauthorized logging. By processing live feeds from camera traps, drones, or other surveillance tools, AI can detect unusual activities and send real-time alerts to authorities. Dalton et al. (2024) highlight that AI can identify the spread of invasive species or the destruction of habitats, ensuring that conservation teams can respond swiftly and effectively.

The continuous flow of data provided by AI systems empowers conservationists to manage ecosystems more proactively and make informed decisions. Kumar (2024) argues that by enabling faster and more precise interventions, AI contributes to the protection of vulnerable habitats and supports the long-term health of biodiversity. Advanced technology in habitat monitoring represents a significant step forward in the fight to preserve the planet's ecosystems.

1.4 Enhancing Species Identification and Monitoring

Accurate identification and monitoring of species are fundamental to effective biodiversity conservation. However, traditional approaches to this task often involve extensive fieldwork and manual data analysis, which can be both time-consuming and labor-intensive. The emergence of AI-powered tools, particularly those leveraging computer vision and deep learning (DL), has transformed species that are identified and tracked (Dhanya et al., 2022). These tools can analyze large datasets of images or audio recordings, enabling researchers to conduct biodiversity surveys on an unprecedented scale.

One notable application of AI in conservation is the analysis of images captured by camera traps. These cameras, frequently used in remote or hard-to-reach locations, generate massive volumes of data. Manually reviewing this data to identify species can take considerable time and effort. AI algorithms, however, can process thousands of images rapidly, detecting and accurately identifying animal species. This capability provides critical insights into population dynamics, habitat preferences, and migration patterns. By enabling continuous and detailed monitoring, these systems significantly enhance our understanding of species and their ecosystems.

In addition to image analysis, AI has advanced acoustic monitoring techniques. AI-driven systems can identify species based on unique vocalizations, such as bird songs, frog calls, or marine mammal sounds. Brodie et al. (2020) note that this technology is particularly valuable for studying species in dense forests, underwater habitats, or other environments where direct observation is challenging. Acoustic monitoring expands the reach of biodiversity studies, allowing researchers to collect data.

Automating species identification with AI offers numerous benefits. It minimizes the likelihood of human error in data processing, increases efficiency, and allows for the collection of more detailed and comprehensive datasets. These advancements are crucial for effective conservation planning, as they provide the information needed to develop targeted strategies for protecting endangered species and their habitats. AI has emerged as a tool for conservationists, enhancing their ability to monitor and safeguard biodiversity on a global scale (Kalogiannidis et al., 2024).

2. PREDICTIVE ANALYTICS: FORECASTING ECOLOGICAL CHANGES

Predictive analytics, a critical application of AI in biodiversity conservation, involves using advanced algorithms to analyze complex environmental data and forecast future conditions for species and ecosystems. This approach has extensive datasets that include species population trends, climate variables, land-use changes, and other ecological indicators. By leveraging ML and DL techniques, predictive models can provide detailed insights into how ecosystems and species respond to various environmental pressures (Casanelles-Abella et al., 2021).

One of predictive analytics' strengths is its ability to process and synthesize vast amounts of ecological data that would be overwhelming to analyze manually. For instance, AI can combine historical records of species distributions with current climate data to predict future habitat shifts. Similarly, models can simulate the potential impacts of deforestation, pollution, or climate change on species survival, allowing conservationists to identify areas at higher risk of biodiversity loss. These predictions are particularly valuable for prioritizing conservation efforts, such as habitat restoration or species protection programs.

For example, DL models can analyze patterns in climate fluctuations and link them to changes in species behavior or population dynamics. Such insights enable researchers to anticipate potential outcomes under different scenarios. These models also allow conservationists to test the effectiveness of strategies, such as reforestation or climate adaptation measures, before implementation.

By providing actionable insights, predictive analytics empowers conservationists to make proactive decisions and address ecological challenges before they escalate. These tools enhance the ability to plan for long-term conservation goals and respond effectively to the dynamic nature of ecosystems. Casanelles-Abella et al. (2021) highlight that as the impacts of environmental pressures continue to intensify, predictive analytics serves as an invaluable resource for ensuring the resilience and sustainability of biodiversity.

2.1 Predictive Analytics Plays a Crucial Role in Several Key Areas of Biodiversity Conservation

2.1.1 Species Distribution Shifts

Predictive analytics plays a pivotal role in biodiversity conservation by helping forecast changes in species distributions due to environmental shifts such as climate change, habitat loss, or land-use changes (Barras et al., 2021). By integrating AI technologies with current climate models and land-use data, conservationists can predict how habitats may evolve and respond to these changes.

For instance, if global temperatures rise or precipitation patterns shift, certain species may find that existing habitats no longer provide the conditions necessary for survival. These changes may force species to migrate to more suitable environments or adapt to the altered conditions (Fontúrbel, 2021). AI-driven models can analyze such scenarios, forecasting which areas are likely to become critical for the survival of specific species. These insights allow conservationists to identify regions that should be prioritized for protection or restoration efforts (Casanelles-Abella et al., 2021; Mkala et al., 2023).

By anticipating shifts in species distributions, conservationists can take pre-emptive action to mitigate the risks associated with habitat loss or fragmentation. For example, these predictions inform the creation of wildlife corridors that enable safe migration or the establishment of new protected areas in regions predicted to become biodiversity hotspots. Such interventions help to maintain ecological balance but reduce the likelihood of species extinction.

Furthermore, AI can evaluate how different conservation strategies might influence species survival under various scenarios. By simulating potential outcomes, conservationists can understand species distribution shifts through predictive analytics, an invaluable tool for addressing the challenges posed by rapid environmental change, ensuring the resilience of ecosystems, and protecting biodiversity for future generations.

2.1.2 Impact of Climate Change on Ecosystems

Weiskopf et al. (2020) note that climate change is one of these factors, as it profoundly affects ecosystems and species worldwide. AI plays a crucial role in addressing these challenges by providing advanced tools to simulate and predict the effects of climate variables, such as temperature changes, altered precipitation patterns, and extreme weather events, on ecosystems (Secci et al., 2023).

Akter et al. (2024) explain that AI-powered models analyze extensive datasets to forecast how these climatic shifts influence ecosystems and species dynamics. For example, rising temperatures might alter the availability of food sources for herbivores, leading to cascading effects across food chains. Similarly, AI can predict how coastal ecosystems are threatened by rising sea levels, which threaten habitats such as mangroves, coral reefs, and wetlands (Raj & Pasfield-Neofitou, 2024). These predictions enable conservationists to identify vulnerable ecosystems and prioritize interventions by offering detailed insights into potential climate change impacts; AI serves as an early warning system. It helps conservationists understand where detrimental effects are likely to occur, providing critical lead time to develop and implement adaptive management strategies. Weiskopf et al. (2020) describe these strategies, which include habitat restoration, assisted migration of species, or the establishment of climate-resilient protected areas.

Furthermore, AI supports scenario testing, allowing conservation planners to evaluate the effectiveness of different mitigation strategies under various climate conditions. This capability ensures that resources are efficient and that interventions are where they will have the most significant positive impact. Olawade et al. (2024A) argue that AI empowers stakeholders to combat and protect biodiversity for future generations by integrating predictive analytics into conservation planning.

2.1.3 Population Viability Analysis

One of the most vital applications of predictive analytics in biodiversity conservation is assessing the viability of species populations under varying environmental conditions. This process, known as population viability analysis (PVA), key factors influencing a species' long-term survival, such as genetic diversity, birth and death rates, and habitat quality (Desbiez et al., 2020). With these variables, AI provides insights into which species are most at risk of extinction and identifies the conservation strategies likely to be most effective.

Leins et al. (2021) explain that AI-powered models enhance PVA by simulating how environmental changes might affect species populations. For instance, changes in habitat quality due to deforestation, food availability caused by climate change, or the effects of human interventions such as conservation breeding programs. These simulations allow conservationists to predict how a species will respond to specific challenges, enabling them to make efforts and allocate resources to the most vulnerable species (Rather et al., 2024).

Moreover, AI enables conservationists to test the outcomes of different management strategies under various future scenarios. For example, models can estimate the impact of establishing wildlife corridors, implementing habitat restoration projects, or controlling invasive species. These insights ensure that conservation plans are data-driven and highly targeted, increasing their likelihood of success.

By incorporating predictive analytics into PVA, conservationists gain a powerful tool for proactive decision-making. AI enhances their ability to anticipate and mitigate risks, ensuring that efforts to preserve biodiversity are practical and efficient (Kamyab et al., 2023). This approach is essential for safeguarding species in a rapidly changing world and contributes significantly to global conservation goals.

2.1.4 Optimizing Resource Allocation and Intervention Timing

Predictive analytics provides invaluable support to conservationists by enabling the strategic allocation of resources. Pachar et al. (2023) suggest that AI models forecast where and when species and ecosystems face significant challenges. These insights allow conservation efforts toward areas requiring immediate attention, ensuring resources.

For example, AI can analyze climate trends, land-use patterns, and vulnerabilities to identify regions expected to experience severe environmental threats such as prolonged droughts, habitat loss, or other ecological pressures (Shivaprakash et al., 2022). Conservationists can use these high-risk zones for protective measures, such as establishing conservation corridors, implementing anti-deforestation policies, or enhancing local water management systems.

Timing is another critical factor that proves invaluable. AI can help determine the optimal moments for specific conservation actions, such as habitat restoration, reforestation, or the relocation of vulnerable species (Buchelt et al., 2024). Silvestro et al. (2022) suggest that habitat restoration efforts are most effective if timed before the onset of seasonal weather changes or anticipated ecological disruptions. Similarly, species relocation efforts can coincide with favorable environmental conditions, increasing the likelihood of success.

By enabling precise forecasts, conservation actions are timely and impactful. Instead of spreading resources thinly across various projects, conservationists can focus on interventions that address the most pressing issues. This targeted approach improves the effectiveness of conservation initiatives but also builds resilience in ecosystems, helping them adapt to future challenges.

The ability to predict the plan is essential. Predictive analytics equips conservationists with the tools to make informed decisions, enhancing the long-term success of efforts to protect biodiversity and ecosystems.

3. AI-DRIVEN HABITAT MONITORING: PROTECTING VULNERABLE ECOSYSTEMS

Habitat monitoring is a critical component of understanding the health of ecosystems and their changes over time. This monitoring offers essential insights into various environmental shifts, including habitat loss, species migration, and the impacts of climate change, particularly in remote or hard-to-reach areas (Wang et al., 2024). However, AI technologies have made habitat monitoring faster, more accurate, and more efficient. Popescu et al. (2024) argue that this transformation empowers conservationists to respond more effectively to emerging threats.

AI-powered tools have become invaluable for vast ecological data, enabling conservationists to track environmental changes with unprecedented precision. For example, AI-driven analysis of satellite imagery can detect habitat degradation, allowing researchers to identify and address ecological issues before they escalate, as presented in Figure 2. MLA excels at identifying patterns in species movements, which helps scientists understand how animals respond to their changing environments (Janga et al., 2023).

In addition to satellite data, AI-based systems can monitor climate impacts by processing information from sensors strategically placed across various ecosystems. Alotaibi and Nassif (2024) explain that this continuous data stream provides a detailed overview of environmental conditions, facilitating timely interventions when necessary.

One notable application of AI in habitat monitoring involves drones equipped with advanced sensors. These drones capture detailed images of forest canopies and soil conditions, enabling researchers to cover large areas quickly. AI algorithms process this data to detect early signs of deforestation, invasive species, or changes in vegetation health, even in the most remote forests (Buchelt et al., 2024).

Another impactful example is the application of AI in tracking endangered species through camera traps. These systems can automatically identify and catalog animals in their natural habitats, distinguishing between different species with remarkable accuracy. Fergus et al. (2024) state that this capability ensures that interventions can protect at-risk species.

Overall, AI is reshaping the landscape of habitat monitoring by providing powerful tools that enable precise, real-time tracking of environmental conditions. These innovations facilitate the early detection of potential threats and allow conservationists to make quicker and more informed decisions. As a result, vital ecosystems ensure a better future for biodiversity.

3.1 Remote Sensing and Satellite Imagery

Remote sensing, mainly through satellites, has emerged as one of the most powerful tools for large-scale habitat monitoring (Zhang & Li, 2023). AI algorithms with satellite imagery significant land-use changes, including deforestation, urban expansion, and agricultural encroachment, are presented in Figure 1, Table 1. Gu and Zeng (2024) highlight that these AI systems have intricate patterns related to vegetation loss, forest fragmentation, and alterations in water bodies, which are crucial for assessing the overall health of ecosystems.

By continuously monitoring these changes over time, conservationists can pinpoint areas most at risk, enabling them to focus on those locations for protection or restoration efforts. For instance, if satellite imagery reveals accelerated deforestation rates in a particular region, conservationists can allocate resources quickly to implement protective measures or restoration initiatives.

AI's remarkable capacity to analyze vast amounts of satellite data allows for large-scale environmental trends in near real-time. This rapid processing capability means that emerging threats can be addressed promptly, thereby minimizing their impact on vulnerable ecosystems. Furthermore, the detailed insights enhance the understanding of ecosystem dynamics, including how various factors interact and influence biodiversity.

Overall, the combination of remote sensing and AI represents a significant advancement in conservation efforts, offering a comprehensive approach to habitat monitoring that is both efficient and effective (Chen et al., 2023). By leveraging these technologies, conservationists can ensure strategies are data-driven and responsive, ultimately contributing to essential ecosystems and the biodiversity they support.

3.2 Camera Traps and Image Recognition

Camera traps are invaluable tools in wildlife research, allowing for images of animals in their natural habitats. They provide crucial data regarding animal populations, behaviors, and habitat usage (Nazir & Kaleem, 2021). However, the sheer volume of images generated by these cameras can be overwhelming for human researchers to process manually. Table 1 indicates where AI-powered image recognition technologies come into play, significantly transforming these images.

Figure 1 presents AI Algorithms that excel at rapidly sorting through thousands of photographs, automating counting individual animals, and even tracking their movements. This automation drastically reduces the time and labor required for data processing, allowing researchers to allocate their efforts toward interpreting results and making informed conservation decisions.

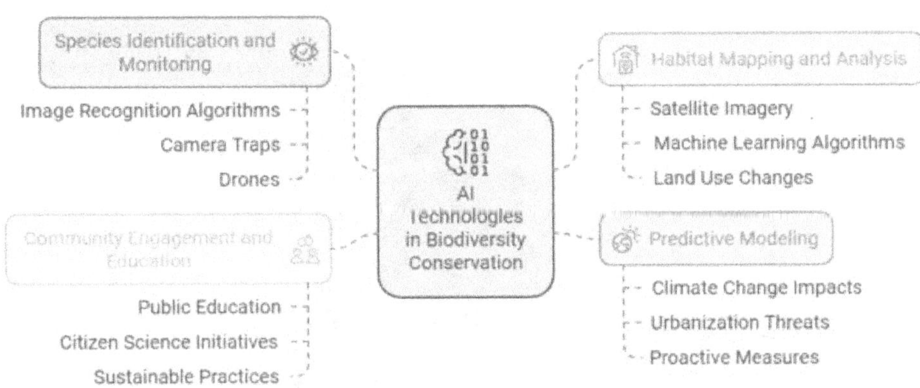

Moreover, AI-driven image recognition is particularly beneficial in identifying elusive or rare species that may be challenging to detect through traditional monitoring methods. By leveraging machine learning (ML), these algorithms can learn from diverse datasets, improving their accuracy and efficiency in recognizing different species. This capability enhances the overall understanding of biodiversity within ecosystems and provides deeper insights into the behaviors and interactions.

3.3 Acoustic Monitoring

Acoustic monitoring is another powerful tool that AI has enhanced for habitat monitoring. Sharma et al. (2023) note that many species, especially amphibians, birds, and marine life, are known for their distinctive vocalizations. AI algorithms can analyze sound recordings from field sensors or remote acoustic monitoring devices to detect these vocalizations, providing valuable data on species presence, population density, and behavioral patterns (Nieto-Mora, 2023). This technology helps monitor populations in remote or inaccessible areas where visual surveys are impossible. AI-driven acoustic analysis can help detect rare or endangered species by identifying their unique calls or sounds, as presented in Figure 2. By automating sound data, AI enables more efficient monitoring of ecosystems, even in challenging environments like dense forests, wetlands, or oceans.

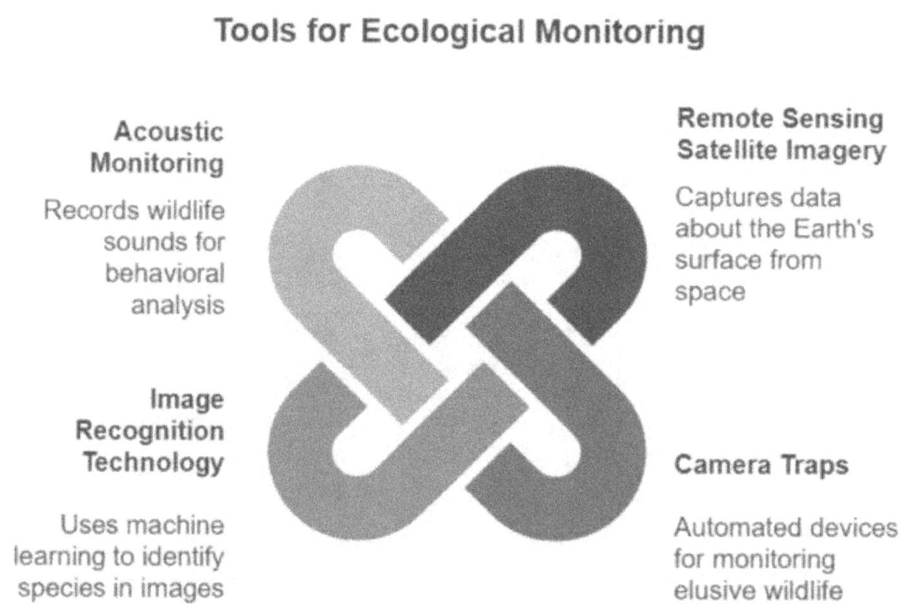

3.4 Continuous Monitoring and Early Threat Detection

AI-powered habitat monitoring offers a significant advantage over traditional methods by enabling real-time observation of ecosystems. While conventional approaches, such as on-site surveys or manual data collection, often provide periodic snapshots of ecological conditions, AI systems can deliver continuous, real-time data on habitat health. Tuia et al. (2022) explain that this capability allows conservationists to detect changes and potential threats.

For instance, AI-driven surveillance systems can monitor protected areas using cameras and drones, swiftly identifying suspicious activities like illegal logging or poaching in real-time, thereby enhancing efforts to protect endangered species (Jiménez et al., 2019).

Moreover, AI can analyze data from various sources, such as satellite imagery and environmental sensors, to predict and detect environmental changes. This proactive approach enables conservationists to anticipate and address issues like habitat degradation, climate change impacts, or the introduction of invasive species before they cause significant ecological damage.

By leveraging AI for real-time habitat monitoring, conservationists can respond more effectively to emerging threats, ensuring a quicker and more informed approach to protecting vulnerable ecosystems and species.

4. AI AND SPECIES CONSERVATION: FROM MONITORING TO PROTECTION

AI is revolutionizing species conservation by enhancing the monitoring of endangered and at-risk species. Traditional methods, such as manual surveys and field monitoring, often struggle to keep pace with rapidly changing ecosystems and the vast, sometimes remote, areas that require attention. Salem et al. (2024) argue that AI addresses these challenges by providing scalability, efficiency, and speed, enabling conservationists to monitor extensive landscapes, track species, and respond promptly to emerging threats.

4.1 Key Applications of AI in Species Conservation

Automated Species Identification and Monitoring: AI-driven image recognition technologies can process large volumes of data from camera traps, drones, and satellites to identify and monitor species. This automation accelerates data analysis, allowing real-time tracking of animal populations and behaviors. For example, AI algorithms can swiftly sort through thousands of photos, accurately identifying species, counting individuals, and even tracking their movements, thereby reducing the time and labor involved in data processing (Jiménez et al., 2019).

Predictive Analytics for Habitat Changes: AI models can analyze complex environmental data to forecast changes in species distributions and habitat conditions. By integrating climate models with current land-use data, AI can predict how species' habitats might shift over time due to climate change, habitat loss, or land-use changes. Casanelles-Abella et al. (2021) and Mkala et al. (2023) emphasize that this proactive approach enables conservationists to anticipate areas that will become increasingly important for species survival, guiding decisions on where to focus conservation efforts and protect vulnerable ecosystems.

Acoustic Monitoring and Communication Analysis: AI decodes animal communications, such as vocalizations of zebra finches, Hawaiian crows, and beluga whales. AI can help researchers understand species' behaviors and needs, facilitating more effective conservation strategies. The Earth Species Project is leveraging advanced AI tools to develop preliminary dictionaries of animal communications, aiming to enhance interspecies understanding and support environmental conservation efforts.

Anti-Poaching and Illegal Activity Detection: AI-powered surveillance systems, including cameras and drones, can monitor protected areas to detect suspicious activities like illegal logging or poaching in real time. By alerting authorities promptly, these systems enable immediate intervention, enhancing efforts to protect endangered species.

Population Viability Analysis: AI can analyze factors such as genetic diversity, birth and death rates, and habitat quality to evaluate the likelihood of a species' long-term survival. This process, known as population viability analysis (PVA), helps identify species at the greatest risk of extinction and determine the most effective conservation strategies. Desbiez et al. (2020) and Rather et al. (2024) explain that by simulating different future scenarios, AI ensures that conservation efforts are data-driven and targeted, increasing the chances of success in preserving biodiversity.

By integrating AI into conservation efforts, researchers and conservationists can enhance their ability to monitor species, predict and mitigate threats, and implement more effective protection strategies. This technological advancement significantly improves the outcomes of conservation initiatives, ensuring the survival of vulnerable species and the preservation of biodiversity.

4.2 Tracking Animal Movements

AI is transforming wildlife conservation by enhancing our ability to monitor and protect animal populations. Traditional methods often involve manual surveys and field observations, which can be time-consuming and limited in scope. AI technologies, however, offer scalable, efficient, and rapid solutions to these challenges.

4.2.1 AI-Enhanced Tracking and Monitoring

AI-powered tracking technologies are revolutionizing the study of animal movements. GPS collars and sensors provide real-time data on animal locations, movements, and behaviors, offering invaluable insights into migration routes, critical habitats, and species-environment interactions (Afzal et al., 2025).

AI models analyze this data to detect trends and predict future movements, helping conservationists identify areas where animals are most vulnerable. Foyet (2024) illustrates that AI can pinpoint migration corridors at risk from human activities like road construction or urban expansion, enabling proactive conservation measures.

Additionally, by monitoring animal behaviors, AI can detect signs of stress or illness, signaling the need for intervention before problems escalate. AlZubi and Al-Zu'bi (2023) highlight that this proactive approach enhances the ability to safeguard vulnerable ecosystems and species, ensuring a quicker, more effective response to environmental challenges.

4.2.2 AI in Habitat Monitoring

AI is also instrumental in habitat monitoring, providing real-time data on environmental conditions. Traditional monitoring methods often offer periodic snapshots, which can delay responses to emerging threats. In contrast, AI-enabled systems deliver continuous data, allowing conservationists to detect changes and potential threats early (Montaser et al., 2024).

AI-driven surveillance systems can monitor protected areas using cameras and drones, swiftly identifying suspicious activities like illegal logging or poaching in real-time. Moreover, AI can analyze data from various sources, such as satellite imagery and environmental sensors, to predict and detect environmental changes. This proactive approach enables conservationists to anticipate and address issues like habitat degradation, climate change impacts, or the introduction of invasive species before they cause significant ecological damage. Reynolds et al. (2024) explain that by leveraging AI for real-time habitat monitoring, conservationists can respond more effectively to emerging threats, ensuring a quicker and more informed approach to protecting vulnerable ecosystems and species.

4.2.3 AI in Species Conservation

AI plays a crucial role in species conservation, particularly for endangered and at-risk species. Traditional conservation methods often struggle to keep up with rapidly changing ecosystems and the large, sometimes remote, areas that need attention. Tuia et al. (2022) note that AI provides the scalability, efficiency, and speed required to monitor vast landscapes, track species, and respond to threats promptly.

Researchers can use AI to determine animal locations, sightings' dates and times, migration patterns, population dynamics, and social environments. This comprehensive data collection enhances our understanding of species' behavior and habitat use, informing more effective conservation strategies (Fergus et al., 2024).

4.2.4 Poaching Detection and Prevention

Poaching continues to pose a significant threat to endangered species worldwide. Traditional anti-poaching efforts often face challenges due to the vastness of protected areas and rugged terrains. Kaur (2023) suggests that AI is emerging as a powerful tool to enhance surveillance and detection capabilities, enabling more efficient and timely responses to poaching activities.

4.2.5 AI-Enhanced Surveillance Systems

AI-driven surveillance systems are transforming wildlife protection by automating suspicious activities. MLA analyzes data from various sources, including cameras, motion sensors, and audio recordings, to identify potential poaching incidents in real-time. Fergus et al. (2024) explain that, for instance, AI can detect unusual patterns of movement or behavior that may indicate illegal activities, allowing authorities to respond promptly.

4.2.6 Predictive Analytics for Poaching Prevention

Beyond real-time detection, AI excels in predictive analytics, assessing historical data to forecast potential poaching hotspots. By identifying areas with a high risk of poaching, conservationists can allocate resources more effectively, focusing patrols on regions most vulnerable to illegal activities (Keskin, 2023). This proactive approach enhances the efficiency of anti-poaching efforts, ensuring that interventions are timely and targeted.

4.2.7 Monitoring Online Wildlife Trade

AI also plays a crucial role in monitoring online platforms for illegal wildlife trade. By analyzing data from social media and online marketplaces, AI systems can detect and track wildlife trafficking activities, enabling law enforcement agencies to dismantle poaching networks more effectively (Rusch & Arora, 2020). Mozer and

Prost (2023) highlight that this capability extends the reach of conservation efforts beyond physical boundaries, addressing the global nature of wildlife trafficking.

4.2.8 Case Study: Vulture Tracking in Uganda

An innovative application of AI in anti-poaching efforts is vultures equipped with AI-infused trackers in Uganda's Murchison Falls National Park. These trackers analyze vulture behavior to provide real-time information on poacher activities, alerting rangers to potential threats. This approach not only aids in poaching detection but also contributes to the resurgence of wildlife populations in the park (Leibniz Institute for Zoo and Wildlife Research [IZW], 2024).

4.2.9 AI-Enhanced Surveillance Systems

AI-driven surveillance systems are transforming wildlife protection by automating suspicious activities. MLA data from various sources, including cameras, motion sensors, and audio recordings, can be used to identify potential poaching incidents in real time. Tuia et al. (2022) suggest that, for instance, AI can detect unusual patterns of movement or behavior that may indicate illegal activities, allowing authorities to respond promptly.

4.3 Genetic Conservation and Breeding Programs

Genetic diversity is essential for the health and survival of endangered species. Reduced genetic variation can lead to inbreeding, resulting in health problems and increased extinction risks. AI genetic research, providing tools, and developing more effective breeding programs.

4.3.1 Identifying Genetic Bottlenecks

AI analyses genetic data to identify genetic bottlenecks—points where species have low genetic diversity, indicating an increased risk of extinction. This information is critical for designing breeding programs aimed at preserving genetic diversity.

4.3.2 Optimizing Breeding Strategies

AI can optimize breeding strategies by identifying compatible individuals to mate with, helping preserve the species' overall health and adaptability. By analyzing genetic data, AI suggests breeding combinations that could lead to a healthier gene pool.

4.3.3 Assisted Reproductive Technologies

AI is also used in assisted reproductive technologies, such as artificial insemination, to diversify gene pools and strengthen populations. Zoos and aquariums have artificial insemination to increase genetic diversity and support conservation efforts.

Through these methods, AI is scientifically informing conservation efforts and helping maintain the genetic health of vulnerable populations.

4.4 Proactive and Targeted Interventions

AI supports conservation efforts by enabling continuous, real-time monitoring of species and ecosystems. This technological advancement empowers conservationists to take a proactive approach to species protection, allowing for early detection of threats and quick responses to mitigate potential harm.

4.5 Real-Time Data Analysis for Early Threat Detection

AI systems can process data from various sources—such as satellite imagery, sensor networks, and camera traps—to identify emerging threats to wildlife and habitats. By analyzing this data in real-time, AI can detect anomalies or patterns indicative of environmental stressors, such as habitat degradation or climate change impacts. This capability enables conservationists to implement timely interventions, such as habitat restoration or species relocation, to prevent irreversible damage.

4.5.1 Predictive Modeling for Proactive Conservation

Beyond immediate threat detection, AI excels in predictive modeling, assessing historical and real-time data to forecast future environmental changes and species behaviors. For example, AI can predict shifts in animal migration patterns due to climate change, allowing for preemptive actions to protect critical habitats or corridors. This foresight is crucial for adapting conservation strategies to dynamic environmental conditions.

4.5.2 Prioritizing Conservation Efforts

AI also aids in prioritizing conservation initiatives by analyzing data to identify species or habitats most at risk. This data-driven approach ensures that limited resources are utilized effectively, focusing efforts on areas where they can have the most significant impact. By highlighting regions with high conservation value and vulnerability, AI supports informed decision-making in resource management. Table 1 indicates the AI tools and techniques to monitor, analyze, and protect biodiversity.

Table 1. The AI tools and techniques to monitor, analyze, and protect biodiversity.

AI Technology	Function	Application in Biodiversity Conservation	Examples
Satellite Imagery	AI analyzes high-resolution satellite imagery to monitor land use, habitat changes, and deforestation.	It provides real-time insights into ecosystems, identifying environmental risks such as deforestation, habitat fragmentation, and the impacts of climate change.	Tracking deforestation in the Amazon rainforest using AI to analyze satellite data and detect illegal logging activities (Olawade et al., 2024B).
Environmental Sensor Data	AI analyses data from environmental sensors such as temperature, humidity, and pollution levels.	Monitors the health of ecosystems by tracking changes in environmental conditions.	Sensors deployed in coral reefs to monitor temperature changes and potential bleaching events, helping to protect marine life (González-Rivero et al., 2020)
Camera Trap Footage	AI-powered computer vision processes camera trap images to identify and monitor animal populations.	Tacks and monitors endangered species, their movements, and populations in remote or difficult-to-access locations.	Using AI to automatically identify and count rare species, such as tigers or snow leopards, from camera trap images in wildlife reserves (Velasco-Montero et al., 2024).
Drones	Drones equipped with cameras and sensors collect environmental data in hard-to-reach areas.	Surveys large areas of land or forest quickly, providing detailed views of the ecosystem complex for humans to assess.	Drones monitor forest canopies and detect illegal activities, such as poaching or logging in protected areas (Buchelt et al., 2024).
ML	ML algorithms analyze large datasets to recognize patterns and predict future environmental changes.	Predict trends in habitat health, species movements, and climate impacts, enabling proactive conservation strategies.	Using ML algorithms to predict the spread of invasive species in a specific region allows for early intervention (Fernandes et al., 2020)
DL	A subset of ML that processes large and complex datasets, such as images or sound recordings, for deeper insights.	Enhances the ability to analyze complex data from cameras, sensors, and acoustic recordings, improving species identification and monitoring.	Identifying animal calls in audio recordings, such as those from endangered birds, to monitor their presence and population trends (Taye, 2023).
Predictive Analytics	AI leverages historical data to predict future trends, such as ecological changes and species decline	. It also anticipates habitat alterations, species migration patterns, and emerging threats like climate change, aiding in developing long-term conservation strategies.	Predicting how rising temperatures might affect the migration patterns of polar bears in the Arctic (Abdelaal et al., 2024).

continued on following page

Table 1. Continued

AI Technology	Function	Application in Biodiversity Conservation	Examples
Remote Sensing	AI processes data from remote sensing technologies to assess large-scale environmental conditions.	Used for habitat mapping, land cover classification, and monitoring ecological changes across vast areas.	Remote sensing of wetlands to assess their health and biodiversity status, enabling targeted conservation efforts (Rana et al., 2024).
Acoustic Monitoring	AI analyses audio data from environmental sensors to track the sounds of species, often in difficult-to-reach areas.	Tracks the presence of species through their vocalizations, providing an efficient way to monitor wildlife populations without direct observation.	Monitoring forest soundscapes to identify the presence of specific animal species, such as elephants or frogs, through their calls (Alexandre, 2024).
GPS Collars and Sensors	GPS collars and sensors provide real-time location data for tracking animal movements and behavior.	Tracks the movement patterns and behavior of endangered species, aiding in understanding their range and habitat use.	To track the migration of endangered species like wildebeests in the Serengeti, providing critical data for conservation planning (English et al., 2024).

5. CHALLENGES AND ETHICAL CONSIDERATIONS IN AI FOR BIODIVERSITY CONSERVATION

While AI presents significant potential to advance biodiversity conservation, its integration brings challenges and ethical concerns to ensure responsible and equitable use.

5.1 Data Quality and Availability

AI's effectiveness in biodiversity conservation heavily relies on the quality of data, which is often lacking or incomplete. Many ecosystems, especially in remote or poorly studied regions, suffer from data gaps, making it difficult to build accurate AI models (Galaz et al., 2021). Inconsistent or biased data can lead to unreliable outcomes, undermining conservation efforts. Additionally, environmental variables used in AI models may not always accurately reflect the complex realities of ecosystems, potentially leading to faulty predictions. Data collection must be improved, focusing on high-resolution, accurate, and comprehensive datasets. Addressing interoperability between diverse datasets is essential to ensure that AI can work effectively across different conservation contexts.

5.2 Cost and Accessibility

Although AI holds transformative potential, its implementation can be financially burdensome, particularly for conservation organizations operating in low-resource environments. The cost of advanced infrastructure, software, and specialized personnel required to use AI can be prohibitive, creating a disparity between well-funded and under-resourced access to cutting-edge tools that could help address biodiversity loss. It should focus on developing cost-effective solutions such as open-source platforms and fostering international collaboration. Supporting partnerships and funding initiatives can help reduce costs, making AI more accessible to conservation projects globally, regardless of their financial capacity.

5.3 Ethical Considerations

Using AI in biodiversity conservation raises important ethical issues, particularly in surveillance, privacy, and animal welfare. GPS tracking can provide valuable data on wildlife movement and cause stress or alter natural behaviors. Similarly, AI-powered surveillance systems that detect illegal activities like poaching can raise concerns about privacy violations, mainly if used to monitor human activities around protected areas. Clear ethical guidelines are necessary to mitigate these risks, emphasizing transparency, informed consent, and privacy protections for humans and wildlife. Conservation efforts should always prioritize animal welfare and respect for local communities, ensuring that the benefits of AI technologies align with ethical standards and do not infringe on fundamental rights.

5.4 Dependency on Technology

While AI can significantly enhance conservation efforts, excessive reliance on technology could overshadow traditional, community-driven conservation practices. Biodiversity conservation is not only about data analysis—it also involves fieldwork, local knowledge, and active community participation. AI should serve as a complement to, rather than a replacement for, these time-tested methods. Striking a balance and ensuring AI tools support local conservation practitioners and integrate traditional ecological knowledge is crucial. By combining advanced technology with local wisdom, AI can contribute to more holistic and culturally respectful conservation strategies that reflect the needs of both ecosystems and the communities that protect them.

5.5 Comparative Analysis of AI and Traditional Methods

While AI offers powerful tools for conservation, its application may not always be ideal or cost-effective, especially in regions with limited resources or infrastructure. Traditional conservation methods often involve deep, local knowledge and a nuanced understanding of ecosystems that AI models may not fully replicate. For example, local communities may have insights into species behaviors or environmental changes that AI cannot capture through data alone. A hybrid approach that combines AI with traditional methods can provide a more comprehensive solution, integrating technological innovation with hands-on, context-specific conservation practices. Evaluating when to use AI and when to rely on traditional methods is essential for optimizing conservation efforts and ensuring long-term sustainability.

6. AI IN ECOLOGICAL RESTORATION

Ecological restoration involves the recovery of degraded or damaged ecosystems, which often face challenges such as identifying intervention priorities, selecting appropriate species, and determining the most effective actions. AI offers great potential to streamline and optimize these processes, improving the efficiency and impact of restoration efforts.

- **Analyzing Ecosystem Health:** AI can process vast amounts of environmental data, such as satellite imagery and field reports, to assess ecosystem health. By analyzing historical and real-time data, AI can track changes in biodiversity and habitat conditions, helping conservationists identify at-risk ecosystems and prioritize areas for restoration.
- **Recommending Restoration Interventions:** AI can suggest specific restoration actions based on the current state of ecosystems. For example, AI can recommend planting native species suited to the local climate, restoring water sources vital for wildlife, or controlling invasive species threatening local biodiversity. By analyzing ecological factors such as soil composition, rainfall patterns, and temperature, AI ensures that restoration measures are appropriate for the environment.
- **Optimizing Species Reintroductions:** AI can help with species reintroductions by modeling ideal conditions for species to thrive in restored habitats. For example, AI can simulate environmental scenarios to determine the best areas for reintroducing endangered species, optimizing food availability, predator presence, and habitat suitability, thereby increasing the likelihood of successful reintroductions.

- **Identifying Priority Areas:** AI can help conservationists pinpoint areas needing immediate restoration. For example, it can identify ecosystems at high risk of collapse due to climate change, pollution, or overexploitation, prioritizing regions based on biodiversity hotspots and ecological stability.

By incorporating AI into ecological restoration, efforts can be more targeted, informed, and efficient, improving the long-term success of conservation initiatives.

6.1AI-Powered Citizen Science

AI has the potential to transform how the public engages with biodiversity conservation, enabling widespread participation in monitoring and data collection through citizen science.

- **Facilitating Species Identification:** AI-powered mobile apps can assist users in identifying species from photos or video feeds using deep learning algorithms. These tools quickly match observations to species databases, providing immediate feedback even to non-experts. For example, AI can help identify a bird species from a photo and provide information about its behavior, habitat, and conservation status.
- **Real-Time Data Collection and Analysis:** AI enables real-time data collection and analysis, improving the efficiency of biodiversity monitoring. For instance, AI can process data from environmental sensors, camera traps, and acoustic devices, allowing for faster and more accurate species identification. AI algorithms can flag unusual sightings, such as rare species or novel behavioral patterns, helping researchers focus on areas of interest.
- **Filling Data Gaps in Remote Areas:** AI-powered citizen science can help fill significant data gaps in remote or under-researched regions, where scientific monitoring is often limited. By collecting data from local communities, tourists, or nature enthusiasts, AI can process large datasets, increasing the geographic and temporal coverage of biodiversity monitoring.
- **Building Awareness and Empowering Communities:** Engaging communities in AI-powered citizen science helps foster a deeper understanding of biodiversity and encourages individuals to participate actively in conservation. This grassroots involvement promotes greater ownership of local environmental challenges and amplifies the reach and impact of conservation efforts.

6.2 Collaboration Across Sectors

The integration of AI into biodiversity conservation will require collaboration across sectors. Governments, NGOs, research institutions, and the private sector bring essential expertise and resources. AI can facilitate partnerships that lead to more holistic and effective conservation strategies.

- **Cross-Sector Data Sharing:** Governments, private companies, academic institutions, and conservation organizations often collect valuable data on biodiversity and ecosystem health. AI can enable seamless integration of these datasets, creating a more comprehensive picture of biodiversity health. For example, satellite data from government agencies could be combined with drone technology from private companies and field data from NGOs, enabling AI to identify trends and inform policy decisions.
- **Public-Private Partnerships:** Collaboration between the public and private sectors will be essential for scaling AI-driven conservation solutions. Private companies with AI and ML learning expertise can provide the tools to develop these technologies, while governments and NGOs can offer infrastructure, funding, and regulatory oversight. For example, private companies might develop AI-driven tools for monitoring deforestation or detecting poaching, while public entities can ensure their deployment and proper regulation in the field.
- **Global Collaboration:** Given the global scope of biodiversity loss, international cooperation will be critical. AI can facilitate cross-border data sharing and international partnerships, enabling joint conservation efforts. For example, AI could track migratory species across different countries, ensuring conservation efforts across regions. Pooling resources and knowledge in global collaborations can address the complex challenges of biodiversity loss on a larger scale.

7. CONCLUSION

In conclusion, integrating AI into biodiversity conservation represents a pivotal advancement in understanding and protecting our planet's ecosystems. The challenges posed by habitat destruction, climate change, poaching, and invasive species are vast and complex, necessitating innovative approaches to safeguard endangered species and their habitats. AI's capacity for predictive analytics and real-time habitat monitoring enables conservationists to make informed decisions based on data-driven insights, ultimately enhancing the effectiveness of their efforts.

Through MLA, AI can identify patterns and trends that might be overlooked through traditional methods, allowing for more proactive management of endangered species. For instance, by analyzing historical data and environmental variables, AI models can forecast potential population declines or habitat degradation, empowering conservationists to implement timely interventions. Additionally, AI-driven habitat monitoring tools, such as remote sensing and drone technology, provide comprehensive and continuous assessments of ecosystems, enabling real-time responses to emerging threats.

The case studies highlighted in this chapter underscore AI's practical applications in various conservation contexts, demonstrating its potential to transform the landscape of wildlife protection. From tracking poaching activities to monitoring species migration patterns, these applications illustrate AI's versatility and emphasize the collaborative efforts required among researchers, technologists, and conservation practitioners.

The future of AI in biodiversity conservation appears promising. Continued technological advancements and a growing awareness of the importance of preserving biodiversity will likely lead to even more innovative solutions. As we move forward, it will be essential to foster interdisciplinary collaboration, engage local communities, and ensure that ethical considerations guide the implementation of AI in conservation efforts.

Ultimately, embracing AI as a tool for biodiversity conservation is not just about technology; it is about a shared responsibility to protect our planet for future generations. By harnessing the power of AI, we can pave the way for a more sustainable and resilient coexistence with nature, ensuring that the rich tapestry of life on Earth continues to thrive.

REFERENCES

Abdelaal, M., AL-Huqail, A. A., Alghanem, S. M. S., Alhaithloul, H. A. S., Al-Robai, S. A., Abeed, A. H. A., Dakhil, M. A., El-Barougy, R. F., & Yahia, A. A. (2024). Population status, habitat preferences, and predictive current and future distributions of three endangered *Silene* species under changing climate. *Frontiers in Plant Science*, *15*, 1336911. DOI: 10.3389/fpls.2024.1336911 PMID: 38966141

Afzal, N., Rehman, M., Seneviratne, L., & Hussain, I. (2025). The convergence of AI and animal-inspired robots for ecological conservation. *Ecological Informatics*, *85*, 102950. DOI: 10.1016/j.ecoinf.2024.102950

Akter, S., Babu, M. M., Hani, U., Sultana, S., Bandara, R., & Grant, D. (2024). Unleashing the power of AI for climate action in industrial markets. *Industrial Marketing Management*, *117*, 92–113. DOI: 10.1016/j.indmarman.2023.12.011

Alotaibi, E., & Nassif, N. (2024). AI in environmental monitoring: In-depth analysis. *Discover AI*, *4*, 84. DOI: 10.1007/s44163-024-00198-1

AlZubi, A. A., & Al-Zu'bi, M. (2023). Application of AI in monitoring of animal health and welfare. *Indian Journal of Animal Research*, *57*(11), 1550–1555.

Aslam, R. W., Shu, H., Javid, K., Pervaiz, S., Mustafa, F., Raza, D., Ahmed, B., Quddoos, A., Al-Ahmadi, S., & Hatamleh, W. A. (2024). Danish Raza, Bilal Ahmed, Abdul Quddoos, Saad Al-Ahmadi, Wesam Atef Hatamleh, Wetland identification through remote sensing: Insights into wetness, greenness, turbidity, temperature, and changing landscapes. *Big Data Research*, *35*, 100416. DOI: 10.1016/j.bdr.2023.100416

Ayoola, V. B., Idoko, I. P., Eromonsei, S., & Afolabi, O. (2024). The role of big data and AI in enhancing biodiversity conservation and resource management in the USA. *World Journal of Advanced Research and Reviews*, *23*(2), 1851–1873. DOI: 10.30574/wjarr.2024.23.2.2350

Barras, A. G., Braunisch, V., & Arlettaz, R. (2021). *Predictive models of the distribution and abundance of a threatened mountain species show that the impacts of climate change overrule those of land use change*. Biodiversity Research., DOI: 10.1111/ddi.13247

Bolón-Canedo, V., Morán-Fernández, L., Cancela, B., & Alonso-Betanzos, A. (2024). A review of green AI: Towards a more sustainable future. *Neurocomputing*, *599*, 128096. DOI: 10.1016/j.neucom.2024.128096

Brodie, S., Allen-Ankins, S., Towsey, M., Roe, P., & Schwarzkopf, L. (2020). Automated species identification of frog choruses in environmental recordings using acoustic indices. *Ecological Indicators, 119*, 106852. DOI: 10.1016/j.ecolind.2020.106852

Buchelt, A., Adrowitzer, A., Kieseberg, P., Gollob, C., Nothdurft, A., Eresheim, S., Tschiatschek, S., Stampfer, K., & Holzinger, A. (2024). Exploring AI for applications of drones in forest ecology and management. *Forest Ecology and Management, 551*, 121530. DOI: 10.1016/j.foreco.2023.121530

Casanelles-Abella, J., Chauvier, Y., Zellweger, F., Villiger, P., Frey, D., Ginzler, C., Moretti, M., & Pellissier, L. (2021). Applying predictive models to study the ecological properties of urban ecosystems: A case study in Zürich, Switzerland. *Landscape and Urban Planning, 214*, 104137. DOI: 10.1016/j.landurbplan.2021.104137

Chen, L., Chen, Z., & Zhang, Y.. (2023). AI based solutions for climate change: A review. *Environmental Chemistry Letters, 21*(8), 2525–2557. DOI: 10.1007/s10311-023-01617-y

Dalton, D., Berger, V., Kirchmeir, H., Adams, V., Botha, J., Halloy, S., Hart, R., Švara, V., Torres Ribeiro, K., Chaudhary, S., & Jungmeier, M. (2024). *A framework for monitoring biodiversity in protected areas and other effective area-based conservation measures: Concepts, methods, and technologies* (IUCN WCPA Technical Report Series No. 7, pp. 1-110). IUCN.

Desbiez, A. L. J., Bertassoni, A., & Traylor-Holzer, K. (2020). Population viability analysis as a tool for giant anteater conservation. *Perspectives in Ecology and Conservation, 18*(2), 124–131. DOI: 10.1016/j.pecon.2020.04.004

Dhanya, V. G., Subeesh, A., Kushwaha, N. L., Vishwakarma, D. K., Kumar, T. N., Ritika, G., & Singh, A. N. (2022). Deep learning-based computer vision approaches for smart agricultural applications. *AI in Agriculture, 6*, 211–229. DOI: 10.1016/j.aiia.2022.09.007

English, H. M., Börger, L., Kane, A., & Ciuti, S. (2024). Advances in bio-logging can identify nuanced energetic costs and gains in predators. *Movement Ecology, 12*(1), 7. DOI: 10.1186/s40462-024-00448-y PMID: 38254232

Fergus, P., Chalmers, C., Longmore, S., & Wich, S. (2024). Harnessing AI for wildlife conservation. *Conservation, 4*(4), 685–702. DOI: 10.3390/conservation4040041

Fernandes, A. C. M., Quintero Gonzalez, R., Lenihan-Clarke, M. A., Trotter, E. F. L., & Arsanjani, J. J. (2020). ML for conservation planning in a changing climate. *Sustainability (Basel), 12*(18), 7657. DOI: 10.3390/su12187657

Folliot, A., Haupert, S., Ducrettet, M., Sèbe, F., & Sueur, J. (2022). Using acoustics and AI to monitor pollination by insects and tree use by woodpeckers. *The Science of the Total Environment, 838*(Part 1), 155883. DOI: 10.1016/j.scitotenv.2022.155883 PMID: 35569672

Fontúrbel, F. E., Nespolo, R. F., Amico, G. C., & Watson, D. M. (2021). Climate change can disrupt ecological interactions in mysterious ways: Using ecological generalists to forecast community-wide effects. *Climate Change Ecology, 2*, 100044. DOI: 10.1016/j.ecochg.2021.100044

Foyet, M. A. (2024, March 5). AI in conservation: Where we came from — and where we are heading. *Nature and Biodiversity*. World Economic Forum.

Galaz, V., Centeno, M. A., Callahan, P. W., Causevic, A., Patterson, T., Brass, I., Baum, S., Farber, D., Fischer, J., Garcia, D., McPhearson, T., Jimenez, D., King, B., Larcey, P., & Levy, K. (2021). AI, systemic risks, and sustainability. *Technology in Society, 67*, 101780. DOI: 10.1016/j.techsoc.2021.101741

González-Rivero, M., Beijbom, O., Rodriguez-Ramirez, A., Bryant, D. E. P., Ganase, A., Gonzalez-Marrero, Y., Herrera-Reveles, A., Kennedy, E. V., Kim, C. J. S., Lopez-Marcano, S., Markey, K., Neal, B. P., Osborne, K., Reyes-Nivia, C., Sampayo, E. M., Stolberg, K., Taylor, A., Vercelloni, J., Wyatt, M., & Hoegh-Guldberg, O. (2020). Monitoring of coral reefs using AI: A feasible and cost-effective approach. *Remote Sensing, 12*(3), 489. DOI: 10.3390/rs12030489

Gu, Z., & Zeng, M. (2024). Using AI and satellite remote sensing in land cover change detection: Review and perspectives. *Sustainability, 16*(1), 274. DOI: 10.3390/su16010274

Janga, B., Asamani, G. P., Sun, Z., & Cristea, N. (2023). A review of practical AI for remote sensing in Earth sciences. *Remote Sensing (Basel), 15*(16), 4112. DOI: 10.3390/rs15164112

Jiménez López, J., & Mulero-Pázmány, M. (2019). Drones for conservation in protected areas: Present and future. *Drones (Basel), 3*(1), 10. DOI: 10.3390/drones3010010

Kahrić, A., Kulijer, D., Dedić, N., & Šnjegota, D. (2022). Degradation of ecosystems and loss of ecosystem services. In Prata, J. C., Ribeiro, A. I., & Rocha-Santos, T. (Eds.), *One health* (pp. 281–327). Academic Press.

Kalogiannidis, S., Kalfas, D., Papaevangelou, O., Giannarakis, G., & Chatzitheodoridis, F. (2024). The role of AI technology in predictive risk assessment for business continuity: A case study of Greece. *Risks, 12*(2), 19. DOI: 10.3390/risks12020019

Kamyab, H., Khademi, T., & Chelliapan, S., SaberiKamarposhti, M., Rezania, S., Yusuf, M., Farajnezhad, M., Abbas, M., Jeon, B. H., & Ahn, Y. (. (2023). The latest innovative avenues for the utilization of AI and big data analytics in water resource management. *Results in Engineering*, *20*, 100912. Advance online publication. DOI: 10.1016/j.rineng.2023.101566

Kaur, R., Gabrijelčič, D., & Klobučar, T. (2023). AI for cybersecurity: Literature review and future research directions. *Information Fusion*, *97*, 101804. DOI: 10.1016/j.inffus.2023.101804

Keskin, B. B., Griffin, E. C., Prell, J. O., Dilkina, B., Ferber, A., MacDonald, J., Hilend, R., Griffis, S., & Gore, M. L. (2023). Quantitative investigation of wildlife trafficking supply chains: A review. *Omega*, *115*, 102780. DOI: 10.1016/j.omega.2022.102780

Kumar, H. (2024, November 14). *The future of AI in ecosystem management*. NASSCOM Community. https://community.nasscom.in

Leibniz Institute for Zoo and Wildlife Research (IZW). (2024, November 19). Vultures and artificial intelligence(s) as death detectors: High-tech approach for wildlife research and conservation. *ScienceDaily*.https://www.sciencedaily.com/releases/2024/11/241119-vultures-ai.htm

Leins, J. A., Banitz, T., Grimm, V., & Drechsler, M. (2021). High-resolution PVA along large environmental gradients to model the combined effects of climate change and land use timing: Lessons from the large marsh grasshopper. *Ecological Modelling*, *440*, 109355. DOI: 10.1016/j.ecolmodel.2020.109355

Llewelyn, J., Strona, G., Dickman, C. R., Greenville, A. C., Wardle, G. M., Lee, M. S. Y., Doherty, S., Shabani, F., Saltré, F., & Bradshaw, C. J. A. (2023). Predicting predator-prey interactions in terrestrial endotherms using random forest. *Ecography*, *2023*(9), e06619. DOI: 10.1111/ecog.06619

Metolo, A., & Fovet, A. (2024, March 5). *AI in conservation: Where we came from – and where we are heading*. World Economic Forum. https://www.weforum.org/agenda/2024/03/ai-in-conservation

Mkala, E. M., Mwanzia, V., Nzei, J., Oluoch, W. A., Ngarega, B. K., Wanga, V. O., Oulo, M. A., Mutie, F. M., Kilingo, F. M., Rono, P., Waswa, E. N., Mutinda, E. S., Ochieng, C. O., Mwachala, G., Hu, G.-W., Wang, Q.-F., Katunge, J. K., & Victoire, I. (2023). Predicting the potential impacts of climate change on the endangered endemic Annonaceae species in East Africa. *Heliyon*, *9*(6), e17405. DOI: 10.1016/j.heliyon.2023.e17405 PMID: 37416643

Montaser, N. A. R., Ali, M. A. H., Khoo, S. Y., & Alkhedher, M. (2024). AI-powered IoT and UAV systems for real-time detection and prevention of illegal logging. *Results in Engineering*, *24*, 103277. DOI: 10.1016/j.rineng.2024.103277

Mozer, A., & Prost, S. (2023). An introduction to illegal wildlife trade and its effects on biodiversity and society. *Forensic Science International. Animals and Environments*, *3*, 100064. DOI: 10.1016/j.fsiae.2023.100064

Muluneh, M. G. (2021). Impact of climate change on biodiversity and food security: A global perspective—a review article. *Agriculture & Food Security*, *10*(1), 36. DOI: 10.1186/s40066-021-00318-5

Nazir, S., & Kaleem, M. (2021). Advances in image acquisition and processing technologies are transforming animal ecological studies. *Ecological Informatics*, *61*, 101212. DOI: 10.1016/j.ecoinf.2021.101212

Nieto-Mora, A., Rodríguez-Buritica, S., Rodríguez-Marín, P., Martínez-Vargaz, J. D., & Isaza-Narváez, C. (2023). A systematic review of machine learning methods applied to eco-acoustics and soundscape monitoring. *Heliyon*, *9*(10), e20275. DOI: 10.1016/j.heliyon.2023.e20275 PMID: 37790981

Olawade, D. B., Wada, O. Z., David-Olawade, A. C., Fapohunda, O., Ige, A. O., & Ling, J. (2024A). AI potential for net zero sustainability: Current evidence and prospects. *Next Sustainability*, *4*, 100041. DOI: 10.1016/j.nxsust.2024.100041

Olawade, D. B., Wada, O. Z., Ige, A. O., Egbewole, B. I., Olojo, A., & Oladapo, B. I. (2024B). AI in environmental monitoring: Advancements, challenges, and future directions. *Hygiene and Environmental Health Advances*, *12*, 100114. DOI: 10.1016/j.heha.2024.100114

Pachar, S., Dudeja, D., Batra, N., Tomar, V., Bhimavarapu, J. P., & Singh, A. K. (2023). Recent developments in machine learning predictive analytics for disaster resource allocation. *Engineering Proceedings*, *59*(1), 19. DOI: 10.3390/engproc2023059019

Popescu, S. M., Mansoor, S., Wani, O. A., Kumar, S. S., Sharma, V., Sharma, A., Arya, V. M., Kirkham, M. B., Hou, D., Bolan, N., & Chung, Y. S. (2024). AI and IoT-driven technologies for environmental pollution monitoring and management. *Frontiers in Environmental Science*, •••, 12.

Raj, N., & Pasfield-Neofitou, S. (2024). Assessment and prediction of sea level and coastal wetland changes in small islands using remote sensing and AI. *Remote Sensing*, *16*(3), 551. DOI: 10.3390/rs16030551

Rather, M. A., Ahmad, I., Shah, A., Hajam, Y. A., Amin, A., Khursheed, S., Ahmad, I., & Rasool, S. (2024). Exploring opportunities of Artificial Intelligence in aquaculture to meet increasing food demand. *Food Chemistry: X*, *22*, 101309. DOI: 10.1016/j.fochx.2024.101309 PMID: 38550881

Rawas, S. (2024). AI: The future of humanity. *Discover Artificial Intelligence*, *4*(1), 25. DOI: 10.1007/s44163-024-00118-3

Reynolds, S. A., Beery, S., Burgess, N., Burgman, M., Butchart, S. H. M., Cooke, S. J., Coomes, D., Danielsen, F., Di Minin, E., Durán, A. P., Gassert, F., Hinsley, A., Jaffer, S., Jones, J. P. G., Li, B. V., Mac Aodha, O., Madhavapeddy, A., O'Donnell, S. A. L., Oxbury, W. M., & Sutherland, W. J. (2024). The potential for AI to revolutionize conservation: A horizon scan. *Trends in Ecology & Evolution*. Advance online publication. DOI: 10.1016/j.tree.2024.11.013 PMID: 39694720

Rusch, N., & Arora, P. (2020). AI-based strategies to combat wildlife trafficking and wet markets in Asia: A critical review. *Konrad-Adenauer-Stiftung*.

Salem, A. H., Azzam, S. M., Emam, O. E., & Abohany, A. A. (2024). Advancing cybersecurity: A comprehensive review of AI-driven detection techniques. *Journal of Big Data*, *11*(1), 105. DOI: 10.1186/s40537-024-00957-y

Secci, D., Tanda, M. G., D'Oria, M., & Todaro, V. (2023). Artificial intelligence models to evaluate the impact of climate change on groundwater resources. *Journal of Hydrology, 627*(Part B), 130359. DOI: 10.1016/j.jhydrol.2023.130359

Sharma, S., Sato, K., & Gautam, B. P. (2023). A methodological literature review of acoustic wildlife monitoring using artificial intelligence tools and techniques. *Sustainability (Basel)*, *15*(9), 7128. DOI: 10.3390/su15097128

Shivaprakash, K. N., Swami, N., Mysorekar, S., Arora, R., Gangadharan, A., Vohra, K., Jadeyegowda, M., & Kiesecker, J. M. (2022). Potential for artificial intelligence (AI) and machine learning applications in biodiversity conservation, managing forests, and related services in India. *Sustainability (Basel)*, *14*(12), 7154. DOI: 10.3390/su14127154

Silvestro, D., Goria, S., Sterner, T., & Antonelli, A. (2022). Improving biodiversity protection through artificial intelligence. *Nature Sustainability*, *5*(5), 415–424. DOI: 10.1038/s41893-022-00851-6 PMID: 35614933

Soori, M., Arezoo, B., & Dastres, R. (2023). A review of artificial intelligence, machine learning, and deep learning in advanced robotics. *Cognitive Robotics*, *3*, 54–70. DOI: 10.1016/j.cogr.2023.04.001

Taye, M. M. (2023). Understanding machine learning with DL: Architectures, workflow, applications, and future directions. *Computers*, *12*(5), 91. DOI: 10.3390/computers12050091

Tuia, D., Kellenberger, B., Beery, S., Costelloe, B. R., Zuffi, S., Risse, B., Mathis, A., Mathis, M. W., van Langevelde, F., Burghardt, T., Kays, R., Klinck, H., Wikelski, M., Couzin, I. D., van Horn, G., Crofoot, M. C., Stewart, C. V., & Berger-Wolf, T. (2022). Perspectives in machine learning for wildlife conservation. *Nature Communications*, *13*(1), 792. DOI: 10.1038/s41467-022-27980-y PMID: 35140206

United Nations Environment Programme. (2022, November 7). How artificial intelligence is helping tackle environmental challenges. *UNEP News*. https://www.unep.org/resources/story/how-artificial-intelligence-helping-tackle environmental-challenges

Veiga Branco, V., Correia, L., & Cardoso, P. (2023). The use of machine learning in species threats and conservation analysis. *Biological Conservation*, *283*, 110091. DOI: 10.1016/j.biocon.2023.110091

Velasco-Montero, D., Fernández-Berni, J., Carmona-Galán, R., Sanglas, A., & Palomares, F. (2024). Reliable and efficient integration of AI into camera traps for smart wildlife monitoring based on continual learning. *Ecological Informatics*, *83*, 102815. DOI: 10.1016/j.ecoinf.2024.102815

Wang, Z., Wang, T., Zhang, X., Wang, J., Yang, Y., Sun, Y., Guo, X., Wu, Q., Nepovimova, E., Watson, A. E., & Kuca, K. (2024). Biodiversity conservation in the context of climate change: Facing challenges and management strategies. *The Science of the Total Environment*, *937*, 173377. DOI: 10.1016/j.scitotenv.2024.173377 PMID: 38796025

Weiskopf, S. R., Rubenstein, M. A., Crozier, L. G., Gaichas, S., Griffis, R., Halofsky, J. E., Hyde, K. J. W., Morelli, T. L., Morisette, J. T., Muñoz, R. C., Pershing, A. J., Peterson, D. L., Poudel, R., Staudinger, M. D., Sutton-Grier, A. E., Thompson, L., Vose, J., Weltzin, J. F., & Whyte, K. P. (2020). Climate change affects biodiversity, ecosystems, ecosystem services, and natural resource management in the United States. *The Science of the Total Environment*, *733*, 137782. DOI: 10.1016/j.scitotenv.2020.137782 PMID: 32209235

Zhang, J., & Li, J. (2023). Spacecraft. In Zhang, J., & Li, J. (Eds.), *Spatial cognitive engine technology* (pp. 129–162). Academic Press., DOI: 10.1016/B978-0-323-95107-4.00004-4

Chapter 3
AI Applications in Environmental Monitoring:
Case Studies on Deforestation and Pollution Assessment.

Olawale Theophilus Ogunwumi
University of Lagos, Nigeria

Rumbidzai Matindike
https://orcid.org/0009-0003-1804-2858
University of Johannesburg, South Africa

ABSTRACT

Artificial Intelligence (AI) is transforming environmental monitoring by providing advanced tools to address ecological challenges. This chapter examines Ai applications in monitoring deforestation and pollution, bridging gaps in traditional methods that lack real-time data, precision and scalability. The study explores how Ai technologies enhance environmental assessment and decision-making, particularly in identifying and managing deforestation and pollution. Using a case study approach, the chapter reviews applications of machine learning, remote sensing, and predictive analytics. Methodologies include analyzing satellite imagery to track forest loss and using Ai models to monitor air and water pollution. Key findings reveal that Ai significantly improves the speed and accuracy of detecting environmental changes, delivering actionable insights for policymakers and conservationists. The findings emphasize Ai's potential to close gaps in environmental monitoring and support sustainable resource management, highlighting challenges such as data access and technological disparities.

DOI: 10.4018/979-8-3373-0680-3.ch003

1. INTRODUCTION

What is Environmental Monitoring without the application of Artificial Intelligence (AI) technology? Environmental Monitoring plays a crucial role in managing and mitigating the adverse effects of deforestation and environmental pollution. In recent years AI has been offering methods that are innovative for the collection, analysis, and interpretation of data which has emerged as a transformative tool. AI models have increased in their use since they aid in predicting the impacts of climate change, forecasting the weather, and assessing climate (Alotaibi & Nassif, 2024).

For real-time monitoring of the quality of air and water, technologies that include Machine (ML) and Deep (DL) Learning have proven to be highly effective (Olawade et al., 2024). These tools are technologically advanced and they allow accurate detection of pollution sources and comprehensive monitoring of critical environmental parameters (Bahadur et al., 2023). Stakeholders can identify trends and patterns including predicting analysis which will enable decision-making that is informed to be done which is all through exploiting AI-driven solutions. Strategic planning and policies that are driven by development are powered by AI-powered models (Putra et al., 2024). Human well-being is affected by the degradation of the environment which is caused by deforestation though pollution also poses a threat to ecosystem diversity. To implement mitigation strategies and advance sustainability of the environment there is a need for effective monitoring. However, traditional environmental monitoring methods require more resources, are time-consuming, and are limited in scope.

Artificial Intelligence (AI) offers transformative solutions that enable real-time processing of satellite imagery and sensor data. This facilitates the early detection of activities such as illegal logging, forest fires, and other forms of deforestation (Martinez et al., 2024). Advanced AI models can greatly analyze and classify large areas that are accurate which reduces the need for human involvement in monitoring which can be done by convolutional neural networks (CNNs) and random forests (Anitha et al., 2024). Large amounts of environmental data obtained from drones, satellites, and sensors are processed with machine learning. AI can foresee future trends of deforestation, and this will enable communities to be aware through the implementation of conservation efforts and proactive forest management (Martinez et al., 2024).

Real-time detection of forest fires, illegal logging, and diseases in plants which all are under the theme of deforestation can be achieved through the integration of AI with IoT and remote sensing technologies (Haq et al., 2024). Timely interventions are supported by these technologies which are crucial for biodiversity protection and effective forest conservation. The adoption of AI has increased due to the rise in challenges within the environment to adopt innovative solutions. AI

Technologies create a comprehensive perspective within the environment whereby it plays a pivotal role in the monitoring of various parameters of pollution which include water quality, air quality, and marine pollution which affects the health of the environment (Kumar et al., 2024).

AI provides a very strong alternative for environmental interventions that are proactive, and this is through exploiting predictive analytics and real-time data processing. The spread of pollution can be tracked and this supports the management and mitigation strategies that can be easily identified by the use of AI. Furthermore, early warnings and proactive measures are done by AI to facilitate predictive models, and this enables the forecasting of pollution events (Putra et al., 2024). In urban planning and public health protection, AI models have been used successfully to predict the level of pollution in the air including pinpointing the levels of pollution.

Measuring physicochemical parameters such as temperature, pH, oxygen dissolved, and the concentration of specific contaminants is a way of doing traditional water quality assessment. However, despite offering a description of water quality through these methods, they fail to capture fully complex interactions between living organisms and pollutants (Kumari & Khare, 2018). To address the limitations of traditional environmental monitoring methods is the aim of this study which are often hindered by constraints on resources, data processing which is slow, and the absence of real-time insights. Conventional approaches which are done manually for instance the counting of plants and flowers individually are time-consuming and labour-intensive despite them being accurate making them not appropriate to use on large-scale monitoring (Van Dongen et al., 2017). AI technologies intend to uncover gaps through the use of traditional methodologies which overcome challenges by providing efficient, scalable, and precise solutions for environmental monitoring. AI-driven monitoring of the environment aligns closely with efforts of the globe to achieve the Sustainable Development Goals (SDGs), particularly those focusing on the action of climate, life on land, clean water and air.

Water sources that are contaminated contribute to high mortality rates from waterborne diseases due to poor hygiene and sanitation in the environment where SDG 6 (Clean Water and Sanitation) aims to reduce water pollution by enhancing water quality and improving sanitation facilities (Kim, 2017) which comes in as a solution. AI technologies play an important role in providing real-time monitoring of changes that occur in the environment and deliver precise accurate data on air and water quality (Putra et al., 2024) which is important for pollution control. Deforestation is believed to be the main cause of climate change and biodiversity loss, and additionally opposes SDG 15 (Life on Land) which seeks to promote sustainable forest management, fight desertification, and end biodiversity loss (Cao & Wyatt, 2020). Initiatives such as the REDD+ program are designed to reduce emissions from deforestation and forest degradation (Siril et al., 2022) while promoting sustainable forest practices. Innovative, data-driven solutions are offered by AI which can significantly advance SDGs related to pollution control and deforestation within the environment (Jain & Mitra, 2024).

Addressing challenges such as ethical implementation and equitable access to AI technologies is important to fully realize their potential for environmental sustainability. Collaborating and sharing views and perceptions among policymakers, researchers, and practitioners are essential to control AI effectively for sustainable development (Yadav, 2023). This chapter discusses how AI is applied in environmental monitoring, with a particular focus on case studies addressing deforestation and pollution assessment in the environment.

This chapter discussed the applications of AI in environmental monitoring, with a particular focus on case studies addressing deforestation and pollution assessment. Through these case studies, the potential of AI to revolutionize environmental management and promote sustainability is explored. The chapter essentially sought to investigate *how AI is currently being utilized to monitor deforestation and assess environmental pollution, what best practices in AI applications for environmental monitoring are highlighted in research case studies, and what strengths/weaknesses are inherent in AI technologies deployed for deforestation and environmental pollution assessment.* The research objectives are to explore the role of AI in deforestation and pollution assessment, to analyze case studies/best practices in AI applications, and to identify the benefits, challenges, and limitations of AI use in environmental monitoring.

2. BACKGROUND

Artificial Intelligence (AI) is vital for improving the gathering of data, analyzing it, and understanding the available environmental data that you would have collected. This whole process and stages are made possible by the availability of AI-powered technologies and approaches that highly boost the efficiency, precision, and breadth of the environment's observation and management. Collection of environmental data is done through sensor networks and IoT integration then analysed and interpreted. AI influences sensor networks to monitor various environmental parameters, including air and water quality, biodiversity, and climate indicators (Dwivedi et al., 2024). These networks provide continuous data streams that AI systems can analyze for immediate insights (Dahake et al., 2024). Additionally, under the collection of environmental data comes IoT Integration where there is integration with the Internet of Things (IoT) to collect real-time environmental data (Arabelli et al., 2024). IoT devices, such as computers, sensors, and cameras, gather extensive amounts of data, which AI algorithms then process and analyze (Zhang et al., 2021). Before the interpretation of the available environmental data, analysis of environmental data is first done where machine learning and predictive modeling are deployed. AI utilizes machine learning (ML) and deep learning models to analyze complex environmental datasets (Arabelli et al., 2024). These models can identify patterns, predict trends, and detect differences in the data, which is crucial for environmental monitoring and conservation efforts (Alloghani, 2024).

AI uses a variety of technologies to improve the precision, effectiveness, and timeliness of forest management initiatives and has emerged as a key instrument in the monitoring and fight against deforestation. When AI is combined with remote sensing it enables satellite images to analyze satellite images to detect changes in forest cover within the environment (Haq et al., 2024). In determining patterns of deforestation and monitoring the environment, this approach is highly effective in identifying patterns which makes this approach greatly effective (Mehdawi & Bin Ahmad, 2021). U-Net, DeepLab V3, and ResNet are deep learning models that are advanced and are used to process imagery of satellites which are efficient in the precision of deforestation detection (Md Jelas et al., 2024).

Machine learning algorithms include networks that are neutral and also support vector machines that are used to classify and foretell activities of deforestation (Adarme et al., 2020). These models can process great amounts of data to identify trends and anomalies in forest cover within the environment (Rodrigues et al., 2024). Attention-based neural networks, which utilize synthetic aperture radar (SAR) images, are particularly effective in tropical regions where cloud cover can obstruct optical satellite imagery (Karaman et al., 2023).

Through learning from large datasets and determining complex patterns, AI models improve the accuracy of deforestation detection (Md Jelas et al., 2024). This however reduces the reliance on manual fieldwork and data processing which is time-consuming (Keskes & Ni ă, 2024). Potential deforestation hotspots can be forecasted for potential deforestation through the use of AI-driven predictive models that are proactive measures to prevent forest loss (Haq et al., 2024). Historical data and current trends are used to predict future deforestation activities within the environment in a specific period (Jain & Mitra, 2024). The quality and availability of data found will determine how accurately the AI model will work. Monitoring of deforestation may be inaccurate if data is inconsistent or lacking and this might occur during the collection of data. However, these technologies can only be effectively used and made essential for resolving ethical and data quality issues within the environment.

In the context of pollution assessment, AI has become a crucial instrument that provides solutions that are creative for a range of issues in the environment. By combining all the information that has been collected from several sensors and satellite photos, AI technologies allow air quality monitoring in real-time. The concentration of pollutants is predicted together with hazards in the environment and health hazards are evaluated using AI and machine learning (ML) models. These models enable researchers to combine huge datasets and estimate exposure levels across several communities which is essential for public health research together with the fight against environmental injustice (Krupnova et al., 2022).

When monitoring the environment, AI applications extend to disaster forecasts, pollution source detection, and comprehensive air and water quality monitoring (Olawade et al., 2024). The understanding, prediction, and mitigation of environmental risks increases these technologies. Despite these promising applications, challenges such as accessibility of data, control, and privacy, especially in regions with developing technological infrastructure, need to be addressed (Olawade et al., 2024). To protect information that is sensitive proactive data governance measures are recommended. For AI there is a variety of applications in pollution assessment that provide answers that are creative in several environmental fields. Including AI in these fields not only enhances monitoring and forecasting skills but also assists in the creation of long-term control plans for efficient pollution.

2.1 The Quest for Environmental Monitoring

Monitoring environmental pollution, land degradation, air quality, and biodiversity loss holds some benefits and co-benefits for the global community (Alotaibi and Nassif, 2024). Direct benefits of environmental monitoring include assisting stakeholders and policymakers in making informed/evidence-based decisions through the provision of real-time data on air/water quality; allowing timely preparations for disasters and evacuation of persons through early-warning systems thus, enhancing loss prevention (Hong et al. 2025); enabling sustainable practices via natural resource management made feasible by AI tracking of natural resources; assisting in pollution control by identifying pollutants and their impacts; enabling biodiversity conservation through ecosystem threat identification (Hong et al. 2025); improving public health; creating public awareness and involvement in conservations efforts by sharing data on environmental conditions; assisting scientific research and innovations by making necessary data available (Alotaibi and Nassif, 2024; Hong et al. 2025). On the other hand, some of the co-benefits of environmental monitoring are technological advancement, educational/social/economic benefits, environmental policy development, and climate change mitigation (Alotaibi and Nassif, 2024; Hong et al. 2025).

Environmental monitoring has been conducted throughout human history and efforts during the ancient civilization periods were primarily qualitative as they relied on human perceptions to track land/soil conditions, water levels, and weather patterns (Nexsens, 2025). During the early civilizations, the Egyptians monitored the concentration of nutrient-rich sediments the Nile River would deposit on its basin by building shafts called nilometers which assisted in measuring the water level and clarity. These measurements were necessary for forecasting agricultural productivity in a planting season (Nexsens, 2025). The Romans built hydraulic systems or aqueducts for irrigation, mining, and water supply (Nexsens, 2025; Fernandez-Lozano et al., 2025; Paul Kessener, 2005). All these efforts led to more precise measurements of water currents and winds at the dawn of nautical exploration (between the 15th and 17th centuries) with the advent of environmental observatory techniques such as putting a floating device in water to measure its currents (Nexsens, 2025). The 17th century brought about standardization in meteorological observations with the invention of the first device for measuring atmospheric pressure (barometer). This was followed by the mercury-in-glass thermometer for measuring ambient temperature in 1714 (Nexsens, 2025). By the 18th and 19th centuries, advancements in the Industrial Revolution brought about innovative sensors for measuring water and wind currents, and the mechanical tensiometer and penetrometer for soil water content, texture, compaction, and density measurements. These innovations led to improvements in agricultural practices and weather predictions.

Much of the early 20th century saw the incorporation of electrical components into sensors (Paul Kessener, 2005) with improvements in devices like the resistance temperature detector (first invented in 1871) and the development of electrochemical and acoustic sensing devices like the pH electrode for determining soil/water alkalinity or acidity, the dissolved oxygen electrode used in assessing water quality, multi-parameter instruments for measuring combined environmental parameters, and the Acoustic Doppler Current Profiler (ADCP) for measuring water current at varying depths in a column, etc. (Nexsens, 2025). The sensor technology for environmental monitoring enhances the integration of computing and data logging into the field. The invention of microprocessors in the mid-20th century allowed continuous environmental observation with little human intervention at automated weather sites that are equipped with data loggers for large-scale data storage (Nexsens, 2025). Advancements in software engineering enhanced data analysis and computer modeling as researchers can execute large datasets rapidly with high accuracy, fostering timely weather forecasts and predictions (Nexsens, 2025). Real-time environmental monitoring was feasible in the 21st century as sensors can now be linked using wireless technologies like cellular communication, radio frequency (RF), and satellite, to data-logging equipment to allow on-the-spot data acquisition even from inaccessible, remote sites. Integrating environmental sensors and wireless technologies with cloud computing and IoT (Internet of Things) capabilities enhances data processing and accessibility, allowing informed ecological management (Nexsens, 2025). Today, further technological innovations with smaller, more efficient, reliable, and affordable sensors and rapid advancements in software engineering have led to the deployment of AI and ML technologies in environmental monitoring with the capability to provide in-depth predictions, deeper insights, and analysis of larger datasets (Nexsens, 2025).

2.2 Environmental Monitoring Using AI Technology

The rapid innovations in computer software engineering and environmental sensor technology, the need for continuous ecological observations, the large volume of datasets generated during monitoring, and the need for accurate environmental forecasts have all led to the adoption of AL and ML technologies in environmental monitoring.

2.2.1 The Concept of Smart Environmental Monitoring

Smart Environmental Monitoring or SEM focuses on building a sustainable environment by monitoring water pollution, soil health, and air quality through information. Knowledge obtained from the real-time, continuous observation, collection, and analysis of environmental data using a combination of AI, IoT, and environmental sensors (Liberata and Sinha, 2020). The SEM system's framework structure has smart sensors, IoT, and wireless smart networks (WSN) as its main support, and, with intercommunication, these devices enable AI-controlled environmental monitoring, enabling smart ecological observations. While ecological data is obtained by IoT hardware devices and smart sensors during a typical monitoring session, the WSN provides network connectivity for the data to be recorded, monitored, and analyzed for various environmental applications (Liberata and Sinha, 2020).

Cloud computing makes SEM systems more robust, allowing data scalability (efficient for handling large volumes of data continuously generated by monitoring systems), real-time data generation and processing, and information accessibility (Quinn Jones, 2023; GaoTek, 2025; Song, 2022). Figures 1 and 2 are two cloud-based SEM architectures. In Figure 1, a cloud-based system structure connects appropriate sensors and IoT hardware, allowing monitoring and control of water contamination. This cloud-SEM system, along with the AI/ML capability of the IoT devices, monitors the water condition (level of contamination), generating and analyzing data that an operator accesses to obtain information on the water quality (Liberata and Sinha, 2020).

Figure 1. Cloud-SEM system structure monitoring contaminated water pollution

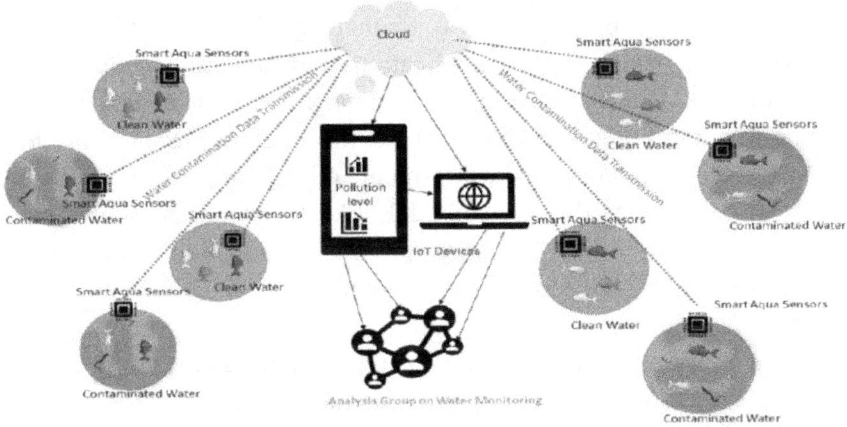

(Liberata and Sinha, 2020)

Also, Figure 2, shows a multi-sensor cloud-SEM system deployed for smart city monitoring of a wide range of environmental parameters including radiation, humidity, UV light signal, particulate matter, and temperature. These parameters are critical for maintaining a sustainable environmental condition (Liberata and Sinha, 2020; Orchard et al., 2025; Sahoo et al., 2024)

Figure 2. A cloud-SEM system deployed for a Smart City's environmental monitoring

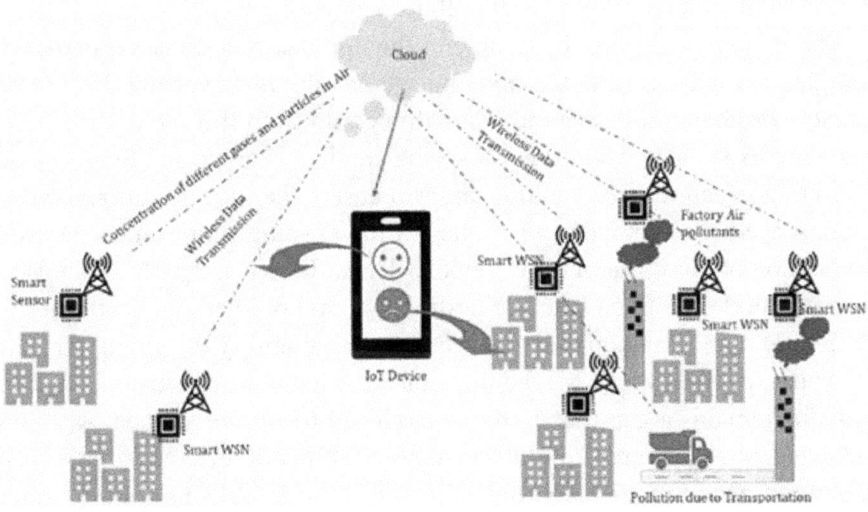

(Liberata and Sinha, 2020)

2.2.3 AI/ML Applications in Environmental Monitoring

Environmental monitoring systems can access AI model capabilities for forecasting, prediction, and analysis of environmental conditions via IoT devices. With different AI models having varying applications and strengths, models are chosen based on individual use cases and areas of deployment. For instance, while ML models like Random Forests and Support Vector Machines (SVMs) are ideal for applications such as biodiversity assessment, climate modeling, and air/water pollution monitoring, deep learning models like Convolution Neural Networks (CNNs) and Recurrent Neural Networks (RNNs) are applied in remote sensing applications and signal processing (Alotaibi and Nassif, 2024; Wani et al., 2024; Chisom et al., 2024).

Researchers have applied AI models for specific environmental monitoring use cases. Li et al. (2020), utilized the Artificial Neural Networks (ANN) ML model to analyze remotely sensed image data to predict wastewater pollution in a Lagoon. The authors reported that the model showed a good linear correlation between water quality parameters like total solids and the reflected electromagnetic spectrum of the water sample (i.e. the blue band), a low mean squared error (MSE), and a high coefficient of determination (R^2), all of which are indicative of a high accurate model. The hybrid ML model, Differential Search Algorithm-Extreme Learning Machine (DSA-ELM) was deployed and evaluated by Yan et al. (2017) for the classification of various pollutants in water in water quality research. Also, Shetty and co-researchers in an air pollution assessment study, utilized the SVM ML model, gas sensors, and IoT devices to capture and analyze air pollutants (Shetty et al., 2020). These studies are a few examples of the applicability of AI/ML technologies in environmental monitoring and pollution assessment.

There is an increasing frequency in the application/utilization of AI/ML technologies in the field of environmental monitoring however, the lack of adequate datasets for model training and validation could limit the accuracy of these technologies, especially when deep learning models (known for their characteristic data-hungry nature) that require larger datasets than the traditional ML models are deployed. This explains why many of the existing studies have focused on the application of conventional ML models. Thus, there is a dearth in the use of deep learning models with advanced classification and prediction capabilities for environmental monitoring applications. The non-availability of bibliometric literature analysis to synthesize AI/ML application knowledge across several environmental domains remains a gap that could be explored for future studies. Also, there is a lack of adequate research around the interoperability of different smart sensors used in environmental monitoring, this is important to establish standard protocols for data capturing and compatibility in monitoring applications.

4. MATERIALS AND METHODS

A systematic review of existing studies was conducted to achieve the research objectives and answer key questions. This is to allow transparency (and enhance reproducibility) of the research process, supporting the assessment of relevant articles.

4.1 Article Source and Search Strategy

The literature search/selection for the current study was based on the PRISMA guidelines for systematic reviews. Articles indexed on Scopus Preview were mainly collected, these were supported by a collection from other databases, including the Web of Science, JSTOR, and PubMed. The search query was structured to collect all the articles containing such search terms as artificial intelligence or AI, machine learning or ML, environmental monitoring, case studies, deforestation, and pollution assessment in their title, abstract, and keywords, combined with Boolean operators.

4.2 Article Selection: Inclusion and Exclusion Criteria

Studies were initially collected based on a set of inclusion/exclusion criteria (highlighted below, Fig. 3) from the databases, after which they were assessed on their title and abstract to ascertain relevance to the current research. A full-text screening was then conducted on each article to determine their quality and appropriateness for the research. Sections like the article's title, abstract, keywords, introduction, methods, results, discussion, and conclusion were examined.

4.2.1 Inclusion Criteria

- Articles focusing on the current research topic, containing search terms like artificial intelligence, machine learning, environmental monitoring, case study, deforestation, and pollution assessment in their title, abstract, and keywords, are included in the research.
- Studies conducted in English were collected.
- Studies conducted between 2018 and 2024 were included for the most up-to-date insights on the research topic.

4.2.2 Exclusion Criteria

- Studies examining the research topic but not in the context of deforestation and pollution assessment were excluded from the current study.
- Articles not written in the English Language were not collected.

- Studies conducted before 2017 were also excluded from the research.

Figure 3. Literature selection process.

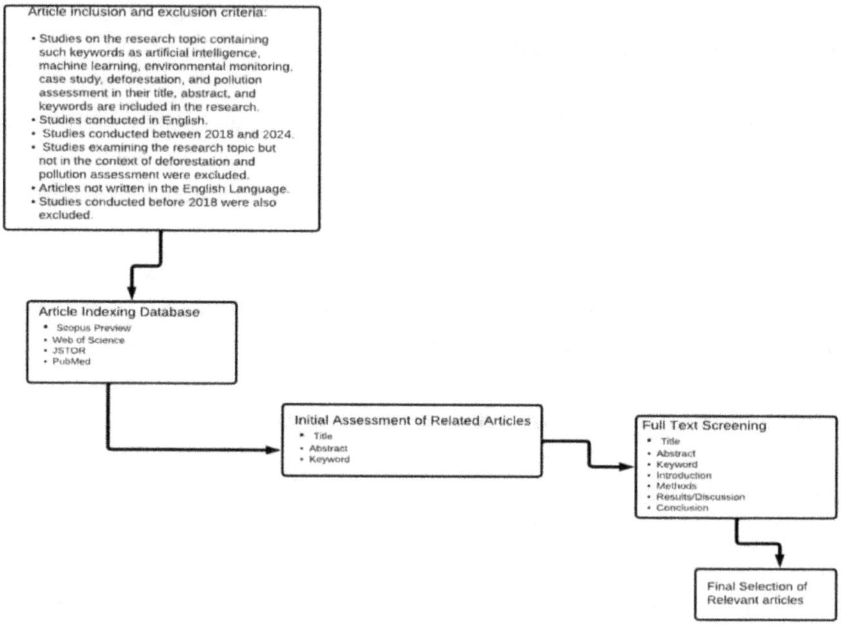

5. RESULTS AND DISCUSSION

5.1 Results

The research selected and analyzed 77 relevant studies from 258 related articles collected from article databases. Thirty-five (35) case studies (Table 1) on AI/ML applications in deforestation and pollution assessment were examined from the selected articles.

Research findings revealed a growing interest in AI applications in environmental monitoring in the areas of deforestation and pollution assessment among researchers, with more research conducted on deforestation than on pollution assessment within the period reviewed, i.e., 2018 -2024 (Fig. 4). The literature reviewed also showed that researchers frequently deployed certain AI algorithms in their research than others. These algorithms are Neural Networks (NNs), Random Forests (RFs),

Support Vector Machines (SVMs), Naïve Bayes (NBs), and Decision Trees (DTs), with NN being mostly utilized (Fig. 5).

An examination of the case studies collected for the research showed a global application of AI technology in Environmental Monitoring with a majority of the application cases in deforestation monitoring concentrated in countries of South, West (Middle East), and East Asia including India, Iran, Turkiye, and China respectively; Africa including Nigeria, Cameroon, and Ethiopia; and South American including Brazil and Colombia. A few of the case studies on pollution assessment were conducted in Pakistan, Taiwan, Ukraine, and Saudi Arabia. The application cases on pollution assessment were focused on water quality assessment and classification in streams, rivers, and reservoirs, real-time, atmospheric air quality monitoring and predictions, and evaluation of air pollutant variability, while deforestation case studies are focused on the assessment of forest cover changes, monitoring of land use/land cover changes, and detection and monitoring of selective/illegal logging. Many of the studies examined deployed models for predictions and forecasts of pollutant concentrations and water quality parameters, assessment and classification of tree cover loss and deforestation probability, and forest surveillance and monitoring.

Figure 4. Number of studies conducted on environmental monitoring activities between 2018 and 2024.

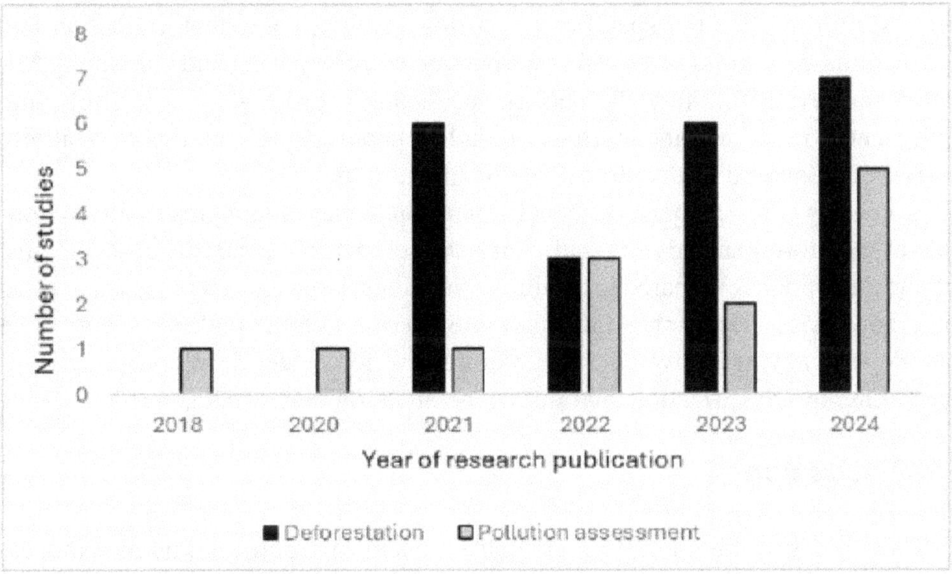

Figure 5. Frequency of application of AI/ML models for environmental monitoring between 2018 and 2024.

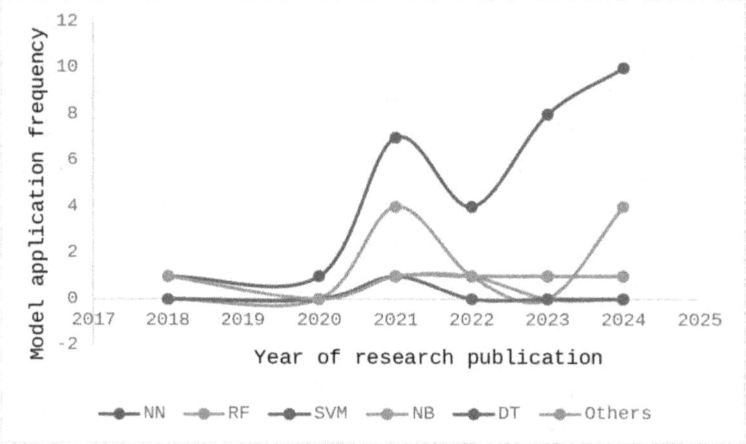

The studies reviewed highlighted best practices in the use of AI algorithms for environmental monitoring as ensuring availability and usage of high-quality, cleaned, pre-processed input data, identifying and utilizing appropriate, representative parameters for the specific monitoring activity, consistent validation and update of models, tuning models' hyperparameters, comparing and combining ML algorithms, evaluating models' performance using appropriate metrics, selecting optimal number of network neurons, validating models using specificity/sensitivity analysis, and ground truth data, and defuzzification models, etc.

According to researchers, the benefits and challenges of applying AI in environmental monitoring are high classification and prediction accuracy, cost-effective monitoring options compared to traditional methods, large dataset processing, good integration with Geographic Information Systems (GISs), models' performance data dependency, intensive resource requirement for model development, risk of overfitting, misclassification, and poor generalization to new data, etc.

5.2 Discussion

There is a growing interest in AI applications in environmental monitoring, especially in deforestation monitoring and pollution assessment, with more research focusing on deforestation monitoring, forest cover change assessment, and land use/land cover classification. The global spread of AI applications in these areas cuts across countries in Asia, Africa, the Middle East, Europe, and South America.

AI models are employed in environmental monitoring for prediction, forecast, assessment, and classification tasks, with Neural Networks including Convolution Neural Network (CNN), Artificial Neural Networks (ANN), and Recurrent Neural Networks (RNN) being the most deployed algorithms. For AI models to achieve reliable performance and high accuracy, some best practices must be considered, and these are cleaning and pre-processing data for high quality before feeding into the models to ensure optimum predictive capability, identifying and using appropriate input parameters/variables to eliminate less significant variables and thus, improve model's accuracy, consistently validating model with different dataset to increase reliability, prevent overfitting, and improve generalizability to new data, and comparing and combining models to ascertain their effectiveness and leverage the strength of individual model respectively. Table 1 extensively discusses these best practices and their benefits.

AI systems do show strength in environmental monitoring applications, including high predictive and classification accuracy, low-cost environmental monitoring and surveillance, good integration with Geographic Information Systems (GISs), and large data processing. However, despite their benefits, these systems have certain drawbacks, such as performance data dependency, high resource consumption for model development and fine-tuning, risks of misclassification, overfitting, and poor generalizability to new data.

Table 1. AI application in environmental monitoring case studies on deforestation and pollution assessment.

S/N	Case study	Ecological Application	AI/ML Model Used	AI/ML Model Role	Best Practices	Benefits/ Limitations/Challenges	Reference
1	Water quality assessment at the Olukdere stream and Lake Iznik (Turkiye)	Heavy metal	ANN	Prediction of environmental pollutants	1. Availability of sufficient high-quality data to prevent ANN model underfitting. 2. Identification and utilization of appropriate input parameters. 3. Consistent validation of the model using a different dataset improves reliability and performance, ensuring the model adjusts well to new data. 4. Using updated/advanced software will ensure prediction accuracy. 5. Environmental media continuous monitoring and model updates can increase prediction performance.	Benefits: 1. High pollutant prediction accuracy. 2. Cost-effective environmental monitoring option compared to the expensive, labor-intensive, traditional method. 3. The ANN model's ability to process multiple input parameters allows a holistic understanding of a wide range of factors that can impact pollutant concentrations in the environment. Limitation/Challenges: 4. The ANN model prediction accuracy is largely dependent on data quantity and quality. 5. Model tuning/training, which accounts for its optimal performance, is usually time-consuming and complex. 6. The model may overfit from learning on noise instead of data pattern, necessitating continuous updates and validations.	Berna Kırıl Mert and Deniz Kasapoğulları, 2023

continued on following page

Table 1. Continued

S/N	Case study	Ecological Application	AI/ML Model Used	AI/ML Model Role	Best Practices	Benefits/ Limitations/Challenges	Reference
2	Air Quality Index Prediction (Punjab, India)	Air quality prediction	Hybrid model- RF model and Genetic Algorithm	Prediction/ forecast of air quality	1). Identification of appropriate input variables via feature selection by correlation analysis, which can eliminate insignificant variables, thereby improving model accuracy. 2. The RF model's accuracy improves on a cleaned, normalized dataset whose missing values are supplied, achieved through data preprocessing. 3. Model parameter optimization or tuning can increase predictive performance and improve accuracy. 4. Comparison of the RF model's performance with other ML models, such as ANN and SVM, helps to ascertain its effectiveness. 5. Combining ML algorithms (e.g., combining an RF algorithm with a generic algorithm) can greatly improve prediction results.	Benefits: 1. The non-parametric nature of RF models makes them ideal for air quality index (AQI) forecasts since they can handle several input data without assumptions. 2. RF prediction accuracy can be improved via feature selection by correlation analysis. 3. The RF model can handle the non-linear relationship between numerical and categorical data, which is crucial for accurate air quality forecasts. Limitations/Challenges: 4. The decision-making process leading to forecasts by RF models can be incomprehensible when the number of trees increases, making the models complex. 5. The RF model consumes high computer resources in resource-limited ecosystems, especially when processing large datasets.	Singh et al., 2024

continued on following page

Table 1. Continued

S/N	Case study	Ecological Application	AI/ML Model Used	AI/ML Model Role	Best Practices	Benefits/ Limitations/Challenges	Reference
3	Water quality assessment and classification of Monsoonal rivers in the Selangor river basin (Malaysia)	Water quality assessment	ANN, SVM	Prediction of seasonal river classification	1). Identification and optimization of appropriate input parameter (hyperparameter) via trial-and-error method and necessary for accurate model prediction. 2). Comparison of performance, strength, and weakness of multiple models helps to select the most appropriate model for an application - seasonal river classification. 3. Ensuring a balanced dataset and consideration of class distribution during model training allows effective training, as class imbalance may lead to biased forecasts. 4). Data- and environmental condition-based continuous model refinements and adjustments can increase predictive accuracy with time.	Benefits: 1). High Accuracy in river classification applications. 2). The models' application to seasonal river classification contributes to water resource management. Limitations/ Challenges: 3). Class Imbalance in the training dataset may cause biased predictions. 4). Selection of hidden neurons in ANN and kernel parameters in SVM may lead to underfitting or overfitting and cause poor predictions. 5). Model performance variability can be caused by different input combinations and environmental conditions, indicative of the need for model adaptability for accurate predictions.	Wong et al., 2021.

continued on following page

Table 1. Continued

S/N	Case study	Ecological Application	AI/ML Model Used	AI/ML Model Role	Best Practices	Benefits/ Limitations/Challenges	Reference
4	Water quality assessment by Nitrate concentration prediction of the Feitsui Reservoir (Taiwan)	Water quality assessment	ANN	Prediction pollutant (nitrate) concentration.	1). Selection of an optimal number of neurons ensures that the ANN model is neither too simple nor complex, increasing prediction accuracy. 2). Utilization of advanced ANN training algorithms such as the Levenberg-Marquardt Algorithm (LMA) allows the model to converge faster and reduces error rates. 3). Utilization of multiple input parameters allows the ANN model to handle complex interactions among those parameters, leading to accurate predictions. 4). Evaluation of the model's performance using correlation coefficients helps to validate the ANN model's accuracy and ensures good generalization with the dataset.	Benefits: 1). High predictive accuracy for pollutant (nitrate) concentrations. 2). Good generalization to new data due to optimum neuron selection. 3). Efficient model training from advanced algorithms reduces error and increases convergence rates. Limitations/Challenges: 4). Dataset class imbalance may lead to biased predictions. 5). Overfitting may occur when training data is non-representative, or the model is too complex. 6). ANN performance variability may occur due to environmental conditions and input parameter combinations, underscoring the need for model adaptability.	Latif et al., 2020.

continued on following page

Table 1. Continued

S/N	Case study	Ecological Application	AI/ML Model Used	AI/ML Model Role	Best Practices	Benefits/ Limitations/Challenges	Reference
5	Forest and biodiversity management-Water quality assessment	Forest water quality assessment	ANN	Prediction of water quality parameters impacting forest ecosystems.	1). Ensuring quality input data through data cleaning and preprocessing is essential for an effective ANN model. 2). Selection of an optimum number of neurons helps the ANN model to achieve a balance between performance and complexity. 3). Utilization of advanced model training algorithms ensures an efficient training process. 4). Monitoring of the model's performance through validation techniques is necessary to prevent overfitting.	Benefits: 1). The ANN model can predict different forest ecosystem parameters from complex data patterns, thus enhancing forest management decision-making. 2). The model has high predictive accuracy. Limitations/Challenges: 3). The ANN model is heavily reliant on data quality and quantity; a lack of these may lead to poor predictions. 4). Model overfitting may occur when the model's performance on training data is better than unseen data due to a lack of generalizability, necessitating good validation techniques. 5). High resource consumption due to computational complexity. 6). ANN model development requires high computer software and AI expertise.	Asif Raihan, 2023.

continued on following page

Table 1. Continued

S/N	Case study	Ecological Application	AI/ML Model Used	AI/ML Model Role	Best Practices	Benefits/ Limitations/Challenges	Reference
6	Forest management-classification of Mangrove forest	Monitoring deforestation	RF, SVM, DT, and object-based NN*	1). Random Forest (RF): The RF model classified mangrove forests and assessed tree cover loss (deforestation). 2). Decision Tree (DT): The DT model analyzed tree cover loss classifications. Nearest Neighbor (NN*): The object-based nearest neighbor model classified mangrove forests using orthophotography.	1). Usage of high-resolution images to enhance classification allows differentiation between ecosystem features. 2). RF and DT models' performance can be enhanced by utilizing a clean, structured dataset via a data preprocessing step. 3). Utilization of a 10-fold cross-validation technique helps to prevent model overfitting and ensures good generalizability with new, unseen datasets. 4). Selection of relevant environmental features and input parameters is necessary for accurate model prediction. 5). Comparing different model algorithms' performance for an application is necessary to select the most appropriate.	Benefits: 1). The RF model can process large datasets and has high accuracy. 2). The DT model allows interpretation of the various features influencing ecosystem classification. 3). With a high-resolution image, the NN model can classify ecosystem features when deployed in an object-based environment. Limitations/Challenges: 4). All the models deployed are dependent on the quality and quantity of input data. 5). The models may be limited in scalability to different environmental contexts. 6). Improper validation may lead to model overfitting, which may prevent good generalization on new, unseen data.	Dawod et al., 2021.

continued on following page

Table 1. Continued

S/N	Case study	Ecological Application	AI/ML Model Used	AI/ML Model Role	Best Practices	Benefits/ Limitations/Challenges	Reference
7	Assessment of forest cover changes in the Zagros Mountains (Iran)	Monitoring deforestation	CNN	Classification of forest cover.	1). Usage of a multi-temporal dataset allows accurate detection of land cover by the CNN model. 2). Utilizing layer stacking for the model's feature extraction allows the collection of low- and high-level features from input data, which is necessary for differentiating between forest and land cover classes. 3). Utilizing CNN's adaptive learning rate prevents model overfitting, ensuring good generalization with new, unseen datasets. 4). CNN model classification accuracy can be improved by utilizing a group of CNN models with varying numbers of neurons at the same time. 5). Applying a change detection analytic process (such as the ICD-image classification difference) after a classification step helps in analyzing changes in ecological features and provides information on environmental dynamics. 6). Verification and validation of classification results with ground truth data ensures the reliability of the output.	Benefits: 1). High environmental feature classification accuracy. 2). CNN models can process large datasets. 3). CNNs can automatically extract low- and high-level features from the dataset, allowing easy detection of patterns in the data. Limitations/Challenges: 4). Intensive computer resource requirements may impact environments with limited access to resources. 5). Poor validation may cause model overfitting, leading to poor generalization with unseen data and unreliable classification. 6). CNN application in heterogeneous environments may be limited by the inability to capture broad data patterns due to the current pixel-level-based classification. 7. CNN's classification accuracy is largely dependent on data quality and quantity.	Elahe Moradi and Alireza Sharifi, 2023.

continued on following page

Table 1. Continued

S/N	Case study	Ecological Application	AI/ML Model Used	AI/ML Model Role	Best Practices	Benefits/ Limitations/Challenges	Reference
8	Water quality assessment of the Verkhno-Ivachivsk Reservoir (Ternopil, Western Ukraine)	Water quality assessment	Fuzzy logic and NN	Evaluation of hydrochemical indicators of water reservoir	1). Constant adaptation of the neural network's knowledge base allows it to stay relevant to the ecological application, ensuring accurate evaluation. 2). Integrating expert Knowledge into fuzzy logic models increases the reliability and effectiveness of the model in real-world situations. 3). Utilizing computer software such as Matlab to develop neural networks enhances the real-time monitoring and self-learning capabilities of the model, ensuring accurate assessment of water quality. 4). Usage of fuzzy linguistic equations allows a small representation of water quality indicators, which enhances decision-making by formalizing data into insights. 5). Defuzzificating neural networks for interpretable, clear model outputs.	Benefits: 1). Fuzzy logic is flexible and can realistically assess water conditions by handling uncertainty and imprecision in water quality data. 2). Timely, real-time monitoring of water quality parameters is enhanced by the integration of fuzzy logic and neural networks. 3). The model is relevant and accurate over a long period with self-learning and good generalization with new data. Limitations/Challenges: 4). Model development and fine-tuning are usually complex. 5). Model accuracy and effectiveness are largely reliant on data quality and quantity. 6) There are challenges in Interpreting the result of fuzzy systems, which may hamper actionable decisions.	Hrubinko et al., 2022.

continued on following page

Table 1. Continued

S/N	Case study	Ecological Application	AI/ML Model Used	AI/ML Model Role	Best Practices	Benefits/ Limitations/Challenges	Reference
9	Deforestation assessment at the Jaldapara National Park, Eastern Himalayan foothills (India)	Monitoring deforestation	SVM, NB, RF, DT, and ANN	Deforestation probability assessment.	1). Deployment of several ML algorithms in one application enhances deforestation probability assessment, allowing cross-validation and increasing accuracy. 2). Utilizing natural breaks classification in deforestation probability assessment improves result interpretation by minimizing and maximizing variance within and between probability classes, respectively. 3). Usage of methods like specificity/sensitivity analysis and area under the curve to validate the models are necessary to ensure effective and reliable prediction of deforestation probability. 4). Including predictors (explanatory factors) in models to improve capability and effectiveness, enhancing the comprehension of variables that influence deforestation. 5). Regular model updates and adaptation to changing environmental conditions and new data are necessary for accurate assessment over time.	Benefits: 1). The models have high predictive capability for deforestation probability assessment. 2). Using multiple ML models enhances the assessment of deforestation probability. Limitations/Challenges: 3). Model development and fine-tuning require significant expertise. 4). Model accuracy and effectiveness is dependent on data quality and quantity. 5). Interpreting model results can be challenging and may impact the decision-making process. 6). Models' validation requires intensive resource consumption, which may be strenuous for environments with limited resources.	Saha et al., 2022.

S/N	Case study	Ecological Application	AI/ML Model Used	AI/ML Model Role	Best Practices	Benefits/ Limitations/Challenges	Reference
10	Real-time air quality monitoring in Islamabad, Lahore, Peshawar, and Karachi (Pakistan)	Air quality monitoring	CNN	Estimation and evaluation of real-time, daily pollutant concentration (particulate matter) in the air.	1). Carrying out data preprocessing by standardizing/normalizing image data via image scaling and pixel value adjustment allows the CNN model to achieve quick convergence. 2). Randomizing input data prevents biases as this ensures the model is not affected by data sequence. 3). Usage of the CNN model's layer architecture, such as the convolutional, pooling, and fully connected layers for feature extraction, dimension reduction, and classification, respectively, allows the model to learn from the complexity in data patterns. 4). Iterating model training through forward and backward propagation procedures for data transmission and weight updates will optimize the performance of the model. 5). Utilizing dropout layers in the CNN model's architecture helps to avoid overfitting, improving the model's generalization to unseen data. 6). Regular assessment of the model's performance using methods like the mean absolute percentage error (MAPE), root mean square error (RMSE), and mean absolute error (MAE) evaluates the model's effectiveness.	Benefits: 1). High pollutant prediction. 2). CNNs have good data extraction features necessary for assessing accurate pollutant concentration. Limitations/Challenges: 3). Model's performance data dependence. 4). Huge resource in terms of computing, time, and power requirements for model continual validation. 5). The black box, the non-transparent nature of how inputs are transformed into outputs by CNN models, prevents easy interpretation of results and hinders actionable decisions. 6). To maintain accuracy over time, the model requires regular validation, which consumes huge resources.	Ahmed et al., 2022.

CNN=Convolution Neural Network, RF= Random Forests, DT= Decision Trees, NN= Neural Network, NN* = Nearest Neighbors, SVM= Support Vector Machines, RNN= Recurrent Neural Network, NB= Naïve Bayes, MLP= Multi-Layered Perceptron, GMM= Gaussian Mixture Model, MLR= Multiple Linear Regression (The continuation of Table 1 can be found in the Appendix).

6. FURTHER RESEARCH

Future research efforts should focus on the establishment of a standard protocol for data collection and compatibility and the interoperability among smart sensors used for environmental monitoring applications. This has several benefits including allowing easy comparison and analysis of the data obtained from different sensors, ensuring reliability and consistency; lowering the efforts and time in managing several systems by streamlining data collection, ensuring quality input data, and allowing monitoring system's scalability through network expansion at no additional costs. When hardware sensors' interoperabilities are standardized, integration and communication among different devices become seamless, enhancing data sharing. Also, the non-availability of bibliometric literature analysis to synthesize AI/ML application knowledge across several environmental domains remains a gap that could be explored for future studies. The impact of the lack of this literature analysis on research in the field of environmental monitoring can be far-reaching, possibly causing knowledge fragmentation, research duplicity, loss of interdisciplinary/collaborative research opportunities, inconsistent methodologies, and less impactful policy implications.

7. MANAGERIAL IMPLICATIONS

AI systems are ideal for real-time surveillance and monitoring of natural ecosystems and forest reserves and are low-cost with high accuracy when compared to traditional forest monitoring techniques. These systems should, therefore, be deployed by stakeholders and policymakers for monitoring and detecting illegal logging and protecting government-reserved spaces.

8. CONCLUSION

The current research explored the application of AI in environmental monitoring, focusing on case studies in deforestation monitoring and pollution assessment. Research on AI applications in forest cover change detection and air and water quality assessment gradually increased between 2018 and 2024, with the highest research output generated in 2024. Among the several AI models employed for environmental monitoring, the Neural Network is mostly utilized. Models are deployed for forecasts, predictions, classifications, detections, and assessments of environmental conditions. Some of the strengths of AI models include high predictive and classification accuracy, low-cost monitoring, easy integration with other environmental surveillance systems (e.g., GIS), and large data processing, while their drawbacks are resource-intensive model tuning and development, data-dependent performance, misclassification, and overfitting risks. Some best practices for building high-performing models include utilizing high-quality datasets, identifying and using appropriate input parameters, combining and comparing AI algorithms in an application, and consistently validating models.

REFERENCES

Adarme, M. O., Feitosa, R. Q., Happ, P. N., Almeida, C. A. D., & Gomes, A. R. (2020). Evaluation of deep learning techniques for deforestation detection in the Brazilian Amazon and Cerrado biomes from remote sensing imagery [Article]. *Remote Sensing (Basel)*, *12*(6), 910. Advance online publication. DOI: 10.3390/rs12060910

Ahmed, M., Xiao, Z., & Shen, Y. (2022). Estimation of Ground PM2.5 Concentrations in Pakistan Using Convolutional Neural Network and Multi-Pollutant Satellite Images. *Remote Sensing (Basel)*, *14*(7), 1735. https://www.mdpi.com/2072-4292/14/7/1735. DOI: 10.3390/rs14071735

Al Mazroa, A., Maray, M., Alashjaee, A. M., Alotaibi, F. A., Alzahrani, A. A., Alkharashi, A., Alotaibi, S. D., & Alnfiai, M. M. (2024). Computer Vision with Explainable Artificial Intelligence for Visual Pollution Detection in the Kingdom of Saudi Arabia. *IEEE Access : Practical Innovations, Open Solutions*, *12*, 193014–193027. DOI: 10.1109/ACCESS.2024.3513696

Alloghani, M. A. (2024). Using AI to Monitor Marine Environmental Pollution: Systematic Review. In Signals and Communication Technology (Vol. Part F1802, pp. 87-97). DOI: 10.1007/978-3-031-45214-7_5

Alotaibi, E., & Nassif, N. (2024). Artificial intelligence in environmental monitoring: In-depth analysis [Article]. *Discover Artificial Intelligence*, *4*(1), 84. Advance online publication. DOI: 10.1007/s44163-024-00198-1

Alsamhi, S. H., Rowan, N. J., Hawbani, A., Nashwan, F., & Brolchain, N. (2024). AI-Empowered Peatland Management for Climate-Resilient and Sustainable Future: Challenges and Opportunities. 4th International Conference on Emerging Smart Technologies and Applications, eSmarTA 2024.

Andrei, A.-T., & Grigore, O. (2022). *Combating Deforestation Using Different AGNES*. Approaches., DOI: 10.1109/COMM54429.2022.9817217

Andrei, A.-T., & Grigore, O. (2024). Development of a Very Low-Cost Deforestation Monitoring System Based on Aerial Image Clustering and Compression Techniques. *Advances in Electrical and Computer Engineering*, *24*(2), 73–84. DOI: 10.4316/AECE.2024.02008

Anitha, C., Devi, S., Nassa, V. K., Mahaveerakannan, R., Das Baksi, K., & Suganthi, D. (2024). Development of Image Processing and AI Model for Drone Based Environmental Monitoring System [Article]. *Journal of Machine and Computing*, *4*(1), 221–229. DOI: 10.53759/7669/jmc202404021

Arabelli, R., Boddepalli, E., Buradkar, M., Goriparti, N. V. S., & Chakravarthi, M. K. (2024). IoT-Enabled Environmental Monitoring System Using AI. Proceedings - 3rd International Conference on Advances in Computing, Communication and Applied Informatics, ACCAI 2024. DOI: 10.1109/ACCAI61061.2024.10602131

Aziz, G., Minallah, N., Saeed, A., Frnda, J., & Khan, W. (2024). Remote Sensing-Based forest cover classification using machine learning. *Scientific Reports*, *14*(1), 69. DOI: 10.1038/s41598-023-50863-1 PMID: 38167902

Bahadur, F. T., Shah, S. R., & Nidamanuri, R. R. (2023). Applications of remote sensing vis-à-vis machine learning in air quality monitoring and modeling: A review [Review]. *Environmental Monitoring and Assessment*, *195*(12), 1502. Advance online publication. DOI: 10.1007/s10661-023-12001-2 PMID: 37987882

Ball, J. G. C., Petrova, K., Coomes, D. A., & Flaxman, S. (2022). Using deep convolutional neural networks to forecast spatial patterns of Amazonian deforestation. *Methods in Ecology and Evolution*, *13*(11), 2622–2634. DOI: 10.1111/2041-210X.13953

Cajamarca-Carrazco, D., Marcelo, E. S. S., Danilo, F. F. V., & Peredes-Godoy, M. (2024). Artificial Intelligence for Predicting WateR Dynamics in Andean Paramo: A Comprehensive Systematic Review. *Data and Metadata*, *3*, 391. 10.56294.

Cao, A. N., & Wyatt, T. (2020). The Sustainable Development Goals Link to Human Security: An Exploration of Illegal Logging in Vietnam. In *The Emerald Handbook of Crime, Justice and Sustainable Development* (pp. 513-532). DOI: 10.1108/978-1-78769-355-520201027

Chaurasia, K., Garg, P. K., Arun, P. V., and Yan, Y. (2024). Spatial intelligence for a greener planet: AI applications in environmental pollution mapping, analysis, and mitigation. .DOI: 10.1201/9781032718323

Chisom, O. N., & Biu, P. W.. (2024). Reviewing the Role of AI in Environmental Monitoring and Conservation: A Data-driven Revolution for our Planet. *World Journal of Advanced Research and Reviews*, *21*(1), 161–171. DOI: 10.30574/wjarr.2024.21.1.2720

Dahake, R., Metre, K., Kale, N., Dange, B. J., & Mahankale, N. (2024). AI-Powered Environmental Monitoring and Conservation Strategies [Article]. *International Journal of Engineering Trends and Technology*, *72*(8), 1–7. DOI: 10.14445/22315381/IJETT-V72I8P101

Dawod, A., & Sharafuddin, M. A. (2021). Assessing mangrove deforestation using pixel-based image: A machine learning approach. *Bulletin of Electrical Engineering and Informatics, 10*(6), 3178–3190. DOI: 10.11591/eei.v10i6.3199

De Andrade, R. B., Mota, G. L. A., & da Costa, G. A. O. P. (2022). Deforestation Detection in the Amazon Using DeepLabv3+ Semantic Segmentation Model Variants. *Remote Sensing (Basel), 14*(19), 4694. https://www.mdpi.com/2072-4292/14/19/4694. DOI: 10.3390/rs14194694

Debus, A., Beauchamp, E., Acworth, J., Ewolo, A., Kamga, J., Verhegghen, A., Zébazé, C., & Lines, E. R. (2024). A labelled dataset to classify direct deforestation drivers from Earth Observation imagery in Cameroon. *Scientific Data, 11*(1), 564. DOI: 10.1038/s41597-024-03384-z PMID: 38821976

Dewi, D. A. (2024). Water Quality Prediction using Random Forest Algorithm and Optimization. *Journal of Applied Data Sciences, 5*(3), 1354–1362. DOI: 10.47738/jads.v5i3.348

Dwivedi, D. N., Mahanty, G., & Dwivedi, V. N. (2024). Intelligent conservation: A comprehensive study on AI-enhanced environmental monitoring and preservation. In The Convergence of Self-Sustaining Systems With AI and IoT (pp. 215-226). https://doi.org/DOI: 10.4018/9798369317020.ch011

Fetene, D., Lohani, T., & Mohammed, I. A. (2023). LULC change detection using support vector machines and cellular automata-based ANN models in Guna Tana watershed of Abay basin, Ethiopia. *Environmental Monitoring and Assessment, 195*(11), 1329. Advance online publication. DOI: 10.1007/s10661-023-11968-2 PMID: 37848752

GaoTek. Applications of Edge Computing for IoT in Environmental Monitoring. 2025. GaoTek. Available from: https://gaotek.com/applications-of-edge-computing-for-iot-in-the-environmental-monitoring/. Accessed: 18-01-2025.

Haq, B., Jamshed, M. A., Member, S., Ali, K., Kasi, B., Arshad, S., Kasi, M. K., Ali, I., Shabbir, A., Abbasi, Q. H., & Ur-Rehman, M. (2024). Tech-Driven Forest Conservation: Combating Deforestation With Internet of Things, Artificial Intelligence, and Remote Sensing [Article]. *IEEE Internet of Things Journal, 11*(14), 24551–24568. DOI: 10.1109/JIOT.2024.3378671

Hong, A., Choi, D.-A., ... (2025). Environmental and Health Monitoring in Smart Cities: A Systematic Review. Available from SSRN – https://ssrn.com/abstract=5071014. Accessed: 10-01-2025. DOI: 10.2139/ssrn.5071014

Hrubinko, V., Halyna, H., Humeniuk, V., Andrusushyn, T., Khomenchuk, V., Harmatiy, N., & Chen, I. (2023). Assessment of the hydro-ecological situation of the Verkhno-Ivachivsk Reservoir in Ternopil using the fuzzy logic apparatus. *Journal of Geology. Geography and Geoecology, 32*(2), 254–265. DOI: 10.15421/112324

Ibam, E. O., & Olowokere, V. (2021). A Smart Deforestation Monitoring and Control System Model. *International Journal of Advanced Trends in Computer Science and Engineering, IJATSCE, 10*, 1901–1913.

Ijaware, V. A. (2024). GIS and Artificial Intelligence Application in Smart Forest Ecosystem Sustainability Evaluation of Olokemeji Forest Reserve, Ogun State, Nigeria. *American Journal of Geospatial Technology, 3*(1), 9–16. DOI: 10.54536/ajgt.v3i1.2621

Jain, V., & Mitra, A. (2024). Integrative innovations: Harnessing artificial intelligence for enhanced environmental sustainability. In *Maintaining a Sustainable World in the Nexus of Environmental Science and AI* (pp. 365-382). DOI: 10.4018/979-8-3693-6336-2.ch014

Jamalani, A., Ku Yusof, K., Azid, A., Samsudin, M., Amin, S., Abd Rani, N., & Abdullah Sani, M. S. (2018). The evaluation on artificial neural networks (ANN) and multiple linear regressions (MLR) models over particulate matter (PM10) variability during haze and non-haze episodes: A decade case study. *Malaysian Journal of Fundamental and Applied Sciences, 15*(2), 164–172. Advance online publication. DOI: 10.11113/mjfas.v15n2.1004

Javier Fernandez-Lozano. Ivan Gonzale et al. (2025) Absent Voices and Unwarranted Presence: A Combined Multi-Approach to Mapping the Roman Hydraulic System Supplying Las Medulas Gold Mine (NW Iberia). Geosciences, 15(1):13. DOI: https://doi.og/10.3390/geosciences15010013

Karaman, K., Garnot, V. S. F., & Wegner, J. D. (2023). Deforestation detection in the Amazon with sentinel-1 sar image time series. *ISPRS Annals of the Photogrammetry, Remote Sensing and Spatial Information Sciences, X-1/W1-2023*, 835–842. DOI: 10.5194/isprs-annals-X-1-W1-2023-835-2023

Keskes, M. I., & Ni ă, M. D. (2024). Developing an AI tool for forest monitoring: introducing sylvamind AI [Article]. *Bulletin of the Transilvania University of Brasov, Series II: Forestry, Wood Industry, Agricultural Food Engineering, 17(66)*(2), 39-54. .DOI: 10.31926/but.fwiafe.2024.17.66.2.3

Kessener, P. (2005). Roman Hydraulic Technology and Problems in Modern Pressurized Pipeline Systems. In Coopey, R., Fahlbusch, H., Hatcho, N., & Janskey, L. (Eds.), *A History of Water Issues* (pp. 217–237). United Nations University Publishing.

Kim, K. H. (2017). Implementation status and monitoring of an environmental health-related index to Sustainable Development Goals [Article]. *Journal of the Korean Medical Association*, *60*(8), 662–671. DOI: 10.5124/jkma.2017.60.8.662

Krupnova, T. G., Rakova, O. V., Bondarenko, K. A., & Tretyakova, V. D. (2022). Environmental Justice and the Use of Artificial Intelligence in Urban Air Pollution Monitoring [Article]. *Big Data and Cognitive Computing*, *6*(3), 75. Advance online publication. DOI: 10.3390/bdcc6030075

Kuck, T. N., Silva Filho, P. F. F., Sano, E. E., Bispo, P. C., Shiguemori, E. H., & Dalagnol, R. (2021). Change Detection of Selective Logging in the Brazilian Amazon Using X-Band SAR Data and Pre-Trained Convolutional Neural Networks. *Remote Sensing (Basel)*, *13*(23), 4944. https://www.mdpi.com/2072-4292/13/23/4944. DOI: 10.3390/rs13234944

Kumar, R., Goyal, M. K., Surampalli, R. Y., & Zhang, T. C. (2024). River pollution in India: Exploring regulatory and remedial paths [Review]. *Clean Technologies and Environmental Policy*, *26*(9), 2777–2799. DOI: 10.1007/s10098-024-02763-9

Kumari, K., & Khare, A. (2018). Integration of Biomarker Approach in Pollution Monitoring Programme of Aquatic Ecosystem. In *Energy* (pp. 331–354). Environment, and Sustainability., DOI: 10.1007/978-981-10-7434-9_18

Latif, S. D., Azmi, M. S. B. N., Ahmed, A. N., Fai, C. M., & El-Shafie, A. (2020). Application of Artificial Neural Network for Forecasting Nitrate Concentration as a Water Quality Parameter: A Case Study of Feitsui Reservoir, Taiwan. *International Journal of Design & Nature and Ecodynamics*, *15*(5), 647–652. DOI: 10.18280/ijdne.150505

Li, Y., & Wang, X.. (2020). Lagoon Water Quality Monitoring Based on Digital Image Analysis and Machine Learning Estimators. *Water Resources*, *172*, 115471. PMID: 32032913

Liberata, S., & Sinha, G. R. (2020). Advances in Smart Environment Monitoring Systems. *Sensors (Basel)*, *20*(11), 3113. DOI: 10.3390/s20113113 PMID: 32486411

Lin, Y., Zhang, T., Ye, Q., Cai, J., Wu, C., Khirni Syed, A., & Li, J. (2021). Long-term remote sensing monitoring on LUCC around Chaohu Lake with new information on algal bloom and flood submerging [Article]. International Journal of Applied Earth Observation and Geoinformation, 102. *Article*, *102413*, 102413. DOI: 10.1016/j.jag

Martinez, J. A. C., da Costa, G. A. O. P., Messias, C. G., Soler, L. D. S., de Almeida, C. A., & Feitosa, R. Q. (2024). Enhancing deforestation monitoring in the Brazilian Amazon: A semi-automatic approach leveraging uncertainty estimation [Article]. *ISPRS Journal of Photogrammetry and Remote Sensing*, *210*, 110–127. DOI: 10.1016/j.isprsjprs.2024.03.004

McCallum, I., Walker, J., Fritz, S., Grau, M., Hannan, C., Hsieh, I. S., Lape, D., Mahone, J., McLester, C., Mellgren, S., Piland, N., See, L., Svolba, G., & De Villiers, M. (2023). Crowd-Driven Deep Learning Tracks Amazon Deforestation. *Remote Sensing (Basel)*, *15*(21), 5204. DOI: 10.3390/rs15215204

Md Jelas, I., Zulkifley, M. A., Abdullah, M., & Spraggon, M. (2024). Deforestation detection using deep learning-based semantic segmentation techniques: A systematic review [Review]. *Frontiers in Forests and Global Change*, *7*, 1300060. Advance online publication. DOI: 10.3389/ffgc.2024.1300060

Mehdawi, A. A., & Bin Ahmad, B. (2012). Classification of forest change by integration of remote sensing data with neural network techniques. *Proceedings of the 2012 International Conference on System Engineering and Technology, ICSET 2012*. DOI: 10.1109/ICSEngT.2012.6339319

Mert, B. K., & Kasapoğulları, D. (2023). A Case Study of Using Artificial Neural Networks to Predict Heavy Metal Pollution in Lake Iznik. *Research Square* Platform LLC. DOI: 10.21203/rs.3.rs-2934605/v1

Moradi, E., & Sharifi, A. (2022). Assessment of forest cover changes using multi-temporal Landsat observation. *Environment, Development and Sustainability*, *25*(2), 1–10. DOI: 10.1007/s10668-021-02097-2

Nexsens. Historical Context and Modern Innovations in Environmental Monitoring. 2025. Nexsens Technology. Available from: https://www.nexsens.com/handbook/1-1-historical-context-and-modern-innovations-in-environmental-monitoring.htm. Accessed: 10-01-2025.

Niyongabo, A., Zhang, D., Guan, Y., Wang, Z., Imran, M., Nicayenzi, B., Guyasa, A. K., & Hatungimana, P. (2024). Predicting Urban Water Consumption and Health Using Artificial Intelligence Techniques in Tanganyika Lake, East Africa. *Water (Basel)*, *16*(13), 1793. https://www.mdpi.com/2073-4441/16/13/1793. DOI: 10.3390/w16131793

Olawade, D. B., Wada, O. Z., Ige, A. O., Egbewole, B. I., Olojo, A., & Oladapo, B. I. (2024). Artificial Intelligence in Environmental Monitoring: Advancements, Challenges, and Future Directions. *Hygiene and Environmental Health Advances*, *12*, 100114. Advance online publication. DOI: 10.1016/j.heha.2024.100114

Orchard, S., Fitzpatrick, B. M., Shah, M. A. R., & Andrade, A. (2025). Impact Assessment Framework for Nature-Based Climate Solutions: A Review of Contemporary Approaches. *Sustainability (Basel), 17*(2), 677. DOI: 10.3390/su17020677

Putra, G. A. A., Wiraatmaja, P. P., & Adhiswari Wedagama, D. A. T. (2024). The Application of Artificial Intelligence (AI) in Addressing Environmental Issues. 2024 10th International Conference on Smart Computing and Communication, ICSCC 2024.

Quinn Jones. IoT-Based Environmental Monitoring: Types and Use Cases. 2023. DIGIXBee, DIGI International. Available from: https://www.digi.com/blog/post/iot-based-environmental-monitoring. Accessed: 18-01-2025.

Raihan, A. (2023). *Artificial intelligence and machine learning applications in forest management and biodiversity conservation.* Natural Resources Conservation and Research., DOI: 10.24294/nrcr.v6i2.3825

Rodrigues, J., Dias, M. A., Negri, R., Hussain, S. M., & Casaca, W. (2024). A Robust Dual-Mode Machine Learning Framework for Classifying Deforestation Patterns in Amazon Native Lands [Article]. *Land (Basel), 13*(9), 1427. Advance online publication. DOI: 10.3390/land13091427

Rodrigues, J., Dias, M. A., Negri, R., Hussain, S. M., & Casaca, W. (2024). A Robust Dual-Mode Machine Learning Framework for Classifying Deforestation Patterns in Amazon Native Lands [Article]. *Land (Basel), 13*(9), 1427. Advance online publication. DOI: 10.3390/land13091427

Sahoo, P. K., Datta, R., Rahman, M. M., & Sarkar, D. (2024). Sustainable Environmental Technologies: Recent Development, Opportunities, and Challenges. *Applied Sciences (Basel, Switzerland), 14*(23), 10956. DOI: 10.3390/app142310956

Schürholz, D., Castellanos-Galindo, G. A., Casella, E., Mejía-Rentería, J. C., & Chennu, A. (2023). Seeing the Forest for the Trees: Mapping Cover and Counting Trees from Aerial Images of a Mangrove Forest Using Artificial Intelligence. *Remote Sensing (Basel), 15*(13), 3334. https://www.mdpi.com/2072-4292/15/13/3334. DOI: 10.3390/rs15133334

Shetty, D., Sowmya, B. J., Seema, S., ... (2020). Air Pollution Control Model Using Machine Learning and IoT Techniques (1st ed.). Elsevier Inc., Netherlands, 117. ISBN 9780128187562. DOI: 10.1016/bs.adcom.2019.10.006

Singh, R., Kaur, P., & Kaur, A. (2024). Air Quality Index Prediction in Punjab Using Genetic Algorithm and Artificial Intelligence. *International Journal of Advanced Multidisciplinary Research Studies, 4*(2), 309–317.

Siril, S., Singh, M., Shahina, N. N., Bhat, J. A., Shukla, G., Panwar, P., & Chakravarty, S. (2022). Role of REDD+ in Reducing Land Degradation and Achieving SDGs. In *Land Degradation Neutrality: Achieving SDG 15 by Forest Management* (pp. 303-313). DOI: 10.1007/978-981-19-5478-8_16

Song, C. H. (2022). A Hybrid SEM and ANN Approach to Predict the Individual Cloud Computing Adoption Based on the UTAUT2. *International Journal of Information Technology : an Official Journal of Bharati Vidyapeeth's Institute of Computer Applications and Management, 14*(7), 3539–3553. DOI: 10.1007/s41870-022-00936-7

Tarazona, Y., Zabala, A., Pons, X., Broquetas, A., Nowosad, J., & Zurqani, H. A. (2021). Fusing Landsat and SAR Data for Mapping Tropical Deforestation through Machine Learning Classification and the PVts-β Non-Seasonal Detection Approach. *Canadian Journal of Remote Sensing*, *47*(5), 677–696. DOI: 10.1080/07038992.2021.1941823

Van Dongen, W. F. D., San Martin, R., Guay, P. J., & Weston, M. A. (2017). A comparison of the effectiveness and time efficiency of traditional and photographic environmental monitoring techniques [Article]. *Journal of Environmental Management*, *193*, 64–69. DOI: 10.1016/j.jenvman.2017.01.050 PMID: 28189930

Wang, Z., Liu, D., Liao, X., Pu, W., Wang, Z., & Zhang, Q. (2023). SiamHRnet-OCR: A Novel Deforestation Detection Model with High-Resolution Imagery and Deep Learning. *Remote Sensing (Basel)*, *15*(2), 463. https://www.mdpi.com/2072-4292/15/2/463. DOI: 10.3390/rs15020463

Wani, A. K., Rahayu, F., Ben Amor, I., Quadir, M., Murianingrum, M., Parnidi, P., Ayub, A., Supriyadi, S., Sakiroh, S., Saefudin, S., Kumar, A., & Latifah, E. (2024). Environmental Resilience through Artificial Intelligence: Innovations in Monitoring and Management. *Environmental Science and Pollution Research International*, *31*(12), 18379–18395. DOI: 10.1007/s11356-024-32404-z PMID: 38358626

Wong, Y. J., Shimizu, Y., Kamiya, A., Maneechot, L., Bharambe, K. P., Fong, C. S., & Nik Sulaiman, N. M. (2021). Application of artificial intelligence methods for monsoonal river classification in Selangor river basin, Malaysia. *Environmental Monitoring and Assessment*, *193*(7), 438. DOI: 10.1007/s10661-021-09202-y PMID: 34159431

Yadav, A. (2023). Leveraging artificial intelligence for sustainable development and environmental resilience. In *Exploring Ethical Dimensions of Environmental Sustainability and Use of AI* (pp. 140-165). DOI: 10.4018/979-8-3693-0892-9.ch008

Yan, H., & Liu, Y.. (2017). An Evaluation Model of Water Quality Based on DSA-ELM Method. *Proceedings of the 16th International Conference on Optical Communications and Network (ICOCN)*, Wuzhen, China. p. 1-3. DOI: 10.1109/ICOCN.2017.8121280

Yuan, X., Wang, S., Fan, F., Dong, Y., Li, Y., Lin, W., & Zhou, C. (2022). Spatio-temporal dynamics and anthropologically dominated drivers of chlorophyll-a, TN and TP concentrations in the Pearl River Estuary based on retrieval algorithm and random forest regression. *Environmental Research*, *215*, 114380. Advance online publication. DOI: 10.1016/j.envres.2022.114380 PMID: 36162468

Zhang, X., Shu, K., Rajkumar, S., & Sivakumar, V. (2021). Research on deep integration of application of artificial intelligence in environmental monitoring system and real economy [Article]. *Environmental Impact Assessment Review*, *86*, 106499. Advance online publication. DOI: 10.1016/j.eiar.2020.106499

Chapter 4
AI-Driven Solutions for Water Quality Assessment

Muhammad Huzaifa Mahmood
https://orcid.org/0000-0001-7303-1500
University of Turin, Italy

Zuhaib Nishtar
China Three Gorges University, Yichang City, China

Usman Basharat
Jilin University, China & Ministry of Education, China

Rehan Mehmood Sabir
https://orcid.org/0009-0007-4711-8304
University of Agriculture, Faisalabad, Pakistan

Mohammed Saleh Al Ansari
https://orcid.org/0000-0001-9425-0294
University of Bahrain, Bahrain

Muntaha Munir
https://orcid.org/0009-0005-3404-3636
University of the Punjab, Lahore, Pakistan

Naima Nawaz
University of Agriculture, Faisalabad, Pakistan

Nalain E Muhammad
University of Agriculture, Faisalabad, Pakistan

Awn Abbas
University of Agriculture, Faisalabad, Pakistan

Maida Maqsood
Shenzhen University, China

ABSTRACT

The chapter of the book delves into the revolutionary impact of artificial intelligence (AI) on water quality evaluation. This chapter offers a fundamental comprehension of water quality criteria, conventional techniques, and the possibilities of artificial intelligence. The primary components encompassed within this framework consist

DOI: 10.4018/979-8-3373-0680-3.ch004

of acquiring data through sophisticated sensors and Internet of Things (IoT) devices, using data preparation methodologies, and utilizing artificial intelligence (AI) algorithms in predictive modeling. The chapter additionally explores the interplay between artificial intelligence (AI) and other technological advancements, delving into the ethical implications of this integration. This chapter discusses the challenges and constraints associated with AI-driven solutions. It also presents a vision for the future of AI, which includes the development of autonomous monitoring systems and the provision of real-time decision support. This chapter holds significant importance for scholars, practitioners, policymakers, and environmentalists.

1. INTRODUCTION AND HISTORICAL EVOLUTION

Water is an essential resource vital for all forms of life. Water pollution diminishes water quality, adversely impacting marine health and, therefore, human health for those who depend on it. Consequently, it is essential to assess water quality (WQ) and ensure the survival of aquatic species (Jain, Shah, & Mehta, 2021). To guarantee the safety and affordability of drinking water, there is a growing necessity for enhanced water management and quality control. Organizational monitoring issues, disposal techniques, and freshwater assessments must be meticulously analyzed to resolve these challenges (Hu, Zhang, & Zhao, 2019). The method of forecasting water quality entails anticipating the characteristics of variations in the condition of a water system at a specific moment. The planning and regulation of water quality are significantly influenced by the outcomes of watershed assessments. Enhancing methods for preventing and managing water pollution can be achieved by forecasting future water purity under varying pollution levels and formulating effective pollution prevention and control strategies. Water diversion plans should evaluate the overall integrity of the water supply. A substantial quantity of water is conveyed to tackle prevalent hydration issues. Consequently, research must be conducted on methods to forecast water quality today (Zhou et a., 2018).

Currently, security issues necessitate the utilization of artificial intelligence (AI) and machine learning (ML) advancements, which delineate the correlation between system inputs and their results, rather than focusing on complex operational methodologies (Waqas et al., 2022). Traditional approaches to assessing water quality, including laboratory analysis and calculations based on measurable criteria, can be costly, labor-intensive, and constrained in their capacity to deliver real-time information. (Ahmed et al., 2020).

Predictive modeling offers an alternate method that employs AI techniques to assess water quality based on existing data. Utilizing forecasted water quality categories, decision-makers can focus on automobile water quality monitoring, analysis, and the optimization of treatment alternatives. Predictive models can serve as the foundation for developing early detection systems for water quality issues (Lowe et al., 2022). The domains of AI and ML-based water quality monitoring and automation are presently merging at a promising and dynamic juncture. Artificial intelligence is employed for various creative applications, including the development of predictive models for water quality, the construction of neural networks to detect specific contaminants like nitrogen and phosphorus, and the automation of water treatment operations through AI-driven decision-making tools. The domains of AI-driven water quality monitoring and treatment are presently converging at a promising and dynamic juncture. Artificial intelligence is employed for various innovative applications, including the development of predictive models for water processes, the construction of neural networks to detect specific pollutants such as heavy metals and pathogens, and the automation of water treatment processes through AI-driven decision-making tools (Godavarthi et al., 2017).

The application of AI and ML in water quality treatment and automation exemplifies the significant advancements in technology. Moreover, it signifies a dedication to promoting environmental sustainability and safeguarding public health, thereby foreshadowing a future when technology and nature coexist happily. Water quality assessment facilitates compliance with regulations, enables early detection of contamination, and fosters the development of effective management strategies. Global regulatory entities establish water quality standards to guarantee that water bodies meet quality criteria for designated uses. Consistent monitoring and evaluation facilitate the identification of trends, the assessment of pollution control methods, and the guidance of water resource management and conservation strategies.

Figure 1. AI in water treatment

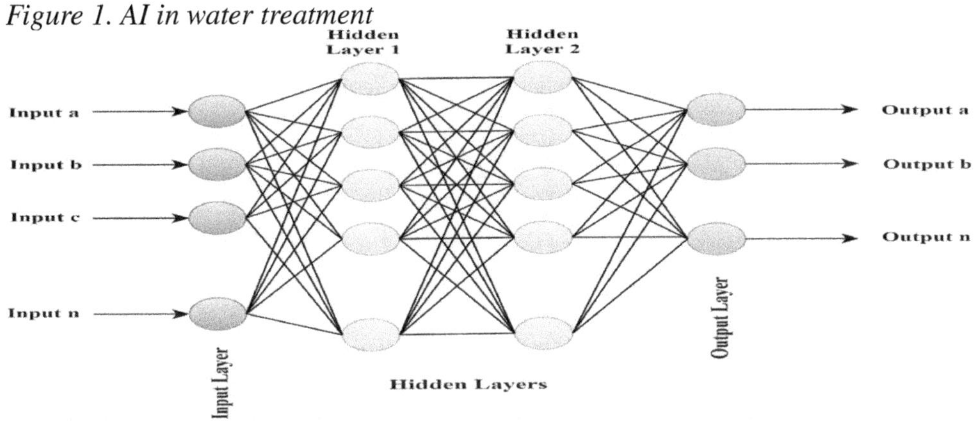

1.1 Significance of Water Quality Assessment

Water quality monitoring is crucial in many ways:

a. Protecting Human Health
b. Preserving Aquatic Ecosystems
c. Supporting Sustainable Water Use
d. Monitoring Pollution Trends

1.2 Parameters of Water Quality Assessment

Several commonly used parameters include:

✓ **Physical Parameters:** Temperature, turbidity, color, odor, conductivity.
✓ **Chemical Parameters:** Dissolved oxygen, pH, nutrients (nitrogen, phosphorus), heavy metals, and organic pollutants.
✓ **Biological Parameters:** The presence and abundance of aquatic organisms are indicators of water quality (e.g., algae, macroinvertebrates).

1.3 Water Quality Standards

Water quality standards are implemented to specify the permissible thresholds of different parameters for distinct water applications, such as potable water or irrigation water.

Table 1. Drinking water quality standards as recommended by FEPA and WHO.

pH 6-9 6.5-9.2 **Total Hardness - 300** **Total Dissolved Solid 2000 500** **Electrical conductivity - 300** **c** **Total Coliform Count (100ml) 0 0** **Sulphate 20 200** **Sodium - 200** **Ammonium 0.01 1.5** **Zinc 5.0 5.0** **Iron 0.05 0.3** **Lead 0.01 0.05** **Cadmium 0.05 0.01** ***Parameter FEPA** **a** **Standards WHO** **b** **standards** ***Parameter**	**FEPA[a] Standards**	**WHO[b] Standards**
pH	6-9	6.5-9.2
Total Hardness	-	300
Total Dissolved Solid	2000	500
Electrical conductivity	-	300[c]
Total Coliform Count (100ml)	0	0
Sulphate	20	200
Sodium	-	200
Ammonium	0.01	1.5
Zinc	5.0	5.0
Iron	0.05	0.3
Lead	0.01	0.05
Cadmium	0.05	0.01

2. METHODOLOGY (AI DRIVEN TECHNIQUES)

Figure 2.1 illustrates the most often utilized AI-based methodologies for water quality assessment. These techniques are widely utilized to oversee wastewater treatment processes, facilitate water reuse, conserve water, and reduce costs through prediction, diagnosis, evaluation, and simulation (Zhao et al., 2020).

Figure 2. Artificial intelligence techniques

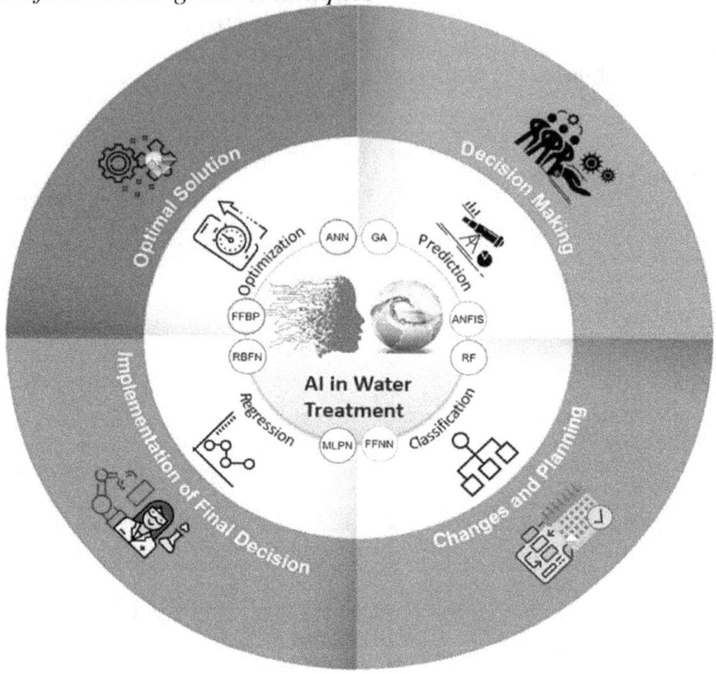

2.1 k-Nearest Neighbor (k-NN)

k-NN is a straightforward machine learning method employed for regression and classification tasks. k-NN retains all current data and classifies new data points based on similarity. For instance, examine a classification problem with two categories, W and Z, as illustrated in Fig. 2.2. In the event of a new data point arising with a placement dilemma about the W and Z categories, it should be assigned to an appropriate category by computing the Euclidean distance. Consequently, the new point will be assigned to category Z, which possesses the highest number of neighbors. k-NN is the most prevalent method employed for classification tasks (J. Han et al., 2012).

Figure 3. k-NN technique before and after a classification problem

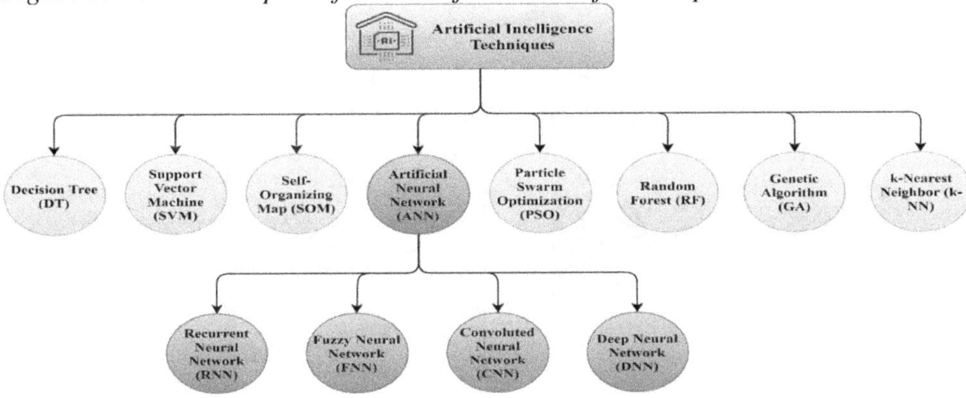

2.2 Decision Tree (DT)

The decision tree technique is primarily employed by AI specialists for classification and regression tasks. The primary objective of decision trees (DT) is to create a training model for class prediction by incorporating the formulation of straightforward decision rules. The structure resembles a tree, where each tree comprises nodes that signify attributes or features of the data, edges that indicate potential solutions to a query, and leaf nodes that reflect the actual output or class label. This technique is predominantly preferred because of its superior accuracy and straightforward implementation. As illustrated in Fig. 2.3, the procedure may yield numerous potential solutions. In a decision tree technique, all elements of an issue are evaluated from the root to the leaf node to identify the ideal solution based on specified criteria.

Figure 4. DT architecture.

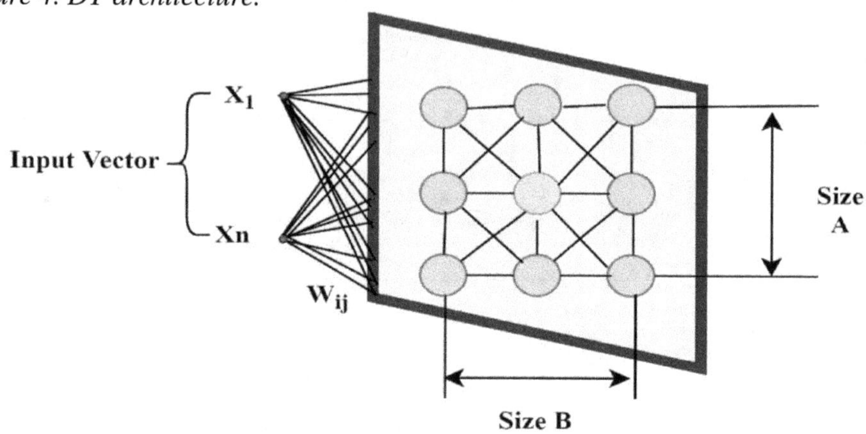

2.3. Random Forest (RF)

Random Forest is employed for both classification and regression tasks. Like a forest, an increased number of decision trees enhances the robustness of the Random Forest. It generates decision trees from data samples, subsequently predicts outcomes for each tree, and ultimately selects the ideal answer through a voting system. The advantage of employing Random Forest is its ability to mitigate the overfitting of Decision Trees through the aggregation of their outcomes. As seen in Fig. 2.4, random samples are selected from a specified dataset, and a decision tree is constructed for each sample. Subsequently, the outcome of each decision tree is derived. The subsequent step involves executing the voting procedure for each anticipated outcome and determining the most often voted projected result as the outcome (Robin et al., 2010).

Figure 5. Typical working procedure of RF technique.

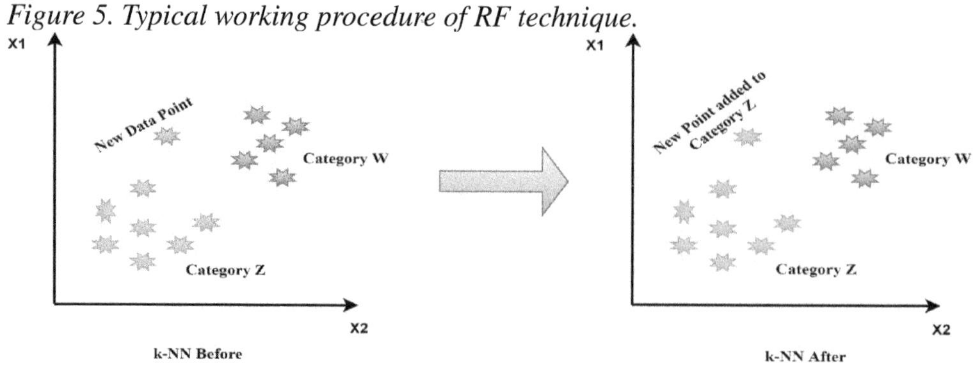

2.4. Artificial Neural Networks (ANNs)

Artificial Neural Networks (ANNs) are statistical models based on biological neurons in the human brain, used for pattern recognition and nonlinear interactions between inputs and outputs. They consist of multiple layers, with the input layer being the input layer, the output layer being the output layer, and hidden layers as intervening layers. Adjustable weights between neurons can be modified using machine learning methods, enhancing the model's effectiveness. An increase in hidden layers can create intricate models that can be trained to improve ANNs' efficiency. A basic design of ANNs includes the input layer, hidden layers, and output layer.

Figure 6. ANNs with four layers: an input layer, two hidden layers and an output layer

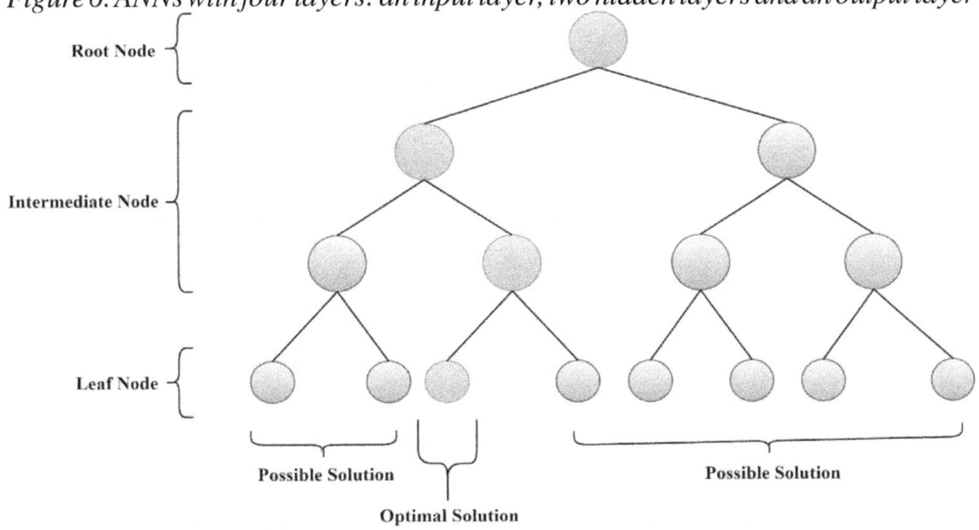

2.4.1. Fuzzy Neural Network (FNN)

FNN is an artificial intelligence methodology that integrates fuzzy logic with neural networks to identify fuzzy system parameters, including fuzzy sets and rules, through neural network approximation methods. It is mostly utilized for pattern recognition, regression analysis, and density estimation.

2.4.2. Convoluted Neural Network (CNN)

CNNs, also known as Feed Forward Neural Networks (FFNN), are artificial neural networks that use convolution instead of matrix multiplication. They are commonly used for image and video recognition, classification, financial time series analysis, and natural language processing. CNNs consist of hidden layers for feature extraction, and classification layers for decision-making based on parameters from the preceding layers. These principles help reduce training parameters and feature size.

Figure 7. Basic CNN architecture.

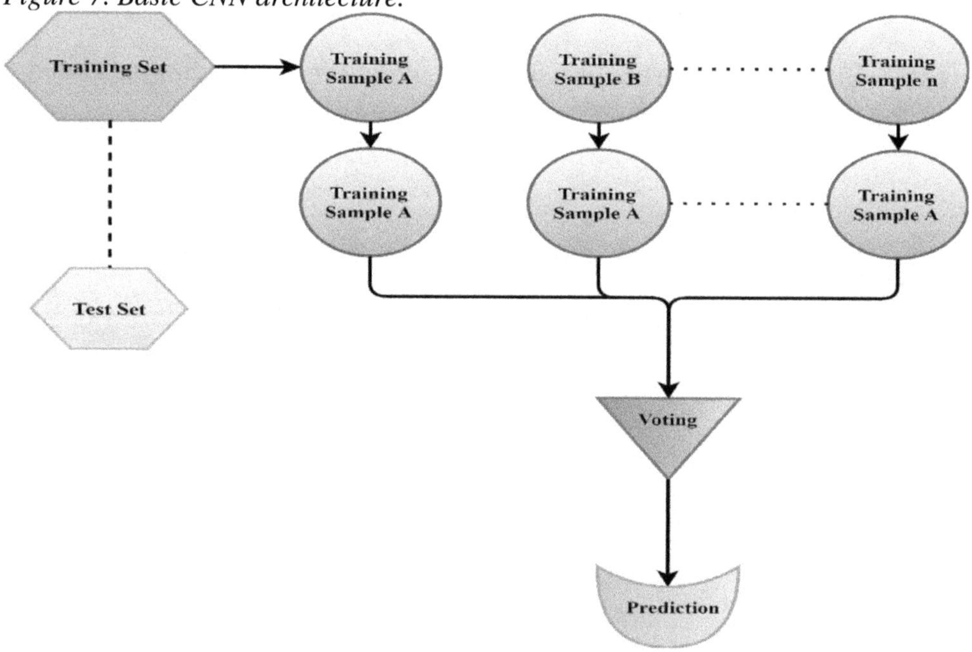

2.4.3. Deep Neural Network (DNN)

DNN comprises several concealed layers in addition to input and output layers (I. Goodfellow et al., 2016; J. Schmidhuber, 2015) as illustrated in Fig. 2.7. Deep Neural Networks (DNN) are frequently employed for the learning of intricate models and the processing of high-dimensional input, incorporating additional hidden layers and neurons. However, DNN requires supplementary computational resources and increases training complexities. In comparison to other artificial neural networks, deep neural networks (DNN) deliver superior performance when the datasets are sufficiently large (G. Pandey, A. Dukkipati, 2014).

Figure 8. DNN architecture

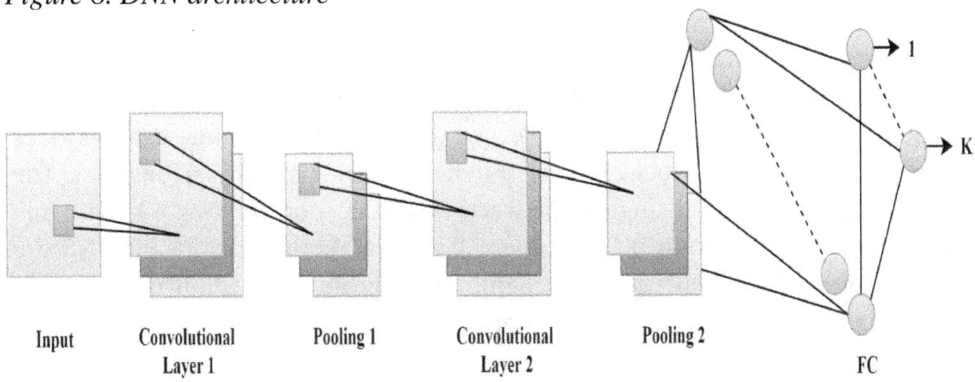

2.4.4. Recurrent Neural Network (RNN)

Recurrent neural networks (RNNs) are feedforward neural networks that use their internal state to process sequences of inputs of various lengths. They incorporate an additional memory state, allowing neurons to exchange identical parameters. The most common RNN is the long short-term memory (LSTM), which uses three gates to compute the hidden state. A basic example of a RNN is a single-layer RNN with parameters X, Y, and Z.

Figure 9. A typical RNN architecture

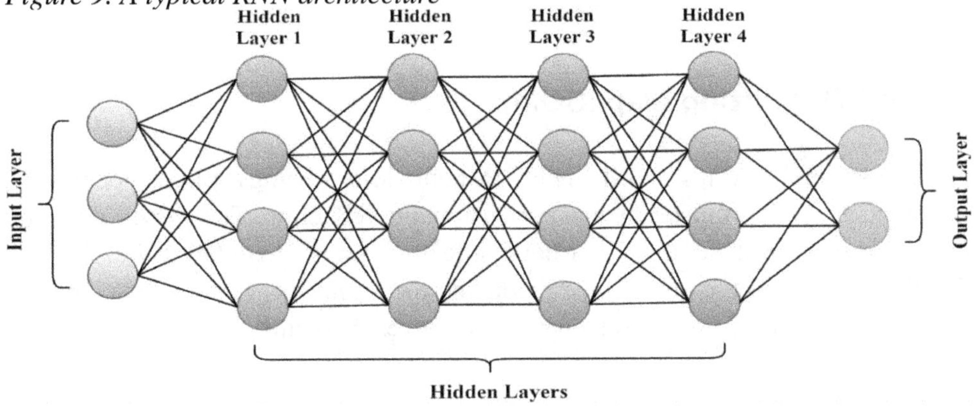

2.5. Support Vector Machine (SVM)

Support Vector Machine (SVM) is an artificial intelligence methodology employed for classification and regression tasks, necessitating labeled training data for each category. It transforms input vectors into high-dimensional feature spaces utilizing kernel functions such as linear, polynomial, and radial basis functions. The primary objective is to distinguish between classes by constructing a hyperplane that maximizes the margin between them. Support Vector Machines (SVM) are predominantly utilized in pattern recognition, where classification involves two distinct classes separated by a single support vector.

Figure 10. SVM classification with a linear hyperplane

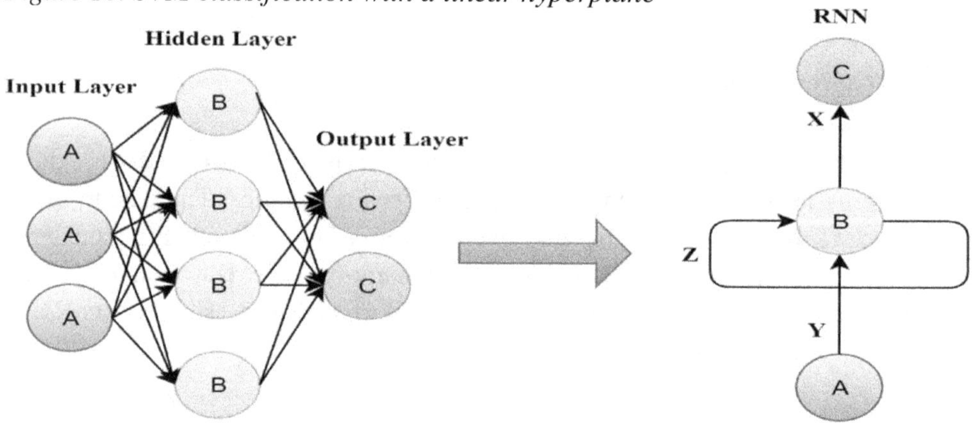

2.6. Self-Organizing Map (SOM)

Self-Organizing Maps (SOM) are an AI methodology using artificial neural network models for data clustering and dimensionality reduction. They have input and output layers, with the output layer being a feature map or map layer. Weights are directly allocated to the output layer, reducing the number of features in a dataset, which simplifies predictive modeling and increases difficulties.

Figure 11. SOM dimensionality reduction.

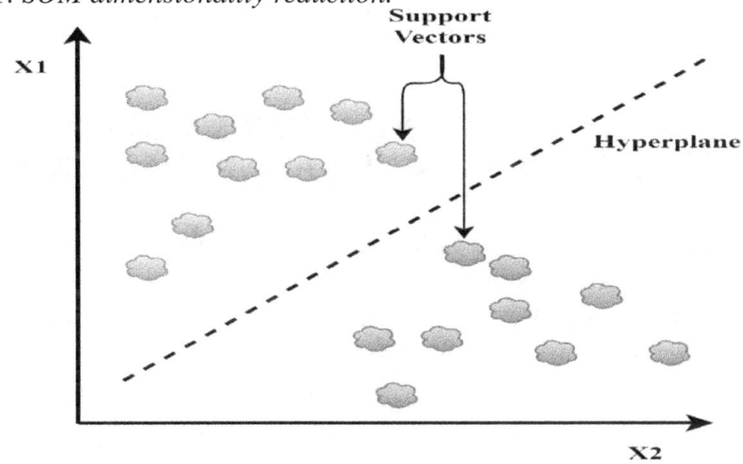

2.7. Genetic Algorithm (GA)

GA is a heuristic search algorithm that employs a recursive method to identify the optimal answer among various alternatives. It employs a gene encoded with alphanumeric sequences, produces novel solutions via mutation, and subsequently merges two solutions through recombination. Genetic Algorithms are predominantly employed to explore the solution space and identify the optimal solution for problem-solving.

2.8. Particle Swarm Optimization (PSO)

Particle Swarm Optimization (PSO) is a widely utilized AI-driven method for optimization challenges, owing to its iterative process that enhances solutions depending on a specified quality metric. In Particle Swarm Optimization (PSO), particles navigate the search space by considering their velocity and position. In the search space, each particle's movement is directed towards the most optimal known point, with its position and velocity being updated over time (T. Cura, 2012). Each particle seeks the optimal position inside the search space by adjusting its velocity in accordance with the established rule (Bansal et al., 2019).

3. RESEARCH OBJECTIVES

- To explore AI's potential in predicting water contamination.

- To assess the efficiency of AI over conventional methods.
- To identify challenges in deploying AI solutions globally.

4. APPLICATIONS OF AI AND ML IN AUTOMATING WATER QUALITY ASSESSMENT, ANALYSIS AND MANAGEMENT

- **Data Gathering:** Real-time gathering of water quality data is conducted using IoT devices and sensors.
- **Data Preprocessing:** The cleaning and manipulation of raw data to eliminate noise and mitigation of inconsistency to do accurate analysis.
- **Anomaly Detection:** Detect irregular variations in water quality indicators via machine learning algorithms, including Support Vector Machines and Isolation Forest.
- **Predictive Modeling:** Anticipate water quality trends utilizing historical data, environmental factors, and predictive variables.
- **Water Quality Assessment**: Automated evaluation of adherence to regulatory criteria in water quality using AI, with deviations identified.
- **Optimization Algorithms:** Enhancement of water treatment procedures for efficiency and cost-effectiveness via genetic algorithms and neural networks.
- **Pattern Recognition:** Artificial intelligence has been employed to identify specific patterns in water quality data linked to distinct pollutants or other pertinent environmental conditions.
- **Decision Support Systems:** Real-time advice and decision-making in water management are conducted based on the current conditions and forecasts generated by AI.
- **Risk Assessment:** Utilizing machine learning models of previous incidences correlated with environmental parameters to evaluate hazards associated with pollution events.
- **Resource Allocation:** Implementing priority-driven resource allocation for water monitoring and management utilizing AI-based technologies.
- **Remote Sensing Integration**: Assessing the condition of extensive water bodies by satellite imagery and incorporating remote sensing data with artificial intelligence algorithms for the identification of distant pollution sources.
- **Maintenance Scheduling:** It can forecast maintenance needs of the water infrastructure by analyzing usage patterns, environmental variables, and deterioration models with AI algorithms.
- **Natural Language Processing:** Examination of reports, social media, and citizen response to comprehend public concerns over water quality.

- **Visualization and Reporting:** Create visual dashboards and reports that include trends, anomalies, and the status of water quality compliance among stakeholders via AI-driven solutions explain in fig. 4.1.

Figure 12. Applications of AI and ML in water quality assessment

Please provide missing figure by **proofing deadline**.

New files can be uploaded to the proofing system via comment attachment or emailed to **bookproofing@igi-global.com**

5. THEORETICAL FRAMEWORKS AND FOUNDATIONS

Artificial intelligence (AI) models for predictive analytics in water quality utilize sophisticated algorithms to anticipate contamination concerns, identify abnormalities, and enhance treatment operations. These models utilize machine learning (ML) methodologies and techniques, to evaluate historical and real-time water quality data. Through the identification of patterns and correlations in intricate information, AI can forecast harmful algal blooms, chemical spills, or bacterial epidemics prior to their escalation, facilitating preventive interventions. The theoretical basis of these models is their capacity to learn from extensive environmental data, hence enhancing accuracy as new information is acquired.

The amalgamation of AI with big data and cloud computing has transformed water quality monitoring through the facilitation of scalable, real-time analysis. Big data platforms consolidate information from sensors, satellites, and municipal records, and cloud computing supplies the computational capacity required for efficient data processing. AI models implemented on cloud infrastructures can manage rapid data streams from IoT-enabled sensors in rivers, treatment facilities, and distribution networks. This collaboration guarantees that predicted insights are both prompt and available to stakeholders, including government agencies and utility operators, promoting data-driven decision-making.

Numerous case studies demonstrate the effective application of AI in water management, especially inside smart water systems. Cities such as Singapore and Barcelona have implemented AI-driven systems to monitor water quality in real time, use predictive analytics to avert contamination and minimize waste. These intelligent grids amalgamate sensor networks with artificial intelligence algorithms to identify leaks, optimize chemical dosing in treatment facilities, and predict demand variations. These applications illustrate the conversion of theoretical AI frameworks into actual solutions, improving both efficiency and sustainability in water resource management.

In addition to urban uses, AI has been employed in natural water bodies to identify pollution sources and evaluate ecosystem health. Researchers have been using deep learning algorithms to examine satellite photos and forecast sediment flow in agricultural watersheds. These case studies highlight the adaptability of AI in tackling various water quality issues, ranging from urban infrastructure to environmental preservation. AI-driven systems are establishing new benchmarks for precision and responsiveness in water quality evaluation by integrating theoretical breakthroughs with empirical data.

The ongoing development of AI models, coupled with progress in big data and cloud technologies, ensures enhanced accuracy in water quality prediction and management. Future advancements may encompass federated learning for decentralized data analysis or explainable AI (XAI) to enhance transparency in decision-making. As these technologies advance, their incorporation into global water management systems will be crucial for preserving this essential resource for future generations.

6. IMPORTANCE OF AI IN WATER QUALITY GOVERNANCE

Access to potable water is a fundamental human right; nonetheless, millions globally continue to confront hazards of water contamination due to pollution, industrial effluents, and deteriorating infrastructure. Conventional water quality monitoring techniques, typically dependent on manual sampling and laboratory analysis, are inefficient, expensive, and incapable of delivering real-time information. Artificial intelligence (AI) emerges as a disruptive force in water administration. Utilizing AI-driven data analytics, predictive modeling, and automated sensing technologies, authorities can swiftly identify contaminants, anticipate possible risks, and execute prompt interventions guaranteeing safer water for communities.

Artificial intelligence improves water quality regulation by analyzing extensive data from sensors, satellites, and historical records to detect trends that traditional approaches cannot reveal. Machine learning algorithms can forecast pollution incidents, identify contamination sources, and provide appropriate remediation solutions. AI-driven technologies in smart cities continuously oversee water networks, promptly identifying irregularities such as chemical spills or bacterial proliferation. This proactive strategy not only protects public health but also saves long-term expenses by averting large-scale disasters before they intensify.

In addition to enhancing efficiency, AI democratizes water management by rendering modern monitoring accessible to underprivileged areas. Portable AI-driven sensors and cloud analytics facilitate water safety monitoring in rural villages and underdeveloped nations without the need for costly laboratory installations. Moreover, AI promotes transparency by producing actionable insights for politicians, facilitating evidence-based legislation and equitable resource allocation. With the escalation of climate change and urbanization leading to increased water stress, AI-driven governance is essential to guarantee the sustainability of clean water for all. The application of conventional techniques for assessing water quality has produced significant insights. Nonetheless, the limitations inherent in these procedures underscore the necessity for the advancement of more complex and authentic methods. Temporal monitoring systems. Such methodologies would yield a more dynamic and comprehensive understanding of water quality.

Table 2. Comparative chart on traditional and current monitoring techniques

Aspect	Traditional Monitoring Techniques	Current Monitoring Techniques
Data Collection Method	Manual sampling and laboratory analysis	Real-time sensor networks and IoT integration
Sampling Frequency	Periodic and infrequent sampling	Continuous and high-frequency monitoring
Parameter Monitoring	Limited parameters (e.g., pH, turbidity)	Multi-parameter monitoring (e.g., pH, DO, EC)
Data Processing and Analysis	Manual data entry and analysis	Automated data aggregation and AI-based analysis
Detection and Alerting	Reactive response to contamination events	Predictive modeling for early detection
Cost and Resource Requirements	Labor-intensive and costly	Cost-effective and resource-efficient

continued on following page

Table 2. Continued

Aspect	Traditional Monitoring Techniques	Current Monitoring Techniques
Scalability and Coverage	Limited spatial coverage and resolution	Broad spatial coverage and high resolution
Integration with Treatment Systems	Disconnected from treatment processes	Integration with AI-enabled treatment systems
Adaptability to Environmental Changes	Limited adaptability to dynamic conditions	Adaptive monitoring and response to changes

7. ADVANTAGES OF AI-DRIVEN SOLUTIONS

The integration of machine learning and artificial intelligence in water quality monitoring and management systems has significantly enhanced operational efficiency and minimized waste. These technologies can autonomously execute routine monitoring tasks with low personnel requirements and reduced costs. It examines extensive datasets to detect inefficiencies in water treatment processes and distribution networks, enabling rectification through operational optimization utilizing AI algorithms (Achite et al., 2024). Predictive maintenance models can foresee equipment failure prior to its occurrence, hence reducing downtime and averting water loss linked to an unanticipated failure event. The AI systems will optimize chemical dosing at treatment plants in real-time via dynamic control, preventing under- or over-application, hence minimizing costs and environmental effects. Real-time data analytics facilitate the identification of leakages and unlawful water usage inside the distribution network, thereby substantially reducing water loss and enhancing operational efficiency. To attain sustainability in water usage, machine learning and artificial intelligence conserve resources and prolong infrastructure lifespan.

Table 3. Advantages of AI-driven solutions

Application Area	Description
Early Warning Systems	Predictive capabilities facilitate early warning systems that notify authorities of imminent water quality concerns, enabling prompt corrective actions to safeguard public health and ecosystems.
Resource Allocation	Precise forecasts enable effective resource distribution by identifying areas with heightened risk of pollution or degradation, so ensuring focused monitoring and remediation efforts.
Preventive Measures	Predictive insights facilitate preemptive measures prior to the escalation of issues. For instance, if an algal bloom is anticipated, authorities might mitigate nutrient runoff from agricultural fields, so averting the occurrence of the bloom.
Regulatory Compliance	Predictive models facilitate adherence to environmental standards by forecasting potential infractions, enabling industries and governments to modify operations to prevent surpassing pollution discharge limitations.
Public Health Protection	Forecasting alterations in water quality that impact drinking water sources facilitates proactive management to protect public health, including prompt advisories and the improvement of treatment systems.
Sustainable Water Management	Predictive analytics facilitates sustainable practices by offering long-term perspectives on water quality changes, assisting in the formulation of plans that reconcile human requirements with ecological well-being, such as improving agricultural water usage. Utilizing or maintaining reservoirs for ecological flow (Alshami, Ali, Elsayed, Eltoukhy, and Zayed, 2024).
Climate Change Adaptation	Predictive models assist in adapting management techniques to evolving conditions induced by climate change, such as forecasting more frequent or severe droughts and planning water conservation measures accordingly.
Community Engagement and Education	Supplying communities with predictive water quality information elevates public knowledge and involvement, fostering support and participation in water protection programs.

8. LIMITATIONS, CHALLENGES AND IMPLICATIONS

A significant impediment is the accessibility and quality of data. Despite the extensive data collected from sensors and other sources, ensuring their reliability and relevance for accurate analysis may prove challenging. Automotive water systems can be intricate, encompassing several factors such as flow rates, pressure, temperature, and contaminants. The intricacies of these systems may pose challenges for ML models to effectively reflect. Deep learning models are often perceived as "black boxes," complicating the comprehension of their conclusions. The absence of interpretability might be a significant disadvantage, especially in critical applications like water quality monitoring. Scalability challenges may occur while implementing ML/AI solutions in automotive water systems, particularly in large automobile manufacturing facilities. Machine learning algorithms may struggle to adapt in real-time to changing environmental conditions or newly recognized contaminants, requiring frequent updates and recalibration. The ramifications of these discoveries are significant for both environmental sustainability and public health. Enhanced water quality monitoring and treatment procedures can result in safer drinking water and less environmental contamination, hence improving public health outcomes. The application of AI and ML in water management highlights the capacity of technology innovations to tackle significant environmental issues. These findings underscore the imperative for sustained investment in research and development to enhance AI and ML technologies, broaden their applications, and guarantee their efficient incorporation into water management methods. These developments signify a substantial progression towards more efficient, reliable, and sustainable water management systems.

9. ETHICAL CONSIDERATIONS

The incorporation of AI in water quality evaluation presents significant potential, although it also engenders crucial ethical dilemmas that require attention. A significant concern is bias in AI training data, which may result in regional discrepancies in water quality estimates. If the data utilized to train AI models predominantly originates from well-regulated urban locales, rural or underdeveloped places may be neglected, leading to erroneous evaluations and inequitable resource allocation. Ensuring varied and representative datasets is crucial to mitigate biases and foster equitable solutions.

A significant issue is privacy in IoT-enabled surveillance systems, which aggregate extensive environmental and, at times, personal data. In the absence of adequate controls, this data may be exploited, resulting in violations of individual or communal privacy. Moreover, accountability in AI-driven decision-making poses a challenge when an AI system errs in water quality evaluations, who bears responsibility. Robust governance structures must be developed to delineate culpability, guarantee transparency, and uphold public trust in these nascent technologies. Harmonizing innovation with ethical accountability will be essential for the sustainable implementation of AI in water management.

10. IMPLEMENTATION OF REGULATORY FRAMEWORKS

The implementation of AI in water management necessitates strong legislative frameworks to guarantee ethical utilization, data protection, and accountability. Standards must be created to regulate AI implementation, focusing on algorithmic openness, bias reduction, and compatibility with current water monitoring systems. These rules must be formulated in conjunction with policymakers, technologists, and environmental specialists to reconcile innovation with public safety. Moreover, explicit standards for data gathering, dissemination, and validation are crucial to uphold trust in AI-driven solutions. In the absence of such criteria, the likelihood of misuse or unexpected repercussions such as erroneous decision-making stemming from biased datasets may compromise the prospective advantages of AI in the protection of water resources.

Nonetheless, the implementation of these systems poses considerable legal and policy obstacles. Regulatory agencies frequently find it challenging to keep up with swift breakthroughs in AI, resulting in deficiencies in oversight. Issues about liability specifically, who bears responsibility if an AI system fails to identify contamination remain unaddressed. Moreover, varying regional and national policies generate discrepancies, hindering cross-border water management initiatives. Case examples illustrate both achievements, such as the EU's AI Act advocating risk-based regulation, and shortcomings, including unregulated AI implementations that intensified disparities in water access. Gaining insights from these experiences is essential for formulating adaptable, egalitarian rules that promote innovation while safeguarding communities and ecosystems.

11. SUMMARY OF KEY FINDINGS

This chapter examined the incorporation of machine learning and artificial intelligence in automotive water quality control systems. The principal conclusions encompass:

1. It is essential to assess all these reasons considering that Auto ML necessitates minimal effort commitment and academic expertise. Auto ML is generally regarded as an advanced tool. Auto ML is less successful in dynamic contexts than static ones.
2. It has been noted that a limited number of research articles explicitly focus on the application of machine learning algorithms in contamination detection and source identification, despite the significant potential for control and prediction offered by these technologies.
3. Establishing and maintaining intelligent water systems necessitates closing the knowledge gap between traditional water management methods and advanced technologies. Furthermore, fostering interdisciplinary collaboration is crucial for cultivating a workforce adept at managing the complexities of technology innovation and water management.

12. SUGGESTIONS AND FUTURE RESEARCH DIRECTIONS

The integration of machine learning and artificial intelligence can improve real-time monitoring capabilities of IoT sensors, enabling proactive management of water quality issues. Hybrid models that combine machine learning techniques with physics-based models can enhance accuracy and interpretability. Explainable artificial intelligence (XAI) research can improve interpretability and transparency in machine learning models, enhancing decision-making and trust. Future research should focus on developing robust machine learning models that can maintain performance over time. Collaboration among automotive manufacturers, water management experts, ML/AI researchers, and regulatory bodies can drive progress in this industry. Multidisciplinary approaches can provide comprehensive solutions for technological and legal challenges. However, temporal intervals between process steps in wastewater treatment plants (WWTP) processes are often overlooked or addressed with insufficient rigor in models (Wang, et al., 2021).

13. CONCLUSION

This chapter delineates the diverse issues in water quality smart quality assessment. AI and ML tackle difficulties related to real-time water quality monitoring and prediction, as well as the identification of contaminants and their sources. This chapter also delineates the application of automated water quality monitoring systems, data-driven decision-making. The integration of AI and ML with the technologies and their viability for water quality control. This chapter examines the limitations and future potential for the advancement of WQ through the utilization of sophisticated computational techniques. The implementation of machine learning and artificial intelligence in automotive water quality assessment, analysis, and management provides substantial advantages in efficiency, cost reduction, and dependability. These technologies provide real-time monitoring, predictive maintenance, and automated management, guaranteeing elevated water quality requirements and efficient resource management. As AI and ML technologies progress, their incorporation into automobile WQ systems is expected to become increasingly complex and influential.

REFERENCES

Achite, M., Farzin, S., Elshaboury, N., Valikhan Anaraki, M., Amamra, M., & Toubal, A. K. (2024). Modeling the optimal dosage of coagulants in water treatment plants using various machine learning models. *Environment, Development and Sustainability*, *26*(2), 3395–3421. DOI: 10.1007/s10668-022-02835-0

Bansal, J. C. (2019). Particle swarm optimization. In *Evol. Swarm Intell. Algorithms* (pp. 11–23). Springer., DOI: 10.1007/978-3-319-91341-4

Cura, T. (2012). A particle swarm optimization approach to clustering. *Expert Systems with Applications*, *39*(1), 1582–1588. DOI: 10.1016/j.eswa.2011.07.123

Genuer, R., Poggi, J.-M., & Tuleau-Malot, C. (2010). Variable selection using random forests. *Pattern Recognition Letters*, *31*(14), 2225–2236. DOI: 10.1016/j.patrec.2010.03.014

Godavarthi, B., Nalajala, P., & Ganapuram, V. (2017). Design and implementation of the vehicle navigation system in urban environments using internet of things (IoT). *IOP Conference Series. Materials Science and Engineering*, *225*(1), 012262. DOI: 10.1088/1757-899X/225/1/012262

Goodfellow, I., Bengio, Y., Courville, A., & Bengio, Y. (2016). Deep learning, MIT press Cambridge. https://mitpress.mit.edu/books/deep-learning

Han, J., Pei, J., & Kamber, M. (2012). *Data mining: concepts and techniques*. Elsevier., DOI: 10.1016/C2009 0-61819-5

Hu, Z., Zhang, Y., Zhao, Y., Xie, M., Zhong, J., Tu, Z., & Liu, J. (2019). A water quality prediction method based on the deep LSTM network considering correlation in smart mariculture. *Sensors (Basel)*, *19*(6), 1420. DOI: 10.3390/s19061420 PMID: 30909468

Jain, D., Shah, S., Mehta, H., ... (2021). A machine learning approach to analyze marine life sustainability. In Proceedings of international conference on intelligent computing, information and control systems (pp. 619 632). Springer. DOI: 10.1007/978-981-15-8443-5_53

Lowe, M., Qin, R., & Mao, X. (2022). A review on machine learning, artificial intelligence, and smart technology in water treatment and monitoring. Water, 14(9), 1384. Lu, H., & Ma, X. (2020).

Pandey, G., & Dukkipati, A. (2014). Learning by stretching deep networks, in: Proc. 31 St Int. Conf. Mach. Learn. Beijing, China. pp. 1719–1727. http://proceedings .mlr. press/v32/pandey14.pdf.

Schmidhuber, J. (2015). Deep learning in neural networks: An overview. *Neural Networks*, *61*, 85–117. DOI: 10.1016/j.neunet.2014.09.003 PMID: 25462637

Wang, G., Jia, Q. S., Zhou, M., Bi, J., & Qiao, J. (2021). Soft-sensing of wastewater treatment process via deep belief network with event-triggered learning. *Neurocomputing*, *436*, 103–113. DOI: 10.1016/j.neucom.2020.12.108

Waqas, S., Harun, N. Y., Sambudi, N. S., Arshad, U., Nordin, N. A. H. M., Bilad, M. R., Saeed, A. A. H., & Malik, A. A. (2022). SVM and ANN modelling approach for the optimization of membrane permeability of a membrane rotating biological contactor for wastewater treatment. *Membranes*, *12*(9), 821. DOI: 10.3390/membranes12090821 PMID: 36135840

Zhao, L., Dai, T., Qiao, Z., Sun, P., Hao, J., & Yang, Y. (2020). Application of artificial intelligence to wastewater treatment: A bibliometric analysis and systematic review of technology, economy, management, and wastewater reuse. *Process Safety and Environmental Protection*, *133*, 169–182.

Zhou, J., Wang, Y., Xiao, F., Wang, Y., & Sun, L. (2018). Water quality prediction method based on IGRA and LSTM. *Water (Basel)*, *10*(9), 1148. DOI: 10.3390/w10091148

Chapter 5
AI's Role in Environmental Health and Tackling Ecological Impacts Through Multidimensional Views

Kholis Ernawati
https://orcid.org/0000-0002-3582-5294
Universitas YARSI, Indonesia

Dian Mardhiyah
Universitas YARSI, Indonesia

Rifda Wulansari
Universitas YARSI, Indonesia

Rifqatussaadah Rifqatussaadah
Universitas YARSI, Indonesia

Sri Utami
Universitas YARSI, Indonesia

ABSTRACT

Ecological change is a global issue impacting environmental health, ecosystems, and human well-being, driven by deforestation, urbanization, and fossil fuel use, leading to air and water pollution and zoonotic diseases. Addressing these challenges requires innovative solutions like artificial intelligence (AI), which offers real-time

DOI: 10.4018/979-8-3373-0680-3.ch005

monitoring, predictive analysis, and resource management. AI aids in identifying risk patterns, providing early warnings, and creating evidence-based strategies while enhancing efficiencies through automation and fostering green technology. Despite its promise, ethical considerations such as data privacy, equitable access, and sustainability remain critical. The article emphasizes integrating technological, economic, and ethical dimensions, calling for policies, collaboration among sectors, and community engagement to maximize AI's potential. It highlights the need for further research on integrating AI with technologies like blockchain to create a sustainable and transparent environmental management system.

INTRODUCTION

The increasing complexity of global ecological changes has posed serious challenges to environmental health and the sustainability of human life. This chapter explores the strategic role of artificial intelligence (AI) in addressing these challenges through a multidimensional approach encompassing technological, economic, and ethical aspects. This chapter also emphasizes the importance of cross-sector collaboration in implementing this technology by highlighting how AI can be used to monitor, analyze, and manage environmental risks more efficiently and sustainably. Through a systematic literature review and a conceptual model, the discussion in this chapter aims to provide practical and academic insights for policymakers, researchers, and practitioners focused on environmental health and social justice in the digital era.

BACKGROUND

Ecological change has become a significant challenge with direct impacts on environmental health and global quality of life. Human activities, such as deforestation, uncontrolled urbanization, fossil fuel consumption, and poor waste management, have accelerated environmental degradation, creating problems such as air pollution, water pollution, climate change, and biodiversity loss (Costanza, 2012; Achmadi & Ernawati, 2017; Wang et al., 2018). Healthy ecosystems are at the core of sustainable development, providing vital services that support human life. However, urbanization pressures continue to threaten ecosystem health, reducing their capacity to support ecological balance and human health (Peng et al., 2015; Wu et al., 2021).

Ecosystem degradation poses serious public health risks, including waterborne zoonotic diseases and respiratory issues from air pollution (Gan et al., 2012; Harada et al., 2018). This highlights the need for integrated approaches that link ecological balance with human well-being. The WHO's "Healthy Cities" concept promotes urban planning centered on ecosystem health as a basis for public health (J. Wang et al., 2018; Mishbahuddin et al., 2021; Ernawati et al., 2022).

Artificial Intelligence (AI) is vital in tackling environmental health issues through advanced data analysis. AI is particularly useful in predicting and monitoring air pollution by analyzing large datasets from satellites and ground sensors. These insights enable timely public health actions and enhance the accuracy of air quality forecasting (Guo et al., 2022). The integration of AI in environmental health strategies not only improves predictive capabilities but also supports sustainable urban planning. AI can optimize resource management in urban environments, such as waste management, energy use, and transportation systems. AI modeling urban dynamics helps urban planners design cities more resilient to environmental change and public health challenges (Denecke et al., 2019).

AI also supports understanding the health impacts of air pollution. Research shows ambient air pollution is a significant risk factor for various health conditions, including cardiovascular disease (Kumari et al., 2024) Using AI to analyze the relationship between air pollution exposure and health impacts, researchers can uncover the mechanisms underlying these factors, thereby aiding public health policies to reduce pollution-related health risks. AI can also identify vulnerable populations and geographic areas most affected, supporting more targeted interventions (Robertson et al., 2018).

AI can support real-time environmental monitoring, efficient resource management, and early detection and response to environmental disasters. For example, machine learning models can predict air pollution patterns or optimize agricultural irrigation to save water (Pardo et al., 2021). However, the application of AI faces challenges, including the need for significant initial investments, the availability of quality data, and ethical considerations such as data privacy and fairness of access (Lu et al., 2015; Ernawati et al., 2022).

In the face of growing ecological challenges, a holistic approach is needed to integrate AI technologies with economic and ethical perspectives. These innovations improve the efficiency of environmental management and promote long-term sustainability, placing human health and ecological integrity as the top priorities (Costanza, 2012; Lu et al., 2015; Tundjungsari et al., 2020).

AI adoption in developing countries faces economic barriers, including high technology, infrastructure, and data costs. Ethical concerns over equitable access and data privacy also pose challenges. Strong data governance and inclusive, participatory approaches are needed to ensure responsible and effective use. Community involvement can enhance trust and ethical implementation (Denecke et al., 2019).

This background emphasizes the importance of AI as a revolutionary tool in addressing environmental health challenges due to ecological change. However, this potential can only be realized if the accompanying economic and ethical challenges are appropriately managed. In this article, further discussion will explore how AI can be effectively integrated to mitigate the impacts of ecological change while considering accessibility, social justice, and sustainability.

PURPOSE AND FOCUS OF ARTICLE

This article explores the potential of artificial intelligence (AI) to provide innovative solutions to environmental health challenges, highlighting the integration of these technologies alongside economic and ethical approaches to create sustainable solutions. In particular, the article focuses on how AI can effectively monitor, manage, and mitigate the impacts of ecological change while ensuring social equity and long-term sustainability. This multidisciplinary approach combining technology, economics, and ethics is designed to provide relevant insights for readers, including policymakers, academics, environmental practitioners, and the general public, to support more integrated strategic planning and policy implementation.

This article aims to:

a. Specifically, identify the role of artificial intelligence (AI) in reducing the impact of ecological changes on environmental health through real-time monitoring and big data analytics by reviewing recent studies (2016–2024).
b. Analyze the economic implications of AI integration in environmental health based on a review of empirical data related to operational cost reduction and productivity improvement.
c. Address ethical challenges in AI implementation, particularly concerning equity of access and data privacy, by providing evidence-based strategies that are achievable and applicable across diverse socioeconomic contexts.
d. Deliver recommendations to stakeholders (government, academia, industry) to develop inclusive, ethical, and sustainable AI-based interventions that can be implemented gradually in the short to medium term.

CONCEPTUAL MODEL

The conceptual model used in this article adopts a multidisciplinary approach that includes three main dimensions: technology, economics, and ethics. Within this framework, artificial intelligence (AI) technology is positioned as a key pillar that provides innovative solutions to environmental health challenges and significantly affects economic and social implications.

Technology as the Main Pillar

AI is increasingly recognized as an essential competency for digital citizens as it shapes how individuals interact with information and technology. Sengsri emphasized the importance of understanding AI's capabilities and ethical implications, especially in education, where future generations need to learn to deal with the complexities of AI responsibly. This understanding is crucial, as misuse of AI technologies can lead to adverse impacts, including criminal activities that capitalize on these advancements.

AI models such as LSTM and CNN are essential for real-time environmental monitoring—LSTM predicts air quality trends, while CNN analyzes satellite images to detect land-use changes. Combined with IoT, algorithms like Random Forest and SVM enhance data processing speed and accuracy. Recent tools like Vision Transformer (ViT) further improve spatial analysis for identifying pollution and at-risk areas, highlighting the powerful synergy between AI and IoT in tackling environmental challenges (Alganci et al., 2020; Watanabe et al., 2020).

AI technology, in the context of environmental health, plays a central role as a catalyst for change. These technologies enable process automation, large-scale data analysis, and evidence-based decision-making quickly and accurately. Some of the technology components that are the focus of this framework include:

a. Real-Time monitoring and prediction

AI enables continuous environmental monitoring (e.g., air quality, temperature) and early detection of changes using machine learning. This real-time capability allows faster public health responses and more accurate forecasting, with over 90% accuracy when tailored to local contexts using IoT, satellite, and demographic data (Mahalisa et al., 2023; Chisom et al., 2024).

b. Big Data Analytics for Ecosystems

AI processes complex datasets from satellites, weather, and field reports to reveal patterns in ecological degradation, such as deforestation and urbanization. These insights enhance environmental decision-making and support sustainable resource management (Vagin et al., 2022)

c. Application of IoT Technology Integrated with AI

Integrating IoT sensors (e.g., for water and air quality) with AI enables automated, accurate environmental monitoring. This improves data collection efficiency and helps detect pollution early, allowing timely interventions in response to complex environmental challenges (Labazanova et al., 2024).

d. AI for Waste and Pollution Management

Recent studies show AI enhances pollution detection and waste segregation through smart biosensors and automated sorting. Combined with blockchain, AI improves traceability and governance in recyclable waste systems. IoT-AI synergy supports scalable, efficient environmental solutions across diverse settings (Nwokediegwu et al., 2024).

Economic Dimensions: Costs and Benefits

Economically, AI has significant and complex effects. It creates new job opportunities but also leads to challenges like wage inequality and unemployment. Policymakers must balance AI's benefits for productivity and innovation with its risks. AI enhances business efficiency but raises concerns about job loss and the need for workforce retraining (Shamim, 2024). Addressing these issues is crucial for adapting to shifting labor markets and reducing inequality.

The use of AI in environmental health impacts technology and has significant economic implications. The economic dimensions in this conceptual model include:

a. Cost efficiency in resource management

Implementing AI can reduce operational costs through efficiencies in the environmental management process. For example, AI technology can help prioritize resource allocation to address areas with the highest pollution levels.

b. New economic opportunities

AI-based innovation opens up opportunities to create new jobs in the green technology sector. In addition, investment in developing AI technologies for environmental health can improve a country's economic competitiveness in the global market.

c. Long-term savings

While the initial investment for implementing AI technologies is high, the resulting long-term benefits, such as reduced environmental damage and decreased healthcare costs due to pollution, can provide significant economic value.

Ethical Dimensions: Justice and Sustainability

Ethically, the application of AI in various sectors, including healthcare, requires a responsible governance framework that considers the ethical, legal, social, and economic implications of AI technologies. Jobin and Ienca highlight the importance of building public trust in AI, which is essential for ethical governance; however, they caution that relying too much on trust may reduce oversight and accountability among AI producers (Jobin et al., 2019). This perspective aligns with the views of Obasa and Palk, who advocate a multidisciplinary approach to AI governance in healthcare, emphasizing the need for collaboration to ensure ethical considerations are adequately addressed (Obasa et al., 2023).

Such an approach is essential for navigating the complexities of integrating AI into society and ensuring the benefits are fairly distributed. The use of AI for environmental health must consider ethical dimensions involving aspects of social justice, sustainability, and impacts on vulnerable communities. These dimensions include:

a. Social justice in technology access

AI technologies are often more accessible to developed countries, leaving developing countries with limited capabilities. Therefore, this conceptual model emphasizes the importance of international collaboration to ensure equitable access to technology globally.

b. Transparency and accountability

AI should be applied with high transparency in collecting, analyzing, and using environmental data. Accountability in AI-based decision-making is also important to ensure that the results align with ethical values and environmental sustainability.

c. Long-term sustainability

AI implementations should be designed to support environmental sustainability holistically. This includes reducing the carbon footprint of AI computing processes, managing e-waste, and protecting natural ecosystems.

Figure 1. Diagram of sustainability analysis of the field of artificial intelligence (AI) based on the five dimensions of sustainability (Khakurel et al., 2018)

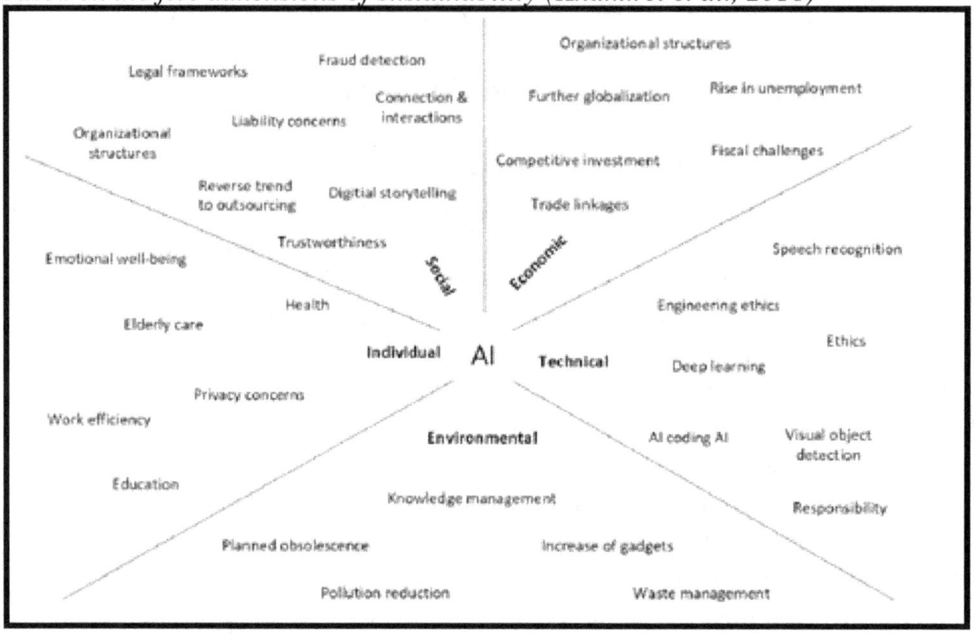

Artificial intelligence (AI) plays a role in five dimensions of sustainability (Figure 1): individual, social, economic, technical, and environmental. On the individual dimension, AI improves work efficiency and access to education but raises privacy concerns. The social dimension includes strengthening communities through digital storytelling, although it requires a legal framework to ensure fairness. In economics, AI boosts productivity but risks increasing unemployment among less-skilled workers. The technical dimension enables innovations such as deep learning but faces ethical challenges and technical errors. Meanwhile, regarding the environmental dimension, AI helps reduce pollution and accelerates e-waste production. A holistic approach is needed to utilize AI sustainably (Khakurel et al., 2018).

Integration of Technology, Economics, and Ethics Dimensions

This conceptual model integrates all three dimensions in a holistic approach. AI technology is used as a foundation to develop economically efficient and ethically responsible solutions. For example, AI-based prediction algorithms not only lower environmental monitoring costs but can also be directed to prioritize areas with vulnerable populations to ensure social equity. The framework also supports data-driven decision-making concerning cost-benefit economic principles and sustainability values. With this approach, the benefits of AI in environmental health can be maximized while minimizing potential social and ethical risks.

Figure 2. Conceptual model integrating three main dimensions (technology, economics, and ethics) in the application of artificial intelligence (AI) for environmental health

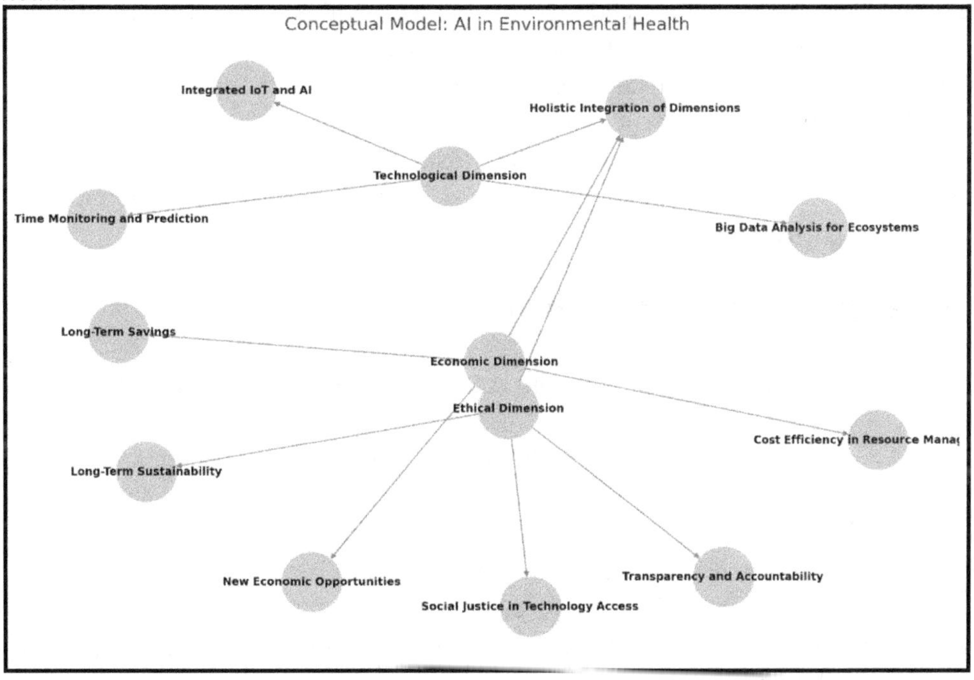

METHODOLOGY

This research uses a systematic literature review method to understand the role of artificial intelligence (AI) in environmental health by considering technological, economic, and ethical perspectives. This methodology was chosen as it allows the collection of relevant and verified secondary data from various high-quality, peer-reviewed sources. The steps in this method are described in detail below:

Data Source

This study utilized literature from Scopus-indexed journals, supplemented by Web of Science and PubMed articles, focusing on publications from 2016–2024 to ensure relevance to recent advancements in AI and environmental health. The review examined the use of various AI models—including deep learning, machine learning, predictive analytics, and IoT integration—highlighting specific algorithms such as neural networks, support vector machines, and random forests. Articles were selected based on topic relevance and AI methodology, with analysis centered on model effectiveness, data types, prediction accuracy, and real-world application in environmental health systems, offering insights into practical implementation, technical challenges, and potential solutions from technological, economic, and ethical perspectives.

Inclusion and Exclusion Criteria

The inclusion criteria applied in the article selection process included:

- Main topic: Articles should address the application of AI in environmental health or ecological impact.
- Time relevance: Articles published within the last five years.
- Type of research: Empirical research-based articles, systematic reviews, or case studies with concrete findings.
- Source credibility: Articles are from Scopus-indexed journals with high-impact factors.

Exclusion criteria include:

- Articles that only discuss AI without linking it to environmental health.
- Publications take the form of opinions or comments not supported by empirical data.
- Articles not available in full text or languages other than English.

Search Procedure

Article searches were conducted through the Scopus portal using keywords such as:

- "AI and environmental health"
- "AI for pollution control"
- "AI and climate change adaptation"

- "AI ethical implications in environment"
- "AI and cost efficiency in environmental management"

This combination of keywords was designed to cover technological, economic, and ethical dimensions. The advanced search feature filters articles by year of publication, field of study, and document type.

Artificial Intelligence (AI) is increasingly vital in environmental health monitoring by improving prediction and management strategies. Models like Random Forest, SVM, CNN, and RNN are used to analyze large environmental datasets, with performance assessed through metrics such as accuracy, sensitivity, specificity, and AUC. These AI tools support data preprocessing and threat detection via image recognition, enabling researchers to forecast climate impacts and develop risk mitigation strategies. This integration marks a significant step forward in tackling environmental health challenges (Adefemi et al., 2023; Bianchi et al., 2024).

Data Analysis

Articles that met the inclusion criteria were analyzed using a thematic approach. Each article was evaluated based on the following:

- Contribution of AI to mitigation or adaptation of ecological change.
- The economic impact of AI implementation, including cost efficiency and potential profits.
- Ethical challenges were identified, such as fairness of access and data privacy.

LITERATURE REVIEW

AI Technology for Environmental Health

Artificial intelligence (AI) has emerged as a transformational technology in environmental health, mainly through its applications in big data management, predictive analysis, and automation. AI facilitates better decision-making in environmental management by providing actionable data-driven insights, as demonstrated by Vagin et al. (2022) and Adefemi et al. (2023). These studies highlight the role of AI in improving the efficiency of environmental business practices and formulating effective strategies to reduce environmental risks.

Here are some of the major applications of AI in environmental health:

a. Air pollution monitoring

A study by Zhang et al. (2023) demonstrated the effectiveness of AI's profound learning algorithms' effectiveness in monitoring air pollution by analyzing data from over 10 million IoT sensors across Chinese cities. This approach significantly improved the accuracy of identifying pollution sources compared to conventional methods, which often rely on limited data and simple models. By processing high-quality sensor data and training deep learning models, the researchers achieved better predictive performance validated against historical data. This advancement highlights the crucial role of AI in enhancing environmental health monitoring and underscores the need for continued research and investment to address growing air pollution challenges (Köse, 2023).

b. Water quality analysis

Integrating artificial intelligence (AI) in water quality analysis has emerged as a transformative approach to monitoring and improving the safety of water sources. AI technology, especially with advanced sensor systems, enables real-time detection of contaminants, which is critical for water management in urban and rural areas.

Some applications of AI in water quality-related research are:

- Rabus et al. (2020). demonstrated that combining machine learning with acoustic sensors improves accuracy and speed in detecting water contaminants, reducing health risks.
- Højris et al. (2016) developed an online optical bacteria sensor for drinking water, offering real-time monitoring every ten minutes and detecting changes in particle size that signal pollution.
- Li et al. (2023) applied UV-visible spectroscopy with advanced algorithms for real-time anomaly detection, improving sensitivity in identifying contamination.
- Zhao et al. (2022) introduced a triboelectric nanogenerator capable of detecting oil and solids at 100 ppm, enabling faster, low-concentration contaminant detection.
- Priya et al. (2022) emphasized smart, in-pipe sensors for real-time detection of physical, chemical, and biological pollutants, supporting EPA standards.
- Hall et al. (2022) showed that AI-enhanced sensors help classify contaminants like oil and heavy metals, improving response strategies and long-term water management.
- Besmer et al. (2017) highlighted the importance of real-time detection by simulating microbial contamination, showing its role in preventing public health crises.

- González-González et al. (2023) advocated for advanced biosensors to detect emerging pollutants efficiently, supporting adaptable monitoring frameworks.

The integration of AI and advanced sensors in water quality analysis enables real-time detection of contaminants, improving response times and supporting sustainable resource management. Continued innovation is essential to tackle the growing complexities of environmental monitoring.

Machine learning models like Support Vector Machine (SVM) and Random Forest (RF) play key roles in applying AI to environmental health. SVM excels in classifying air and water quality with limited data, while RF is effective in predicting pollutants and identifying sources due to its resilience to complex and noisy datasets (Adefemi et al., 2023a; Ohalete et al., 2024) (Adefemi et al., 2023; Ohalete et al., 2024). These AI tools enhance monitoring accuracy and support public health by addressing data and ethical challenges (Hedayet et al., 2024).

Climate Change Prediction

Artificial intelligence (AI) has emerged as an important tool in climate data analysis, especially in predicting changes in complex ecosystems such as global rainfall patterns. The integration of machine learning techniques, particularly neural networks, has shown significant potential in improving the accuracy of predictive models over traditional statistical methods. For example, Crane-Droesch highlighted the effectiveness of machine learning in crop yield prediction and climate change impact assessment, showing that these methods can account for a wide range of environmental factors that affect agricultural yields, such as growing degree days (GDDs) and soil moisture levels (Crane-Droesch, 2018). This capability is critical for developing effective mitigation policies in the agriculture and water management sectors, as accurate predictions can inform management strategies and resource allocation.

The role of AI is not just limited to prediction; it also facilitates the development of adaptive strategies in agriculture. AI can provide insights into optimal irrigation practices and crop management techniques by utilizing data from various sources, including IoT devices and remote sensing technology (Qazi et al., 2022). This is particularly important in regions facing water scarcity, where efficient water management is crucial for sustainable agricultural practices (García et al., 2020). AI's ability to analyze large data sets and identify patterns allows stakeholders to make informed decisions aligned with agricultural productivity and environmental (Sharma et al., 2021).

Economic Implications of AI Technology in Environmental Management

Integrating artificial intelligence (AI) technologies into various sectors brings significant economic impacts, especially in reducing operational costs, increasing productivity, driving technological innovation, and the accompanying economic challenges. This section will explore these dimensions in depth, supported by empirical evidence and academic research.

a. Operating cost reduction

Investments in AI technology have been proven to reduce operational costs significantly in various industries. For example, a study on water pollution monitoring systems in factories showed that AI applications could save up to 30% of operational costs. This reduction is mainly due to AI's ability to detect problems earlier than traditional methods, thus minimizing the need for costly human intervention and other reactive measures. The efficiencies generated by AI technologies enable process optimization, better resource allocation, and waste reduction, contributing to reduced operational costs. The efficiencies generated by AI technologies enable process optimization, better resource allocation, and waste reduction, which overall contribute to the reduction of operational costs (Chen et al., 2023).

In addition, the role of AI in improving operational efficiency is also seen in other sectors, such as healthcare. AI-based systems have demonstrated the ability to significantly reduce administrative costs by automating routine tasks and improving accuracy in billing and coding processes (Kilanko, 2023). This automation reduces labor costs and improves overall operational efficiency, allowing organizations to allocate resources more effectively.

b. Increased economic productivity

AI technologies contribute to long-term improvements in economic productivity by supporting more efficient environmental management practices. In the agriculture sector, for example, AI algorithms have been instrumental in predicting weather patterns and managing water resources, resulting in significant increases in crop yields. In addition to improving food security, these technologies promote sustainability by reducing production waste.

AI-driven economic productivity is also seen in various other sectors. For example, AI applications have been shown to optimize manufacturing production processes, resulting in higher output at a lower cost per unit (Chen et al., 2023). Analyzing large amounts of data in real time enables businesses to make informed decisions, improving operational efficiency and productivity. This trend is supported by the *Global Innovation Index*, which shows that sectors utilizing AI are experiencing substantial growth, with investments in AI-based startups reaching $15 billion in the past five years.

c. Encouraging technological innovation

The application of AI in environmental management has been driving technological innovation across sectors. The startup ecosystem is experiencing rapid growth in developing AI-based solutions to address environmental challenges, such as e-waste management and biodiversity monitoring. The growth of this ecosystem reflects a broader trend where AI catalyzes innovation, creating new business models and approaches to sustainability.

According to the Global Innovation Index report, the AI sector has recorded a 20% growth in the last five years, highlighting the importance of AI in driving technological progress. AI-based startups utilize advanced technologies such as machine learning, computer vision, and data analytics to create effective and scalable solutions. For example, an AI-based platform for biodiversity monitoring allows organizations to track species populations and habitat changes in real time, supporting better conservation efforts.

d. Economic challenges

Despite AI technology's many benefits, its implementation presents significant economic challenges, especially for developing countries. The initial investment required, including hardware procurement, software development, and workforce training, can be prohibitive (Chen & Jin, 2023). This financial burden is often more pronounced in regions with limited access to capital and technological infrastructure, widening the gap between developed and developing countries.

Collaboration between governments, non-governmental organizations (NGOs), and the private sector is essential to overcome these challenges. Initiatives such as technology subsidies and fiscal incentives can facilitate the adoption of AI technologies in economically disadvantaged regions. For example, governments can provide grants or tax incentives to organizations investing in AI-based environmental management solutions, alleviating some of the initial cost burdens. In addition to the initial investment, ongoing maintenance costs, including algorithm updates and hardware maintenance, are also challenging for organizations, especially in resource-constrained environments (Ruamviboonsuk et al., 2021). A balance between the cost of AI implementation and its long-term benefits is crucial to ensure sustainable growth.

Ethical Perspective

Integrating artificial intelligence (AI) technologies in environmental health requires a solid ethical framework prioritizing fairness, data privacy, environmental sustainability, and community engagement. Each aspect ensures that AI technologies are applied fairly, inclusively, and beneficially to all parties involved.

a. Equity in technology access

The digital divide remains a significant obstacle to the equitable distribution of AI technologies, especially in developing countries and marginalized communities. Moreover, the ethical implications of unequal access to technology go beyond mere availability; they include issues of justice and human dignity. Russell et al. (2023) emphasize that an ethical framework for digital technologies must consider the balance of power and rights of individuals in disadvantaged positions. By creating an environment that prioritizes equitable access to AI technologies, society can move towards social justice that empowers all individuals regardless of their socio-economic status.

b. Data privacy and security

Ethical challenges related to data privacy and security are particularly prominent in the context of AI applications in environmental health. Collecting and analyzing large datasets, often containing sensitive personal information, raises significant concerns about potential data breaches and misuse. A study by Katirai (2023) found that more than 15% of AI-based devices are vulnerable to security breaches, putting individuals' privacy rights at risk.

A strict regulatory framework is needed to ensure data is managed responsibly and transparently. This includes implementing strict security measures, regular audits, and a culture of accountability among AI developers and users. In addition, engaging the public in discussions about data privacy can increase trust and facilitate a more informed dialog about the ethical implications of AI technologies (Couture et al., 2023). By prioritizing data privacy and security, risks can be minimized, and a safer environment for individuals whose information is used can be created (Lal et al., 2024)

c. Environmental sustainability

A paradox often occurs when AI technologies are designed to promote sustainability while contributing to environmental degradation. Developing energy-efficient algorithms and using renewable energy sources for data centers are important steps to reduce applications' carbon footprint. The urgency of environmental sustainability should be integrated into the ethical considerations of AI development. Richie (2022) articulated that the environmental consequences of AI implementation cannot be ignored, as they have far-reaching impacts on human and planetary health. With a holistic approach that prioritizes sustainability, AI technologies can contribute positively without exacerbating ecological challenges

d. Community engagement

The success of AI applications in environmental health largely depends on the level of community engagement during the technology implementation process. Local communities should be actively involved in every stage, from design to evaluation, to ensure that the solutions developed are relevant to their needs and context. This participatory approach increases the effectiveness of AI technologies and fosters a sense of ownership among community members.

Involving the community in developing and applying AI technologies can also help identify unintended consequences and mitigate negative impacts. Couture et al. (2023) argue that the ethical governance of AI in public health requires public engagement to ensure the technology is aligned with community values and priorities. By prioritizing community engagement, stakeholders can create AI solutions that are not only effective but also ethically appropriate and socially responsible.

RESULTS AND DISCUSSION

Improving Economic Efficiency in Resource Management

Integrating artificial intelligence (AI) in resource management has become a transformative force that increases economic efficiency across sectors. By optimizing energy consumption, improving agricultural practices, and improving waste management, AI contributes to environmental sustainability while providing substantial economic benefits. This multifaceted approach emphasizes the critical role of AI in establishing a balance between economic growth and ecological preservation.

The economic implications of AI-based resource management are profound. By reducing operational costs through improved efficiency in energy consumption, agricultural practices, and waste management, organizations can allocate resources more effectively, ultimately improving profitability and sustainability. In addition, AI enables organizations to streamline operations, improve decision-making processes, and enhance overall performance (Davianto, 2022).

a. Energy optimization

AI technology plays an important role in reducing energy consumption inefficiencies. Predictive algorithms can analyze historical energy usage patterns to forecast future demand in the industrial and residential sectors. This allows energy providers to adjust production and distribution strategies in real-time, minimizing waste and operational costs (Rojek et al., 2023). In addition, AI-based energy management systems can identify inefficiencies and project future energy demand, which is crucial for reducing energy waste and optimizing resource allocation (Hanafi et al., 2024).

b. Resource optimization in agriculture

In the agricultural sector, AI optimizes water, fertilizers, and pesticide-critical inputs in crop production. Through advanced data analytics, farmers can determine the optimal timing and amount of applications, resulting in increased crop yields while promoting environmental sustainability. Predictive models incorporating weather data and soil conditions allow farmers to make more informed decisions, thereby reducing the overuse of resources and increasing overall efficiency (Hanafi et al., 2024).

This precision farming approach maximizes productivity and reduces the ecological impact of agricultural practices. Furthermore, using AI in agricultural resource management also helps create a more sustainable farming environment by reducing waste and improving resource conservation.

c. Increased efficiency in waste management

AI makes a significant contribution to improving the efficiency of the waste management process through the automation of waste sorting using computer vision technology. The system can accurately identify and separate recyclable materials from the general waste stream, reducing reliance on manual labor and increasing recycling rates (Sharma et al., 2021). This results in better resource recovery and decreased landfill utilization, crucial for sustainable waste management practices.

Improving Social Justice Through Community-Based Data Analytics

Integrating artificial intelligence (AI) in environmental health is important in improving social justice, mainly through community-based data analysis. AI's ability to process large amounts of geospatial and demographic data enables the identification of vulnerable areas disproportionately affected by environmental hazards. This focused approach not only improves the efficiency of resource distribution but also addresses systemic inequalities that have marginalized certain groups.

a. Local community empowerment

Empowering local communities through AI is also important to promoting social justice. AI-based applications can provide communities real-time information on environmental conditions, such as air and water quality. With this access, residents can take proactive steps to protect their health (Lohani, 2024). Access to this timely and relevant information increases public awareness and engagement in environmental issues and encourages a sense of ownership and responsibility among community members.

By integrating AI into community-based environmental management, this technology acts as a mitigation tool and a means of empowerment. With accurate and timely information, communities can make better decisions, advocate for their rights, and actively participate in shaping a healthier and more equitable environment (Gatla, 2019)

b. Benefits of data analysis for policy and planning

The implications of AI in improving social justice go beyond identifying vulnerable areas and empowering communities. Using AI for data analysis, stakeholders can develop comprehensive vulnerability assessments that inform policy and planning decisions. For example, methodologies such as the DRASTIC model have been used to evaluate groundwater vulnerability, providing important insights that can guide environmental management strategies (Hendriks et al., 2019; Al-Ozeer et al., 2021). Such assessments are critical to understanding the complexity of environmental risks and implementing effective interventions according to the unique needs of each community.

c. AI for sustainability and AI sustainability

The role of AI in environmental sustainability must be considered. It facilitates resource management optimization, particularly in renewable energy systems. Using predictive analytics and optimization methodologies, AI can improve the efficiency of renewable energy utilization, contributing to a low-carbon energy future (van Wynsberghe, 2021; Rathor, 2024). This collaboration between AI and sustainability affirms the potential of technology to not only address current environmental challenges and promote long-term ecological resilience and social justice.

AI applications in environmental health represent a transformative opportunity to improve social justice through community-based data analysis. By identifying vulnerable populations and empowering local communities with actionable insights, AI fosters an inclusive approach to environmental health that ensures equitable access to resources and opportunities. Integrating AI in environmental management practices addresses systemic inequalities and promotes a sustainable future where all communities can thrive.

Challenges in AI Implementation for Environmental Health

Integrating artificial intelligence (AI) in environmental health presents various challenges that must be overcome to maximize its potential. These challenges include high costs for developing and maintaining AI technologies, inadequate infrastructure, ethical and privacy concerns, and a shortage of skilled labor.

First, the cost burden for the implementation of AI technologies is enormous. Developing countries, in particular, face budget constraints that hinder investment in advanced AI systems These costs include initial investment, maintenance, and system updates, which are often too expensive for countries with limited financial resources (Adefemi et al., 2023). This financial disparity widens the digital divide, where rich countries can utilize AI for environmental health initiatives while poor countries are left behind. This lack of funding hinders innovation and adoption of AI solutions that can reduce environmental health risks, such as air and water quality monitoring, predictive models for disease outbreaks, and resource management (Ayoubi et al., 2023).

In addition, inadequate infrastructure is a significant barrier to implementing AI in environmental health. Many rural and less developed areas need more stable internet access and advanced computing devices for AI technologies. These shortcomings exacerbate health access and outcomes inequalities, with communities without adequate infrastructure missing out on critical information that AI can provide. These challenges are further compounded by the rapid pace of technological development, often outpacing the ability of local systems to adapt (Badawy, 2023).

Ethical and privacy issues are also significant concerns regarding the use of AI in environmental health. AI technologies often involve collecting and analyzing large amounts of data, which raises questions about data ownership, consent, and potential misuse (Alshawi et al., 2022). A lack of clear guidelines on data privacy can lead to using personal information without explicit consent, undermining public trust in AI systems. Algorithmic bias in AI systems can result in inequalities in health outcomes, especially for groups that have experienced systemic discrimination. A transparent and accountable regulatory framework is needed (Murdoch, 2021).

More skilled labor is another important challenge in developing and using AI. Successful implementation of AI technologies requires a workforce that is not only proficient in AI and machine learning but also knowledgeable about environmental health issues (Naik et al., 2022). Adequate training programs to develop these skills still need to be improved. This gap hinders the adoption of AI solutions as organizations need help finding a qualified workforce to manage these technologies.

Addressing these challenges requires a comprehensive approach. Targeted investments in infrastructure, especially in less developed regions, can help reduce the technology access gap. Strict privacy policies and ethical guidelines are needed to protect individuals' data and build trust in AI systems (Fiske et al., 2019). Comprehensive workforce training programs should be developed to ensure a workforce ready for the AI-based era. By addressing these challenges holistically, AI can contribute significantly to improving environmental health and social justice globally.

Detailed Issues in Implementing AI Models

Implementing Artificial Intelligence (AI) in environmental health presents various challenges, primarily around algorithm bias, interoperability issues, predictive accuracy, scalability, and the requisite infrastructural capacities, especially in developing nations. Addressing these challenges is critical to harness the full potential of AI while ensuring fairness, efficiency, and practical relevance in interventions tailored to diverse environmental health needs.

One of the foremost challenges is algorithm bias and model interpretability. Biases in the training data—historical or due to unrepresentative datasets—can lead to skewed algorithm decisions, potentially exacerbating inequalities within public health interventions. The ramifications are severe in environmental health, where the equitable distribution of resources is crucial for effective public health responses. Moreover, without a clear understanding of how AI systems arrive at specific decisions, stakeholders may find it challenging to trust these systems, which can further impede their adoption in health contexts (Petersson et al., 2022). Interoperability is another significant hurdle.

Various Internet of Things (IoT) devices and AI systems are often developed by different manufacturers, leading to difficulties in seamless integration, thereby hindering effective data collection and analysis. To mitigate this, there is a pressing need for standardized interoperability frameworks that can accommodate diverse technologies, thus bridging the gaps between disparate systems and promoting coherent data flow (Svedberg et al., 2022). Such standardizations will facilitate more efficient utilization of AI tools by ensuring they can function collectively rather than in isolation. Predictive accuracy and model validation further complicate the deployment of AI. AI models must be continuously validated against real-world data to maintain their accuracy and relevance, particularly given the volatility of environmental conditions. Dynamic validation methods need to be adopted to adapt to changing environmental contexts in real time, thus ensuring that AI tools can provide actionable insights that are both relevant and timely (Lammons et al., 2023).

The scalability and adaptability of AI models face significant hurdles, mainly when tools from high-income countries are applied in developing regions. Sociocultural and economic differences require significant adjustments, often needing substantial resources. Health challenges vary by region, making it hard to scale narrow AI solutions without losing local relevance. Poor internet access and low digital literacy in developing areas further limit AI use. Without investments in infrastructure and human capacity, the benefits of AI may remain out of reach. Successful AI implementation demands improved tech infrastructure, workforce training, and strategies to overcome algorithm bias, interoperability, and validation issues—ensuring equitable and impactful use in environmental health.

Solutions to Overcome Challenges

Addressing the challenges of implementing artificial intelligence (AI) technologies, particularly in environmental health, requires a comprehensive approach. This includes government co-funding and subsidies, technology infrastructure development in less developed regions, transparent privacy policies, and thorough training and education programs. These measures aim to ensure that AI technologies are accessible, ethical, and effective in addressing environmental health issues.

a. Co-funding and Government Subsidies

Co-funding and subsidies are key to promoting AI adoption, especially in resource-limited areas. Public-private partnerships can ease financial barriers for institutions by supporting AI projects, notably in healthcare, where high initial costs hinder implementation. Such funding supports AI R&D in environmental health, encouraging cross-sector collaboration (Teixeira et al., 2015).

b. Technology Infrastructure Development

Strong infrastructure—like internet access and digital literacy—is vital for equitable AI access, particularly in rural areas. Enhancing connectivity and tech training helps communities apply AI in health monitoring and decision. (Wood et al., 2021).

c. Transparent Privacy Policy

Ethical AI uses transparent privacy policies to ensure responsible data handling and public trust. Data collection, consent, and use regulations must involve multisector stakeholder input (Cantu et al., 2016).

d. Training and Education Programs

Workforce capacity can be improved through AI and environmental health education. Community-based training and AI literacy in curricula prepare professionals to effectively use AI to address environmental and health challenges (Cantu et al., 2016).

FURTHER RESEARCH

In the context of the rapid development of artificial intelligence (AI) technology, further research is needed to ensure that the application of AI in environmental health is practical but inclusive and sustainable.

Some of the pressing research areas are:

a. Integration of AI with other technologies, such as blockchain, for environmental data transparency

Blockchain is a digital ledger-based technology that enables transparent and secure data recording. Integrating AI and blockchain can improve environmental data's accuracy, transparency, and security. AI can analyze real-time environmental data, such as air pollution, water quality, and land change. Blockchain, in turn, can record and store such data transparently, enabling monitoring by various stakeholders.

The study by Zhang et al. (2023) showed that AI and blockchain can be used to manage e-waste more efficiently. AI identifies locations of waste accumulation that require immediate management, while blockchain ensures that waste management companies, governments, and communities can track the waste data. Future research could focus on developing AI and blockchain-based platforms for various environmental sectors, including sustainable agriculture, forest monitoring, and toxic waste management.

b. Data-driven policy development for AI use in the environmental health sector

Currently, many environmental policies need to be made utilizing the great potential offered by AI-based data analysis. Further research is needed to develop data-driven policy models that utilize AI to support strategic decision-making in environmental health. These policy models may include:

- Predictive modeling

AI can be used to predict the impact of air pollution on respiratory disease incidence in a given region so that policies can be directed towards reducing industrial emissions in the most severely affected areas.

- Real-time risk analysis

AI can assist governments in identifying urgent environmental health risks, such as chemical leaks or pollution spikes, and respond immediately through emergency policies.

- Improved regulatory efficiency

AI can analyze a company's compliance with environmental regulations, such as carbon emissions or waste management, and provide automated reports to regulators.

c. Interdisciplinary collaboration in research

Further research should involve interdisciplinary collaboration, including environmental science, public health, information technology, economics, and ethics. This multidisciplinary approach will ensure that research covers all important aspects, from technical to socio-economic impacts and equity. Research may include engaging with local communities to understand their needs and perspectives. By listening to community voices, AI-based solutions can be designed in a more inclusive and relevant manner.

MANAGERIAL IMPLICATIONS

Integrating artificial intelligence (AI) in environmental health requires a comprehensive managerial approach that includes education, an ethical framework, cross-sector collaboration, and a robust implementation strategy. This approach is essential for effective, sustainable, and ethically sound AI implementation. Collaboration between policymakers, industry leaders, and the scientific community is essential to optimize the benefits of AI while addressing the challenges inherent in its use in environmental health.

Investment in education and training is a key cornerstone in improving the utilization of AI for environmental health. Formal education programs need to be designed to integrate curricula covering big data analysis, ecosystem change prediction, and machine learning models for environmental risk mitigation (Adefemi et al., 2023; Chisom et al., 2024). In addition, industry-led training initiatives can improve employees' competencies using AI-based tools, creating a workforce ready to face the technological challenges in the sector (Ayoubi et al., 2023). Collaboration between academia and industry can also drive innovation, creating a dynamic ecosystem that supports technological advancement in environmental (Chisom et al., 2024). On the other hand, improving the technological literacy of the general public is also important to encourage active participation in AI-based environmental health initiatives (Murphy et al., 2021).

An ethical framework is needed to ensure equitable AI access and ecological sustainability. This includes the development of technological infrastructure in underserved regions to bridge the digital divide and foster inclusivity (Khakurel et al., 2018; Murphy et al., 2021). Transparency in data usage and algorithm development is also crucial; stakeholders should know how data is collected and used and the impact of AI-based decisions (Morgenstern et al., 2021). AI system designs that support sustainability, such as low-carbon waste management solutions, can significantly contribute to environmental health goals. Regulation, including national standards and incentives for ethical AI practices, is needed to ensure that the application of AI does not exacerbate inequalities or damage the environment (Murphy et al., 2021; Adefemi et al., 2023) .

Collaboration between the government, private sector, and academia is essential to maximize the potential of AI in environmental health. The government can act as a facilitator through funding, building digital infrastructure, and creating incentives for private sector investment in AI technologies. The private sector contributes to developing AI products and services and provides technical training. Meanwhile, academia plays a role in basic research and workforce capacity building. Cross-sector collaborations, such as the Global Partnership on Artificial Intelligence (GPAI), have shown that these joint efforts can lead to high-impact solutions in environmental health (Morgenstern et al., 2021; Chisom et al., 2024).

Practical implementation strategies include mapping critical needs, implementing pilot projects, and continuous evaluation to refine approaches. Challenges such as limited technological infrastructure, resistance to change, and budget constraints can be addressed through targeted interventions, including government investments, awareness campaigns, and public-private partnerships (Adefemi et al., 2023; Ayoubi et al., 2023). Pilot projects can be the basis for testing the effectiveness and scalability of AI applications before they are widely implemented. Regular evaluations are also important to ensure that AI applications remain relevant and effective in addressing environmental health issues (Gama et al., 2022; Chisom et al., 2024).

By adopting this managerial strategy, stakeholders can harness the transformative potential of AI to address environmental health challenges while ensuring inclusivity, equity, and long-term sustainability. AI integration offers innovative solutions to monitor and mitigate environmental risks and encourages a proactive approach to public health that is responsive to the complexities of our changing environment. As AI technologies continue to evolve, stakeholders need to remain mindful of ethical considerations and ensure that the benefits of AI are equitably distributed across sectors of society.

CONCLUSIONS

This article highlights that artificial intelligence (AI) has great potential in addressing the increasingly complex environmental health challenges posed by ecological change. With advanced data analysis capabilities and evidence-based approaches, AI can provide innovative solutions to monitor, analyze, and manage environmental risks more efficiently. The technology enables real-time monitoring of air, water, and soil quality, prediction of natural disasters, and better resource management. Its predictive capabilities, powered by machine learning algorithms and smart sensors, facilitate timely interventions, reducing human and ecosystem health risks.

From an economic standpoint, AI significantly benefits operational cost savings, waste reduction, and improved efficiency. AI-based automation helps organizations manage resources more effectively, supporting environmental sustainability and creating new economic opportunities in the green technology sector. For example, AI can optimize energy consumption, increase agricultural yields through precision farming, and support more efficient waste management. However, economic challenges, including initial investment costs and technology infrastructure requirements, remain key barriers that must be overcome through cross-sector collaboration and supporting policies.

The ethical dimension is equally important in implementing AI for environmental health. The use of this technology must consider data privacy, fairness of access, and sustainability. The risk of privacy violations due to large-scale data collection and unequal access to technology between developed and developing countries poses significant global challenges. Therefore, an ethical framework is needed that ensures AI is used in an inclusive, transparent, and responsible manner. This approach includes developing equitable policies, transferring technology to developing countries, and strengthening data privacy regulations and algorithm transparency regulations.

Cross-sector collaboration is key to maximizing the benefits of AI. The government, academia, private sector, and society must work together to create an ecosystem that supports adopting this technology. The government can provide funding and supportive policies, the private sector can develop AI-based solutions, and academia can participate in research and workforce training. The involvement of local communities is also important to ensure that AI technologies are relevant and beneficial to vulnerable groups who are often the most affected by environmental changes.

As a way forward, further research is needed to explore the integration of AI with other technologies, such as blockchain, which can improve data transparency and security. Case studies in developing countries can help understand the specific needs of vulnerable communities and create more inclusive solutions. With a multidimensional approach that combines technology, economics, and ethics, AI can be a catalyst for creating an environmental health system that is more equitable, sustainable, and responsive to future challenges.

REFERENCES

Achmadi, U. F., & Ernawati, K. (2017). Climate Change Globalization and Emerging Disease, Challenges and Good Pracitces for Environmental Health Specialist. *International Journal of Public Health and Clinical Sciences*, *4*(2), 2289–7577.

Adefemi, A., Ukpoju, E. A., Adekoya, O., Abatan, A., & Adegbite, A. O. (2023a). Artificial intelligence in environmental health and public safety: A comprehensive review of USA strategies. *World Journal of Advanced Research and Reviews*, *20*(3), 1420–1434. DOI: 10.30574/wjarr.2023.20.3.2591

Adefemi, A., Ukpoju, E. A., Adekoya, O., Abatan, A., & Adegbite, A. O. (2023b). Artificial intelligence in environmental health and public safety: A comprehensive review of USA strategies. *World Journal of Advanced Research and Reviews*, *20*(3), 1420–1434. DOI: 10.30574/wjarr.2023.20.3.2591

Al-Ozeer, A. Z., & Al-Abadi, A. M. (2021). Groundwater Vulnerability Evaluation in the Nineveh Plain, Northern Iraq, using a GIS-based DRASTIC Model. [INJES]. *Iraqi National Journal of Earth Science*, *21*(2), 78–92. DOI: 10.33899/earth.2021.170392

Alganci, U., Soydas, M., & Sertel, E. (2020). Comparative Research on Deep Learning Approaches for Airplane Detection from Very High-Resolution Satellite Images. *Remote Sensing (Basel)*, *12*(3), 458. DOI: 10.3390/rs12030458

Alshawi, A., Al-Razgan, M., AlKallas, F. H., Bin Suhaim, R. A., Al-Tamimi, R., Alharbi, N., & AlSaif, S. O. (2022). Data privacy during pandemics: A systematic literature review of COVID-19 smartphone applications. *PeerJ. Computer Science*, *7*, e826. DOI: 10.7717/peerj-cs.826 PMID: 35111915

Ayoubi, H., Tabaa, Y., & El kharrim, M. (2023). Artificial Intelligence in Green Management and the Rise of Digital Lean for Sustainable Efficiency. *E3S Web of Conferences*, *412*, 01053. DOI: 10.1051/e3sconf/202341201053

Badawy, M. (2023). Integrating Artificial Intelligence and Big Data into Smart Healthcare Systems: A Comprehensive Review of Current Practices and Future Directions. *Artificial Intelligence Evolution*, 133–153. DOI: 10.37256/aie.4220232980

Besmer, M. D., Sigrist, J. A., Props, R., Buysschaert, B., Mao, G., Boon, N., & Hammes, F. (2017). Laboratory-Scale Simulation and Real-Time Tracking of a Microbial Contamination Event and Subsequent Shock-Chlorination in Drinking Water. *Frontiers in Microbiology*, *8*, 1900. Advance online publication. DOI: 10.3389/fmicb.2017.01900 PMID: 29085343

Bianchi, O., & Putro, H. P. (2024). Artificial Intelligence in Environmental Monitoring: Predicting and Managing Climate Change Impacts. [ITALIC]. *International Transactions on Artificial Intelligence*, *3*(1), 85–96. DOI: 10.33050/italic.v3i1.650

Cantu, A., Graham, M. A., Millard, A. V., Flores, I., Mugleston, M. K., Reyes, I. Y., & Carbajal, E. S. (2016). Environmental Justice and Community-Based Research in Texas Borderland Colonias. *Public Health Nursing (Boston, Mass.)*, *33*(1), 65–72. DOI: 10.1111/phn.12187 PMID: 25787846

Chen, Y., & Jin, S. (2023a). Artificial Intelligence and Carbon Emissions in Manufacturing Firms: The Moderating Role of Green Innovation. *Processes (Basel, Switzerland)*, *11*(9), 2705. DOI: 10.3390/pr11092705

Chen, Y., & Jin, S. (2023b). Artificial Intelligence and Carbon Emissions in Manufacturing Firms: The Moderating Role of Green Innovation. *Processes (Basel, Switzerland)*, *11*(9), 2705. DOI: 10.3390/pr11092705

Chisom, O. N., Biu, P. W., Umoh, A. A., Obaedo, B. O., Adegbite, A. O., & Abatan, A. (2024). Reviewing the role of AI in environmental monitoring and conservation: A data-driven revolution for our planet. *World Journal of Advanced Research and Reviews*, *21*(1), 161–171. DOI: 10.30574/wjarr.2024.21.1.2720

Costanza, R. (2012). Ecosystem health and ecological engineering. *Ecological Engineering*, *45*, 24–29. DOI: 10.1016/j.ecoleng.2012.03.023

Couture, V., Roy, M.-C., Dez, E., Laperle, S., & Bélisle-Pipon, J.-C. (2023). Ethical Implications of Artificial Intelligence in Population Health and the Public's Role in Its Governance: Perspectives From a Citizen and Expert Panel. *Journal of Medical Internet Research*, *25*, e44357. DOI: 10.2196/44357 PMID: 37104026

Crane-Droesch, A. (2018). Machine learning methods for crop yield prediction and climate change impact assessment in agriculture. *Environmental Research Letters*, *13*(11), 114003. DOI: 10.1088/1748-9326/aae159

Davianto, H. (2022). The Advantages of Artificial Intelligence in Operational Decision Making. *Hasanuddin Economics and Business Review*, *6*(1), 24. DOI: 10.26487/hebr.v6i1.5082

Denecke, K., Gabarron, E., Grainger, R., Konstantinidis, S. Th., Lau, A., Rivera-Romero, O., Miron-Shatz, T., & Merolli, M. (2019). Artificial Intelligence for Participatory Health: Applications, Impact, and Future Implications. *Yearbook of Medical Informatics*, *28*(01), 165–173. DOI: 10.1055/s-0039-1677902 PMID: 31022749

Ernawati, K., Nugroho, B. S., Suryana, C., Riyanto, A., & Fatmawati, E. (2022a). advantages of digital applications in public health services on automation era. *International Journal of Health Sciences*, *6*(1), 174–186. DOI: 10.53730/ijhs.v6n1.3684

Ernawati, K., Nugroho, B. S., Suryana, C., Riyanto, A., & Fatmawati, E. (2022b). Advantages of Digital Applications in Public Health Services on Automation Era. *International Journal of Health Sciences*, *6*(1), 174–186. DOI: 10.53730/ijhs.v6n1.3684

Fiske, A., Henningsen, P., & Buyx, A. (2019). Your Robot Therapist Will See You Now: Ethical Implications of Embodied Artificial Intelligence in Psychiatry, Psychology, and Psychotherapy. *Journal of Medical Internet Research*, *21*(5), e13216. DOI: 10.2196/13216 PMID: 31094356

Gama, F., Tyskbo, D., Nygren, J., Barlow, J., Reed, J., & Svedberg, P. (2022). Implementation Frameworks for Artificial Intelligence Translation Into Health Care Practice: Scoping Review. *Journal of Medical Internet Research*, *24*(1), e32215. DOI: 10.2196/32215 PMID: 35084349

García, L., Parra, L., Jimenez, J. M., Lloret, J., & Lorenz, P. (2020). IoT-Based Smart Irrigation Systems: An Overview on the Recent Trends on Sensors and IoT Systems for Irrigation in Precision Agriculture. *Sensors (Basel)*, *20*(4), 1042. DOI: 10.3390/s20041042 PMID: 32075172

Gatla, T. R.Teja Reddy Gatla. (2019). A Cutting-Edge Research on AI Combating Climate Change: Innovations and Its Impacts. *International Journal of Innovations in Engineering Research and Technology*, *6*(9), 1–8. DOI: 10.26662/ijiert.v11i3.pp1-8

González-González, R. B., Flores-Contreras, E. A., González-González, E., Torres Castillo, N. E., Parra-Saldívar, R., & Iqbal, H. M. N. (2023). Biosensor Constructs for the Monitoring of Persistent Emerging Pollutants in Environmental Matrices. *Industrial & Engineering Chemistry Research*, *62*(11), 4503–4520. DOI: 10.1021/acs.iecr.2c00421

Guo, Q., Ren, M., Wu, S., Sun, Y., Wang, J., Wang, Q., Ma, Y., Song, X., & Chen, Y. (2022). Applications of artificial intelligence in the field of air pollution: A bibliometric analysis. *Frontiers in Public Health*, *10*, 933665. Advance online publication. DOI: 10.3389/fpubh.2022.933665 PMID: 36159306

Hall, R., Pal, D., & Ariya, P. A. (2022). Novel Dynamic Technique, Nano-DIHM, for Rapid Detection of Oil, Heavy Metals, and Biological Spills in Aquatic Systems. *Analytical Chemistry*, *94*(32), 11390–11400. DOI: 10.1021/acs.analchem.2c02396 PMID: 35929664

Hanafi, A., Moawed, M., & Abdellatif, O. (2024). Advancing Sustainable Energy Management: A Comprehensive Review of Artificial Intelligence Techniques in Building. *Engineering Research Journal (Shoubra)*, *53*(2), 26–46. DOI: 10.21608/erjsh.2023.226854.1196

Hedayet, H., & Haseen, F. (2024). Artificial Intelligence in Public Health: A Review Article. *Bangladesh Journal of Bioethics*, *15*(2), 15–19. DOI: 10.62865/bjbio.v15i2.108

Hendriks, C. M. J., Gibson, H. S., Trett, A., Python, A., Weiss, D. J., Vrieling, A., Coleman, M., Gething, P. W., Hancock, P. A., & Moyes, C. L. (2019). Mapping Geospatial Processes Affecting the Environmental Fate of Agricultural Pesticides in Africa. *International Journal of Environmental Research and Public Health*, *16*(19), 3523. DOI: 10.3390/ijerph16193523 PMID: 31547208

Højris, B., Christensen, S. C. B., Albrechtsen, H.-J., Smith, C., & Dahlqvist, M. (2016). A novel, optical, on-line bacteria sensor for monitoring drinking water quality. *Scientific Reports*, *6*(1), 23935. DOI: 10.1038/srep23935 PMID: 27040142

Jobin, A., Ienca, M., & Vayena, E. (2019). The global landscape of AI ethics guidelines. *Nature Machine Intelligence*, *1*(9), 389–399. DOI: 10.1038/s42256-019-0088-2

Katirai, A. (2023). The ethics of advancing artificial intelligence in healthcare: Analyzing ethical considerations for Japan's innovative AI hospital system. *Frontiers in Public Health*, *11*, 1142062. Advance online publication. DOI: 10.3389/fpubh.2023.1142062 PMID: 37529426

Khakurel, J., Penzenstadler, B., Porras, J., Knutas, A., & Zhang, W. (2018). The Rise of Artificial Intelligence under the Lens of Sustainability. *Technologies*, *6*(4), 100. DOI: 10.3390/technologies6040100

Kilanko, V. (2023). Leveraging Artificial Intelligence for Enhanced Revenue Cycle Management in the United States. *International Journal Of Scientific Advances*, *4*(4). Advance online publication. DOI: 10.51542/ijscia.v4i4.3

Köse, A. (2023). *Artificial Intelligence in Health and Applications*., DOI: 10.4018/978-1-6684-8337-4.ch002

Kumari, G., Srivastava, A., & Hansraj, B. (2024). *The Utilization of Artificial Intelligence in the Health Care Sector*., DOI: 10.4018/979-8-3693-3140-8.ch013

Labazanova, S., Aygumov, T., & Mustafaev, T. (2024). How artificial intelligence can fight environmental issues. *E3S Web of Conferences, 515*, 02021. DOI: 10.1051/e3sconf/202451502021

Lal, S., Raghav, A., Arora, M. K., Mishra, S., & Singh, B. (2024). *Impact of Artificial Intelligence in the Health Sector.*, DOI: 10.4018/979-8-3693-5976-1.ch005

Lammons, W., Silkens, M., Hunter, J., Shah, S., & Stavropoulou, C. (2023). Centering Public Perceptions on Translating AI Into Clinical Practice: Patient and Public Involvement and Engagement Consultation Focus Group Study. *Journal of Medical Internet Research, 25*, e49303. DOI: 10.2196/49303 PMID: 37751234

Li, Q., Shao, X., Cui, H., Wei, Y., & Shang, Y. (2023). Dynamic Multivariate Outlier Detection Algorithm Using Ultraviolet Visible Spectroscopy for Monitoring Surface Water Contamination With Hydrological Fluctuation in Real-Time. *Applied Spectroscopy, 77*(12), 1371–1381. DOI: 10.1177/00037028231206191 PMID: 38010873

Lohani, N. (2024). AI-based Environmental Sustainability: Transforming Conservation Efforts. *International Journal For Multidisciplinary Research, 6*(2), 16997. Advance online publication. DOI: 10.36948/ijfmr.2024.v06i02.16997

Lu, Y., Wang, R., Zhang, Y., Su, H., Wang, P., Jenkins, A., Ferrier, R. C., Bailey, M., & Squire, G. (2015). Ecosystem health towards sustainability. *Ecosystem Health and Sustainability, 1*(1), 1–15. DOI: 10.1890/EHS14-0013.1

Mahalisa, G., & Nurarminarahmah, N. (2023). Air Pollution Standard Index (APSI) Detection Application Based on the Flask Model. *Brilliance: Research of Artificial Intelligence, 3*(2), 270–274. DOI: 10.47709/brilliance.v3i2.3194

Mishbahuddin, A. A., Fadli, Justicia, A. K., & Ernawati, K. (2021). Understanding islamic commandments to maintain people's health and environment: A critical review on literature. In International Journal of Health Sciences (Vol. 5, Issue 2, pp. 202–211). Universidad Tecnica de Manabi. doi: DOI: 10.29332/ijhs.v5n2.1427

Morgenstern, J. D., Rosella, L. C., Daley, M. J., Goel, V., Schünemann, H. J., & Piggott, T. (2021). "AI's gonna have an impact on everything in society, so it has to have an impact on public health": A fundamental qualitative descriptive study of the implications of artificial intelligence for public health. *BMC Public Health, 21*(1), 40. DOI: 10.1186/s12889-020-10030-x PMID: 33407254

Murdoch, B. (2021). Privacy and artificial intelligence: Challenges for protecting health information in a new era. *BMC Medical Ethics, 22*(1), 122. DOI: 10.1186/s12910-021-00687-3 PMID: 34525993

Murphy, K., Di Ruggiero, E., Upshur, R., Willison, D. J., Malhotra, N., Cai, J. C., Malhotra, N., Lui, V., & Gibson, J. (2021). Artificial intelligence for good health: A scoping review of the ethics literature. *BMC Medical Ethics, 22*(1), 14. DOI: 10.1186/s12910-021-00577-8 PMID: 33588803

Naik, N., Hameed, B. M. Z., Shetty, D. K., Swain, D., Shah, M., Paul, R., Aggarwal, K., Ibrahim, S., Patil, V., Smriti, K., Shetty, S., Rai, B. P., Chlosta, P., & Somani, B. K. (2022). Legal and Ethical Consideration in Artificial Intelligence in Healthcare: Who Takes Responsibility? *Frontiers in Surgery*, *9*, 862322. Advance online publication. DOI: 10.3389/fsurg.2022.862322 PMID: 35360424

Obasa, A. E., & Palk, A. C. (2023). Responsible application of artificial intelligence in health care. *South African Journal of Science*, *119*(5/6). Advance online publication. DOI: 10.17159/sajs.2023/14889 PMID: 39328370

Oluwatoyin Ayo-Farai, , Paschal Maduka, , & Temidayo Olorunsogo, Nzubechukwu Chukwudum OhaleteTolulope O Olorunsogo. (2024). AI-Driven Environmental Health Disease Modeling: A Review of Techniques and Their Impact on Public Health in The USA and African Contexts. *International Medical Science Research Journal*, *4*(1), 51–73. DOI: 10.51594/imsrj.v4i1.737

Pardo, M. de C., Azevedo, J. C., & Fernandez, P. (2021). *Ecosystem Services, Sustainable Rural Development and Protected Areas*. MDPI., DOI: 10.3390/books978-3-0365-2294-4

Peng, J., Liu, Y., Wu, J., Lv, H., & Hu, X. (2015). Linking ecosystem services and landscape patterns to assess urban ecosystem health: A case study in Shenzhen City, China. *Landscape and Urban Planning*, *143*, 56–68. DOI: 10.1016/j.landurbplan.2015.06.007

Petersson, L., Larsson, I., Nygren, J. M., Nilsen, P., Neher, M., Reed, J. E., Tyskbo, D., & Svedberg, P. (2022). Challenges to implementing artificial intelligence in healthcare: A qualitative interview study with healthcare leaders in Sweden. *BMC Health Services Research*, *22*(1), 850. DOI: 10.1186/s12913-022-08215-8 PMID: 35778736

Qazi, S., Khawaja, B. A., & Farooq, Q. U. (2022). IoT-Equipped and AI-Enabled Next Generation Smart Agriculture: A Critical Review, Current Challenges and Future Trends. *IEEE Access : Practical Innovations, Open Solutions*, *10*, 21219–21235. DOI: 10.1109/ACCESS.2022.3152544

Rabus, D., Friedt, J.-M., Arapan, L., Lamare, S., Baqué, M., Audouin, G., & Chérioux, F. (2020). Subsurface H_2S Detection by a Surface Acoustic Wave Passive Wireless Sensor Interrogated with a Ground Penetrating Radar. *ACS Sensors*, *5*(4), 1075–1081. DOI: 10.1021/acssensors.0c00013 PMID: 32202415

Rathor, K. (2024). Exploring The Challenges and Oppurtunities of Implementing Artificial Intelligence in Supply Chain Management: A Survey-Based Study in Asian Manufacturing Sector. *International Research Journal of Modernization in Engineering Technology and Science*. Advance online publication. DOI: 10.56726/IRJMETS49724

Richie, C. (2022). Environmentally sustainable development and use of artificial intelligence in health care. *Bioethics*, *36*(5), 547–555. DOI: 10.1111/bioe.13018 PMID: 35290675

Robertson, S., & Miller, M. R. (2018). Ambient air pollution and thrombosis. *Particle and Fibre Toxicology*, *15*(1), 1. DOI: 10.1186/s12989-017-0237-x PMID: 29298690

Rojek, I., Mroziński, A., Kotlarz, P., Macko, M., & Mikołajewski, D. (2023). AI-Based Computational Model in Sustainable Transformation of Energy Markets. *Energies*, *16*(24), 8059. DOI: 10.3390/en16248059

Ruamviboonsuk, P., Chantra, S., Seresirikachorn, K., Ruamviboonsuk, V., & Sangroongruangsri, S. (2021). Economic Evaluations of Artificial Intelligence in Ophthalmology. *Asia-Pacific Journal of Ophthalmology*, *10*(3), 307–316. DOI: 10.1097/APO.0000000000000403 PMID: 34261102

Russell, R. G., Lovett Novak, L., Patel, M., Garvey, K. V., Craig, K. J. T., Jackson, G. P., Moore, D., & Miller, B. M. (2023). Competencies for the Use of Artificial Intelligence–Based Tools by Health Care Professionals. *Academic Medicine*, *98*(3), 348–356. DOI: 10.1097/ACM.0000000000004963 PMID: 36731054

Shamim, M. I. (2024). Impact of Artificial Intelligence (AI) on Existing Businesses and the Global Economy. *Global Mainstream Journal*, *1*(1), 1–6. DOI: 10.62304/ijbm.v1i1.120

Sharma, A., Jain, A., Gupta, P., & Chowdary, V. (2021). Machine Learning Applications for Precision Agriculture: A Comprehensive Review. *IEEE Access : Practical Innovations, Open Solutions*, *9*, 4843–4873. DOI: 10.1109/ACCESS.2020.3048415

Svedberg, P., Reed, J., Nilsen, P., Barlow, J., Macrae, C., & Nygren, J. (2022). Toward Successful Implementation of Artificial Intelligence in Health Care Practice: Protocol for a Research Program. *JMIR Research Protocols*, *11*(3), e34920. DOI: 10.2196/34920 PMID: 35262500

Teixeira, S., & Krings, A. (2015). Sustainable Social Work: An Environmental Justice Framework for Social Work Education. *Social Work Education*, *34*(5), 513–527. DOI: 10.1080/02615479.2015.1063601

Tundjungsari, V., Ernawati, K., & Mutia, N. (2020). Public Health Surveillance to Promote Clean and Healthy Life Behaviours Using Big Data Approach (An Indonesian Case Study). *Lecture Notes in Networks and Systems*, *80*(February), 761–775. DOI: 10.1007/978-3-030-23162-0_69

Vagin, S. G., Klimenko, V. A., Telegina, Z. A., & Aleksashina, T. V. (2022). Improving environmental decision-making in environmental business-management using big data and AI. *Frontiers in Environmental Science*, *10*, 951306. Advance online publication. DOI: 10.3389/fenvs.2022.951306

van Wynsberghe, A. (2021). Sustainable AI: AI for sustainability and the sustainability of AI. *AI and Ethics*, *1*(3), 213–218. DOI: 10.1007/s43681-021-00043-6

Wang, J., Huang, B., Zhang, T., Wong, H., & Huang, Y. (2018). Impact of housing and community conditions on multidimensional health among middle- and low-income groups in Hong Kong. *International Journal of Environmental Research and Public Health*, *15*(6), 1132. Advance online publication. DOI: 10.3390/ijerph15061132 PMID: 29857544

Wang, Z., Tang, L., Qiu, Q., Chen, H., Wu, T., & Shao, G. (2018). Assessment of Regional Ecosystem Health—A Case Study of the Golden Triangle of Southern Fujian Province, China. *International Journal of Environmental Research and Public Health*, *15*(4), 802. DOI: 10.3390/ijerph15040802 PMID: 29671817

Watanabe, S., Sumi, K., & Ise, T. (2020). Identifying the vegetation type in Google Earth images using a convolutional neural network: A case study for Japanese bamboo forests. *BMC Ecology*, *20*(1), 65. DOI: 10.1186/s12898-020-00331-5 PMID: 33246473

Wood, E. A., Ange, B. L., & Miller, D. D. (2021). Are We Ready to Integrate Artificial Intelligence Literacy into Medical School Curriculum: Students and Faculty Survey. *Journal of Medical Education and Curricular Development*, *8*, 23821205211024078. Advance online publication. DOI: 10.1177/23821205211024078 PMID: 34250242

Wu, F., Wang, X., & Ren, Y. (2021). Urbanization's Impacts on Ecosystem Health Dynamics in the Beijing-Tianjin-Hebei Region, China. *International Journal of Environmental Research and Public Health*, *18*(3), 918. DOI: 10.3390/ijerph18030918 PMID: 33494446

Zamathula, Q. S. N., Ugwuanyi, E. D., Dada, M. A., Majemite, M. T., & Obaigbena, A. (2024). AI-Driven Waste Management Systems: A Comparative Review of Innovations in The USA and Africa. *Engineering Science & Technology Journal*, *5*(2), 507–516. DOI: 10.51594/estj.v5i2.828

Zhang, B., Zhang, D., Feng, Z., Zhang, L., Zhang, M., Fu, R., & Wang, Z. (2023). Assessment of the Potential of Indirect Measurement for Sap Flow Using Environmental Factors and Artificial Intelligence Approach: A Case Study of Magnolia denudata in Shanghai Urban Green Spaces. *Forests*, *14*(9), 1768. DOI: 10.3390/f14091768

Zhao, J., Wang, D., Zhang, F., Pan, J., Claesson, P., Larsson, R., & Shi, Y. (2022). Self-Powered, Long-Durable, and Highly Selective Oil–Solid Triboelectric Nanogenerator for Energy Harvesting and Intelligent Monitoring. *Nano-Micro Letters*, *14*(1), 160. DOI: 10.1007/s40820-022-00903-8 PMID: 35930162

Chapter 6
The Transformative Potential of Artificial Intelligence in Building Sustainable Agricultural Systems: Innovations, Challenges, and Future Directions

Ananya Pandey
https://orcid.org/0000-0002-2419-6314
Christ University, Bangalore, India

ABSTRACT

Artificial Intelligence is revolutionizing agricultural practices by enhancing efficiency, productivity, and sustainability. Through precision agriculture, AI enables data-driven decision-making, optimizing resource use, crop management, and yield prediction. Innovations such as machine learning algorithms and IoT integration facilitate real-time monitoring of soil, weather patterns, and pest control. These technologies address critical challenges like food security and environmental conservation. However, the adoption of AI in agriculture faces obstacles, including high implementation costs, data privacy, and limited access to technology. Future directions involve advancing AI capabilities to support climate-resilient farming, promoting digital inclusion, and developing policy frameworks to address socio-economic concerns. By leveraging AI, agriculture can transition toward sustainable practices, ensuring food security while minimizing environmental impact. In light of the above perspective, this chapter highlights AI's potential to transform agriculture

DOI: 10.4018/979-8-3373-0680-3.ch006

amidst ongoing challenges.

INTRODUCTION

Global leaders and important stakeholders have turned their attention to conserving natural resources in the face of accelerating climate change and environmental deterioration. Sustainable agriculture is gaining traction in this quest for answers because it offers ideas, methods, and resources that seek to combine supplying the world's food needs with protecting the environment for coming generations (Umesha et al., 2018). However, significant players like the Sustainable Agriculture Initiative Platform have broadened the definition of sustainable agriculture to encompass advancing the social and economic welfare of farmers and rural communities in addition to environmental preservation. An environmental beneficial strategy that lessens dependency on finite natural resources is sustainable agriculture (Al-Agele et al., 2021). Crop rotation, intercropping, and companion planting are some of its meticulous agricultural techniques that assist in preserving soil health and water quality while removing the need for harmful practices (Hobbs et al., 2008). Since they understand and accept that soil is a limited and non-renewable resource, farmers that practice sustainable farming practically treat the soil like an egg. Crop yields in the present and the future may be at risk if its nutrient content is reduced or if its structure is compromised (Timsina, 2018). In order to replenish the soil's nutrients, these farmers prioritize techniques including composting, cover crops, and less tillage. The best method for farmers to save and conserve water is still sustainable farming. Farmers may save a lot of water while maintaining or even increasing crop yields by implementing sustainable agriculture techniques like drip irrigation and soil moisture monitoring (Tiwari et al., 2023). Adopting sustainable farming methods is even more advantageous in dry areas where water scarcity is a major problem because it guarantees agriculture's long-term viability and safeguards water ecosystems.

Utilizing effective farming techniques is essential to boosting output on the finite amount of land due to the expanding global population and rising food demand (Viana et al., 2022). AI-based technologies are improving the current farming system, and AI is becoming more and more common in agriculture every day. Numerous factors, such as crop rotation, rainfall, temperature, moisture content, and soil nutrient level, affect agriculture. These factors can be used by artificial intelligence-based products to monitor crop productivity. Industries are looking to artificial intelligence technologies to enhance a variety of agriculture-related operations across the whole food supply chain (Javaid et al., 2023). AI-based agricultural applications and solutions have been developed to help farmers with precise and controlled farming by providing them with the best guidance on water management, crop rotation, timely harvesting, crop type, optimal planting, pest attacks, and nutrition management (Sharma, 2021). AI-enabled systems use information such as temperature, precipitation, wind speed, and sun radiation along with satellite and drone photos to forecast the weather, track agricultural sustainability, and evaluate farms for the presence of pests or illnesses and undernourished plants (Zha, 2020). Farmers without connectivity can immediately benefit from AI with as little as an SMS-enabled phone and the Sowing App. Meanwhile, farmers with Wi-Fi connectivity can use AI apps to receive a continuously AI-tailored farm plan. With the use of IoT and AI-driven technology, farmers can sustainably improve output and earnings while meeting the rising demand for food without depleting valuable natural resources (Liu, 2020). Heat, precipitation, wind, and solar radiation are examples of climate variables. Keeping in view the above, this chapter adapts doctrinal methodology with analytical, critical and exploratory methods to determine the role of AI in sustainable agriculture. In light of this perspective the chapter explores the importance of sustainable agriculture and the need of AI in enhancing sustainability. Furthermore, this chapter examines the current global trends, innovations, challenges and future directions, where AI plays a crucial role in sustainable agriculture.

FOUNDATIONS OF AI IN AGRICULTURE

With the advancement of AI, agriculture has also undergone a revolutionary change. Crop yield has been shielded by this technology from a number of concerns, including population expansion, climate change, joblessness, and issues with food security (Talaviya et al., 2020). With the aid of sensors and other tools built into robots and drones, artificial intelligence is finding a wide range of uses in agriculture, including irrigation, weeding, and spraying (Gandhi et al., 2020). These technologies reduce the need for excessive amounts of water, pesticides, and herbicides, preserve soil fertility, aid in the effective use of labor, increase productivity, and enhance quality (Kakkad et al., 2019). The modern agricultural system has advanced to a new level thanks to AI-based machinery and equipment. Crop output has increased along with real-time harvesting, processing, marketing, and monitoring because of this technology. The agro-based industry has greatly benefited from the most recent automated system innovations that include drones and farm robots. Numerous advanced computer-based systems are made to measure a variety of crucial factors, such as crop quality, production, and weed detection, among many other methods (Liakos et al., 2018).

To cut down on waste and environmental damage, farmers can optimize their pesticide use, fertilizer application rates, and irrigation schedules with the aid of machine learning. It helps farmers make better choices regarding fertilizer application schedules, irrigation techniques, and planting times (Ghazal et al., 2024). Predictive analytics technology, a branch of data science, can be applied to agriculture to help farmers forecast crop yields, estimate demand for certain crops, and maximize fertilizer and irrigation use (Sabu & Kumar, 2020). Predictive analytics may help farmers make better decisions about when to plant, how to care for their crops, and how much to charge for their produce by examining historical data patterns. Furthermore, early warning indicators of agricultural diseases and pests can be found using predictive analytics, enabling farmers to take precautions and minimize possible harm (Johnraja et al., 2024). Computer vision may be used to assess attributes, track movement, and identify things. Precision irrigation, automated farm equipment, yield prediction, and crop monitoring are just a few of the many uses for it in smart agriculture. One of the most significant uses of computer vision in agriculture is crop health monitoring (Tian et al., 2020). Farmers can identify early indicators of disease or stress that might not be apparent to the human eye by examining photos of plants. When and how to apply pesticides or take other corrective measures can be advised by this information (Shoaib et al., 2023). A branch of machine learning that uses artificial neural networks is called deep learning. Deep learning algorithms are similar to people in how they learn from data. In addition to this, they are able to forecast data that is too complicated for conventional machine learning methods (Sarkar, 2021). In smart farming, deep learning can be used to identify pests and illnesses, forecast crop yields, and improve irrigation systems. Additionally, new crop types that are more climate change resilient can be developed with the aid of deep learning (Panda & Bhatnagar, 2020).

Farmers may also monitor soil conditions using drone cameras, deep learning networks, and ground sensors by combining IoT systems with computer vision. These devices are capable of conducting tests like pH assessments, soil erosion monitoring, and root health evaluations (Rajak et al., 2023). Unmanned aerial vehicles (UAVs) and some IoT-enabled devices are helpful for precise monitoring of insect attacks and related diseases in agricultural vegetation. IoT-based technologies are used in precision farming to track agricultural variables. For improved environmental protection and ongoing agricultural production sustainability, IoT offers real-time critical data about crop, soil, water, and air conditions (Velusamy et al., 2021). In order to speed up the irrigation process, irrigation can be converted into a smart irrigation system. UAVs are a single device of equipment that is used to quickly gather vital information about essential agricultural aspects such as pest assault, soil quality, and water availability over a wide area (Bhagat et al., 2019). Even in remote locations, UAVs equipped with spectrum analysis technology can detect pest infestations in farms by taking high-resolution pictures that are saved in cloud server databases for additional analysis. With the least amount of human involvement, smart greenhouses are expanding the possibilities for production (Kreuze et al., 2022). For improved yield, it uses sensor-based continuous monitoring of soil moisture, temperature, humidity, and illumination. These automated sensors aid in decision-making so that more appropriate actions can be taken to prevent damage to agriculture (Subeesh & Mehta, 2021). An integrated approach that combines AI and IoT is the farm management system (FMS). In order to improve the amount and quality of agricultural production, it provides an improved tracking system for analyzing different physical, chemical, and biological agricultural elements (Alahmad, 2023). With real-time data processing capabilities, it requires fewer personnel. Thus, with less time and effort, FMS can support the sustainability and quality of agricultural production (Abiri et al., 2023).

INNOVATIONS IN AI FOR SUSTAINABLE AGRICULTURE

Utilizing technology, precision agriculture maximizes crop yields and optimizes farming methods while using the fewest resources possible. By automating procedures and offering data-driven insights, AI is essential to enhancing these practices. AI innovations are revolutionizing sustainable agriculture by enabling precision farming, optimizing resource use, and improving crop management through technologies like crop health monitoring, automated weeding, and intelligent spraying

Crop Monitoring and Management

Crop monitoring is the process of routinely assessing the health of crops in order to make better judgments about their maintenance. This includes monitoring growth rates, identifying illnesses, and evaluating the condition of the soil (Pretty, 2008). By gathering and evaluating vast volumes of data from many sources, including drones, satellites, and Internet of Things sensors, artificial intelligence significantly aids in crop monitoring. These tools offer crucial data on crop health, temperature, and soil moisture (Khan et al., 2021). After that, AI systems examine this data to identify trends and problems. Farmers can increase crop output by using this real-time information to make informed decisions about fertilizer application and irrigation schedule changes (Giller et al., 2021). By guaranteeing that resources are used precisely where and when they are needed, AI systems support sustainable farming. Over time, this targeted approach improves the health of ecosystems by reducing waste and the environmental impact of farming (Young, 2020). AI is also capable of forecasting hazards associated with market trends, pest issues, and weather variations. This makes it possible for farmers to act proactively to protect their operations (Delfani et al., 2024).

Soil and Water Management

AI is used to improve the management of the food distribution network by encouraging the production of healthier crops, controlling pesticides and soil conditions, and analyzing the needs of farmers (Javaiad et al., 2023). In order to increase crop yields, pesticides are frequently administered to agricultural regions, whether they be in greenhouses or open-air fields. These days, farmers use machine learning to decide which agrichemicals to use, when to apply them, and where (Durai & Shamili, 2022). To increase product values and decrease waste, crop quality attributes must be accurately identified and classified. Data-driven machines are able to identify and uncover novel features that have a major impact on crop quality (Musannase et al., 2023). By providing a long-term solution to the pressing problems of water scarcity and environmental degradation, smart irrigation systems are transforming the management of water resources. These systems minimize environmental effect, eliminate waste, and optimize water usage by utilizing machine learning algorithms and sensor data (Obaideen et al., 2022). The ability of smart irrigation systems to adjust to changing environmental conditions is one of their main advantages. Precise water delivery, which guarantees plants receive the right amounts, is made possible by monitoring variables such as soil moisture levels and meteorological conditions (Bhavsar et al., 2023). This preserves precious water resources by reducing water runoff and evaporation and guaranteeing that plants receive the proper amount of water at the right time. Additionally, smart irrigation systems learn continuously and increase their efficiency over time by integrating machine learning algorithms (Wei et al., 2024). In order to identify leaks or failures, forecast future water requirements, and optimize irrigation schedules, these algorithms examine data patterns and trends. Smart irrigation systems maximize sustainability and water efficiency by continuously improving their operations (Javaid et al., 2023).

Pest and Disease Control

Algorithms for picture recognition driven by AI can assess the health of plants and spot early indicators of pest infestations or diseases that may be hidden from view. Early identification minimizes crop damage and lessens the demand for broad-spectrum herbicides by enabling prompt responses (Li et al., 2024). AI can estimate the possibility of outbreaks in particular places by analyzing a variety of criteria, including historical data, soil conditions, and weather patterns. This lessens the impact on the environment and the expense of treatments by allowing farmers to concentrate their control efforts on the most vulnerable locations (Subedi et al., 2023). Taking into account variables like the kind of pest or illness, the extent of the infestation, and the surrounding environment, AI algorithms are able to analyze enormous volumes of data and suggest the best control strategies (Domingues et al., 2022). Farmers can use this to improve their pest management techniques and make better-informed decisions. AI has the potential to drastically lessen the need for chemical pesticides by facilitating early identification and focused treatments (Javaid et al., 2023). This helps to maintain a healthy ecosystem and conserve beneficial insects in addition to helping the environment. Farmers now have more access to AI-powered tools and technology, enabling them to make more sustainable and efficient decisions about pest and disease control (Demirel & Kumral, 2021).

Supply Chain and Market Management

With its many advantages for supply chain and market management, artificial intelligence is revolutionizing the agriculture sector. Businesses may improve the quality and safety of agricultural goods while also increasing efficiency, sustainability, and profitability by implementing these technologies (Ben Ayed & Hanana, 2021). To guarantee successful acceptance and optimize the advantages, it is crucial to solve the difficulties and factors related to AI deployment (Rane et al., 2024). Throughout the supply chain, AI can optimize resource allocation, inventory control, and transportation routes. This may result in lower expenses, quicker delivery, and more productivity all around (Perez, 2023). In order to find new opportunities and improve marketing tactics, AI may evaluate market data, including pricing patterns, competition activity, and consumer preferences. Businesses may benefit from this by seeing a rise in sales and market share (Haleem et al., 2022). Agriculture is undergoing a digital revolution brought on by population expansion and climate change. As a result, agri-food supply chains are becoming increasingly sophisticated, generating enormous volumes of data, and expanding the parameters of the conventional food lifecycle. But the use of AI in the broader agri-food supply chain is only getting started (Monterio & Barata, 2021). Intelligent planning for demand uncertainty and end-customer customization, storage optimization, waste reduction in the post-production phase (e.g., distribution and recycling), and boundary-spanning analytics are all necessary to expand the significant advancements of AI in traditional stages of production (Lezoche et al., 2020).

Climate Resilient Agriculture

In addition to providing sustainable production tailored to regional hydro-meteorological circumstances, Climate Resilient Agriculture (CRA) successfully adapts to climate change. By improving agricultural viability through water exchange between surface and groundwater systems, CRA offers resistance to climatic variability (Sahoo et at., 2024). CRA provides solutions to present and future food security issues in water-stressed nations like India, where the demand for water from home, industrial, and agricultural uses is rising (Angom & Viswanathan, 2023). CRA can offer a viable option to increase resilience, production, and carbon sequestration by fusing old knowledge with contemporary methods. Case studies from throughout the world show how effective the CRA is in enhancing water management and food security (Sekeran et al., 2021). Kim et al. (2019), for instance, used generalized linear models (GLMs) in South Korea to investigate how land cover affects flooding, highlighting the significance of irrigation systems maintained by farmers and guided by traditional ecological knowledge.

BENEFITS OF AI IN PROMOTING SUSTAINABILITY

Farmers may improve soil fertility and lessen their environmental impact by using AI-powered mobile soil analysis. Real-time, on-site assessments of soil nutrient levels are made possible by this technique. By combining state-of-the-art technologies and data-driven tactics, precision agriculture transforms conventional agricultural methods in order to maximize resource use and reduce environmental effect (Junior et al., 2024). The accurate monitoring and control of inputs like water, fertilizer, and pesticides using cutting-edge technologies like remote sensing, GPS-guided equipment, and complex data analytics platforms is essential to its use (Choudhary et al., 2022). In addition to increasing crop yields, this granular method reduces resource waste, which lowers the agricultural sector's carbon impact. Precision farming not only conserves resources but also makes it easier to implement ecologically friendly methods that encourage energy and carbon sequestration (Getahun et al., 2024). Based on information about the soil and environment, practices including crop rotation, conservation tillage, and cover crops enhance soil health and increase carbon storage. Farmers may reduce greenhouse gas emissions and improve soil health and resilience at the same time by using fewer chemical inputs and making better use of their property (Busari et al., 2015). Additionally, by offering useful data for carbon accounting and emissions monitoring, precision agriculture supports larger carbon and energy management initiatives. Farmers may precisely estimate their carbon footprint and make well-informed decisions to reduce emissions thanks to the comprehensive insights provided by precision agricultural technologies (Guan et al., 2023). In addition to facilitating regulatory compliance, this data-driven strategy enables farmers to take part in carbon markets and incentive schemes that promote sustainable farming methods. In general, precision agriculture signifies a change in farming paradigms toward a more resource-efficient and ecologically aware approach (Getahun et al., 2024). Precision agriculture may address climate change, resource scarcity in food production, and sustainable development by utilizing technology and data to optimize every facet of agriculture. In many agricultural domains, precision farming improves crop yield and environmental sustainability (Khan et al., 2021).

CHALLENGES IN IMPLEMENTING AI IN AGRICULTURE

Precision agriculture promises major improvements in agricultural productivity and sustainability, but it also faces a number of global difficulties.

Technological Barrier

Despite the fact that AI is essential for increasing agricultural efficiency, sustainability, and productivity, many farmers have trouble acquiring and using these cutting-edge technologies. AI infrastructure, hardware, and software are significantly expensive for smallholder farmers in poor nations (Balaska et al., 2023). The use of AI technologies, such as computer vision weeding or robotic harvesters, necessitates greater investment and experience, even for larger farms in developed nations (Agarwal & Bangale, 2023). AI usage in research initiatives and agriculture activities is also hindered by fundamental technological shortcomings. In distant locations, technological access is limited by unreliable electricity, poor internet connectivity, malfunctioning sensors, and a lack of technical help (Dixit et al., 2023). The majority of AI applications today rely on big data analytics and cloud computing, which necessitate constant internet access (Atapattu et al., 2024). Real-time AI is impossible in barns and fields without connectivity. Technology use is further hampered by farmers' lack of fundamental digital skills, even in the presence of infrastructure. If the training data does not represent various farms, AI algorithms are also hampered by data and societal prejudice (Kakani et al., 2020). Today's agricultural AI runs the danger of failing or ignoring smaller farms with fewer resources because the majority of it is created using data from huge industrial farms (Bali et al., 2023). Gathering field data in a range of settings is necessary to create AI tools that are suited to smallholder farming systems. However, the difficulties brought on by inadequate technology accessibility have made it troublesome (Mohsan et al., 2023).

Data Availability

One of the main obstacles to the application of AI in agriculture is the scarcity of high-quality data. Large, high-quality datasets are necessary for AI systems to efficiently program algorithms (Jha et al., 2019). It is challenging and costly to gather agricultural data over wide geographic areas and seasons. Because agricultural production takes place outdoors in intricate natural systems, these statistics are extremely variable and inconsistent due to factors like weather, soil conditions, crop genetics, and management approaches (Matthews et al., 2002). Farmers can get high-resolution data on a variety of aspects using precision agricultural techniques, but this data is frequently insufficient (Khanal et al., 2020). It's possible that important data like weather, fertilizer applications, watering schedules, and insect pressure won't be noted or connected to the sensor data. It is challenging to design AI models to generate precise predictions and suggestions in the absence of these data (Zhang et al., 2019). Data quality is further diminished by the variation in agricultural techniques. The consistency of data across farms is diminished if different producers employ disparate preventive strategies for the same illness (Kuma et al., 2021). Common concepts must be shared in order to align and integrate disparate datasets. However, privacy, proprietary practices, and competitiveness concerns frequently make agricultural stakeholders reluctant to exchange data (Atapattu et al., 2024).

Social and Cultural Barriers

There are significant social impediments to the deployment of AI technologies in agriculture. Due to worries about job losses, privacy issues, and a lack of knowledge or confidence in the technology, many farmers and other agricultural workers are hesitant to embrace AI (Cubric, 2020). The possible effect of AI technology on jobs in agriculture is one significant obstacle. Even while AI can increase productivity and efficiency, it would result in job losses in a variety of occupations, such as those involving machinery operation, agricultural labor, and food processing (Lassoued et al., 2021). This could make the financial difficulties that rural communities now face worse. Addressing unemployment issues requires building confidence between agricultural communities and tech corporations (Soma & Nuckchady, 2021). Agricultural laborers' lack of internet knowledge and accessibility is another obstacle. Introducing cutting-edge technologies necessitates a fundamental understanding of computers and technology. Training initiatives can aid in improving the usability and accessibility of AI systems for pertinent stakeholders (Shaw et al., 2006). Concerns exist around ethical AI methods, transparency, and data privacy. Without knowing exactly how their data is being utilized, farmers are hesitant to embrace AI technologies (Kerr, 2004). Strong cybersecurity safeguards, equitable, responsible

AI algorithms, and transparent communication on data collection, processing, and storage are all necessary. Another difficulty is cultural attitudes and views (Rose et al., 2016). The agricultural sector is typically antiquated and resistant to quickly evolving technologies. Generally speaking, younger farmers are more receptive to advances than older ones. Over time, creating intuitive AI interfaces and implementing educational initiatives can help to progressively lessen unfavorable perceptions (Atapattu et al., 2024).

Policy Barriers

Both farmers and IT businesses are impacted by the policy issues surrounding the use of AI in agriculture. Concerns about privacy, security, and ethics have made agricultural stakeholders reluctant to embrace AI technologies due to unclear laws and guidelines (Ghaffarianhoseini et al., 2017). As of right now, there are no particular regulations controlling the collection, use, or sharing of farmer data by agricultural AI systems. Because of worries about handling data responsibly, this prevents people from adopting new technologies (Dhillon & Moncur, 2023). The absence of explicit guidelines for accountability for mistakes, prejudices, or damages brought about by agricultural AI is another significant problem (Mohammed & Tejay, 2017). In the absence of robust frameworks, farmers would be reluctant to depend on AI systems, and developers might refrain from implementing potentially dangerous technology (Atapattu et al., 2024). Legislators must provide fair liability guidelines for ethical AI development. Another obstacle is the lack of standardized certification or testing procedures for agricultural AI. Despite the potential advantages, farmers view AI tools as unreliable in the absence of official validation of their safety, security, and effectiveness (Marda, 2018). Creating standardized supervision procedures can contribute to increased technological trust. Furthermore, farmers' access to AI systems that rely on cloud computing and data transfer is limited by connectivity issues (Galaz et al., 2021). AI can be made more accessible to the agriculture industry through policies aimed at closing the digital divide in rural areas through subsidies and infrastructure investment (Rotz et al., 2019).

FUTURE DIRECTIONS AND SUGGESTIONS

By tackling important issues like food security, resource efficiency, and climate resilience, AI has the potential to completely transform sustainable agriculture. Predictive analytics, smart irrigation, and precision farming are all made possible by the combination of AI and cutting-edge technologies, which helps to maximize agricultural output while reducing environmental effects (Javaid et al., 2023). Using

AI-powered sensors and drones to monitor crop conditions, pest infestations, and soil health in real time is one of the newest trends. By using these tools, farmers may optimize their usage of pesticides, fertilizers, and water by making data-driven decisions (Sharma et al., 2023). In order to prevent losses, machine learning models are being used in conjunction with remote sensing data to forecast weather patterns, evaluate crop yields, and predict the likelihood of natural disasters. Agricultural methods are being further transformed by advancements in automation and robotics (Subeesh & Mehta, 2021). In order to alleviate labor shortages and save operating expenses, AI-driven robots are being created to carry out labor-intensive tasks including planting, weeding, and harvesting. AI-assisted controlled environment agriculture and vertical farming are becoming viable answers to the problems associated with urban farming (Moshayedi et al., 2024). With the help of AI algorithms that track and modify light, temperature, and nutrient levels, these systems maximize growing conditions and guarantee year-round crop production while using the fewest resources possible. Another exciting direction is the use of AI in genomics (Smith, 2018). Finding hardy crop varieties that can resist harsh weather, pests, and illnesses is being made easier by AI algorithms. Breeding crops that are both high-yielding and environmentally friendly requires this ingenuity. By tracking products from farm to fork, blockchain technology combined with artificial intelligence is being investigated to improve supply chain traceability and transparency, guaranteeing ethical standards and lowering food waste (Demestichas et al., 2020). AI use in agriculture is not without its difficulties, though. Smallholder farmers, who make up a sizable share of the agricultural industry, frequently do not have access to the tools, resources, and expertise needed to fully utilize AI. The digital divide between rural and urban areas and the high cost of AI tools continues to be major obstacles (Božić, 2023). Concerns about data security and privacy also surface since AI systems need enormous volumes of data to function properly. Governments, businesses, and non-governmental organizations must work together to address these issues by developing infrastructure, growing capacity, and offering reasonably priced AI solutions. Looking ahead, developments in edge computing and the IoT will probably influence how AI is used in sustainable agriculture (Villegas-Ch & García-Ortiz, 2023). It is anticipated that Edge AI, which handles data locally instead of depending on centralized servers, will improve the effectiveness and real-time capabilities of AI tools in remote farming regions. Furthermore, self-sustaining agricultural systems can be developed by combining AI with renewable energy sources, which will lessen reliance on non-renewable resources (Majeed et al., 2023). AI has the potential to revolutionize sustainable agriculture by promoting creativity, increasing productivity, and tackling environmental issues. The agriculture industry may progress toward a more resilient and sustainable future by removing present obstacles and encouraging equitable access to AI technologies. The convergence of sustainable practices and AI developments will be crucial (Ziesche et al., 2023).

CONCLUSION

A vital component of the human community is agriculture. In order to boost agricultural output, it has been improved over time. In order to better monitor plant development, illnesses, and insect populations, modern technology has also been used in horticulture, forestry, and agriculture. There are several emerging applications of AI and IoT in agriculture to address problems such as soil conditions, weather, pest detection, harvesting schedule determination, and more. To optimize agricultural productivity using few resources. Deep Technology, Innovations like AI and ML, robots, temperature and moisture sensors, aerial photos, and GPS let farmers cultivate crops even in desert regions with high resource efficiency. The precision and response time of the system are crucial components of AI. Even behavioral changes in field crops brought on by modifications to the microclimate can be accurately and quickly studied. But given how quickly technology is evolving in this digital age and how quickly changing devices and sensors in tandem with advancements in adopted technologies won't be cost-effective for India's small-scale farmers, worries about the durability of AI technologies may deter farmers from adopting them. If appropriate efforts are made to pursue the potential of AI-based technology in the agriculture sector, not only from the perspective of the user and customer but also from the governance and policy side, it is undeniable that digital innovation may revolutionize sustainable agriculture. AI in agriculture has the potential to revolutionize conventional farming methods. The advantages are obvious, ranging from predictive analytics to precision farming. Still, there are obstacles like exorbitant prices and integration problems. Investigating AI technologies is essential for farmers and agribusiness owners to maintain their competitiveness. Data-driven, effective, and sustainable farming is the way of the future.

REFERENCES

Abiri, R., Rizan, N., Balasundram, S. K., Shahbazi, A. B., & Abdul-Hamid, H. (2023). Application of digital technologies for ensuring agricultural productivity. *Heliyon*, *9*(12), e22601. DOI: 10.1016/j.heliyon.2023.e22601 PMID: 38125472

Agrawal, K. N., & Bangale, R. A. (2023). Mechanizing Indian Agriculture with precision farming technologies: Challenges and perspective. *RASSA Journal of Science for Society, 5*(2and3), 129-138.

Al-Agele, H. A., Nackley, L., & Higgins, C. W. (2021). A pathway for sustainable agriculture. *Sustainability (Basel)*, *13*(8), 4328. DOI: 10.3390/su13084328

Alahmad, T., Neményi, M., & Nyéki, A. (2023). Applying IoT sensors and big data to improve precision crop production: A review. *Agronomy (Basel)*, *13*(10), 2603. DOI: 10.3390/agronomy13102603

Angom, J., & Viswanathan, P. K. (2023). Climate-Smart Agricultural Practices and Technologies in India and South Africa: Implications for Climate Change Adaptation and Sustainable Livelihoods. In *The Palgrave Handbook of Socio-ecological Resilience in the Face of Climate Change: Contexts from a Developing Country* (pp. 161–195). Springer Nature Singapore. DOI: 10.1007/978-981-99-2206-2_12

Atapattu, A. J., Perera, L. K., Nuwarapaksha, T. D., Udumann, S. S., & Dissanayaka, N. S. (2024). Challenges in Achieving Artificial Intelligence in Agriculture. In *Artificial Intelligence Techniques in Smart Agriculture* (pp. 7–34). Springer Nature Singapore. DOI: 10.1007/978-981-97-5878-4_2

Balaska, V., Adamidou, Z., Vryzas, Z., & Gasteratos, A. (2023). Sustainable crop protection via robotics and artificial intelligence solutions. *Machines (Basel)*, *11*(8), 774. DOI: 10.3390/machines11080774

Bali, A., Chouhan, S. S., Kumar, G., Kumar, R., & Singh, U. P. (2023). Adaptive fault-tolerant control for pure-feedback stochastic nonlinear systems with sensor and actuator faults. *Circuits, Systems, and Signal Processing*, *42*(10), 5838–5867. DOI: 10.1007/s00034-023-02366-7

Ben Ayed, R., & Hanana, M. (2021). Artificial intelligence to improve the food and agriculture sector. *Journal of Food Quality*, *2021*(1), 5584754. DOI: 10.1155/2021/5584754

Bhagat, M., Kumar, D., & Kumar, D. (2019). Role of Internet of Things (IoT) in smart farming: A brief survey. *2019 Devices for Integrated Circuit (DevIC)*, 141-145.

Bhavsar, D., Limbasia, B., Mori, Y., Aglodiya, M. I., & Shah, M. (2023). A comprehensive and systematic study in smart drip and sprinkler irrigation systems. *Smart Agricultural Technology*, *5*, 100303. DOI: 10.1016/j.atech.2023.100303

Božić, V. (2023). Risks of digital divide in using artificial intelligence (AI).

Busari, M. A., Kukal, S. S., Kaur, A., Bhatt, R., & Dulazi, A. A. (2015). Conservation tillage impacts on soil, crop and the environment. *International Soil and Water Conservation Research*, *3*(2), 119–129. DOI: 10.1016/j.iswcr.2015.05.002

Choudhury, A., Biswas, A., Singh, T. P., & Ghosh, S. K. (Eds.). (2022). *Smart Agriculture Automation Using Advanced Technologies: Data Analytics and Machine Learning, Cloud Architecture, Automation and IoT*. Springer Nature.

Cubric, M. (2020). Drivers, barriers and social considerations for AI adoption in business and management: A tertiary study. *Technology in Society*, *62*, 101257. DOI: 10.1016/j.techsoc.2020.101257

Delfani, P., Thuraga, V., Banerjee, B., & Chawade, A. (2024). Integrative approaches in modern agriculture: IoT, ML and AI for disease forecasting amidst climate change. *Precision Agriculture*, *25*(5), 1–25. DOI: 10.1007/s11119-024-10164-7

Demestichas, K., Peppes, N., Alexakis, T., & Adamopoulou, E. (2020). Blockchain in agriculture traceability systems: A review. *Applied Sciences (Basel, Switzerland)*, *10*(12), 4113. DOI: 10.3390/app10124113

Demirel, M., & Kumral, N. A. (2021). Artificial intelligence in integrated pest management. In *Artificial intelligence and IoT-based technologies for sustainable farming and smart agriculture* (pp. 289–313). IGI Global. DOI: 10.4018/978-1-7998-1722-2.ch018

Dhillon, R., & Moncur, Q. (2023). Small-scale farming: A review of challenges and potential opportunities offered by technological advancements. *Sustainability (Basel)*, *15*(21), 15478. DOI: 10.3390/su152115478

Dixit, S., Bohre, K., Singh, Y., Himeur, Y., Mansoor, W., Atalla, S., & Srinivasan, K. (2023). A Comprehensive review on AI-enabled models for Parkinson's disease diagnosis. *Electronics (Basel)*, *12*(4), 783. DOI: 10.3390/electronics12040783

Domingues, T., Brandão, T., & Ferreira, J. C. (2022). Machine learning for detection and prediction of crop diseases and pests: A comprehensive survey. *Agriculture*, *12*(9), 1350. DOI: 10.3390/agriculture12091350

Durai, S. K. S., & Shamili, M. D. (2022). Smart farming using machine learning and deep learning techniques. *Decision Analytics Journal, 3,* 100041. DOI: 10.1016/j.dajour.2022.100041

Galaz, V., Centeno, M. A., Callahan, P. W., Causevic, A., Patterson, T., Brass, I., Baum, S., Farber, D., Fischer, J., Garcia, D., McPhearson, T., Jimenez, D., King, B., Larcey, P., & Levy, K. (2021). Artificial intelligence, systemic risks, and sustainability. *Technology in Society*, *67*, 101741. DOI: 10.1016/j.techsoc.2021.101741

Gandhi, M., Kamdar, J., & Shah, M. (2020). Preprocessing of non-symmetrical images for edge detection. *Augmented Human Research*, *5*(1), 10. DOI: 10.1007/s41133-019-0030-5

Getahun, S., Kefale, H., & Gelaye, Y. (2024). Application of Precision Agriculture Technologies for Sustainable Crop Production and Environmental Sustainability: A Systematic Review. *TheScientificWorldJournal, 2024*(1), 2126734. DOI: 10.1155/2024/2126734 PMID: 39421732

Ghaffarianhoseini, A., Tookey, J., Ghaffarianhoseini, A., Naismith, N., Azhar, S., Efimova, O., & Raahemifar, K. (2017). Building Information Modelling (BIM) uptake: Clear benefits, understanding its implementation, risks and challenges. *Renewable & Sustainable Energy Reviews*, *75*, 1046–1053. DOI: 10.1016/j.rser.2016.11.083

Ghazal, S., Munir, A., & Qureshi, W. S. (2024). *Computer vision in smart agriculture and precision farming: Techniques and applications*. Artificial Intelligence in Agriculture.

Giller, K. E., Delaune, T., Silva, J. V., Descheemaeker, K., van de Ven, G., Schut, A. G., van Wijk, M., Hammond, J., Hochman, Z., Taulya, G., Chikowo, R., Narayanan, S., Kishore, A., Bresciani, F., Teixeira, H. M., Andersson, J. A., & van Ittersum, M. K. (2021). The future of farming: Who will produce our food? *Food Security*, *13*(5), 1073–1099. DOI: 10.1007/s12571-021-01184-6

Guan, K., Jin, Z., Peng, B., Tang, J., DeLucia, E. H., West, P. C., Jiang, C., Wang, S., Kim, T., Zhou, W., Griffis, T., Liu, L., Yang, W. H., Qin, Z., Yang, Q., Margenot, A., Stuchiner, E. R., Kumar, V., Bernacchi, C., & Yang, S. J. (2023). A scalable framework for quantifying field-level agricultural carbon outcomes. *Earth-Science Reviews*, *243*, 104462. DOI: 10.1016/j.earscirev.2023.104462

Haleem, A., Javaid, M., Qadri, M. A., Singh, R. P., & Suman, R. (2022). Artificial intelligence (AI) applications for marketing: A literature-based study. *International Journal of Intelligent Networks*, *3*, 119–132. DOI: 10.1016/j.ijin.2022.08.005

Hobbs, P. R., Sayre, K., & Gupta, R. (2008). The role of conservation agriculture in sustainable agriculture. *Philosophical Transactions of the Royal Society of London. Series B, Biological Sciences*, *363*(1491), 543–555. DOI: 10.1098/rstb.2007.2169 PMID: 17720669

Javaid, M., Haleem, A., Khan, I. H., & Suman, R. (2023). Understanding the potential applications of Artificial Intelligence in the Agriculture Sector. *Advanced Agrochem*, *2*(1), 15–30. DOI: 10.1016/j.aac.2022.10.001

Johnraja, J. I., Leelipushpam, P. G. J., Shirley, C. P., & Princess, P. J. B. (2024). Impact of Cloud Computing on the Future of Smart Farming. In *Intelligent Robots and Drones for Precision Agriculture* (pp. 391–420). Springer Nature Switzerland. DOI: 10.1007/978-3-031-51195-0_18

Júnior, M. R. B., de Almeida Moreira, B. R., dos Santos Carreira, V., de Brito Filho, A. L., Trentin, C., de Souza, F. L. P., & Shiratsuchi, L. S. (2024). Precision agriculture in the United States: A comprehensive meta-review inspiring further research, innovation, and adoption. *Computers and Electronics in Agriculture*, *221*, 108993. DOI: 10.1016/j.compag.2024.108993

Kakani, V., Nguyen, V. H., Kumar, B. P., Kim, H., & Pasupuleti, V. R. (2020). A critical review on computer vision and artificial intelligence in food industry. *Journal of Agriculture and Food Research*, *2*, 100033. DOI: 10.1016/j.jafr.2020.100033

Kakkad, V., Patel, M., & Shah, M. (2019). Biometric authentication and image encryption for image security in cloud framework. *Multiscale and Multidisciplinary Modeling. Experiments and Design*, *2*(4), 233–248.

Kerr, D. (2004). Factors influencing the development and adoption of knowledge based decision support systems for small, owner-operated rural business. *Artificial Intelligence Review*, *22*(2), 127–147. DOI: 10.1023/B:AIRE.0000045503.74951.7a

Khan, N., Ray, R. L., Sargani, G. R., Ihtisham, M., Khayyam, M., & Ismail, S. (2021). Current progress and future prospects of agriculture technology: Gateway to sustainable agriculture. *Sustainability (Basel)*, *13*(9), 4883. DOI: 10.3390/su13094883

Khanal, S., Kc, K., Fulton, J. P., Shearer, S., & Ozkan, E. (2020). Remote sensing in agriculture—Accomplishments, limitations, and opportunities. *Remote Sensing (Basel)*, *12*(22), 3783. DOI: 10.3390/rs12223783

Kim, G., Kang, W., Lee, D., Vaswani, R. T., & Chon, J. (2019). A spatial approach to climate-resilient infrastructure in coastal social-ecological systems: The case of dumbeong in Goseong County, South Korea. *Environment International*, *131*, 105032. DOI: 10.1016/j.envint.2019.105032 PMID: 31374442

Kreuze, J., Adewopo, J., Selvaraj, M., Mwanzia, L., Kumar, P. L., Cuellar, W. J., ... & Blomme, G. (2022). Innovative digital technologies to monitor and control pest and disease threats in root, tuber, and banana (RT&B) cropping systems: Progress and prospects. *Root, Tuber and Banana Food System Innovations: Value Creation for Inclusive Outcomes*, 261-288.

Kumar, G., Basri, S., Imam, A. A., Khowaja, S. A., Capretz, L. F., & Balogun, A. O. (2021). Data harmonization for heterogeneous datasets: A systematic literature review. *Applied Sciences (Basel, Switzerland)*, *11*(17), 8275. DOI: 10.3390/app11178275

Lassoued, R., Macall, D. M., Smyth, S. J., Phillips, P. W., & Hesseln, H. (2021). Expert insights on the impacts of, and potential for, agricultural big data. *Sustainability (Basel)*, *13*(5), 2521. DOI: 10.3390/su13052521

Lezoche, M., Hernandez, J. E., Díaz, M. D. M. E. A., Panetto, H., & Kacprzyk, J. (2020). Agri-food 4.0: A survey of the supply chains and technologies for the future agriculture. *Computers in Industry*, *117*, 103187. DOI: 10.1016/j.compind.2020.103187

Li, W. Q., Han, X. X., Lin, Z. B., & Rahman, A. (2024). Enhanced pest and disease detection in agriculture using deep learning-enabled drones. [ATAIML]. *Acadlore Transactions on AI and Machine Learning*, *3*(1), 1–10. DOI: 10.56578/ataiml030101

Liakos, K. G., Busato, P., Moshou, D., Pearson, S., & Bochtis, D. (2018). Machine learning in agriculture: A review. *Sensors (Basel)*, *18*(8), 2674. DOI: 10.3390/s18082674 PMID: 30110960

Liu, S. Y. (2020). Artificial intelligence (AI) in agriculture. *IT Professional*, *22*(3), 14–15. DOI: 10.1109/MITP.2020.2986121

Majeed, Y., Khan, M. U., Waseem, M., Zahid, U., Mahmood, F., Majeed, F., Sultan, M., & Raza, A. (2023). Renewable energy as an alternative source for energy management in agriculture. *Energy Reports*, *10*, 344–359. DOI: 10.1016/j.egyr.2023.06.032

Marda, V. (2018). Artificial intelligence policy in India: A framework for engaging the limits of data-driven decision-making. *Philosophical Transactions. Series A, Mathematical, Physical, and Engineering Sciences*, *376*(2133), 20180087. DOI: 10.1098/rsta.2018.0087 PMID: 30323001

Matthews, R., Stephens, W., Hess, T., Middleton, T., & Graves, A. (2002). Applications of crop/soil simulation models in tropical agricultural systems.

Mohsan, S. A. H., Othman, N. Q. H., Li, Y., Alsharif, M. H., & Khan, M. A. (2023). Unmanned aerial vehicles (UAVs): Practical aspects, applications, open challenges, security issues, and future trends. *Intelligent Service Robotics*, *16*(1), 109–137. DOI: 10.1007/s11370-022-00452-4 PMID: 36687780

Monteiro, J., & Barata, J. (2021). Artificial intelligence in extended agri-food supply chain: A short review based on bibliometric analysis. *Procedia Computer Science*, *192*, 3020–3029. DOI: 10.1016/j.procs.2021.09.074

Moshayedi, A. J., Khan, A. S., Yang, Y., Hu, J., & Kolahdooz, A. (2024). Robots in Agriculture: Revolutionizing Farming Practices. *EAI Endorsed Transactions on AI and Robotics, 3*.

Musanase, C., Vodacek, A., Hanyurwimfura, D., Uwitonze, A., & Kabandana, I. (2023). Data-driven analysis and machine learning-based crop and fertilizer recommendation system for revolutionizing farming practices. *Agriculture*, *13*(11), 2141. DOI: 10.3390/agriculture13112141

Obaideen, K., Yousef, B. A., AlMallahi, M. N., Tan, Y. C., Mahmoud, M., Jaber, H., & Ramadan, M. (2022). An overview of smart irrigation systems using IoT. *Energy Nexus*, *7*, 100124. DOI: 10.1016/j.nexus.2022.100124

Panda, C. K., & Bhatnagar, R. (2020). Social internet of things in agriculture: an overview and future scope. *Toward Social Internet of Things (SIoT): Enabling Technologies, Architectures and Applications: Emerging Technologies for Connected and Smart Social Objects*, 317-334.

Perez, J. (2023). How automation drives business growth and efficiency. *Harvard Business Review*.

Pretty, J. (2008). Agricultural sustainability: Concepts, principles and evidence. *Philosophical Transactions of the Royal Society of London. Series B, Biological Sciences*, *363*(1491), 447–465. DOI: 10.1098/rstb.2007.2163 PMID: 17652074

Rajak, P., Ganguly, A., Adhikary, S., & Bhattacharya, S. (2023). Internet of Things and smart sensors in agriculture: Scopes and challenges. *Journal of Agriculture and Food Research, 14*, 100776. DOI: 10.1016/j.jafr.2023.100776

Rane, N., Choudhary, S. P., & Rane, J. (2024). Acceptance of artificial intelligence: Key factors, challenges, and implementation strategies. *Journal of Applied Artificial Intelligence*, *5*(2), 50–70. DOI: 10.48185/jaai.v5i2.1017

Rose, D. C., Sutherland, W. J., Parker, C., Lobley, M., Winter, M., Morris, C., Twining, S., Ffoulkes, C., Amano, T., & Dicks, L. V. (2016). Decision support tools for agriculture: Towards effective design and delivery. *Agricultural Systems*, *149*, 165–174. DOI: 10.1016/j.agsy.2016.09.009

Rotz, S., Gravely, E., Mosby, I., Duncan, E., Finnis, E., Horgan, M., LeBlanc, J., Martin, R., Neufeld, H. T., Nixon, A., Pant, L., Shalla, V., & Fraser, E. (2019). Automated pastures and the digital divide: How agricultural technologies are shaping labour and rural communities. *Journal of Rural Studies*, *68*, 112–122. DOI: 10.1016/j.jrurstud.2019.01.023

Sabu, K. M., & Kumar, T. M. (2020). Predictive analytics in Agriculture: Forecasting prices of Areca Nuts in Kerala. *Procedia Computer Science*, *171*, 699–708. DOI: 10.1016/j.procs.2020.04.076

Sahoo, S., Singha, C., Govind, A., & Moghimi, A. (2024). Review of Climate-Resilient Agriculture for Ensuring Food Security: Sustainability Opportunities and Challenges of India. *Environmental and Sustainability Indicators*, •••, 100544.

Sarker, I. H. (2021). Deep learning: A comprehensive overview on techniques, taxonomy, applications and research directions. *SN Computer Science*, *2*(6), 420. DOI: 10.1007/s42979-021-00815-1 PMID: 34426802

Sekaran, U., Lai, L., Ussiri, D. A., Kumar, S., & Clay, S. (2021). Role of integrated crop-livestock systems in improving agriculture production and addressing food security–A review. *Journal of Agriculture and Food Research*, *5*, 100190. DOI: 10.1016/j.jafr.2021.100190

Sharma, A., Sharma, A., Tselykh, A., Bozhenyuk, A., Choudhury, T., Alomar, M. A., & Sánchez-Chero, M. (2023). Artificial intelligence and internet of things oriented sustainable precision farming: Towards modern agriculture. *Open Life Sciences*, *18*(1), 20220713. DOI: 10.1515/biol-2022-0713 PMID: 37854322

Sharma, R. (2021, May). Artificial intelligence in agriculture: a review. In *2021 5th international conference on intelligent computing and control systems (ICICCS)* (pp. 937-942). IEEE. DOI: 10.1109/ICICCS51141.2021.9432187

Shaw, B., Gustafson, D. H., Hawkins, R., McTavish, F., McDowell, H., Pingree, S., & Ballard, D. (2006). How underserved breast cancer patients use and benefit from eHealth programs: Implications for closing the digital divide. *The American Behavioral Scientist*, *49*(6), 823–834. DOI: 10.1177/0002764205283803

Shoaib, M., Shah, B., Ei-Sappagh, S., Ali, A., Ullah, A., Alenezi, F., Gechev, T., Hussain, T., & Ali, F. (2023). An advanced deep learning models-based plant disease detection: A review of recent research. *Frontiers in Plant Science*, *14*, 1158933. DOI: 10.3389/fpls.2023.1158933 PMID: 37025141

Smith, M. J. (2018). Getting value from artificial intelligence in agriculture. *Animal Production Science*, *60*(1), 46–54. DOI: 10.1071/AN18522

Soma, T., & Nuckchady, B. (2021). Communicating the benefits and risks of digital agriculture technologies: Perspectives on the future of digital agricultural education and training. *Frontiers in Communication*, *6*, 762201. DOI: 10.3389/fcomm.2021.762201

Subedi, B., Poudel, A., & Aryal, S. (2023). The impact of climate change on insect pest biology and ecology: Implications for pest management strategies, crop production, and food security. *Journal of Agriculture and Food Research*, *14*, 100733. DOI: 10.1016/j.jafr.2023.100733

Subeesh, A., & Mehta, C. R. (2021). Automation and digitization of agriculture using artificial intelligence and the internet of things. *Artificial Intelligence in Agriculture*, *5*, 278–291. DOI: 10.1016/j.aiia.2021.11.004

Talaviya, T., Shah, D., Patel, N., Yagnik, H., & Shah, M. (2020). Implementation of artificial intelligence in agriculture for optimisation of irrigation and application of pesticides and herbicides. *Artificial Intelligence in Agriculture*, *4*, 58–73. DOI: 10.1016/j.aiia.2020.04.002

Tian, H., Wang, T., Liu, Y., Qiao, X., & Li, Y. (2020). Computer vision technology in agricultural automation—A review. *Information Processing in Agriculture*, *7*(1), 1–19. DOI: 10.1016/j.inpa.2019.09.006

Timsina, J. (2018). Can organic sources of nutrients increase crop yields to meet global food demand? *Agronomy (Basel)*, *8*(10), 214. DOI: 10.3390/agronomy8100214

Tiwari, A. K., Mishra, H., Nishad, D. C., & Pandey, A. (2023). Sustainable water management in agriculture: irrigation techniques and water conservation. *Dr. Ajay B. Jadhao, 53*.

Umesha, S., Manukumar, H. M., & Chandrasekhar, B. (2018). Sustainable agriculture and food security. In *Biotechnology for sustainable agriculture* (pp. 67–92). Woodhead Publishing. DOI: 10.1016/B978-0-12-812160-3.00003-9

Velusamy, P., Rajendran, S., Mahendran, R. K., Naseer, S., Shafiq, M., & Choi, J. G. (2021). Unmanned Aerial Vehicles (UAV) in precision agriculture: Applications and challenges. *Energies*, *15*(1), 217. DOI: 10.3390/en15010217

Viana, C. M., Freire, D., Abrantes, P., Rocha, J., & Pereira, P. (2022). Agricultural land systems importance for supporting food security and sustainable development goals: A systematic review. *The Science of the Total Environment*, *806*, 150718. DOI: 10.1016/j.scitotenv.2021.150718 PMID: 34606855

Villegas-Ch, W., & García-Ortiz, J. (2023). Toward a comprehensive framework for ensuring security and privacy in artificial intelligence. *Electronics (Basel)*, *12*(18), 3786. DOI: 10.3390/electronics12183786

Wei, H., Xu, W., Kang, B., Eisner, R., Muleke, A., Rodriguez, D., ... & Harrison, M. T. (2024). Irrigation with Artificial Intelligence: Problems, Premises, Promises. *Human-Centric Intelligent Systems*, 1-19.

Young, S. (2020). The future of farming. *Harvard International Review*, *41*(1), 45–47.

Zha, J. (2020, December). Artificial intelligence in agriculture. [). IOP Publishing.]. *Journal of Physics: Conference Series*, *1693*(1), 012058. DOI: 10.1088/1742-6596/1693/1/012058

Zhang, J., Huang, Y., Pu, R., Gonzalez-Moreno, P., Yuan, L., Wu, K., & Huang, W. (2019). Monitoring plant diseases and pests through remote sensing technology: A review. *Computers and Electronics in Agriculture*, *165*, 104943. DOI: 10.1016/j.compag.2019.104943

Ziesche, S., Agarwal, S., Nagaraju, U., Prestes, E., & Singha, N. (2023). Role of artificial intelligence in advancing sustainable development goals in the agriculture sector. In *The ethics of artificial intelligence for the sustainable development goals* (pp. 379–397). Springer International Publishing. DOI: 10.1007/978-3-031-21147-8_21

Chapter 7

Empowering Sustainable Agriculture Harnessing AI for Enhanced Yields, Resource Efficiency, and Eco-Friendly Farming Practices

Ankur Bhardwaj
https://orcid.org/0000-0003-0687-8810
Shri Vaishnav Vidyapeeth Vishwavidyalaya, Indore, India

Surendra Prakash Gupta
https://orcid.org/0000-0002-4873-6346
Shri Vaishnav Vidyapeeth Vishwavidyalaya, Indore, India

ABSTRACT

The present chapter delves into the revolutionary role of AI technologies in promoting environmentally responsible and economically viable farming practices. The chapter highlights case studies where AI-driven tools, like, drones, sensors, etc., have successfully enabled real-time monitoring of crop health, pest control, and yield prediction. Socioeconomic implications of AI adoption, emphasizing AI potential to bridge the gap between smallholder farmers and large-scale agricultural enterprises are also discussed. This chapter anticipates future trends, including AI's role in developing resilient crops, mitigating climate risks, and facilitating circular agricultural systems. The ethical and environmental considerations of AI deployment in agriculture are critically analysed to ensure sustainable and equitable outcomes.

DOI: 10.4018/979-8-3373-0680-3.ch007

This comprehensive analysis provides a roadmap for harnessing AI to achieve the United Nations Sustainable Development Goals (SDGs) related to agriculture, food security, and environmental conservation.

1. INTRODUCTION

Sustainable agriculture is increasingly recognized as a foundation for addressing global food security while preserving the environment. It aims to balance the growing demand for food, fibre, and bioenergy with the need to conserve natural resources and ensure long-term ecological health (Kwakye et al., 2024). However, achieving this balance presents formidable challenges, including resource scarcity, climate change, pest management, soil degradation, and the need for economic viability for farmers. These challenges necessitate innovative approaches to enhance productivity while minimizing environmental impact (Mondal and Palit, 2022).

The evolution from traditional to digital to AI-powered agriculture marks a significant shift in farming practices. Traditional agriculture relied heavily on manual labour and basic tools, with practices passed down through generations (Haded et al., 2023). The advent of digital agriculture introduced technologies like GPS, drones, and sensors, enabling precision farming and data-driven decision-making. This phase improved efficiency and yield prediction, optimizing resource use and reducing environmental impact (Giller et al., 2021). Now, AI-powered agriculture integrates advanced algorithms and machine learning to analyse vast amounts of data in real-time. AI enables predictive analytics for crop health, pest management, and optimal harvesting times, enhancing productivity and sustainability (Sharma and Shivandu, 2024). Future trends may include autonomous farming equipment and robotics, further revolutionizing how farms operate. This evolution underscores a transition towards more sustainable, efficient, and resilient agricultural systems, capable of meeting growing global food demand while minimizing environmental footprint.

Artificial Intelligence (AI) has emerged as a transformative force in agriculture, offering tools to optimize farming practices, predict and mitigate risks, and streamline resource use. Integrating AI-driven insights into decision-making, farmers can achieve precision in planting, irrigation, pest control, and harvests, thereby reducing waste and improving yields (Thangamani et al., 2024).

1.1 Overview of Sustainable Agriculture and its Challenges

Sustainable agriculture is a farming approach designed to meet present and future food, fibre, and fuel needs while preserving the environment, promoting economic stability, and supporting social equity (Tian et al., 2021). It emphasizes practices that maintain soil health, conserve water, protect biodiversity, and minimize greenhouse gas emissions, ensuring agricultural productivity over the long term (Bhatti et al., 2024). This holistic framework aligns ecological stewardship with the economic well-being of farming communities, making it a foundation stone of global food security and environmental sustainability.

However, achieving sustainable agriculture is fraught with challenges. Resource scarcity is a significant issue, as arable land and freshwater supplies are limited and under increasing pressure from urbanization, industrial demands, and population growth. Climate change exacerbates these problems, leading to unpredictable weather patterns, extreme events, and shifts in pest and disease dynamics, which threaten crop yields and livestock health (Skendžić et al., 2021). Intensive farming practices, deforestation, and chemical overuse have depleted soil nutrients and caused erosion, reducing agricultural productivity. In parallel, the widespread use of chemical fertilizers and pesticides has raised concerns about water pollution and biodiversity loss (Musa et al., 2024).

Economic and social challenges also impede sustainable agriculture. Smallholder farmers, who produce a significant portion of the world's food, often lack access to modern technologies, financing, and markets, making it difficult for them to adopt sustainable practices (Giller et al., 2021). Furthermore, global supply chains can create inequities, leaving producers in developing regions vulnerable to price volatility and market instability (Gikunda, 2024).

1.2 Role of AI in Transforming Agriculture

AI-powered technologies enable precision agriculture, where real-time data from sensors, drones, and satellite imagery are analysed to guide decisions on planting, irrigation, fertilization, and pest control (Gupta and Bhardwaj, 2024). These insights allow farmers to apply inputs only where and when they are needed, reducing waste and environmental impact while improving crop yields (Sharma and Shivandu, 2024). For instance, machine learning algorithms can predict optimal planting times based on weather patterns and soil conditions, ensuring better harvest outcomes (Elbasi et al., 2023).

AI models analyse historical and real-time data (predictive analytics) to forecast weather changes, pest outbreaks, and disease risks, allowing farmers to take proactive measures. This not only minimizes potential losses but also reduces reliance on harmful chemicals (Ali et al., 2024).

AI also plays a critical role in autonomous farming systems, such as robots for planting, weeding, and harvesting. These systems enhance operational efficiency and reduce labour costs while maintaining consistent quality. Additionally, AI-driven market analytics help farmers make informed decisions about crop selection and pricing, improving their profitability and resilience to market fluctuations (Das, 2024).

In livestock management, AI tools monitor animal health and behaviour, ensuring timely interventions to improve productivity and welfare. Similarly, AI facilitates supply chain optimization by enhancing logistics, reducing food waste, and improving traceability, which is crucial for meeting consumer demands for sustainability and transparency (Zhang et al., 2024a). AI integration into farming practices empowers producers to meet global food demands while aligning with environmental and social sustainability goals (Atapattu et al., 2024a).

1.3 Objectives and Scope of the Chapter

The primary objective of the present chapter is to explore the transformative potential of AI in addressing the challenges of sustainable agriculture. The chapter aims to provide a comprehensive understanding of how AI-driven innovations can optimize farming practices, improve resource efficiency, and foster environmentally sustainable and economically viable agricultural systems.

The prominent objectives of the chapter are:

- To identify and discuss the critical challenges facing sustainable agriculture, including resource scarcity, climate change, soil degradation, pest management, and socio-economic barriers.
- To explore how AI technologies can address these challenges through applications such as precision farming, predictive analytics, autonomous systems, and supply chain optimization.
- To demonstrate how AI can enhance crop yields, reduce environmental impact, and support the global transition to eco-friendly farming practices.
- To consider how AI innovations can be made accessible to smallholder farmers, ensuring inclusivity and equity in the adoption of sustainable technologies.

- To discuss the role of supportive policies, research collaborations, and infrastructure in accelerating AI-driven transformations in agriculture.

The scope of this chapter includes an in-depth examination of AI applications in crop and livestock management, resource optimization, and risk mitigation. It also considers the broader implications of integrating AI into agricultural ecosystems, such as improving global food security and mitigating climate change. Through a synthesis of current trends, challenges, and innovations, the chapter aims to provide insights into how AI can be harnessed to achieve the goals of sustainable agriculture on a global scale.

2. THE INTERSECTION OF AI AND SUSTAINABLE AGRICULTURE

The integration of AI with sustainable agriculture is transforming the way food is grown, harvested, and distributed, addressing critical challenges like climate change, resource scarcity, and the growing global population. Tools like drones, IoT sensors, and machine learning models provide real-time insights into soil health, crop conditions, and weather patterns (Zhang et al., 2024a; Das, 2024). These technologies help reduce water usage, optimize fertilizer application, and minimize pesticide use, promoting eco-friendly farming practices.

AI systems monitor soil moisture levels and weather forecasts to recommend precise irrigation schedules. Smart irrigation reduces water consumption and prevents overwatering, protecting soil quality and conserving water resources. Wearable devices for animals track vital signs, enabling early disease detection and reducing the use of antibiotics. This enhances animal welfare and reduces the environmental impact of livestock farming (Neethirajan, 2024).

AI streamlines agricultural supply chains by improving logistics, reducing food waste, and predicting market demands. Advanced algorithms optimize transportation routes and storage conditions, ensuring produce reaches markets efficiently with minimal spoilage. AI helps farmers adapt to climate change by predicting extreme weather events and providing actionable insights for mitigation strategies (Pandey et al., 2023). AI supports the development of resilient crops and farming systems through modelling climate scenarios.

AI applications analyse large datasets to recommend optimal land use strategies, balancing the needs for agriculture, biodiversity conservation, and carbon sequestration. These insights help in preventing deforestation and maintaining soil health. AI optimizes energy use in farming operations, such as through the efficient use of machinery or renewable energy sources like solar panels (Mohammad and

Mahjabeen, 2023). By reducing energy consumption, farms can lower their carbon footprint. While AI holds immense potential for sustainable agriculture, challenges such as high implementation costs, limited data availability in remote regions, and a lack of technical expertise hinder widespread adoption (Onwude et al., 2020). Coupled with emerging technologies like blockchain for transparent supply chains and synthetic biology for advanced crop engineering, AI can be a basis of sustainable, resilient food systems worldwide.

2.1 Definition and Principles of Sustainable Agriculture

Sustainable agriculture is the practice of farming that meets current food, fibre, and fuel needs without compromising the ability of future generations to meet their own needs (Chandel et al., 2024). It focuses on producing agricultural goods in a way that is environmentally sound, economically viable, and socially equitable, ensuring long-term sustainability of farming systems.

Soil management practices (crop rotation, cover cropping, and reduced tillage) maintain soil fertility and prevent erosion (Musa et al., 2024). Efficient irrigation systems and rainwater harvesting to minimize water usage and preserve freshwater resources. Encouragement of diverse ecosystems by planting multiple crops, helps maintaining natural habitats, and reducing monoculture reliance. Reducing or minimizing synthetic fertilizers and pesticides prevent soil and water contamination (Lakhiar, et al., 2024). Adapting farming practices remain economically viable for farmers by increasing yields, reducing input costs, and optimizing resource use. Developing sustainable supply chains support fair trade and equitable access to markets. Diversifying crops and adopting climate-resilient practices reduce financial risks from environmental shocks (Kopeć, 2024).

Efficient labour practices, ensure the well-being, safety, and fair wages of agricultural workers. Collaborating promotions among farmers, local communities, and policymakers could achieve common sustainability goals. Further it also safeguards consistent and equitable access to nutritious food for local and global populations (Tjilen et al., 2024).

To ensure agricultural resilience and sustainability, it is crucial to focus on climate adaptation by developing farming systems capable of withstanding extreme weather events, droughts, and floods. This effort can be enhanced through innovation, incorporating emerging technologies like AI and precision farming to improve efficiency and reduce environmental impact. At the same time, a long-term vision is essential, balancing the need for immediate productivity with strategies to sustain ecosystems and conserve natural resources for future generations.

2.2 Overview of AI Technologies Used in Agriculture

AI technologies are transforming agriculture by improving efficiency, sustainability, and decision-making. Machine Learning (ML) is revolutionizing agriculture through its diverse applications. It enables accurate yield prediction by analysing weather, soil, and crop data, while also detecting crop diseases and pests by identifying patterns in data (Elbasi et al., 2023). Additionally, ML optimizes critical farming practices such as irrigation, fertilization, and pesticide application, ensuring resource efficiency and sustainability. For instance, algorithms trained on historical and real-time data provide predictive analytics, empowering farmers to make informed decisions and enhance overall productivity (Pandey et al., 2023).

Computer vision is transforming modern agriculture by enabling advanced applications that improve efficiency and precision. Through image recognition, it facilitates accurate identification of crops and weeds, aiding in targeted interventions. Robots equipped with cameras and AI can perform automated fruit picking, significantly reducing labour requirements while enhancing productivity (Gupta and Gupta, 2024). For instance, drones or robotic harvesters equipped with sophisticated vision systems can identify ripe produce with exceptional accuracy, optimizing harvests and reducing waste (Thangamani et al., 2024).

Robotics and automation are revolutionizing agriculture by streamlining labour-intensive tasks and promoting sustainable practices. Autonomous tractors are transforming field operations, efficiently handling ploughing, seeding, and spraying with minimal human intervention (Singh et al., 2024a). Robotic weeders, such as those developed by *Naïo Technologies*, mechanically remove weeds while distinguishing them from crops, reducing the reliance on manual labour and chemical herbicides (Upadhyay et al., 2024). In addition, automated greenhouse management systems optimize climate control and watering, ensuring ideal growing conditions for crops (Singh et al., 2024b). These advancements not only enhance productivity but also contribute to environmentally friendly farming practices.

The integration of the Internet of Things (IoT) with AI is transforming agriculture by enabling real-time monitoring and data-driven decision-making. IoT sensors provide continuous updates on soil moisture, temperature, and crop health, creating a comprehensive picture of field conditions (Krishnababu et al., 2024). AI analyses this data to optimize farming practices, reducing waste and enhancing efficiency. For instance, smart irrigation systems use IoT sensors to assess soil conditions and automatically adjust watering schedules, ensuring crops receive the right amount of water while conserving resources (Sharma and Shivandu, 2024).

Natural Language Processing (NLP) is enhancing the way farmers access information and make decisions by enabling more efficient communication and data processing. Virtual assistants powered by NLP provide valuable crop advisory and weather forecasts, helping farmers stay informed and optimize their practices (Marla et al., 2023). For instance, chatbots using NLP can offer guidance on best farming practices or respond to questions in regional languages, making farming knowledge more accessible and tailored to local needs (Kaushal, 2024).

Predictive analytics is revolutionizing agricultural planning and risk management by utilizing data to anticipate future events and improve decision-making. It helps farmers optimize planting and harvesting schedules through accurate weather forecasting, reducing the risk of crop loss (Tantalaki et al., 2019). For instance, AI models that predict locust swarms enable governments and farmers to take preventive measures, minimizing damage and safeguarding food security. This proactive approach to agriculture is helping to mitigate risks and enhance overall productivity (Javaid et al., 2023).

Drones and remote sensing technologies are transforming agriculture by providing valuable insights into crop health and field conditions. Aerial imaging allows for continuous monitoring of crop health and soil analysis, enabling farmers to identify issues like nutrient deficiencies or irrigation needs with precision (Khanpara et al., 2024). Drones equipped with AI-powered cameras capture multispectral images, offering a detailed view of crop conditions that can reveal areas requiring more attention or resources. For instance, these drones can detect nutrient deficiencies, allowing for targeted interventions that improve crop yields while reducing waste and input costs (Guebsi et al., 2024).

Blockchain, when combined with AI, is revolutionizing the agricultural supply chain by enhancing transparency and traceability. This integration allows for the secure tracking of agricultural products from farm to table, ensuring their authenticity and reducing the risk of fraud (Doshi et al., 2024). With the help of AI, blockchain technology can also optimize logistics, making it easier for farmers to manage inventory, track shipments, and improve delivery efficiency. For instance, AI-powered blockchain systems help farmers streamline operations and ensure fair trade practices, giving them greater control over their products and fostering trust with consumers (Oriekhoe et al., 2024).

Digital twins are transforming agricultural management by providing virtual models of farm environments that can simulate various scenarios to optimize operations. These models allow farmers to test different strategies, from crop rotation to irrigation practices, to determine the most effective approaches (Escribà-Gelonch et al., 2024). Additionally, digital twins can model the impacts of climate change on specific crops and regions, helping farmers anticipate challenges and adapt their strategies accordingly. For instance, AI-driven digital twins replicate real-world fields, enabling farmers to make more informed management decisions, improving productivity, and increasing resilience to environmental changes (Dawn et al., 2024).

Genetic analysis and breeding, powered by AI, are revolutionizing the development of crops that are more resilient and productive. AI tools analyse genetic data to identify key traits such as drought resistance or faster growth, enabling the creation of crops that can thrive under challenging conditions (Farooq et al., 2024). This technology accelerates the breeding process, helping scientists develop high-yield crop varieties more quickly. For instance, AI is used to identify genes in crops that contribute to climate adaptability, allowing for the development of varieties that can withstand the impacts of climate change, ensuring food security in the face of environmental challenges (Nguy n et al., 2024).

2.3 Importance of Integrating AI Into Sustainable Farming Practices

AI offers innovative solutions that enhance efficiency, minimize environmental impact, and ensure long-term agricultural sustainability. One of the significant benefits of AI is precision farming, which optimizes resource use by analysing real-time data from IoT sensors, drones, and satellite imagery (Bhardwaj and Gupta, 2024; Agrawal and Arafat, 2024). This reduces water consumption, prevents overuse of fertilizers and pesticides, and conserves soil health, making farming more environmentally friendly. For instance, AI-powered irrigation systems deliver water precisely where and when needed, reducing waste and protecting freshwater resources (Nova, 2023).

Machine learning models and computer vision can identify crop diseases, pest infestations, or nutrient deficiencies early, allowing farmers to take timely action and avoid significant losses. This reduces dependence on chemical interventions, promoting eco-friendly farming practices (Amulothu et al., 2024). Moreover, AI enhances resilience to climate change by providing accurate weather forecasts, predicting extreme weather events, and offering adaptive strategies (Rane et al., 2024). It also aids in developing climate-resilient crops through genetic analysis and modelling.

AI-driven supply chain optimization improves the distribution of agricultural products, reducing food waste and ensuring better market access for farmers. Integrating AI into sustainable farming ensures food security, supports environmental conservation, and boosts the economic viability of farming systems (Vinod Chandra et al., 2024). AI empowers farmers to meet current agricultural needs by enabling smarter resource use and promoting adaptive strategies without compromising the ability of future generations to thrive. This makes AI a cornerstone of sustainable agricultural transformation.

3. AI APPLICATIONS IN SUSTAINABLE AGRICULTURE

AI applications in sustainable agriculture include precision farming, where AI analyses data to optimize crop yield, reduce water usage, and minimize pesticide application. AI-driven models predict weather patterns, monitor soil health, and detect plant diseases, fostering efficient resource management and promoting environmentally friendly farming practices.

3.1 Precision Farming

Precision farming uses advanced technologies like AI, GPS, and IoT to optimize agricultural practices. It involves collecting and analysing data on soil conditions, weather patterns, crop health, and resource use to make informed decisions (Rhoads, 2023). Precision farming applications include soil health monitoring, precise irrigation, targeted fertilization, and automated planting and harvesting. Precision farming helps farmers tailor practices to specific field conditions, minimizing environmental impact while boosting productivity. Ultimately, precision farming fosters more resilient, efficient, and eco-friendly farming operations (Sharma et al., 2024b).

3.1.1 Use of AI in Soil Health Monitoring and Nutrient Management

AI plays a pivotal role in monitoring soil health and optimizing nutrient management. By using sensors, drones, and satellite imaging, AI systems can gather real-time data on soil moisture, pH levels, nutrient content, and microbial activity (Zhang et al., 2024b). Machine learning algorithms analyse this data to provide actionable insights for farmers, enabling precise recommendations for soil amendments and fertilizer application (Lu et al., 2024). This reduces overuse of chemicals, promotes efficient nutrient cycles, and supports sustainable farming practices by improving soil fertility and long-term productivity. AI-driven soil health monitoring ensures the timely detection of imbalances, preventing soil degradation and supporting environmental conservation (Sivakumar et al., 2023).

3.1.2 Optimized Planting, Irrigation, and Harvesting Practices

AI also contributes significantly to optimizing planting, irrigation, and harvesting practices. Analysis of historical climate data, soil conditions, and crop performance, AI models can determine the best planting times, crop varieties, and spacing for maximum yield (Akkem et al., 2023). During harvesting, AI technologies such as machine vision and autonomous robots can identify the right time for harvest and perform tasks like fruit picking with precision, reducing labour costs and minimizing crop loss (Rajendran et al., 2024). Together, these AI applications help streamline agricultural processes, making them more resource-efficient and environmentally sustainable.

3.2 Resource Management

Resource management in sustainable agriculture involves efficiently utilizing water, soil, nutrients, and energy while minimizing waste and environmental impact. AI applications optimize resource use by analysing data from sensors, drones, and satellites (Bhardwaj and Gupta, 2024; Thangamani et al., 2024). For water management, AI systems track soil moisture and weather patterns to automate irrigation, reducing water waste (Lakhiar, et al., 2024). AI also helps in nutrient management by analysing soil health and recommending targeted fertilizer use, preventing over-application. Additionally, AI aids in monitoring crop health and pest control, reducing pesticide use (Sivakumar et al., 2023). Farmers can manage resources more efficiently, enhance productivity, and support environmentally sustainable agricultural practices by leveraging AI.

3.2.1 Smart Irrigation Systems

AI algorithms analyse weather data, soil moisture levels, and crop water requirements to determine the exact amount of water needed, adjusting irrigation schedules in real-time (Sharma and Shivandu, 2024). This reduces water wastage and ensures crops receive optimal hydration, improving water efficiency. AI-enabled systems can also use predictive analytics to forecast future irrigation needs based on environmental variables, further conserving water (Elbasi et al., 2023). These smart systems contribute to sustainable resource management by promoting responsible water use and enhancing crop productivity, all while reducing the environmental footprint of farming.

3.2.2 AI-Driven Pest and Disease Management

AI-driven pest and disease management enhances sustainable farming by enabling early detection and precise control measures. Using machine learning, image recognition, and data analytics, AI systems can identify pests, diseases, and plant stress symptoms from images or sensor data (Amulothu et al., 2024). AI then predicts the spread of pests or diseases, helping farmers apply targeted treatments, such as eco-friendly pesticides or biological controls, only when necessary. This approach reduces the overuse of chemicals, minimizes environmental harm, and prevents crop damage (Kebe et al., 2023).

3.2.3 Minimizing Chemical and Water Usage

Through data analysis from sensors, drones, and satellite imaging, AI systems can monitor soil moisture, crop health, and weather conditions in real-time. This allows for tailored irrigation and nutrient application, reducing the need for excessive water and fertilizers (Nguy n et al., 2024). In terms of chemical use, AI-driven pest and disease detection enables targeted treatments, applying pesticides or herbicides only when and where necessary, rather than blanket spraying (Naveed et al., 2024). AI helps reduce waste, lower costs, and minimize environmental impact, by optimization of these resources, promoting a more sustainable and efficient agricultural system.

3.3 Yield Prediction and Enhancement

Analysis of soil health, weather patterns, pest and disease occurrences, and crop growth stages, AI can predict the optimal harvest time and potential yields. This allows farmers to adjust practices, such as irrigation and fertilization, to maximize productivity (Kebe et al., 2023). Additionally, AI can identify high-performing crop varieties and suggest improvements in planting techniques (Chandel et al., 2024). By integrating AI into yield management, farmers can enhance efficiency, reduce resource waste, and improve sustainability while meeting growing food demands (Table 1).

AI-powered predictive analytics in sustainable agriculture enable farmers to make informed decisions on irrigation, fertilization, and planting schedules, optimizing resources and maximizing productivity (Elufioye et al., 2024). By forecasting yield outcomes, AI allows for proactive adjustments to farming practices, reducing resource wastage and enhancing overall crop productivity, leading to more sustainable and efficient agricultural operations. Using AI image recognition, sensor data, and machine learning, AI systems can monitor crop health and identify early signs of stress, such as wilting, discoloration, or pest infestation. Identification of these issues at an early stage, farmers can take targeted actions, such as applying localized treatments or adjusting environmental conditions, to prevent crop loss. AI's ability to forecast weather anomalies like droughts or storms further supports timely interventions, reducing risks and ensuring stable, enhanced crop yields.

Table 1. Case studies and success stories, with instances and impact analysis of AI applications and farming projects in various regions.

Case Study	Region/s	AI Application	Project Details	Impact Analysis	References
John Deere's AI-Powered Precision Agriculture	USA	Precision Farming (AI in machinery & equipment)	John Deere uses AI in tractors and harvesters, equipping them with sensors and GPS to collect real-time data on soil health, crop performance, and moisture levels for precise planting, fertilizing, and harvesting.	Increased crop yields by up to 20%, reduced water usage by 30%, and lowered operational costs by optimizing resource use. It also enabled farmers to make data-driven decisions, reducing environmental impact.	Karunathilake et al., 2023
AI for Pest Detection in Rice Farming	India	Pest and Disease Management (AI-based detection)	In India, AI-powered drones and cameras are used to detect pests and diseases in rice fields. Machine learning models analyse imagery to identify pests like locusts, enabling targeted pesticide applications.	60% reduction in pesticide use, improved pest control accuracy, and enhanced crop health. The project also reduced the environmental footprint and pesticide-related costs.	Bhoi et al., 2021
SAB Miller's AI-Driven Water Efficiency Project	South Africa	Smart Irrigation Systems	SAB Miller partnered with IBM to implement AI-driven irrigation systems in barley fields. Sensors monitor soil moisture and weather forecasts to optimize irrigation schedules.	Reduced water usage by 50% in barley production, supporting sustainable water management practices. The project helped conserve water resources and improved crop yield quality.	Moorsom et al., 2020
AI-Powered Crop Yield Prediction by Taranis	Argentina	Yield Prediction and Enhancement	Taranis, an AI-powered platform, provides precise crop yield predictions using satellite imagery, IoT sensors, and machine learning algorithms. It helps farmers predict outcomes and optimize harvesting schedules.	25% increase in yield accuracy, enabling farmers to make informed decisions regarding planting, fertilizing, and harvesting. It also minimized the risk of over-fertilization and resource waste.	Krishnababu et a., 2024

continued on following page

Table 1. Continued

Case Study	Region/s	AI Application	Project Details	Impact Analysis	References
AI for Soil Health Monitoring in Europe	European Union	Soil Health Monitoring and Nutrient Management	AI models, using satellite data and on-ground sensors, monitor soil conditions, moisture levels, and nutrient requirements across farms in the EU. These systems give farmers tailored advice for soil management.	Improved soil fertility by 30%, reduced fertilizer use by 20%, and increased crop health. The system promoted eco-friendly farming practices by preventing overuse of fertilizers and chemicals.	Konya, A. Nematzadeh, 2024
Agri-Tech in Kenya: M-Farm and AI Solutions	Kenya	Smart Irrigation & Market Insight	M-Farm, a Kenyan agri-tech startup, integrates AI to provide farmers with real-time data on weather patterns, market prices, and irrigation requirements through mobile apps.	40% reduction in irrigation costs, increased market access for small farmers, and enhanced productivity through better pricing and resource management. It also empowered smallholder farmers.	Bunde, 2024
China's AI for Rice Farming	China	Smart Irrigation and Pest Management	AI-based sensors and drones are deployed in rice paddies across China to monitor pest outbreaks and automate irrigation processes based on weather data and soil moisture levels.	15% increase in rice yields, reduced pesticide use by 25%, and improved water efficiency. The use of AI also helped optimize labour usage and reduced costs associated with manual labour.	Ai et al., 2023
AI-Based Crop and Climate Monitoring in Australia	Australia	Climate and Yield Prediction	Australian farms use AI-based platforms for climate prediction and yield estimation. These tools analyze historical data, weather patterns, and environmental conditions to forecast crop performance.	Enhanced crop resilience to climate change, with up to 30% better yield predictions. Farmers adapted farming practices based on predictive insights, reducing water use and improving soil health.	Haded et al., 2023

continued on following page

Table 1. Continued

Case Study	Region/s	AI Application	Project Details	Impact Analysis	References
AI-Powered Farm Management by IBM Watson	North America, Canada, Europe, UK, France, Australia, Brazil, India, Sub-Saharan Africa	Farm Management System	IBM Watson's AI-powered farm management system integrates weather, soil, and crop data to help farmers manage operations. It offers tailored recommendations for irrigation, fertilization, and crop management.	Increased operational efficiency, reducing fertilizer use by 25%, and improving crop quality. The system's predictive insights allowed farmers to optimize resource use and reduce costs.	Hadeed et al., 2024
AI in Vineyard Management by XyloTech	France	Precision Farming and Yield Prediction	XyloTech uses AI to manage vineyards, using drones and sensors to monitor grapevine health and predict the best harvesting times. Machine learning algorithms analyse climate and soil data.	Improved grape quality, reduced pesticide use by 20%, and optimized harvest timing for better wine production. The system also helped manage labour costs and improve overall sustainability.	Tardaguila et al., 2021

Footnotes:

- Water Efficiency: Many projects focus on smart irrigation, reducing water usage by up to 50%, crucial in water-scarce regions.
- Yield Prediction: AI systems improve yield forecasting accuracy by up to 25%, enabling optimized resource allocation.
- Pest & Disease Control: AI enables early pest detection and targeted interventions, reducing pesticide use by 25-60%.
- Soil & Nutrient Management: AI-driven soil monitoring systems lead to more efficient use of fertilizers and enhanced soil health, reducing environmental impact.
- Global Impact: AI applications are having a measurable impact on crop yields, cost reduction, and environmental sustainability across various regions.

5. SOCIOECONOMIC AND ETHICAL IMPLICATIONS

The integration of advanced technologies like AI into agriculture offers transformative potential, but it also raises significant socioeconomic and ethical considerations. To ensure these innovations empower sustainable agriculture rather than exacerbate disparities, the following points must be addressed:

5.1 Addressing the Digital Divide and Ensuring Equitable Access to AI Technologies

The digital divide, gaps in access to technology and infrastructure, poses a significant challenge to the equitable adoption of AI in agriculture. Infrastructure disparities, including limited access to high-speed internet and reliable electricity in rural and developing regions, pose significant barriers to the widespread implementation of AI-based solutions in agriculture (Samuel-Okon and Abejide, 2024). Smallholder farmers, often lacking the financial resources and technical expertise, struggle to adopt and fully benefit from these advanced tools. To address these challenges, a collaborative effort from governments, NGOs, and the private sector is essential (Dhillon and Moncur, 2023). By subsidizing technology, providing targeted training programs, and supporting local innovations, an environment that fosters inclusive access to AI can be created (Limpamont et al., 2024). This approach will democratize technology, enabling marginalized farmers to harness the power of precision farming, improve yield predictions, and optimize resource use, ultimately driving sustainable agricultural growth.

5.2 Impact on Smallholder Farmers and Rural Communities

AI technologies hold immense empowerment potential for smallholder farmers, offering them access to vital resources such as accurate weather forecasts, pest control strategies, and market trends/tools that are often essential for their survival in competitive markets (Tiwari et al., 2024). However, without targeted interventions, there is a risk of further marginalization, as small-scale farmers may struggle to keep pace with large agricultural enterprises that quickly adopt AI, deepening income inequalities (Villalba et al., 2024). To combat this, AI-driven solutions must prioritize community-based approaches, including shared AI resources and cooperatives, to uplift rural economies and foster collaboration among farmers (Tiwari et al., 2024). Implementing of strategic frameworks that emphasize inclusivity, we can ensure that smallholder farmers and rural communities are not only participants but also beneficiaries of agricultural innovations, ultimately paving the way for more equitable and sustainable agricultural development.

5.3 Ethical Concerns and Environmental Considerations

AI applications in agriculture bring ethical and environmental challenges that must be addressed to align with sustainable development goals. The rapid adoption of AI in agriculture raises crucial ethical and environmental concerns that must be addressed. Data privacy and ownership are significant issues, as many AI systems require vast amounts of data, prompting questions about who owns this data and how it is used, particularly for vulnerable farmers (Saikanth et al., 2024). Moreover, while AI can enhance resource use efficiency, its computational demands and reliance on electronic hardware contribute to increased energy consumption and e-waste, presenting environmental trade-offs (Mallinger and Baeza-Yates, 2024). Another challenge lies in ensuring fairness, poorly designed AI algorithms can perpetuate biases in resource allocation, deepening social inequalities (Khang et al., 2023). To balance innovation with environmental stewardship, it is essential to design low-energy AI systems, promote circular economies in technology manufacturing, and implement transparent policies for data governance (Ziesche et al., 2023). Such measures will ensure that AI-driven agricultural solutions are not only effective but also ethically sound and environmentally responsible.

The adoption of AI in agriculture presents an opportunity to achieve sustainable agricultural practices, but only if socioeconomic and ethical implications are adequately addressed. Proactively bridging the digital divide, empowering smallholder farmers, and integrating ethical and environmental considerations into AI development and deployment are essential steps toward equitable and sustainable agricultural transformations.

6. CHALLENGES IN ADOPTING AI FOR SUSTAINABLE AGRICULTURE

The adoption of AI in agriculture has the potential to transform farming practices and promote sustainability. However, significant challenges must be addressed to ensure widespread implementation. These challenges include issues related to data, costs, and the willingness of farmers to adopt new technologies.

6.1 Data Availability and Quality

Data serves as the backbone of AI applications in agriculture, but various limitations hinder its effective use. In many regions, inadequate data collection systems hinder the ability to gather high-quality agricultural data, such as soil health, crop growth patterns, and weather conditions, limiting the potential for AI-driven solutions. Moreover, existing agricultural data is often fragmented and inconsistent, scattered across multiple sources without standardization, making it difficult to integrate into AI systems effectively. Smallholder farmers and researchers face further barriers, as they often lack access to proprietary datasets, limiting their ability to fully leverage AI technologies. Poor-quality or biased data only exacerbates these challenges, leading to flawed AI predictions that can result in inefficient resource use and even crop failures. To overcome these obstacles, it is essential to invest in robust data collection infrastructure, establish open data platforms, and develop universal standards that ensure data quality and interoperability, paving the way for more accurate and impactful AI applications in agriculture.

6.2 High Costs of Implementation

The financial barriers associated with AI adoption are a significant challenge, especially for smallholder farmers and developing regions. The adoption of AI-powered tools in agriculture is often hindered by significant financial barriers, with high upfront costs for technologies such as drones, sensors, and software deterring many farmers from making the initial investment. Beyond the initial purchase, the ongoing expenses for maintaining AI systems, acquiring updates, and ensuring compatibility with rapidly evolving technologies can place additional strain on already limited budgets. Furthermore, many farmers lack access to affordable financing options, such as loans or subsidies, which are crucial for investing in these advanced solutions. In regions with unstable agricultural markets, the economic risk associated with AI adoption may be perceived as a financial gamble. To overcome these challenges, governments and organizations must step in to provide essential financial support, including subsidies, low-interest loans, and public-private partnerships, to alleviate the financial burden on farmers and make AI-driven solutions more accessible and sustainable.

6.3 Resistance to Technology Adoption Among Farmers

Even when AI technologies are accessible, farmers may resist adopting them due to various social and psychological factors. In many rural areas, a lack of awareness and education about the potential benefits of AI, coupled with limited understanding of how to use it effectively, creates a significant barrier to adoption. Long-standing cultural and traditional farming practices often foster skepticism toward new technologies, especially in communities with deep ties to manual methods. Additionally, the fear of job displacement in labour-intensive farming regions further deters farmers from embracing AI, as they worry it may replace human labour. Trust issues also arise, as farmers may fear the misuse of their data or hidden costs associated with technology adoption. To address these challenges, it is essential to implement targeted awareness campaigns and hands-on training programs that help farmers understand the value of AI. Involving local agricultural communities in co-developing AI solutions that are tailored to their specific needs can also build trust and ensure that these technologies are embraced as tools to enhance, rather than replace, traditional practices.

7. FUTURE TRENDS AND OPPORTUNITIES

The potential for AI to drive sustainable agriculture continues to expand with advancements in technology and a growing emphasis on addressing global challenges. The future promises innovative applications that enhance climate resilience, foster integration with other technologies, and promote circular agricultural systems.

7.1 AI for Climate-Resilient Crop Development

Climate change poses a significant threat to global agriculture, making the development of climate-resilient crops a priority. Advancements in AI and machine learning are revolutionizing agriculture by offering transformative solutions to climate change challenges. Genomic insights powered by AI-driven analysis enable the identification of key traits such as drought tolerance, heat resistance, and pest resilience, accelerating the development of climate-resilient crop varieties. Predictive breeding models further enhance efficiency by forecasting the success of crossbreeding efforts, reducing both time and resources traditionally required for experimentation. Additionally, AI systems can provide tailored, region-specific recommendations for crop selection and planting schedules, utilizing real-time climate data to help farmers adapt to changing environmental conditions. Together, these innovations promise to bolster global food security, reduce agriculture's vulnerability to climate change, and ensure a sustainable future for food production.

7.2 Integration of AI With Other Emerging Technologies

The convergence of AI with other cutting-edge technologies opens new frontiers for sustainable agriculture. The integration of cutting-edge technologies is transforming agriculture into a smarter, more efficient industry with the potential to address pressing challenges in food production. Blockchain, when combined with AI, ensures transparency in supply chains, verifying the authenticity of organic produce and promoting fair trade practices while curbing food fraud. Robotics, guided by AI, enhances precision farming, enabling autonomous robots to perform tasks such as planting, weeding, and harvesting with unmatched accuracy, reducing resource waste and labour costs. AI-powered IoT devices offer real-time monitoring of soil health, crop growth, and weather conditions, supporting precision agriculture practices that optimize resources. Remote sensing and satellite imagery, processed through AI algorithms, help detect pest infestations, nutrient deficiencies, and water stress, providing farmers with actionable insights to improve crop health. The synergy of these technologies is creating more productive and sustainable agricultural systems, shaping a future where food security and environmental stewardship go hand in hand.

7.3 Role of AI in Circular Agricultural Systems

Circular agricultural systems aim to minimize waste and maximize resource efficiency, and AI plays a pivotal role in enabling this transformation. AI is at the forefront of revolutionizing agricultural waste management, offering innovative solutions that can transform the industry's environmental footprint. Through resource recycling, AI tracks nutrient flows and recommends strategies to reuse organic materials, such as converting crop residues into biofertilizers, thereby enriching soil health. AI also plays a crucial role in optimizing water usage in irrigation and energy consumption in farming operations, contributing to greater resource efficiency and reduced environmental impact. Additionally, by analyzing market trends, AI helps align production with demand, minimizing overproduction and its associated waste. Integrating AI into circular agricultural systems represents a transformative shift, moving agriculture from a linear model to a sustainable, regenerative approach that enhances both productivity and environmental stewardship.

The future of AI in sustainable agriculture is rich with possibilities, from developing climate-resilient crops to integrating with emerging technologies and advancing circular agricultural systems. Embracing of these trends, stakeholders can harness AI's potential to build a resilient, equitable, and environmentally sustainable agricultural sector that meets the challenges of the 21st century.

8. RECOMMENDATIONS AND POLICY IMPLICATIONS

To maximize the transformative potential of AI in sustainable agriculture, targeted strategies and policy interventions are essential. These must address barriers to adoption, foster collaboration among stakeholders, and establish robust policy frameworks that align technology with sustainability goals.

8.1 Strategies for Fostering AI Adoption in Agriculture

Successfully integrating AI into agriculture requires strategies tailored to diverse farming contexts, focusing on accessibility, education, and incentives. To foster widespread adoption of AI in agriculture, a comprehensive approach focused on capacity building, scalable solutions, incentivizing adoption, and democratizing data access is essential (Parra-López et al., 2024). Establishing training programs and workshops to enhance farmers' digital literacy and technical skills, in collaboration with local agricultural cooperatives and NGOs, ensures that region-specific AI solutions are accessible and effective. Scalable solutions, such as affordable, user-friendly AI tools tailored for smallholder farmers and the promotion of open-source platforms, reduce reliance on expensive proprietary software and democratize technology access (Atapattu et al., 2024b). Financial incentives, like subsidies and low-interest loans, can offset the initial costs of AI adoption, while recognizing early adopters through certifications or tax benefits encourages peer learning and broader adoption. Centralized, open-access databases for agricultural data, supported by public-private partnerships, ensure that farmers and researchers have the resources needed to make data-driven decisions, ultimately empowering farmers and creating a more equitable agricultural landscape (Vetrivel and Arun, 2025).

8.2 Role of Governments, Private Sectors, and Academic Institutions

For AI adoption to transform agriculture, a collaborative effort between governments, the private sector, and academic institutions is essential. Governments must prioritize agriculture as an important sector for innovation by developing robust infrastructure, such as reliable internet connectivity and electricity in rural areas, to enable AI integration (Majumdar et al., 2024). National AI strategies should also ensure policies that protect farmers' data privacy and ownership rights, fostering trust and security in the digital landscape. The private sector has a critical role to play by investing in affordable, accessible AI tools for both smallholder and large-scale farmers and promoting inclusive business models (shared AI services or subscription plans) which lower entry barriers for smallholders (Aderibigbe et al., 2023). By partnering with governments to pilot AI-driven projects, the private sector can showcase the real-world impact of these technologies. Academic institutions must conduct interdisciplinary research to refine AI models for agriculture, ensuring accuracy and scalability, while collaborating with agricultural extension services to bring AI tools to the grassroots level (Yigitcanlar et al., 2024). Providing training that combines AI expertise with agricultural sciences, universities can build a workforce capable of advancing agri-tech and driving sustainable, technology-driven agricultural practices.

8.3 Policy Frameworks for Promoting Sustainable Farming Through AI

Policy frameworks must ensure that AI adoption aligns with environmental, economic, and social sustainability goals. To harness the full potential of AI in agriculture while safeguarding ethical, environmental, and social standards, it is crucial to establish robust regulatory frameworks and incentivize sustainable practices (Dara et al., 2022). Regulatory standards should set clear guidelines for the ethical use of AI, addressing data ownership, environmental impact, and fair labor practices, while ensuring that AI technologies meet reliability, safety, and environmental compliance (Ryan et al., 2023). Sustainability incentives, such as tax benefits or grants, can promote AI solutions that optimize resource use, like water and fertilizers, and reduce greenhouse gas emissions, while targeted subsidies can support AI-driven regenerative practices, including precision composting and waste management. To ensure that AI adoption aligns with sustainability goals, monitoring and accountability are vital: developing metrics to assess the social, economic, and environmental impacts of AI in agriculture and establishing independent oversight bodies to track AI deployment (Rane et al., 2024). Global collaboration is essential to accelerate AI-driven innovation in sustainable farming, facilitating knowledge exchange, sharing best practices, and promoting harmonized policies to foster cross-border research and trade. By integrating these strategies, we can create a future where AI not only drives agricultural productivity but also contributes to environmental stewardship and social equity (Hacker, 2024).

To empower sustainable agriculture through AI, a holistic approach is required that includes strategies for adoption, active engagement from diverse stakeholders, and robust policy frameworks. By addressing the needs of farmers and aligning AI technologies with sustainability objectives, these recommendations can guide a transition toward a resilient, inclusive, and environmentally sound agricultural future.

9. CONCLUSION

The integration of AI into agriculture is revolutionizing the sector, offering unprecedented opportunities to balance productivity with sustainability. This journey of transformation requires a holistic understanding of its implications, potential, and necessary actions.

AI has proven to be a game-changer in sustainable agriculture through its ability to enhance yields, optimize resource use, and promote eco-friendly practices. AI-driven precision agriculture technologies are reshaping farming by improving crop management, reducing waste, and enabling climate-resilient practices that enhance productivity and resilience. These advanced tools optimize critical resources like water usage, fertilizer application, and pest management, minimizing environmental impact while increasing operational efficiency. However, challenges such as limited data accessibility, high costs, and resistance to adoption underscore the need for inclusive, scalable, and affordable solutions that can benefit all farmers, regardless of size. The collaborative potential between AI and emerging technologies, such as blockchain and robotics, offers the opportunity to create smarter, integrated agricultural systems. Together, these innovations have the transformative power to build a resilient agricultural ecosystem capable of meeting the growing demands of a global population while safeguarding natural resources for future generations.

The future of AI in agriculture is one of inclusivity, innovation, and sustainability. AI envisions to empower smallholder farmers, providing them with accessible tools that bridge the digital divide and ensure their equitable participation in the agricultural transformation. By seamlessly integrating AI with IoT, robotics, and renewable energy systems, we can create circular agricultural models that minimize environmental impact and promote sustainability. AI also plays a crucial role in building climate resilience, enabling farmers to develop adaptive strategies to mitigate the adverse effects of climate change on their communities. This vision is supported by global collaboration, where governments, private enterprises, and other stakeholders unite to advance research, share knowledge, and deploy AI solutions worldwide. By combining technological innovation with human ingenuity, this approach is poised to redefine agriculture as a sustainable, inclusive, and equitable endeavour for future generations.

Achieving this vision requires collective effort and commitment from all stakeholders. A coordinated effort from governments, the private sector, academic institutions, farmers, and the global community is essential. Governments must prioritize investment in infrastructure, establish supportive policy frameworks, and ensure the ethical use of AI in agriculture, creating an environment that fosters innovation while protecting farmers' rights. The private sector plays a pivotal role in driving the development of affordable, scalable, and accessible AI technologies that cater to farmers of all scales, bridging the gap between cutting-edge solutions and practical applications. Academic institutions must lead the charge in refining AI algorithms for agricultural use, while also building capacity through education and training to develop a skilled workforce. Farmers and communities should actively engage in co-developing AI solutions that reflect local realities and cultural contexts, ensuring relevance and sustainability. Lastly, the global community must foster international collaboration to tackle shared challenges like climate change, food insecurity, and resource depletion, working together to create a future where AI-driven agriculture thrives for all.

Harnessing the power of AI for sustainable agriculture is not merely a technological challenge, it is a moral imperative. By working together to implement cutting-edge solutions, we can create a future where agriculture is productive, resilient, and harmonious with the planet. The time to act is now.

REFERENCES

Aderibigbe, A. O., Ohenhen, P. E., Nwaobia, N. K., Gidiagba, J. O., & Ani, E. C.Adebayo Olusegun AderibigbePeter Efosa OhenhenNwabueze Kelvin Nwaobia-Joachim Osheyor GidiagbaEmmanuel Chigozie Ani. (2023). Artificial intelligence in developing countries: Bridging the gap between potential and implementation. *Computer Science & IT Research Journal*, *4*(3), 185–199. DOI: 10.51594/csitrj.v4i3.629

Agrawal, J., & Arafat, M. Y. (2024). Transforming farming: A review of AI-powered UAV technologies in precision agriculture. *Drones (Basel)*, *8*(11), 664. DOI: 10.3390/drones8110664

Ai, J., Hu, L., Xia, S., Xiang, H., & Chen, Z. (2023). Analysis of factors influencing the adoption behavior of agricultural productive services based on logistic—ISM model: A case study of rice farmers in jiangxi province, China. *Agriculture*, *13*(1), 162. DOI: 10.3390/agriculture13010162

Akkem, Y., Biswas, S. K., & Varanasi, A. (2023). Smart farming using artificial intelligence: A review. *Engineering Applications of Artificial Intelligence*, *120*, 105899. DOI: 10.1016/j.engappai.2023.105899

Ali, F., Rehman, A., Hameed, A., Sarfraz, S., Rajput, N. A., & Atiq, M. (2024). Climate Change Impact on Plant Pathogen Emergence: Artificial Intelligence (AI) Approach. In Abd-Elsalam, K. A., & Abdel-Momen, S. M. (Eds.), *Plant Quarantine Challenges under Climate Change Anxiety*. Springer., DOI: 10.1007/978-3-031-56011-8_9

Amulothu, D. V. R. T., Rodge, R. R., Hasan, W., & Gupta, S. (2024). Machine Learning for Pest and Disease Detection in Crops. In *Agriculture 4.0* (pp. 111–132). CRC Press., DOI: 10.1201/9781003570219-6

Atapattu, A. J., Perera, L. K., Nuwarapaksha, T. D., Udumann, S. S., & Dissanayaka, N. S. (2024b). Challenges in Achieving Artificial Intelligence in Agriculture. In Chouhan, S. S., Saxena, A., Singh, U. P., & Jain, S. (Eds.), *Artificial Intelligence Techniques in Smart Agriculture*. Springer., DOI: 10.1007/978-981-97-5878-4_2

Atapattu, A. J., Ranasinghe, C. S., Nuwarapaksha, T. D., Udumann, S. S., & Dissanayaka, N. S. (2024a). Sustainable Agriculture and Sustainable Development Goals (SDGs). In *Emerging Technologies and Marketing Strategies for Sustainable Agriculture* (pp. 1-27). IGI Global. https://doi.org/DOI: 10.4018/979-8-3693-4864-2.ch001

Bhatti, U. A., Bhatti, M. A., Tang, H., Syam, M. S., Awwad, E. M., Sharaf, M., & Ghadi, Y. Y. (2024). Global production patterns: Understanding the relationship between greenhouse gas emissions, agriculture greening and climate variability. *Environmental Research*, *245*, 118049. DOI: 10.1016/j.envres.2023.118049 PMID: 38169167

Bhoi, S. K., Jena, K. K., Panda, S. K., Long, H. V., Kumar, R., Subbulakshmi, P., & Jebreen, H. B. (2021). An Internet of Things assisted Unmanned Aerial Vehicle based artificial intelligence model for rice pest detection. *Microprocessors and Microsystems*, *80*, 103607. DOI: 10.1016/j.micpro.2020.103607

Bunde, A. O. (2024). The Digital Platform Economy Revolution And Entrepreneurship Ecosystem in Kenya. *Available atSSRN* 4954030. https://dx.doi.org/DOI: 10.2139/ssrn.4954030

Chandel, N., Kumar, A., & Kumar, R. (2024). Towards Sustainable Agriculture: Integrating Agronomic Practices, Environmental Physiology and Plant Nutrition. *International Journal of Plant and Soil Science*, *36*(6), 492–503. DOI: 10.9734/ijpss/2024/v36i64651

Dara, R., Hazrati Fard, S. M., & Kaur, J. (2022). Recommendations for ethical and responsible use of artificial intelligence in digital agriculture. *Frontiers in Artificial Intelligence*, *5*, 884192. DOI: 10.3389/frai.2022.884192 PMID: 35968036

Das, S. (2024). Transforming Agriculture: Harnessing Robotics and Drones for Sustainable Farming Solution. *Journal of Experimental Agriculture International*, *46*(7), 219–231. DOI: 10.9734/jeai/2024/v46i72577

Dawn, N., Ghosh, S., Saha, A., Chatterjee, S., Ghosh, T., Guha, S., Sarkar, S., Mukherjee, P., & Sanyal, T. (2024). A Review on Digital Twins Technology: A New Frontier in Agriculture. In *Artificial Intelligence and Applications* (Vol. 2, No. 4, pp. 250-262). https://doi.org/DOI: 10.47852/bonviewAIA3202919

Dhillon, R., & Moncur, Q. (2023). Small-scale farming: A review of challenges and potential opportunities offered by technological advancements. *Sustainability (Basel)*, *15*(21), 15478. DOI: 10.3390/su152115478

Doshi, S., Jangir, S., & Gohil, P. (2024). Role of Blockchain Technology in Enhancing Supplychain Traceability, Transparency and Efficiency. *Journal of Experimental Agriculture International*, *46*(5), 636–653. DOI: 10.9734/jeai/2024/v46i52419

Elbasi, E., Zaki, C., Topcu, A. E., Abdelbaki, W., Zreikat, A. I., Cina, E., Shdefat, A., & Saker, L. (2023). Crop prediction model using machine learning algorithms. *Applied Sciences (Basel, Switzerland)*, *13*(16), 9288. DOI: 10.3390/app13169288

Elufioye, O. A., Ike, C. U., Odeyemi, O., Usman, F. O., & Mhlongo, N. Z. (2024). Ai-Driven predictive analytics in agricultural supply chains: a review: assessing the benefits and challenges of ai in forecasting demand and optimizing supply in agriculture. *Computer Science & IT Research Journal*, *5*(2), 473–497. DOI: 10.51594/csitrj.v5i2.817

Escribà-Gelonch, M., Liang, S., van Schalkwyk, P., Fisk, I., Long, N. V. D., & Hessel, V. (2024). Digital Twins in Agriculture: Orchestration and Applications. *Journal of Agricultural and Food Chemistry*, *72*(19), 10737–10752. DOI: 10.1021/acs.jafc.4c01934 PMID: 38709011

Farooq, M. A., Gao, S., Hassan, M. A., Huang, Z., Rasheed, A., Hearne, S., Prasanna, B., Li, X., & Li, H. (2024). Artificial intelligence in plant breeding. *Trends in Genetics*, *40*(10), 891–908. Advance online publication. DOI: 10.1016/j.tig.2024.07.001 PMID: 39117482

Gikunda, K. (2024). Harnessing Artificial Intelligence for Sustainable Agricultural Development in Africa: Opportunities, Challenges, and Impact. *arXiv preprint* arXiv:2401.06171. https://doi.org//arXiv.2401.06171DOI: 10.48550

Giller, K. E., Delaune, T., Silva, J. V., Descheemaeker, K., van de Ven, G., Schut, A. G., van Wijk, M., Hammond, J., Hochman, Z., Taulya, G., Chikowo, R., Narayanan, S., Kishore, A., Bresciani, F., Teixeira, H. M., Andersson, J. A., & van Ittersum, M. K. (2021). The future of farming: Who will produce our food? *Food Security*, *13*(5), 1073–1099. DOI: 10.1007/s12571-021-01184-6

Guebsi, R., Mami, S., & Chokmani, K. (2024). Drones in Precision Agriculture: A Comprehensive Review of Applications, Technologies, and Challenges. *Drones (Basel)*, *8*(11), 686. DOI: 10.3390/drones8110686

Gupta, N., & Gupta, P. K. (2024). Robotics and Artificial Intelligence (AI) in Agriculture with Major Emphasis on Food Crops. In Priyadarshan, P. M., Jain, S. M., Penna, S., & Al-Khayri, J. M. (Eds.), *Digital Agriculture*. Springer., DOI: 10.1007/978-3-031-43548-5_19

Hacker, P. (2024). Sustainable AI regulation. *Common Market Law Review*, *61*(2), 345–386. Advance online publication. DOI: 10.54648/COLA2024025

Hadeed, M. Z., Ali, A., Malik, A., Raza, A., & Shoaib, M. (2024). Harnessing AI and GIS Technologies for Climate-Resilient Agriculture and Environmental Sustainability. In *Maintaining a Sustainable World in the Nexus of Environmental Science and AI* (pp. 333-364). IGI Global. AI-Powered Farm Management by IBM Watson. https://doi.org/DOI: 10.4018/979-8-3693-6336-2.ch013

Javaid, M., Haleem, A., Khan, I. H., & Suman, R. (2023). Understanding the potential applications of Artificial Intelligence in Agriculture Sector. *Advanced Agrochem, 2*(1), 15–30. DOI: 10.1016/j.aac.2022.10.001

Karunathilake, E. M. B. M., Le, A. T., Heo, S., Chung, Y. S., & Mansoor, S. (2023). The path to smart farming: Innovations and opportunities in precision agriculture. *Agriculture, 13*(8), 1593. DOI: 10.3390/agriculture13081593

Kaushal, S. (2024). Enhancing agricultural feedback analysis through Voice User Interface and Deep Learning Integration (*Doctoral dissertation, Kansas State University*). https://hdl.handle.net/2097/44460

Kebe, A. A., Hameed, S., Farooq, M. S., Sufyan, A., Malook, M. B., Awais, S., Riaz, M., Waseem, M., Amjad, U., & Abbas, N. (2023). Enhancing Crop Protection and Yield through Precision Agriculture and Integrated Pest Management: A Comprehensive Review. *Asian Journal of Research in Crop Science, 8*(4), 443–453. DOI: 10.9734/ajrcs/2023/v8i4225

Khang, A., Rath, K. C., Panda, S., Sree, P. K., & Panda, S. K. (2023). Revolutionizing agriculture: exploring advanced technologies for plant protection in the agriculture sector. In *Handbook of Research on AI-Equipped IoT Applications in High-Tech Agriculture* (pp. 1–22). IGI Global., DOI: 10.4018/978-1-6684-9231-4.ch001

Khanpara, B. M., Patel, B. P., Parmar, N. B., & Mehta, T. D. (2024). Transforming Agriculture with Drones: Applications, Challenges and Implementation Strategies. *Journal of Scientific Research and Reports, 30*(8), 792–802. DOI: 10.9734/jsrr/2024/v30i82299

Konya, A., & Nematzadeh, P. (2024). Recent applications of AI to environmental disciplines: A review. *The Science of the Total Environment, 906*, 167705. DOI: 10.1016/j.scitotenv.2023.167705 PMID: 37820816

Kopeć, P. (2024). Climate Change—The Rise of Climate-Resilient Crops. *Plants, 13*(4), 490. DOI: 10.3390/plants13040490 PMID: 38498432

Krishnababu, M. E., Devi, B. R., Soni, A., Panigrahi, C. K., Sudeepthi, B., Rathi, A., & Shukla, A. (2024). A Review on Precision Agriculture Navigating the Future of Farming with AI and IoT. *Asian Journal of Soil Science and Plant Nutrition, 10*(2), 336–349. DOI: 10.9734/ajsspn/2024/v10i2291

Kwakye, J. M., Ekechukwu, D. E., & Ogundipe, O. B.Jephta Mensah KwakyeDarlington Eze EkechukwuOlorunshogo Benjamin Ogundipe. (2024). Policy approaches for bioenergy development in response to climate change: A conceptual analysis. *World Journal of Advanced Engineering Technology and Sciences*, *12*(2), 299–306. DOI: 10.30574/wjaets.2024.12.2.0292

Lakhiar, I. A., Yan, H., Zhang, C., Wang, G., He, B., Hao, B., Han, Y., Wang, B., Bao, R., Syed, T. N., Chauhdary, J. N., & Rakibuzzaman, M. (2024). A review of precision irrigation water-saving technology under changing climate for enhancing water use efficiency, crop yield, and environmental footprints. *Agriculture*, *14*(7), 1141. DOI: 10.3390/agriculture14071141

Limpamont, A., Kittipanya-ngam, P., Chindasombatcharoen, N., & Cavite, H. J. M. (2024). Towards agri-food industry sustainability: Addressing agricultural technology adoption challenges through innovation. *Business Strategy and the Environment*, *33*(7), 7352–7367. DOI: 10.1002/bse.3871

Lu, K., Yan, W., & Tan, H. (2024). Applications and recent advances in artificial intelligence for soil data analysis. *Advances in Resources Research*, *4*(4), 668–680. DOI: 10.50908/arr.4.4_668

Majumdar, P., Mitra, S., Bhattacharya, D., & Bhushan, B. (2024). Enhancing sustainable 5G powered agriculture 4.0: Summary of low power connectivity, internet of UAV things, AI solutions and research trends. *Multimedia Tools and Applications*, *84*(17), 1–45. DOI: 10.1007/s11042-024-19728-1

Mallinger, K., & Baeza-Yates, R. (2024). Responsible AI in Farming: A Multi-Criteria Framework for Sustainable Technology Design. *Applied Sciences (Basel, Switzerland)*, *14*(1), 437. DOI: 10.3390/app14010437

Marla, A., Paul, R., Saha, A. K., Basha, N. K., & Anandhakrishnan, B. (2023, September). An AgroBot: Natural Language Processing Based Chatbot for Farmers. In *2023 4th International Conference on Smart Electronics and Communication (ICOSEC)* (pp. 1235-1241). IEEE. https://doi.org/DOI: 10.1109/ICOSEC58147.2023.10276356

Mohammad, A., & Mahjabeen, F. (2023). Revolutionizing solar energy: The impact of artificial intelligence on photovoltaic systems. *International Journal of Multidisciplinary Sciences and Arts*, *2*(1). Advance online publication. DOI: 10.47709/ijmdsa.v2i1.2599

Mondal, S., & Palit, D. (2022). Challenges in natural resource management for ecological sustainability. In *Natural Resources Conservation and Advances for Sustainability* (pp. 29–59). Elsevier., DOI: 10.1016/B978-0-12-822976-7.00004-1

Moorsom, T. L., Rao, S., Gengenbach, H., & Huggins, C. (2020). Food security and the contested visions of agrarian change in Africa. *Canadian Journal of Development Studies. Canadian Journal of Development Studies*, *41*(2), 212–223. DOI: 10.1080/02255189.2020.1786356

Musa, I. O., Samuel, J. O., Adams, M., Abdulsalam, M., Nathaniel, V., Maude, A. M., Adedayo, O. A., & Tiamiyu, A. G. T. (2024). Soil Erosion, Mineral Depletion and Regeneration. In: Aransiola, S.A., Babaniyi, B.R., Aransiola, A.B., Maddela, N.R. (eds) *Prospects for Soil Regeneration and Its Impact on Environmental Protection*. Earth and Environmental Sciences Library. Springer, Cham. https://doi.org/ DOI: 10.1007/978-3-031-53270-2_7

Naveed, M., Zahid, N., Fatima, I., Saleem, A., Majeed, M., Abid, A., Javed, K., Wazir, R., & Qasim, A. (2024). Sustainable AI-Driven Applications for Plant Care and Treatment. In: Khamparia, A., Pandey, B., Pandey, D.K., Gupta, D. (eds) *Microbial Data Intelligence and Computational Techniques for Sustainable Computing*. Microorganisms for Sustainability, vol 47. Springer, Singapore. https://doi.org/DOI: 10.1007/978-981-99-9621-6_16

Neethirajan, S. (2024). Artificial intelligence and sensor innovations: Enhancing livestock welfare with a human-centric approach. *Human-Centric Intelligent Systems*, *4*(1), 77–92. DOI: 10.1007/s44230-023-00050-2

Nguy n, C. T., Gammatantrawet, N., Susawaengsup, C., Tandee, K., Ramli, A. N. M., Tongkoom, K., Chatsungnoen, T., Dangtungee, R., & Bhuyar, P. (2024). Drought-ready plant resilience: Harnessing nano-biotechnology techniques for swift screening and selection of organic crop varieties. *South African Journal of Botany*, *169*, 553–566. DOI: 10.1016/j.sajb.2024.05.011

Nova, K. (2023). AI-enabled water management systems: an analysis of system components and interdependencies for water conservation. *Eigenpub Review of Science and Technology,* *7*(1), 105-124. https://studies.eigenpub.com/index.php/erst/article/view/12

Onwude, D. I., Chen, G., Eke-Emezie, N., Kabutey, A., Khaled, A. Y., & Sturm, B. (2020). Recent advances in reducing food losses in the supply chain of fresh agricultural produce. *Processes (Basel, Switzerland)*, *8*(11), 1431. DOI: 10.3390/pr8111431

Oriekhoe, O. I., Ashiwaju, B. I., Ihemereze, K. C., Ikwue, U., & Udeh, C. A. (2024). Blockchain technology in supply chain management: A comprehensive review. *International Journal of Management & Entrepreneurship Research*, *6*(1), 150–166. DOI: 10.51594/ijmer.v6i1.714

Pandey, V. K., Srivastava, S., Dash, K. K., Singh, R., Mukarram, S. A., Kovács, B., & Harsányi, E. (2023). Machine Learning algorithms and fundamentals as Emerging Safety Tools in Preservation of fruits and vegetables: A review. *Processes (Basel, Switzerland)*, *11*(6), 1720. DOI: 10.3390/pr11061720

Parra-López, C., Abdallah, S. B., Garcia-Garcia, G., Hassoun, A., Sánchez-Zamora, P., Trollman, H., Jagtap, S., & Carmona-Torres, C. (2024). Integrating digital technologies in agriculture for climate change adaptation and mitigation: State of the art and future perspectives. *Computers and Electronics in Agriculture*, *226*, 109412. DOI: 10.1016/j.compag.2024.109412

Rajendran, V., Debnath, B., Mghames, S., Mandil, W., Parsa, S., Parsons, S., & Ghalamzan-E, A. (2024). Towards autonomous selective harvesting: A review of robot perception, robot design, motion planning and control. *Journal of Field Robotics*, *41*(7), 2247–2279. DOI: 10.1002/rob.22230

Rane, N., Choudhary, S., & Rane, J. (2024). Artificial intelligence for enhancing resilience. *Journal of Applied Artificial Intelligence*, *5*(2), 1–33. DOI: 10.48185/jaai.v5i2.1053

Rane, N. L., Paramesha, M., Rane, J., & Mallick, S. K. (2024). *Policies and regulations of artificial intelligence in healthcare, finance, agriculture, manufacturing, retail, energy, and transportation industry*. Deep Science Publishing., DOI: 10.70593/978-81-981271-1-2_4

Rhoads, J. (2023). Next-Generation Precision Farming Integrating AI and IoT in Crop Management Systems. *AI, IoT and the Fourth Industrial Revolution Review*, 13(7), 1-9. https://scicadence.com/index.php/AI-IoT-REVIEW/article/view/1

Ryan, M. (2023). The social and ethical impacts of artificial intelligence in agriculture: Mapping the agricultural AI literature. *AI & Society*, *38*(6), 2473–2485. DOI: 10.1007/s00146-021-01377-9

Saikanth, D. R. K., Ragini, M., Tripathi, G., Rajesh, K., Giri, A., Pandey, S. K., & Verma, L. K. (2024). The Impact of Emerging Technologies on Sustainable Agriculture and Rural Development. *International Journal of Environment and Climate Change*, *14*(1), 253–263. DOI: 10.9734/ijecc/2024/v14i13830

Samuel-Okon, A. D., & Abejide, O. O. (2024). Bridging the digital divide: Exploring the role of artificial intelligence and automation in enhancing connectivity in developing nations. *Journal of Engineering Research and Reports*, *26*(6), 165–177. DOI: 10.9734/jerr/2024/v26i61170

Sharma, K., & Shivandu, S. K. (2024). Integrating Artificial Intelligence and Internet of Things (IoT) for Enhanced Crop Monitoring and Management in Precision Agriculture. *Sensors International, 100292*, 100292. Advance online publication. DOI: 10.1016/j.sintl.2024.100292

Singh, K. A. P., Goutam, P. K., Xaxa, S., Pandey, S. K., Panotra, N., & Rajesh, G. M. (2024b). The Role of Greenhouse Technology in Streamlining Crop Production. *Journal of Experimental Agriculture International, 46*(6), 776–798. DOI: 10.9734/jeai/2024/v46i62532

Singh, S., Vaishnav, R., Gautam, S., & Banerjee, S. (2024a). Agricultural Robotics: A Comprehensive Review of Applications, Challenges and Future Prospects. In *2024 2nd International Conference on Artificial Intelligence and Machine Learning Applications Theme: Healthcare and Internet of Things (AIMLA)* (pp. 1-8). IEEE. https://doi.org/DOI: 10.1109/AIMLA59606.2024.10531517

Sivakumar, V. G., Baskar, V. V., Vadivel, M., Vimal, S. P., & Murugan, S. (2023). IoT and GIS Integration for Real-Time Monitoring of Soil Health and Nutrient Status. In *2023 International Conference on Self Sustainable Artificial Intelligence Systems (ICSSAS)* (pp. 1265-1270). IEEE. https://doi.org/DOI: 10.1109/ICSSAS57918.2023.10331694

Skendžić, S., Zovko, M., Živković, I. P., Lešić, V., & Lemić, D. (2021). The impact of climate change on agricultural insect pests. *Insects, 12*(5), 440. DOI: 10.3390/insects12050440 PMID: 34066138

Tantalaki, N., Souravlas, S., & Roumeliotis, M. (2019). Data-driven decision making in precision agriculture: The rise of big data in agricultural systems. *Journal of Agricultural & Food Information, 20*(4), 344–380. DOI: 10.1080/10496505.2019.1638264

Tardaguila, J., Stoll, M., Gutiérrez, S., Proffitt, T., & Diago, M. P. (2021). Smart applications and digital technologies in viticulture: A review. *Smart Agricultural Technology, 1*, 100005. DOI: 10.1016/j.atech.2021.100005

Thangamani, R., Sathya, D., Kamalam, G. K., & Lyer, G. N. (2024). AI Green Revolution: Reshaping Agriculture's Future. In Balasubramanian, S., Natarajan, G., & Chelliah, P. R. (Eds.), *Intelligent Robots and Drones for Precision Agriculture. Signals and Communication Technology*. Springer., DOI: 10.1007/978-3-031-51195-0_19

Tian, Z., Wang, J. W., Li, J., & Han, B. (2021). Designing future crops: Challenges and strategies for sustainable agriculture. *The Plant Journal, 105*(5), 1165–1178. DOI: 10.1111/tpj.15107 PMID: 33258137

Tiwari, S., Bhardwaj, B., Arora, D., & Khatri, S. (2024). Challenges and Barriers to Smart Farming Adaptation: A Technical, Economic, and Social Perspective. *Smart Agritech: Robotics, AI, and Internet of Things (IoT) in Agriculture*, 75-111. https://doi.org/DOI: 10.1002/9781394302994.ch4

Tjilen, A. P., Tambaip, B., Dharmawan, B., Adrianus, A., Riyanto, P., & Ohoiwutun, Y. (2024). Engaging stakeholders in policy decision-making for food security governance: Identification, perception, and contribution. *Corporate Governance and Organizational Behavior Review*, *8*(1), 144–154. DOI: 10.22495/cgobrv8i1p12

Upadhyay, A., Zhang, Y., Koparan, C., Rai, N., Howatt, K., Bajwa, S., & Sun, X. (2024). Advances in ground robotic technologies for site-specific weed management in precision agriculture: A review. *Computers and Electronics in Agriculture*, *225*, 109363. DOI: 10.1016/j.compag.2024.109363

Vetrivel, S. C., & Arun, V. P. (2025). Smart Farming and Precision Agriculture Using AI Technologies. In *Real-World Applications of AI Innovation* (pp. 85–106). IGI Global Scientific Publishing., DOI: 10.4018/979-8-3693-4252-7.ch005

Villalba, R., Joshi, G., Daum, T., & Venus, T. E. (2024). Financing Climate-Smart Agriculture: A case study from the Indo-Gangetic Plains. *Mitigation and Adaptation Strategies for Global Change*, *29*(5), 33. DOI: 10.1007/s11027-024-10127-3

Vinod Chandra, S. S., Hareendran, A., & Albaaji, G. F. (2024). Precision farming for sustainability: An agricultural intelligence model. *Computers and Electronics in Agriculture*, *226*, 109386. DOI: 10.1016/j.compag.2024.109386

Yigitcanlar, T., David, A., Li, W., Fookes, C., Bibri, S. E., & Ye, X. (2024). Unlocking Artificial Intelligence Adoption in Local Governments: Best Practice Lessons from Real-World Implementations. *Smart Cities*, *7*(4), 1576–1625. DOI: 10.3390/smartcities7040064

Zhang, L., Guo, W., Lv, C., Guo, M., Yang, M., Fu, Q., & Liu, X. (2024a). Advancements in artificial intelligence technology for improving animal welfare: Current applications and research progress. *Animal Research and One Health*, *2*(1), 93–109. DOI: 10.1002/aro2.44

Zhang, X., Yang, P., & Lu, B. (2024b). Artificial intelligence in soil management: The new frontier of smart agriculture. *Advances in Resources Research*, *4*(2), 231–251. DOI: 10.50908/arr.4.2_231

Ziesche, S., Agarwal, S., Nagaraju, U., Prestes, E., & Singha, N. (2023). Role of Artificial Intelligence in Advancing Sustainable Development Goals in the Agriculture Sector. In Mazzi, F., & Floridi, L. (Eds.), Philosophical Studies Series: Vol. 152. *The Ethics of Artificial Intelligence for the Sustainable Development Goals*. Springer., DOI: 10.1007/978-3-031-21147-8_21

Chapter 8
Artificial Intelligence-Based Landslide Prediction Using Remote Sensing Data

Subha Vaiapury
https://orcid.org/0009-0005-9800-4660
Pondicherry University, India

V. Uma
https://orcid.org/0000-0002-7257-7920
Pondicherry University, India

ABSTRACT

Disaster Management (DM) is preparing for, responding to, and recovering from disasters to reduce the impact on people and property. Landslides are a significant hydrogeological hazard affecting large parts of India. Remote Sensing provides crucial and timely data about natural disasters through spatial and multi-temporal remote sensing inputs. Geographical Information System plays a crucial role in disaster management by providing support for collecting and processing remote sensing data. Advances in AI models enable accurate landslide identification by processing the RS satellite data using GIS. The successful integration of AI, RS, and GIS in landslide detection has paved the way for a standard change in disaster risk reduction and response. This chapter elucidates the significance of integrating RS, GIS, and Artificial Intelligence in landslide disaster prediction. This chapter explains the characteristics and importance of RS data, processes involved in identifying the landslide causative factors, and AI-based object detection principles useful for target landslide prediction.

DOI: 10.4018/979-8-3373-0680-3.ch008

INTRODUCTION

A disaster is a catastrophe due to natural or artificial sources. India is exposed to both natural and man-made disasters. Natural disasters include floods, landslides, cyclones, drought, and earthquakes. DM involves a continuous and combined process of planning, organizing, coordinating, and implementing measures that are necessary for prevention, mitigation, evacuation, and rehabilitation. A distinctive DM continuum includes six elements; the pre-disaster phase includes prevention, mitigation, and preparedness, while the post-disaster phase includes response, rehabilitation, reconstruction, and recovery. In India, limited resources delay disaster identification, posing a significant challenge to DM.

Landslides are one of the most dangerous natural disasters that affect numerous hilly areas, particularly in hills with high slope angles around the world. In India, landslides occur frequently in the Himalayas, the North-eastern hill ranges, the Western Ghats, the Nilgiris, the Eastern Ghats, and the Vindhyas - resulting in substantial loss of life, property, and the change in terrain. Specifically, 12% of hilly areas are at risk caused due to landslides and avalanches. Landslide-influencing factors are categorized into geological, terrain, geomorphological, hydrological, and anthropogenic factors that can be derived from satellite imagery. Landslide-causing factors are slope, elevation, curvature, aspect, land use/land cover, geology, lineament, rainfall, soil type, soil thickness, and distance from road. In 90's and early 2000's, landslide detection are done with field surveys and statistical models. To overcome the limitation of conventional methods, RS and GIS were incorporated with AI technology. Landslide forecasting techniques have significantly progressed over the last two decades with remarkable RS, GIS, and AI advancements.

Satellites, aircraft, and drones are the RS platforms that provide crucial data and make RS a valuable tool in DM. It enables large-area coverage, monitoring vast areas unreachable by traditional methods. High-resolution imagery includes detailed information on terrain changes, while multi-temporal analysis allows tracking changes over time. RS is also cost-effective compared to the costs associated with field surveys. By integrating multiple data sources like the digital elevation model (DEM), thematic maps, and resources-1&2, RS enables timely warnings and evacuations. RS provides unbiased topographic data without human intervention, while GIS is powerful in processing the data and analysing causative factors of landslides.

GIS is a tool capable of predicting landslides because it is capable of handling large amounts of vector, raster data and analysing, displaying results efficiently. It can be useful in landslide prediction by working on DEM raster data and performing preprocessing task, such as merging, clipping etc. to fit the disaster area. Analysing the landslide causative factors such as slope, aspect, Terrain Ruggedness Index (TRI), Topographic Wetness Index (TWI), etc., can be precisely calculated and classified by GIS. GIS software viz. ArcGIS and QGIS are mostly used for landslide monitoring and prediction. Using GIS software, all conditioning factors can be created as a raster with spatial resolution. Important topographic attributes namely slope angle, aspect, profile curvature and plane curvature can be analysed using GIS tools. Geoprocessed data fed to AI models can analyse this data and predict potential landslide areas.

AI enables machines to achieve human intelligence. ML models namely Support Vector Machine (SVM), Logistic Regression (LR), Random Forest (RF), Ensemble model, and Artificial Neural Network (ANN) learn the processed numerical data features. DL models learn the processed image features. Object detection is carried out using Convolutional Neural Network (CNN), Region-Based Convolutional Neural Network (R-CNN), and U-net. Recurrent Neural Network (RNN) carries out multi-temporal analysis useful for landslide forecasting. The potential solution for landslide detection is also achieved through Transformer models since it allows parallel processing and uses attention mechanisms. Advances in Artificial Intelligence, Machine Learning, and Deep Learning models enable accurate landslide identification by processing the RS data in GIS software, reducing response times and mitigating risks compared to traditional approaches. This integration of AI, RS, and GIS enables Disaster Management authorities to predict and respond promptly to landslides, reducing risks and saving lives.

BACKGROUND

Disaster and Its Types

A disaster is a catastrophe due to natural or artificial sources. India is exposed to both natural and man-made disasters. Natural disasters (ND) include floods, landslides, cyclones, drought, and earthquakes. Man-made disasters include deforestation, setting of fires, epidemics and wars. Figure 1 clearly depicts the different types of disasters. Among these disasters this chapter provides the detailed information about landslides, and landslide analysis (prediction/detection/forecasting) using conventional and AI approaches based on landslide causative factors which directs optimal DM activities.

Figure 1. Disaster and its types

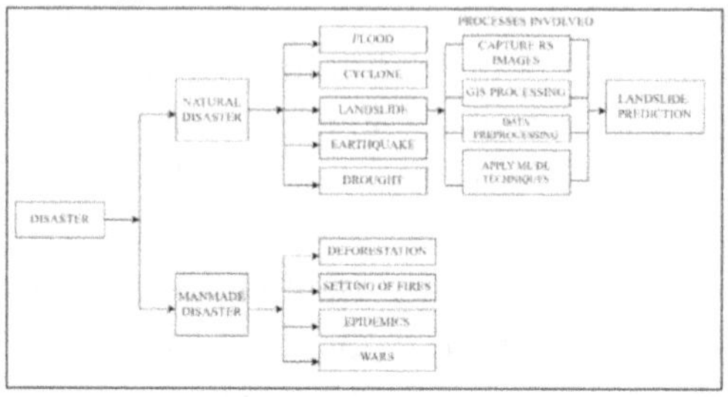

Landslides and Types

Landslides, also known as landslips or rockslides, are the motion of rock or soil down the slopes under the force of gravity, including rockfalls, mudflows, shallow or deep-seated slope failures, and debris flows. Table 1 presents the different types of landslides that occurred in India. Landslide is triggered by torrential/antecedent rainfall, deforestation, climate change, earthquakes, a slope cut to build roads, destabilization of the hill slopes, and cloud bursts.

Table 1. Landslide types

Landslide Types	Description	Occurrence Regions of India
Falls	Occur when debris separates from a steep slope and falls freely	Himalayan region
Topples	Occur due to erosion that make forward rotation of rock around a pivot point	Himalayan region and Western Ghats
Slides	Occur when a block of earth material moves down a slope and it can either be rotational slide or translational slide	Nilgiris, Himalayan region and Northeastern States
Flows	Occur when mass movement of materials (debris, mud, and earthflows)	Kerala, Sikkim and Himachal Pradesh
Creep	Occur due do slow motion of soil/ rock	Western Ghats

Disaster Management

Disaster Management (DM) refers to the process of preparing for, responding to, and recovering from disasters to minimize the impact on people and property. DM involves a consistent and combined process of arranging, marshalling, coordinating, and executing measures that are necessary for prevention, mitigation, evacuation, and rehabilitation.

Disaster management is one of the most demanding applications in the field of emergency assistance and long-term maintenance of refugees and displaced persons (Sara et al., 2015) Disaster management involves the pillars of emergency management: planning and preparation, mitigation, response, and recovery (Bly et al., 2021). Effective DM requires a cyclic approach by integrating these phases in order to reduce disaster risks and impacts, and ensure resilient communities.

A distinctive DM chain consists of six essentials; the pre-disaster phase includes prevention, mitigation, and preparedness, while the post-disaster phase includes response, rehabilitation, reconstruction, and recovery. In India, limited resources delay disaster identification, posing a significant challenge to DM.

In India, limited resources hinder disaster identification posing a significant challenge to DM. To bridge this gap, RS technology plays a vital role, providing crucial and timely data during natural disasters through space and aerial remote sensing inputs. RS is undoubtedly a valuable tool for disaster management, and offers geographical data in various situations. The application of GIS and Spatial Data Management for visualizing and analysing spatial data facilitates informed decision-making and resource allocation during disaster (Khan et al., 2023).

Artificial intelligence offers an innovative approach to enhance DM activities, promising improved disaster prediction, response, and mitigation strategies. By leveraging RS and AI, India can strengthen its disaster management capabilities, ultimately saving lives and reducing damage. AI technology enable rapid and effective action in disaster management activities for disaster-affected countries. The use of AI systems will be effective in minimizing logistical challenges in the intervention and management of natural disasters (Simsek et al., 2023).

The effective DM is accomplished by analysing valid dataset provided by RS. Both the pre-disaster and post-disaster activities of DM for any type of disaster are efficiently accomplished since DM gets the large-scale geospatial information of the specific region as RS data (Lv et al., 2023).

REMOTE SENSING

Remote sensing involves observing an object in the globe or an event that occurs globally through instruments that operate remotely.

Remote Sensing Platforms

There are three types of remote sensing platforms viz. spaceborne platforms, airborne platforms, and ground-based platforms.

- Spaceborne Platforms: Include satellites orbiting the Earth to gather large amounts of remote sensing data. Examples are Earth Observation Satellites and High-Resolution Satellites. The acquired data is helpful for disaster management.
- Airborne Platforms: Include aircraft and drones that work within the troposphere; examples are aircraft and UAVs (Unmanned Aerial Vehicles).
- Ground-Based Platforms: Mobile or stationary systems located on the ground that include handheld sensors, towers, and Ground Penetrating Radar (GPR).

Remote Sensing Data

The RS data must be pre-processed since the raw RS data encompasses numerous misrepresentations data and faulty data. RS device namely satellite/UAV can easily capture the geospatial data. Geospatial information's are available in two formats namely raster data and vector data.

Raster data include numerical aerial photographs, satellite images, DEM, pictures, and thematic maps. RS raster data consists of a pixel (picture elements) organized in grid form; each cell contains a value representing information, such as reflectance, spatial/temporal radiation, temperature, intensity and height values.

Remote sensing data can be procured by a user in the form of a data product. Data products are made available in India to users by the National Data Center (NDC) of the National Remote Sensing Centre (NRSC) of the Indian Space Research Organization, Hyderabad. The remote sensing data products are based on level of processing, output media, and area of coverage.

The digital product formats are LSSOWG (Landsat Ground Station Operators Working Group), Fast Format, GeoTIFF (Geographic Tagged Image File Format), and HDF (Hierarchical Data Format). DEM of the specific study area can be procured from bhukosh portal governed by ISRO.

The shape file and thematic map of particular study can be downloaded from bhuwan portal that is governed by ISRO an USGS international data portal. Rainfall data (annual/monthly/daily) can be procured through India- Water Resource Information System (WRIS) governed by Ministry of Jal Shakti and ISRO.

GIS software namely QGIS and ArcGIS plays a vital role in processing raw RS data by integrating multiple data source such as raster (DEM file), vector (shape file) and thematic map (LULC map).

GEOGRAPHIC INFORMATION SYSTEM

Geographic Information System (GIS) data is geographic information that is referenced to specific locations on Earth. GIS is a high-powered toolbox for real-world problems. It helps in capturing, storing, analysing, managing, manipulating, and visualizing geographically referenced data which can be either spatial or attribute data.

GIS is used to store, retrieve, operate, and analyse geospatial data and output numerical data/images to support decision-making using AI techniques. Shapefile is the standard vector data format used in GIS.

Spatial data falls under two categories, namely, vector and raster. Vector data are further classified into three forms, namely, point, line, and polygon. The vector data are based on (X, Y) coordinates, whereas the raster data comes in pixel format based on grid cells without (X, Y) coordinates.

Vector data is resolution-independent and is not pixel-based. Even enlarged, it does not affect data quality; print vector images in any size SVG (Scalable Vector Graphics), VSTM (Visio Macro-Enabled Drawing Template), and AI (Adobe Illustrator) are some of the vector file formats. Vector data is divided into points, lines (arc), and polygons.

Raster data, also called grid data, represents surface and is cell-based. The resolution of a raster data is based on cell (for example, a 10-meter resolution means that one cell represents a 10 * 10-meter section on the ground) Raster data are stored as a grid of cells, where each cell represents a value.

Raster data is cell-based (DEM, surface temperature, rainfall amounts, and satellite imagery). For instance, each cell in a DEM contains only one value representing surface elevation. Raster graphics are also called bitmaps, which are made up of many small squares or dots/pixels. JPEG (Joint Photographic Experts Group), PNG (Portable Network Graphics), and GIF (Graphics Interchange Format) are some of the raster file formats.

Points are non-adjacent features and have zero dimensions. Point data can show the location of certain features and measure density (number of points per area). Line data represents linear features like roads, highways, and rivers. The line has one dimension and can be used to calculate the length of a feature.

Polygons are used to represent areas like lakes and urban areas, and they have two dimensions that can be used to calculate area.

Different stages in GIS are listed below

- Purpose of study(primary/secondary)
- Data collection
- Data input
- Data correction and verification

- Spatial analysis
- Output and report generation
- Metadata generation and refinement.

The purpose of the study is to define the goal of the GIS project in this primary means original data, and secondary implies existing data. Data Collection is gathering spatial and non-spatial data through RS devices. RS data input, either vector or raster, depending upon the application.

Input data must be verified using a coordinate reference system to check its accuracy. Spatial Analysis to identify patterns and relationships, for instance, include buffer analysis, network analysis, spatial interpolation, etc. The final process generates maps, charts, and tables and documents the dataset details, including source, date, and format.

RS provides the raw spatial data and forms the input for GIS analyses. GIS integrates RS data with other spatial datasets and enables complex spatial analysis functional for landslide prediction by processing different landslide causative factors like slope, aspect, etc.

The output data (numerical and images) from GIS is given as input to the AI model, both machine learning and deep learning, to efficiently predict landslide events in hilly areas. DM, Resource Management, Urban planning, and Environmental monitoring are the applications of GIS. Commonly used GIS software's are ArcGIS, QGIS, and GRASS GIS.

Challenges

- Sometimes the data may be inconsistent and noisy which affect AI model accuracy.
- Processing RS shape file and raster data require high computational resources.
- Domain knowledge both in earth science (RS and GIS) and computer science (AI) is needed for effective implementation of landslide detection.

To efficiently handle the RS data in GIS and for optimal decision-making. Knowing fundamental geospatial information, basic terminologies, and file formats are crucial to correctly analyse the causative factors influencing landslides. The next section of this chapter briefly discusses the geospatial information and landslide causative factors.

GEOSPATIAL INFORMATION

All natural disasters impact geology changes. Geological changes are observed as geospatial information. Geospatial information is acquired using RS technologies and processed in GIS software. Integrated AI and geospatial technologies (GIS and RS) have emerged as powerful tools for DM. Geospatial data (Geodata) defines objects or features (valley/ridges), and events (floods/landslides), with location on the surface of the Earth. It integrates location data (latitude and longitude coordinates) and attribute data (characteristics). Features are characteristics and factors that influences a result.

Basic Geospatial Terminologies

- Geology is derived from Greek words (ge- 'earth' and logia- 'study of'). It defines the structure of the Earth on and below its surface and the evolution of its structure over time.
- Terrain also called as relief or topographical relief is derived from the Latin word (terra - 'earth'). It is a given land surface's measurement and silhouette(shape).
- Term geomorphology is the abstraction of the evolution of terrain and deals with the overall structure of the Earth's surface, particularly the nature, source, growth, arrangement, relationship of landforms, and the history of geologic changes. Term terrain and topography are synonymous.
- Hydrology is the study of water, derived from ancient Greek words (hudor – 'water' and logia- 'study of') is the study of the measure, dispensation, and governance of water on Earth.
- Anthropogenic the term derived from Greek words (anthropo -' having to do with humanity' genic - 'creation' the ecosystem altered by human activities directly or indirectly.
- Geospatial features are the spatial description of a real-world entity and are saved in a spatial database with the following file formats.
    - .idx - indexing file with subtitle index.
    - .prj – projection file, contains projection information with coordinate system.
    - .shp – shape file (vector data) encompasses the geometry for all objects.
    - .sbx and .sbn - contains a spatial index of the objects.
    - .dbf – dBase table file stores feature attributes in a database (table format).
    - . shp.xml - geospatial metadata in xml (eXtensible Markup Language) format.

- .mif – MapInfo Interchange File, stores geometry and attributes for all features.
- .tif/.tiff - Tagged Image File, contains geo location data connected to the image (raster data)
- .cpg – code page, optional text file defines the text character interpreted in shape file.

Landslide Causative Factors

Landslide-causing/causative factors (attributes) that collectively influence landslide hazards are categorized under five geospatial factors: geological factors, terrain factors, geomorphological factors, hydrological factors, and anthropogenic factors as shown in Figure 2.

Figure 2. Geospatial factors and its categories

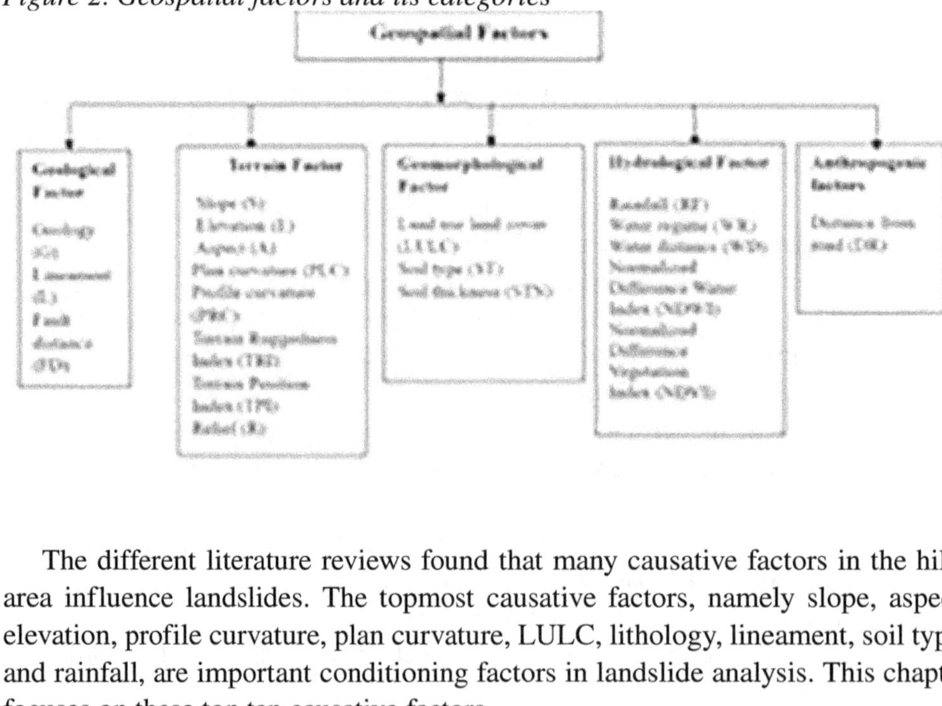

The different literature reviews found that many causative factors in the hilly area influence landslides. The topmost causative factors, namely slope, aspect, elevation, profile curvature, plan curvature, LULC, lithology, lineament, soil type, and rainfall, are important conditioning factors in landslide analysis. This chapter focuses on these top ten causative factors.

Elevation

Elevation is the general altitude of land above the mean sea level. The elevation of a geographic location is its height above or below a fixed reference geoid. Geoid is a mathematical model of the Earth's sea level as an equipotential (potential function is continuous across its entire range) gravitational surface. The angle of elevation is the angle between the line of sight of a viewer and the straight line at their eye level Figure 4. This angle is used to calculate the height of an object.

Elevation is a continuous geographic field measured everywhere, characterized by gradual changes in field values. As shown in the Figure 3 the gradient (slope) can be measured as the change in elevation, calculated as the raise over run or increase in height with respect to distance. This calculation determines the slope of the study area.

Mathematically, elevation of an object is calculated using equation number (1) where, θ is an angle of elevation.

$$\tan \theta = \frac{Opposite\ side}{Adjacent\ side} = \frac{Raise}{Run} \quad \ldots\ldots \quad (1)$$

Figure 3. Elevation

Figure 4. Angle of elevation

Figure 5. DEM of Idukki region

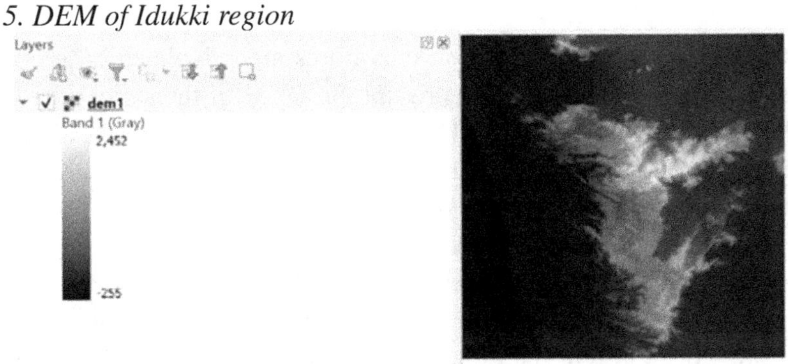

Figure 6. 3×3 moving window

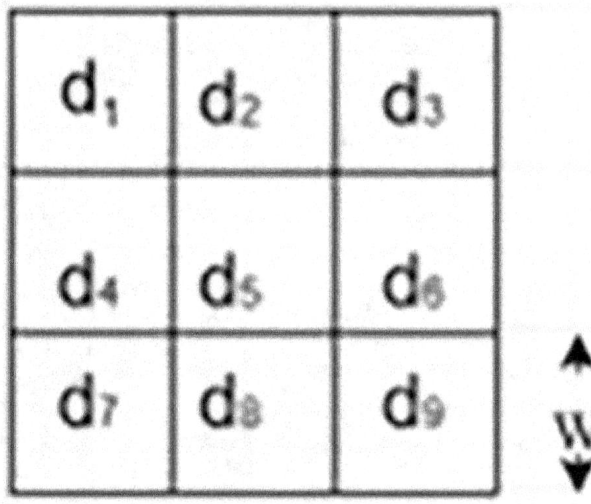

Causative factors like slope, aspect, curvatures are calculated based on DEM (Digital Elevation Model) Figure 5 represent the DEM of the study area which is processed using QGIS software. In a DEM, the grid resolution refers to the spatial resolution or the distance between adjacent elevation points. In Figure 6 the grid resolution(w) depends on the spacing of the elevation points, which is expressed in meters (Example: 10m, or 30m per pixel).

DEM is represented as a raster grid, where each cell has an elevation value. A 3×3 moving window is applied to extract the elevation values surrounding a central point d_c. d_1, d_2...d_9 represent the elevation values at each of the nine positions in the moving window.

Let us consider $d = (c_x, c_y)$ be a position in DEM plane, d represents the elevation at each grid and w denotes spatial resolution (moving window) in DEM.

$$c_x = (d_8 - d_2)/2w \ \ldots\ldots \quad (2)$$

$c_y = (d_6 - d_4)/2w$ ...... (3)

Slope:

The slope of the hilly region is denoted by the symbol '*m*' and represents the percentage change in elevation over a certain distance as shown in Figure 7 and Figure 8. It is also known as grade, gradient, steepness, incline, pitch, and rise. It measures the ratio of vertical change (dy) to horizontal change (dx) and is calculated using equation (3). The slope remains constant anywhere on a straight line.

Figure 7. Slope of hill

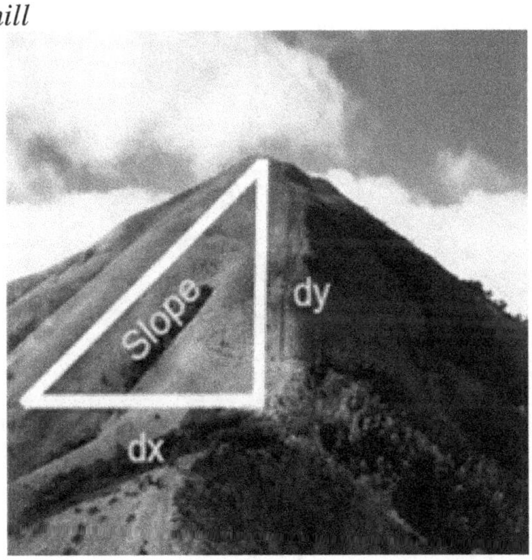

Figure 8. Slope m

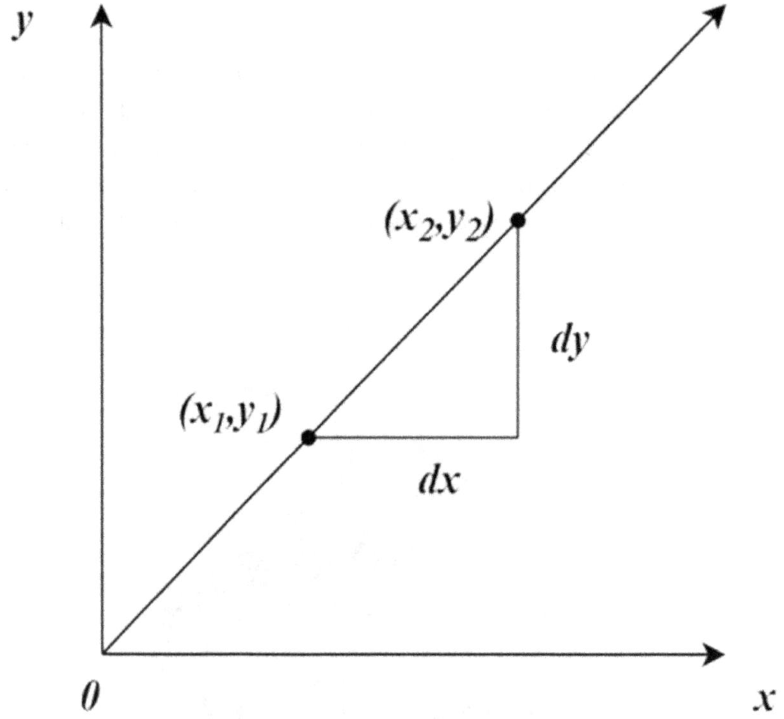

$$m = \frac{(y_2 - y_1)}{(x_2 - x_1)} = \frac{dy}{dx}$$

..... (4)

A significant relationship exists between landslides, geological formation, slope angle, and slope length. Specifically, increasing slope height and length exacerbates slope instability. Investigating the relationships between landslides, geological formation, slope angle, and slope length is essential. Slope stability is calculated based on the slope angle and its strength. Shear force reacts throughout the length of the surface, and it is related to slope; sometimes, the shear force is constant. However, shear strength can vary spontaneously depending on the terrain. Shear strength is directly proportional to slope stability; an increase in shear strength increases slope stability and vice versa, so a decrease in shear strength strongly influences landslides.

Slope angle θ is calculated based on c_x, c_y using (2). Here c_x, $c_{y\,is}$ calculated using equation (2) and (3) which is based on simple difference method.

Slope angle θ = arctan

$$\sqrt{c_x^{\,2} + c_y^{\,2}}$$

...... (5)

here, arctan (tan inverse) is used to measure the angle of a right triangle by using known tangent ratio.

Known tangent ratio is the ratio of the raise to the run of a right triangle. Slope values are calculated based on steepness; the lower slope value, for example, 5°, means the terrain is flatter, making a region less prone to landslide occurrence. If the slope value is higher, for example, 45 ° or above, the terrain becomes steeper, resulting in a region highly prone to landslide occurrence.

Aspect (Direction of the slope)

Aspect is the orientation or direction of a slope measured clockwise from 0° to 360°, with 0°, 90°, 180°, and 270° facing north, east, south, and west, respectively. The slope's orientation is grouped into 45° intervals: North-Northeast (0°-45°), East-Southeast (45°-135°), South-Southwest (135°-225°), and West-Northwest (225°-315°).

Figure 9. Aspect

Figure 10. Aspect orientation

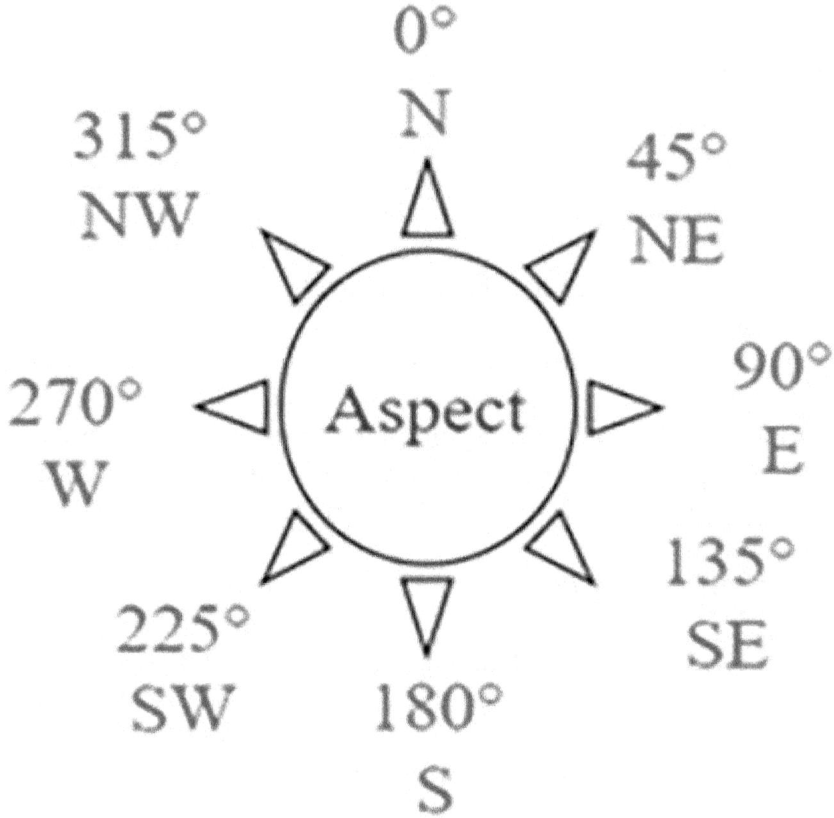

Slope Aspect A =

$270° + \arctan\left(\frac{c_y}{c_x}\right) -$

$90°$

$\frac{c_x}{|c_x|}$

…… (6)

Figure 9 and Figure 10 clearly represent slope aspect of hilly area and its eight orientations. Slope aspect is calculated using equation number (6). From the literature survey it's found that the places with higher aspect values have a relatively large impact on the landslide hazard level.

Among these eight directions, south, southeast, northwest, and west are disposed to a high probability of the occurrence of landslides (Kunwar et al., 2024). Since aspect maps show both the direction and grade of a terrain simultaneously, they are considered as an important factor in the analysis and creation of landslide susceptibility maps.

Curvature

Curvature is the ratio of change in slope and distance. It plays an essential role in slope steadiness by varying stress and strain distribution. Stress refers to the force applied to the hill slope, while strain is a change due to the resultant stress. Curvature is classified in to different types such as plan curvature, planform curvature, profile curvature, tangential curvature, etc.

This chapter focus on plan curvature and profile curvature as depicts in Figure 11.

Profile curvature parallels the slope and controls flow velocity of water, impacting erosion in negative curvature (convex) and deposition in positive curvature (concave). On the other hand, plan curvature is perpendicular to the slope and manages flow of water (merging and separating).

Plan and profile curvature are defined as curvature in a horizontal and vertical plane are calculated using equations (7) and (8). The general curvature is the sum of plan and profile curvature. By measuring plan curvature and profile curvature, we can accurately understand flow patterns across a slope surface, enabling better prediction and management of landslide risks (Alkhasawneh et al., 2013).

Figure 11. Curvature

Plan curvature (hc) and profile curvature (vc) are calculated using equations (7) and (8). Here, d is the set that contains $d_{i,}$ $d = \{d_{1,}\ d_{2\ldots}\ d_9\}$ represents the elevation at each grid and w denotes spatial resolution (moving window) in DEM.

$$\text{Plan Curvature}\left(\text{hc}\right) = \frac{((d_4 + d_6)/2 - d_5)}{2w} \quad \ldots\ldots \quad (7)$$

$$\text{Profile Curvature}\left(\text{vc}\right) = \frac{((d_2 + d_8)/2 - d_5)}{2w} \quad \ldots\ldots \quad (8)$$

Rainfall

In India, the northeastern hilly region and western ghats region are highly prone to landslide occurrence since this region receives antecedent rainfall and torrential rainfall as shown in Figure 12 specifically in monsoon seasons. Predominant wetlands directly affect the slope stability and trigger landslides because of their slope topography.

Figure 12. Kerala's Idukki: Heavy rains have increased the probability of debris flows, soil erosion and landslides

Source:https://maktoobmedia.com/india/13-dead-in-landslide-in-kerala-after-heavy-rain-52-feared-trapped -under-debris

Three categories of rainfall in the hilly region directly trigger landslides: antecedent rainfall, torrential rainfall, and cloudburst rainfall.

- Antecedent rainfall occurs before and during a storm and increases pore pressure in the soil.
- Torrential rainfall refers to hefty rainfall characterized by high intensity (typically >100 mm/h), short duration (minutes to hours), and localized spatial extent.
- Cloudburst is a severe amount of rainfall over a particular region in a short duration, like 50 mm rainfall per minutes.

The rainfall intensity of a region is evaluated using threshold analysis and probabilistic analysis. This threshold analysis is based on the power-law equation, and probabilistic analysis is based on Bayes theorem.

The rainfall threshold is the least amount of rainfall required in a specific time to trigger a landslide. Two basic models are employed to calculate the threshold of landslide-triggering rainfall events: the physical model and the empirical model (Amatya et al., 2022).

For instance, Threshold analysis is measured using equation number (9).

$I(t) = k$

t^{-n}

...... (9)

Where I is the intensity of rainfall, t represents duration, k is the scaling constant which determines the magnitude of the I, and n is an exponent constant (negative value) that represents I decreases as the duration of rainfall increases.

Rainfall data for specific region can be downloaded from India- Water Resource Information System (I-WRIS) portal in which the water data is published by National Water Information Centre (NWIC). Daily rainfall data in mm can be procured from this website based on the Standard Rain Gauge (SRG) located in that specific study area. Rainfall intensity is normally expressed in millimeters per hour.

Land Use/Land Cover

Land use refers to the human activities and purposes for which land is utilized, such as agriculture, urbanization, industrialization etc. Land cover, on the other hand, describes the physical features of the Earth's surface, including forests, water bodies, and wetlands.

Land Use/LandCover (LULC) of the hilly region play a vital role in landslide analysis. Routine human activities like agriculture, road construction and engineering works create direct impact on slope strength. Table 2 show the LULC classes of different study area in India.

LULC map with different classes are generated using Sentinel 2 (Kshetrimayum et al., 2024) and LISS IV (Kaur et al., 2019).

LULC vector data are observed from Resourcesat-2 LISS III (Multi-temporal satellite data) and can be obtained from Bhuvan (Indian Geo platform of ISRO). For instance, the LULC of the Idukki area is broadly classified in to five classes viz. built-up (urban/rural/mining), agriculture (plantation/cropland/fallow), forest (evergreen/deciduous/scrub), barren (unculturable/wastelands/ scrublands) and wetlands (waterbodies/ river/ stream/ reservoir) as shown in the Table 2.

Table 2. LULC class

Class	LULC Class	Area (Sq. Km)
1	Built-up (urban/rural/mining)	162.43
2	Agriculture (plantation/cropland/fallow)	1641.74
3	Forest (evergreen/deciduous/scrub)	2709.34
4	Barren (unculturable/wastelands/ scrublands)	357.91
5	Wetlands (waterbodies/ river/ stream/ reservoir)	147.59

Built-up is modified geomorphological area because of human activities for their sophisticated livelihood by constructing multi-storeyed buildings, resorts, and roads. Buil-up directly disturb the slope stability and trigger landslide in hilly areas.

Agriculture land with proper cultivation process maintains slope stability. On the other side, using pesticides for profitable cultivation cause soil erosion, which directly affects slope stability. The dense vegetation and proper maintenance throughout the cultivation process achieves more slope stability compared to the barren lands. Forest and plantation region are prone to landslide occurrence in western ghats.

Lineament

Lineament is a significant line of landscape caused by joints and faults. It may be continuous or discontinuous and, under certain conditions, observed as the surface index of fault and fracture zones. It is a linear feature of a surface whose part is aligned in a straight-line and probably reflect some surface phenomenon.

Lineament is classified as positive (ridge trend) and negative (river valleys) lineaments. Positive lineament interpreted as dykes and dyke swarms whereas the negative lineament is interpreted as joints, faults and shear zone. Negative lineament is highly prone to landside (Ramli et al., 2010).

In RS, lineaments are represented as lines or linear formation pixels (both lighter and darker) compared to the background pixels. Lineament details can be learned from the RESOURCESAT-1 and 2 with the resolution of 23.5 and 5 meters (Kaur et al., 2019).

Geology

Landslides are caused by internal factors (lithology, geological structure, rock weathering) and external factors (heavy rainfall, earthquakes, human activities). The geology map of the landslide analysis area can be downloaded from the USGS portal and the different classes are classified based on their geological structure [9].

Lithology is the branch of geology that deals with the study of the rocks. Heavy rainfall during monsoon season often erodes the rocks and induces recurrent landslides in the mountainous area.

For example, in the Idukki district geology is classified as Pink granitic gneiss, weathered, Hornblende-biotite gneiss, Charnockite, Biotite gneiss, and Gneissic rock which are highly prone to landslides. In landslide analysis, the geology is evaluated based on the multi class classification method.

Soil Types

Soil genesis/pedogenesis is the study of the origin of soil. Soil development is influenced by five features, namely organisms(biogeochemical nutrient cycling), topography(local/regional scale), climate (temperature/ rainfall), parent material (rock/minerals), and time (temporal outcomes). Figure 13 depicts the general soil development.

Figure 13. Soil formation

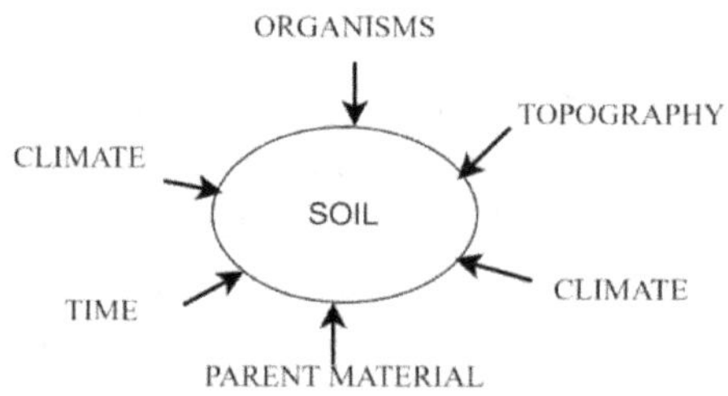

Soils are both penetrable and impenetrable by nature. Some slopes in the hilly regions are characterized by penetrable soils over impenetrable soils. If antecedent rainfall occurs in that hilly region, the top penetrable soil is filled with water. Impenetrable soil generates high water pressure that directly weakens the slope stability and induces shallow landslides.

The topsoil cover on a slope has a significant influence on landslide occurrence. Atterberg limits are values that calculate water content in the soil and is used to regulate soil stability. The soil map of the study area can be generated in GIS from the input thematic map downloaded from government portals like State Disaster Management Authority (SDMA), GSI, or the National Bureau of Soil Survey and Land Use Planning (NBSS&LUP).

Table 3. Soil types and their regions

Regions	Soil types	References
Manipur	Orthic aerisols, Dystric cambisols Ferralic cambisols, Humic cambislos, Dystric gleysols, and Humic gleysols	(Kshetrimayum et al., 2024)
Darjiling	Coarse loamy, fine loamy and loamy	(Chawla et al., 2019)
Sikkim	Clayey soil with naturally formed surface Older well compacted mature soil Sandy soil with naturally formed surface	(Kaur et al., 2019)
Idukki	forest loams, lateritic soils, brown hydromorphic soils and alluvial soils	(District Survey Report Of Minor Minerals. 2016)

The soil thematic map of the study area provides soil parameters like soil types, texture, organic content, pH (Potential of Hydrogen), and depth. Slope stability mainly depends on the soil parameters that influence the landslide in the hilly region. Table 3 provides the different soil types in various hilly areas of India and its references.

RESEARCH OBJECTIVES/HYPOTHESIS

Hypothesis

AI, RS, and GIS integration will improve potential landslide predictions. In achieving the hypothesis the following objectives are framed

1. Acquiring Satellite images using RS.
2. Processing RS data using GIS software and capture landslide causative factors.
3. Predicting and detecting potential landslide using ML and DL models.

CONCEPTUAL MODEL

This study explores landslides/landslips/rockslides, investigating AI-driven solutions for DM to improve monitoring and timely response. Traditional manual methods are time-consuming and unsafe. ML (RF, SVM), and DL (CNN, RNN, YOLO) algorithms enable accurate landslide identification using historical data maps, reducing response times and mitigating risks.

The integration of AI technology and high-resolution RS images procured through the latest technology provide better landslide mapping than the conventional method and overcome its limitations (Yang et al., 2024).

Conventional Approaches

Conventional approaches to landslide detection are field-based approaches, probabilistic likelihood ratio (PLR), fuzzy set model (FM), swarm optimization, etc.

Among these, the *field-based approach* to landslide detection is one of the oldest practices used by the expert group. Since the field-based approach is manual, human intervention is challenging in hilly regions, especially with high pick points, and it may be unsafe for human lives. This approach may be useful for generating statistical reports after the disaster and not for detecting and forecasting landslides.

The FM approach based on Fuzzy Sets was introduced in 1965 by Zadeh. fuzzy relations are distinct parts of L-relations (an abstract algebraic context), where Lis is the unit interval [0,1]. The fuzzy set model was the conventional landslide detection approach through its decision-making abilities. The fuzzy membership (Fm) ranges from 0 to 1, where 1 indicates full importance, and 0 indicates no importance, and its characteristics are entirely different from Boolean logic (Yang et al., 2024).

The fuzzy set model typically includes causative factors (such as slope, rainfall, soil type, aspect, etc.) and assigns them fuzzy membership values to measure the Landslide Risk Index (LRI). For instance, LRI can be calculated using the equation

$$LRI = \sum_{i=1}^{n} w_i * f_i\left(m\right) \quad \ldots\ldots \quad (9)$$

where,

n = number of causative factors

wi = weight assigned to the causative factor

fi(m) = fuzzy membership value of the causative factor, which represents the grade of landslide risk (0 to 1).

Disadvantage

- Fuzzy models do not learn from data; they use static rules and need manual updating.
- Fuzzy models can be computationally expensive for large areas because they require complex rule-based computations.

Swarm optimization is a computational approach similar to the movement of the organisms in an ant colony or bird flock or fish school. Some of the algorithms used for Landslide Detection helps in analysing and classifying satellite images to identify landslide-prone areas are,

- Ant Colony Optimization (ACO),
- Particle Swarm Optimization (PSO),
- Bee Algorithm and Artificial Immune Systems.

(Chawla et al., 2019) employs (PSO-SVM) without feature selection and Ant Colony Optimization (ACO-SVM) approach with feature selection for the classification and prediction of landslide susceptible zones in Darjeeling.

Disadvantage

Swarm Optimization struggles to select the most relevant features because landslide analysis is based on more features (different causative factors).

Parameter tuning is highly sensitive; improper tuning can lead to slow convergence or oscillations in landslide susceptibility mapping compared to the AI model.

To overcome the disadvantages mentioned in the conventional approaches, this chapter focuses on landslide detection with RS data and processing using GIS techniques and ML algorithms.

RESEARCH METHODOLOGY

Artificial Intelligence based LS prediction

Figure 14. AI based model for landslide prediction

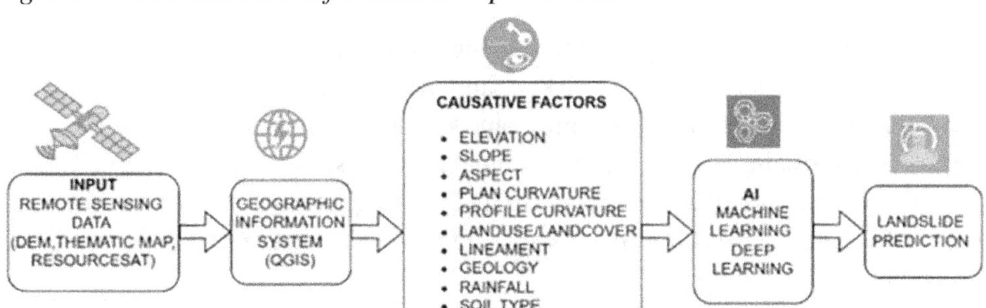

Concepts

RS input data, namely DEM, shape files, and thematic maps can be downloaded from government organizations like ISRO, GSI, and IWRIS. Input data (raster/vector) are pre-processed in GIS, and extract the causative factors of the landslide event from the RS data. For example, terrain attributes like slope, aspect, and curvature are calculated based on DEM's latitude and longitude points.

The output of GIS is numerical data or image data. This output is fed as input to the ML and DL model to predict the landslide. Development in AI techniques (ML and DL) enables accurate landslide identification by processing the RS data in GIS software, reducing response times and mitigating risks compared to traditional approaches.

Figure 14 depicts the integration architecture. AI, RS, and GIS integration enables Disaster Management authorities to predict and respond promptly to landslides, reducing risks and saving lives. AI-powered analysis of RS data with GIS, DM works can shift from reactive to proactive, attenuating the impact of landslides.

Benefits of an integrated approaches are

- Increased accuracy.
- Efficient spatial analysis and pattern recognition.
- Active monitoring and advanced warning.
- Improved data processing.
- Improved Decision-making.

Applications

Applications include real-time monitoring and early warning systems, enabling timely evacuations and mitigating risks. Landslide susceptibility mapping and risk assessment allow for proactive planning and infrastructure development.

Post-disaster damage evaluation and emergency response are also streamlined, facilitating swift relief efforts. By an integrated approach, landslide detection systems can save lives, decrease economic losses, and provision sustainable environmental management.

Machine Learning

Definition:

"Arthur Samuel defined ML as the field of study that allows computers to learn without being explicitly programmed."

"ML, it's the science of getting computers to learn without being explicitly programmed."

Machine learning is primarily used in data Classification and Regression to classify and predict the data.

Three main types of ML are

1. Supervised Machine Learning,
2. Unsupervised Machine Learning and
3. Reinforcement Learning.

Commonly used ML algorithms in landslide prediction are discussed below:

Logistic Regression

Logistic Regression is the single most widely used classification algorithm. An example of binary classification is where there are two possible classes: True/False, Yes/No, or 1/0. Its operation is based on the sigmoid function, which starts very close to zero and slowly builds up or grows to the value of one, and it passes the vertical axis at 0.5.

Sigmoid function $g(s) = \frac{1}{1 + e^{-s}}$

Two steps to compute output using logistic Regression

1. Compute s = w∗x+b
2. Apply sigmoid function "g" to this value s

The logistic Regression makes a prediction depending on s; if wx+b =0, this line is also called the decision boundary because the line is almost neutral. It can learn to fit pretty complex data.

Support vector machine

SVM applications include face detection, image classification, Geo-spatial data-based applications, and text classification. The mathematical insight of SVM is based on dot product (projection of any vector on another vector)

Let us consider vector x vertical to the hyperplane; c refers to the distance of vector w from the origin to the decision boundary. Take projection of x vector on w

- if dot product > c implies the point lies on the right side (positive samples)
- if dot product < c implies the point lies on the left side (negative samples)
- if dot product = c implies the point lies on the decision boundary.

Ensemble learning

Ensemble learning is a powerful method to improve the model's performance and combine the prediction from multiple models. Bagging or Bootstrap Aggregation, Boosting, and Stacking are the different types of ensemble learning.

The main goal of the ensemble model is to combine the predictions of several base evaluators constructed with a given learning procedure to improve strength over unit evaluators.

Random Forest (Supervised Learning)

Random Forest (RF) is based on the Bagging principle and handles overfitting. Sometimes, it increases the bias and reduces variance with less training time. They are called parallel classifiers. Gradient Boosting is based on the Boosting principle and can overfit with reduced bias; they are called sequential classifiers.

RF is a robust tree ensemble algorithm; sampling with replacement procedure causes the RF to explore a lot of small changes to the data and be able to predict future results based on past data.

Deep Learning

Deep learning mimics the human brain, learns to identify objects, and performs complex tasks with increasing accuracy without human intervention. There are several applications for deep learning. In any dataset, each instance is represented using neural network features. Features may be continuous, categorical, or binary. Deep networks have an input, multiple hidden, and output layers.

Each node in the hidden and output layers applies its weight (w), bias (b), and activation function. The Activation function consents the network to train complex relationships between input features and outputs. Sigmoid activation function is the best choice for binary classification problems and softmax activation function for multiclass classification at the output layer.

The hidden layer applies ReLU and tanh activation functions. ReLU, Gradient clipping, batch normalization, and Residual connection solve the vanishing and exploding gradient descent problems during propagation.

For instance, to detect whether the zone has a chance of landslide or not, we consider the input features X by considering landslide conditioning factors like elevation, slope aspect, rainfall, and curvature and adding the two parameters (w) and (b) to predict the output Y. During forward propagation input features are pass to hidden and output layer of neural network and calculate the expected output.

Figure 15. Deep learning architecture

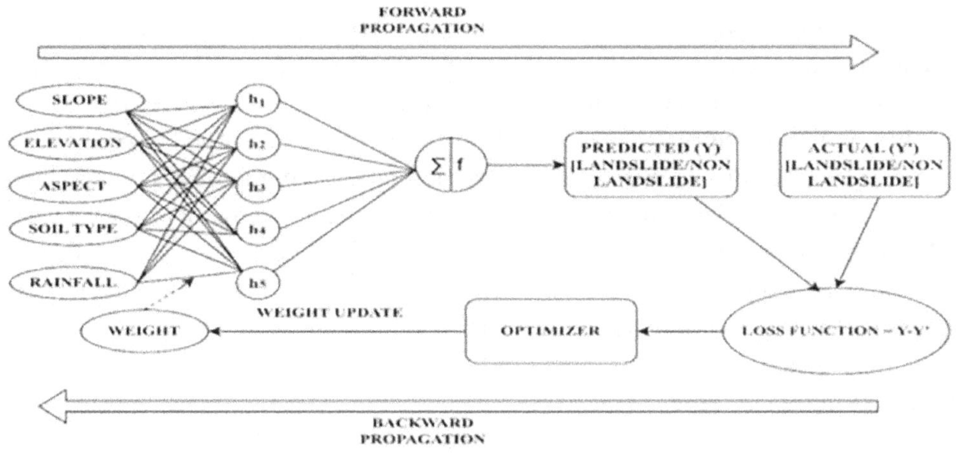

During backward propagation the predicted loss are passed backwards by derivative operation to adjust (w) value to get the true value as shown in Figure 15.

For investigational settings the datasets sample are divided in the ratio of 70%,20% and 10% representing training, validation and testing respectively. Each experiment is to be conducted for several epochs, and the evaluation metrics like F1-score, recall, precision and accuracy are to be calculated using confusion matrix. Epoch constitute the number of times that the entire training dataset is passed the network. Batch size represents number of training samples that are fed to the neural network at once.

Convolutional Neural Network

A typical CNN comprises of four components: convolutional layers, activation layers (sigmoid/ReLU), pooling layers (maximum/average), and fully connected layers. CNN performs the convolutional operations on an input image represented as a grid of pixel values using filters or kernels.

Three key parameters of CNN are kernel size, stride, and padding. The kernel or filter is a small matrix (3*3) used to extract local features from the input image slides across the input and forms an output feature map by scalar product calculation. Stride refers to the step size of the kernel.

Dropout is the regularization process in CNN that arbitrarily turns off neurons during training to avoid overfitting. The two essential characteristics of CNN are sparse connectivity and weight sharing.

Object Detection

The YOLO X model is used in landslide analysis to improve landslide prediction efficiency and accuracy. Object detection: The location of the landslide in the input image is predicted using bounding boxes and class labels. YOLO system detects objects in a single pass and provides fast and precise object detection and segmentation.

A single convolutional network concurrently predicts numerous bounding boxes and class probabilities for those boxes. YOLO model trains on entire images and directly optimizes detection performance. The backbone network consists of convolutional and pooling layers to capture semantic information from images by increasing the number of feature map channels.

The neck includes attention modules to enhance object detection capabilities by integrating the features of the backbone network. The final detection network of YOLOX consists of a sequence of convolutional and activation for extracting information and converting feature maps into object detection.

The U-Net

In 2015, Olaf Ronneberger, Philipp Fischer, and Thomas Brox developed U-Net architecture. The architecture consists of a contracting part (encoder) and an expansive part (decoder), which makes u-shaped architecture.

U-Net architecture is similar to an encoder-decoder model; the contracting path decreases the spatial dimensions of the image, whereas the expanding path aims to up-sample the feature map and produce a pertinent segmentation map. The contracting path works as a convolutional network with recurrent convolution, uniquely followed by the activation function (ReLU- Rectified Linear Unit) and pooling (max pooling) layer.

The contracting path extracts the feature information. It generates high-resolution features and concatenations with the expansive part through a sequence of up-sampling that combines feature and spatial data. The skip connections allow the data to be communicated from the contracting path to the expanding path and preserve high-level and low-level features.

The U-Net model can provide precise and accurate segmentations by up sampling and refining the features learned during the encoding step. The feature extractor or backbone extracts the high-level features from the input.

Recurrent Neural Network: Time Series Analysis

RNN is a class of deep learning model used for time series analysis and processing of sequential data, specifically it processes data across multiple time steps. Recurrent unit is the basic block of RNN, which is updated at each time step based on present input and the preceding hidden state. Hidden layers are the main features of RNN, which remember the sequence pattern for the prediction.

Vanishing and exploding gradient descent problems are the most significant problems in deep neural networks. During backpropagation RNN is unable to hold information of more extended sequence; thus, RNN have short-term memory. Both Long Short-Term Memory (LSTM) and Gated Recurrent Unit (GRU) maintain the relevant information of the model by having gates as an internal mechanism. For instance, GRU and LSTM are particularly effective in forecasting a variety of landslide deformation due to their capacity to capture nonlinear dynamics in time-series data (Ma et al., 2024).

Transformer

The Transformer model, presented in the paper "*Attention is All You Need*" (Vaswani et al., 2017), has reshaped the field of natural language processing (NLP). The Central part of the Transformer's network is the attention mechanism, which allows models to focus on particular parts of the input sequence when producing each element of the output sequence.

The transformer model enables parallel processing. The encoder and decoder constitute the model, and a SoftMax activation function is employed to normalize the output probabilities. Since input is given in parallel with the attention model global feature extraction is possible quickly and predicts more accurate results.

Vision Transformer

ViT, processing an image through a transformer-based architecture to perform tasks like classification, detection, or segmentation. The attention mechanism and embedding (patch and positional) blocks of ViT architecture play a vital role in processing input images and extracting meaningful features.

Basic Operations

Patch splitting is the process of dividing the input images (height×width×channel) into non-overlapping patches, typically of size P×P, and flattening each patch into a 1-dimensional vector. Patch embedding is comprised of linear projection in which each patch is passed through a learnable projection layer and positional encodings to positional information. [CLS] refers to the "classification token," a unique token added to the input sequence to aggregate information for classification tasks. [CLS] token is prepended to the sequence of patch embeddings.

The transformer layers contain a Multi-Head Self-Attention (MHSA) that calculates relationships between all patches in the sequence. Feed-Forward Network (FFN) is applied independently to each embedding, with an anon-linear activation function. MHSA and FFN are wrapped with residual connections and LayerNorm for faster convergence.

After passing through multiple transformer layers, the [CLS] token contains a globally aggregated image representation and is used for downstream tasks. Finally, the [CLS] token output is passed through a Multi-Layer Perceptron (MLP) head for final prediction, which depends on the application, such as classification, regression, and segmentation.

LITERATURE REVIEW

Gong Cheng et al. (2024) employ deep learning recognition models such as AlexNet, YOLO, Mask R-CNN, U-Net, Transformer, and PSPNet in landslide identification with the Bijie landslide dataset, open landslide datasets. The author compares the DL model with ML and concludes that DL has potential landslide recognition by obtaining data from InSAR and LiDAR RS.

Pengyuan Lv et al. (2023) proposed a shape-enhanced vision transformer. The ShapeFormer, a PVT model, extracts multiscale deep features from the original imagery based on a self-attention operation to detect landslides. The PVT model results were compared with DL models on two public datasets, the Bijie and Nepal datasets.

Kushanav Bhuyan et al. (2023) designed standard multi-temporal (MT) landslide inventories using RS data from RapidEye images and generated MT landslide inventories in Nepal. Analysed this dataset in a DL model like U-Net and detected the landslide efficiently. Ghorbanzadeh et al. (2019) employ optical data from RapidEye images and processing using both ML (RF and SVM) and DL (ANN and CNN) models to detect landslides.

Soumik Saha et al. (2023) created the real-time dataset from RS images downloaded from ASTER DEM, Landsat 8 satellite, the GSI website, and Google Earth Pro. Employs RF, SVM, and Neural Network models to precisely describe landslide susceptibility zones of the Garhwal Himalaya with limited datasets.

Table 4. Summary of landslide prediction techniques.

Year [References]	Dataset	Objective	Techniques	Issues	Merits	Results
2024 Gong Cheng	Bijie landslide dataset CAS landslide dataset	To detect and classify the landslide feature using deep learning.	AlexNet, ResNet, YOLO, Mask R-CNN, U-Net, Transformer, EfficientNet, DeeplabV3+ and PSPNet	YOLO optimization model still needs optimized according to real time scenarios.	Improves recognition accuracy and efficiency for active landslides.	The accuracy of the UNet model is higher than the other DL models
2024 Arvindd Kshetrimayum	Bhukosh is a gateway to all geoscientific data of Geological Survey of India. Sentinel datasets.	To compare conventional models with machine learning and deep learning methods to determine classification accuracy and success rate.	Analytical Hierarchy Process (AHP), Frequency Ratio (FR), K-Nearest Neighbors (KNN), RF, SVM and DL models.	Computational complexity (training and testing time) of DL is higher than the KNN, RF, and SVM.	Based on the DL Landslide susceptibility map, it is recommended that the DL be implemented as the overall best strategy for a landslide hazard mitigation plan.	The research found that the DL model, which has an overall accuracy of 97.2%, is the most attractive.
2023 Soumik Saha	Source: ASTER DEM Landsat 8 satellite image GSI website Google earth pro	To precisely delineate landslide susceptibility zones of the Garhwal Himalaya.	SVM RF bagging, ANN DLNN (Deep Learning Neural Network)	Limited datasets.	Support large scale landslide management.	The DLNN technique has high capability (AUC = 0.925) and accuracy for landslide area differentiation.
2023 Pengyuan Lv	Bijie dataset and the Nepal dataset	A shape-enhanced vision transformer (ShapeFormer) model PVT model	Pyramid Vision Transformer (PVT)model ResUNet DeepLabv3+ TransUNet FC-DenseNet PVT-UNet.	Bilinear interpolation is fast and convenient, it will misplace some details during up sampling of the decoder.	The proposed ShapeFormer extracts shape and maintain the boundary information.	DL model attains the highest values in the recall, F1-score, and IoU.
2019 O. Ghorbanzadeh	Satellite imageries RapidEye Data and field survey	To analyse the potential ML and DL models for landslide detection	Artificial neural network (ANN), SVM and RF Convolution neural networks (CNNs)	The topographical information did not improve the results.	CNN recognize complex image patterns and classify landslide efficiently	CNN yielded the best landslide detection results with the highest mIOU value of 78.26%.

continued on following page

Table 4. Continued

Year [References]	Dataset	Objective	Techniques	Issues	Merits	Results
2023 Kushanav Bhuyan	Satellite Imageries Rapideye Images.	To automatically detect and map landslides over time (multi-temporal)	Transfer learning U-Net model	Lack of training data.	The transfer learning model generating annual inventories.	The finest F1-score is only 58% and total landslide detected in 2015 is 108.
2024 Kinh Bac Dang	Both field and Sentinel-2images Google Earth Pro 3Dmodels	To develop DL models capable of automatically identifying landslides from (optical) images	UNet using four S2-bands, UNet3+ using four S2-bands	Limited input samples.	Improved frequency and spatial resolution of landslide detection.	U-Net model attaining an impressive accuracy rate of 99%.
2024 Yong Song	Multi-source geo-environmental data Google Earth images	To provide technical support for mitigating and preventing the impacts of these land slide disasters and to explore the advantages of stacking, voting and meta-learning.	Ensemble technique Stacking, Voting and Meta-learning the Multi-Layer Perceptron RF, Light Gradient Boosting Machine (LGBM), and MLP.	Imbalanced datasets and complex variable interactions.	Provides technical support for disaster mitigation efforts.	Stacking Classifier models, attaining the highest F1 Score of 0.846 and AUC (Area Under ROC Curve) of 0.923,
2022 Byron Guerrero	Google Earth, Satellite Image.	To govern the most influential causative factors in the occurrence of landslides with the use of time windows.	SVM and RF Logistic Regression, Multilayer perceptron (MLP) and Naive-Bayes.	The variables fluctuate in different scales.	Results provide technical support for decision making in DM	The highest accuracy corresponds to MLP with 99.6% is achieved.

Arvind Kshetrimayum et al. (2024) proved that the DL model provides better Landslide susceptibility map results than conventional methods like Frequency Ratio (FR). They created their dataset by getting data from Bhukosh, the Geological Survey of India, and Sentinel. Kinh Bac Dang et al. (2024) generated a dataset using fusion data of Sentinel-1 (SAR) and Sentinel-2 (optical) images employ DL models(U-Net) to detect landslides.

A literature review of landslide research, as listed in the Table 4 shows that the dataset generation for the study area is crucial in landslide research, and datasets are not publicly available. Only limited datasets are there. So, the integration of AI techniques with RS and GIS can provide timely data about landslides to DM.

AI techniques ML and DL have emerged as potential tools for improving the efficiency and accuracy of predicting landslides in hilly regions with limited datasets. RNN provides accurate output with time series data compared to the other DL model. A Vision Transformer (ViT) model with an attention mechanism provides better prediction of RS images.

RESULTS/DISCUSSION

From the literature survey, the U-Net model has achieved better accuracy with an impressive accuracy rate of 99% tested with both field datasets and RS datasets in Vietnam. Followed by 97.2% accuracy obtained using the DL model working with RS data downloaded from the BUKOSH website for Manipur state.

The best score was achieved in landslide detection using the U-Net model in the Rapideye Satellite dataset. The ensemble learning achieved the highest F1 Score of 0.846 and AUC of 0.923 in predicting landslide susceptibility maps using integrated datasets (DEM, Sentinel-2, Geological Map, and Google Earth image) for Poyang country.

CNN model yielded the best landslide detection results with the highest mIOU value of 78.26% using the USGS and CAS datasets. The highest accuracy corresponds to MLP, with 99.6% achieved in RS numerical data in Ecuador.

The integrated AI approach for landslide prediction does not follow the common dataset since the earth's terrain surface varies in latitudes and longitudes. The use of different datasets depends upon the availability and consideration of causative factors. The RS data may be a shape file or a .csv file. Working with numerical data achieved better accuracy than RS image data. From the literature survey it is clear DL models like YOLO and its latest versions, U-Net, and variants of Vision Transformer with attention mechanism provides optimal landslide prediction models.

Research Gap:

1. Landslide repository for specific regions using RS and GIS.
2. Multi-temporal dataset for effective prediction adaptable to all climatic conditions.
3. Inadequate RS image dataset.

FURTHER RESEARCH

Since the landslide prediction depends entirely on the study area's latitude and longitude, a region-specific repository for landslide slide prediction is most needed in DM for planning and mitigation. The diverse climatic conditions in the mountainous region bring unsuspected torrential rainfall, which in turn causes disaster; it may be predicted by using multi-temporal time series data analysed by an AI model. The RS image captured through satellites lacks resolution compared to the RS images captured with UAV. So, choosing the input data depends upon the scenario, such as an emergency during DM, susceptible mapping, or data analysis. The research gap will be resolved by adding variants of ViT networks and layers and incorporating different data resources like optical, microwave, or LiDAR data.

MANAGERIAL IMPLICATIONS

The AI-driven approach offers opportunities for DM to improve decision-making and planning by providing early warning systems for various disasters in hilly regions and potential predictive capabilities. Integrated AI solutions have challenges, including inconsistent data, region specific data, a high-end GPU system to process complex data, and ethical considerations. Addressing these issues is vital in developing a model for landslide prediction.

CONCLUSION

This chapter explored how the integrated AI model based on RS and GIS plays a crucial role in DM. The background study clearly explains the mathematical principle behind the causative factors of landslides and paved the way for a clear idea of how to process the data in GIS after getting it from RS. The variants of different AI models discussed in this chapter provides potential ideas for choosing the correct model for region-specific landslide prediction. This integrated AI approach enables us to create an early warning system before the disaster and helps DM authorities save human lives and provides rehabilitation. For future research, region-specific datasets can be used based on the causative factors of each area, incorporating advanced Deep Learning models for improved landslide prediction and effective disaster management.

REFERENCES

Alkhasawneh, M. Sh., Ngah, U. K., Tay, L. T., Mat Isa, N. A., & Al-batah, M. S. (2013). Mat Isa, N. A., & Al-batah, M. S. (2013). Determination of important topographic factors for landslide mapping analysis using MLP network. *TheScientificWorldJournal, 2013*(1), 415023. Advance online publication. DOI: 10.1155/2013/415023 PMID: 24453846

Bhuyan, K., Meena, S. R., Nava, L., Van Westen, C., Floris, M., & Catani, F. (2023). Mapping landslides through a temporal lens: An insight toward multi-temporal landslide mapping using the U-Net deep learning model. *GIScience & Remote Sensing*, *60*(1), 2182057. DOI: 10.1080/15481603.2023.2182057

Bly, J., Francescutti, L. H., & Weiss, D. (2021). Disaster management: A state-of-the-art review. In Noroozinejad Farsangi, E. (Ed.), *Natural hazards – Impacts, adjustments and resilience*. IntechOpen., DOI: 10.5772/intechopen.94489

Chawla, A., Pasupuleti, S., Chawla, S., Rao, A. C. S., Sarkar, K., & Dwivedi, R. (2019). Landslide susceptibility zonation mapping: A case study from Darjeeling District, Eastern Himalayas, India. *Photonirvachak (Dehra Dun), 47*(3), 497–511. DOI: 10.1007/s12524-018-0916-6

Cheng, G., Wang, Z., Huang, C., Yang, Y., Hu, J., Yan, X., Tan, Y., Liao, L., Zhou, X., Li, Y., Hussain, S., Faisal, M., & Li, H. (2024). Advances in deep learning recognition of landslides based on remote sensing images. *Remote Sensing (Basel)*, *16*(10), 1787. DOI: 10.3390/rs16101787

Ghorbanzadeh, O., Blaschke, T., Gholamnia, K., Meena, S. R., Tiede, D., & Aryal, J. (2019). Evaluation of different machine learning methods and deep-learning convolutional neural networks for landslide detection.

Guerrero-Rodriguez, B., Garcia-Rodriguez, J., Salvador, J., Mejia-Escobar, C., Bonifaz, M., & Gallardo, O. (2022). Landslide prediction with machine learning and time windows. In Ferrández Vicente, J. M., Álvarez-Sánchez, J. R., De La Paz López, F., & Adeli, H. (Eds.), *Bio-inspired systems and applications: From robotics to ambient intelligence* (Vol. 13259, pp. 193–202). Springer., DOI: 10.1007/978-3-031-06527-9_19

Kaur, H., Gupta, S., Parkash, S., Thapa, R., Gupta, A., & Khanal, G. C. (2019). Evaluation of landslide susceptibility in a hill city of Sikkim Himalaya with the perspective of hybrid modelling techniques. *Annals of GIS*, *25*(2), 113–132. DOI: 10.1080/19475683.2019.1575906

Khan, S. M., Shafi, I., Butt, W. H., Diez, I. D. L. T., Flores, M. A. L., Galán, J. C., & Ashraf, I. (2023). A systematic review of disaster management systems: Approaches, challenges, and future directions. *Land (Basel)*, *12*(8), 1514. DOI: 10.3390/land12081514

Kshetrimayum, A. R. H., & Goyal, A. (2024). Exploring different approaches for landslide susceptibility zonation mapping in Manipur: A comparative study of AHP, FR, machine learning, and deep learning models. *Journal of Spatial Science*, •••, 1–30. DOI: 10.1080/14498596.2024.2368156

Lv, P., Ma, L., Li, Q., & Du, F. (2023). ShapeFormer: A shape-enhanced vision transformer model for optical remote sensing image landslide detection. *IEEE Journal of Selected Topics in Applied Earth Observations and Remote Sensing*, *16*, 2681–2689. DOI: 10.1109/JSTARS.2023.3253769

Ma, Z., & Mei, G. (2024). Forecasting landslide deformation by integrating domain knowledge into interpretable deep learning considering spatiotemporal correlations. *Journal of Rock Mechanics and Geotechnical Engineering*. Advance online publication. DOI: 10.1016/j.jrmge.2024.02.034

Ramli, M. F., Yusof, N., Yusoff, M. K., Juahir, H., & Shafri, H. Z. M. (2010). Lineament mapping and its application in landslide hazard assessment: A review. *Bulletin of Engineering Geology and the Environment*, *69*(2), 215–233. DOI: 10.1007/s10064-009-0255-5

Saha, S., Majumdar, P., & Bera, B. (2023). Deep learning and benchmark machine learning based landslide susceptibility investigation, Garhwal Himalaya (India). *Quaternary Science Advances*, *10*, 100075. DOI: 10.1016/j.qsa.2023.100075

Sara, A., & Vikraman, R. (2015). A survey on disaster management systems. *International Journal of Computer Applications*, *128*(10), 27–31. DOI: 10.5120/ijca2015906652

Şimşek, D., Kutlu, İ., & Şık, B. (2023, October). The role and applications of artificial intelligence (AI) in disaster management. In *Proceedings of the 3rd International Civil Engineering & Architecture Conference* Volume 2*: Architecture*. Golden Light Publishing. DOI: 10.31462/icearc.2023.arc992

Song, Y., Song, Y., Wang, C., Wu, L., Wu, W., Li, Y., Li, S., & Chen, A. (2024). Landslide susceptibility assessment through multi-model stacking and meta-learning in Poyang County, China. *Geomatics, Natural Hazards & Risk*, *15*(1), 2354499. DOI: 10.1080/19475705.2024.2354499

Vaswani, A., Shazeer, N., Parmar, N., Uszkoreit, J., Jones, L., Gomez, A. N., & Polosukhin, I. (2017). Attention is all you need. *Advances in Neural Information Processing Systems*, •••, 30.

Wang, H., Zhang, L., Yin, K., Luo, H., & Li, J. (2021). Landslide identification using machine learning. *Geoscience Frontiers*, *12*(1), 351–364. DOI: 10.1016/j.gsf.2020.02.012

Xu, P., Zhu, X., & Clifton, D. A. (2023, May 10). Multimodal learning with transformers: A survey. *arXiv*. https://doi.org//arXiv.2206.06488DOI: 10.48550

Yang, Z. Q., Qi, W. W., Xu, C., & Shao, X. Y. (2024). Exploring deep learning for landslide mapping: A comprehensive review. *China Geology*, *7*(2), 330–350. DOI: 10.31035/cg2024032

Chapter 9
Enhancing Climate Resilience and Sustainable Development Through the Integration of Artificial Intelligence

Shweta Jaiswal
https://orcid.org/0000-0002-7849-3461
CMP Degree College, University of Allahabad, Prayagraj, India

ABSTRACT

AI is emerging as a potentially revolutionary tool in addressing climate-related challenges, optimizing resource management, and facilitating sustainable development. This necessitates the creation of innovative methods to enhance resilience and adaptability. This narrative study explores the application of AI in climate resilience strategies, focusing on predictive analytics, disaster risk management, carbon footprint reduction, and sustainable urban planning. It synthesizes findings from existing literature to showcase AI-driven innovations, including machine learning models for climate forecasting, AI-powered decision support systems for policymakers, and automated monitoring of environmental indicators. Additionally, it examines the ethical considerations, challenges, and opportunities associated with using AI for climate resilience. The study underscores the potential of AI to accelerate the transition to sustainability but emphasizes the need for interdisciplinary collaboration, policy integration, and inclusive governance to fully realize its benefits.

DOI: 10.4018/979-8-3373-0680-3.ch009

1. INTRODUCTION

Climate change effects are becoming more apparent from rising seas to weather extremes, these trends present unprecedented challenges to global societies (Akter, 2024). In this context, the application of advanced technologies, including Artificial Intelligence (AI), presents high potential to facilitate climate resilience and support sustainable development (Kumari & Pandey, 2023). Climate change and its impact on the world are among the world's greatest concerns. Many experts made alarming reports on the destructive impact of climate change. To satisfy the diversified needs of customers in the ever-changing business environment and to minimize carbon emission, banks are increasingly looking at the environmental effects of their products and the optimum use of natural resources (Kang & Nam, 2024). This change triggered the emergence of green banking, promoting eco-friendly operations entered on sustainable development and carbon footprint reduction. Banking is usually considered environmentally friendly however, its carbon footprint is growing due to high energy consumption, scarce spaces and poorly planned buildings (Satish et al., 2023Brevini, 2023). The banking sector is now adopting innovative green products, including digital banking, electronic payment of bills, green mortgages, green credits and support of local environmental initiatives (Selvapriya et al., 2024).

Against the environmental backdrop and the increasing emphasis on green investments, most companies from advanced nations have made environmental protection an integral part of their strategy, offering green services and products (Sahil et al., 2023). Developing nations, including India, however, understand that environmental sustainability is imperative to inclusive growth. Green banking guidelines necessitate those financial institutions employ environmentally sound practices however, the banking sector is expected to assist the government in reducing carbon footprints as responsible corporate citizens. Technological innovations have transformed the way financial institutions conceptualize sustainability (Dubey et al., 2022). Artificial intelligence (AI) plays an important role by processing enormous sets of data to identify good sustainable investment opportunities. Blockchain technology offers transparency and reliability of transactions by providing an immutable and secure record, enabling tracking of the impact of green investments. Moreover, big data analytics offers improved means of measuring environmental impacts and personalization of financial products to suit the requirements of sustainability objectives Although there are obstacles, the continuous development within these areas is stimulating (Habila et al., 2023). The growing challenges of climate change and sustainable development has unleashed the search for new solutions across sectors. The use of Artificial Intelligence (AI) is one of the most exciting strategies, with the potential to drive paradigm shifts in how we plan for, adapt to, and respond to climate impacts (Haldorai et al., 2024).

By exploring the intersections of AI applications like waste reduction programs, energy-efficient automation, and predictive maintenance, the study aims to provide context around how the drive for operational efficiency from these technologies simultaneously contributes to environmental sustainability. The study also examines the Opportunities and challenges of implementing AI in agriculture and manufacturing industries.

2. LITERATURE REVIEW

Climate Resilience through AI refers to the idea that utilizes the capabilities of artificial intelligence (AI) to enhance society's ability to predict, react to and recover from climate challenges (Saad et al., 2022). The project, however, transcends across different sectors, from agriculture and infrastructure and health, among many others. The idea usually involves different elements aimed at building systems' resilience to climate change effects (Shaamala et al., 2024). Data Collection and Observation Climate Data Analytics Artificial intelligence models undertake the analysis of huge amounts of data acquired from satellites, sensors and weather stations, observing environmental conditions including temperature, rain, sea level and extreme weather (Cembrowska-Lech et al., 2023). Interconnected Internet of Things (IoT) sensors monitor ongoing climate conditions, from air quality to soil moisture (Cembrowska-Lech et al., 2023). This allows AI systems to access real-time information to make predictions and initiate interventions. Additionally, AI is able to process people's and local communities' data (e.g., crowd-sourced climate event reports) to calibrate localized predictions, hence enhancing responsiveness (Khan et al., 2022).

But the effectiveness of this system is based on intersectoral collaboration, since it is through intersectoral collaboration that we can effectively counter the threats caused by climate change (Farooq et al., 2024). AI greatly complements traditional climate models by using machine learning algorithms to improve the accuracy of climate forecasts. These innovations include long-term forecasts, as well as forecasts related to extreme weather conditions, such as hurricanes, floods and drought. AI risk models can simulate different climate scenarios, thereby predicting vulnerabilities across different sectors: agriculture, energy and infrastructure (Balogun et al., 2020). Additionally, AI can evaluate the impact of climate disasters and optimize resource allocation following such disasters. In precision agriculture, AI uses optimize the use of water, detect pests and predict crop yields against weather-based estimates; this allows the farmer to react to changing weather conditions (Haldorai et al., 2024). Additionally, AI can be used to design buildings and roads, for instance, which are more resistant to adverse weather conditions. While past occurrence can be analyzed by machine learning algorithms to recommend enhancement, they are also crucial while optimizing the usage of water (Amiri et al., 2024). AI applications can predict leaks and predict the availability of water due to variable rainfalls, drought or floods. Yet, the application of AI in these domains is still to mature and there are hurdles that are still to be crossed (Roychowdhury et al., 2023). Artificial Intelligence (AI) can greatly contribute to energy efficiency across different industries, such as infrastructure, factories and transport (Rane et al., 2024). By means of energy forecasting and the capacity to control smart grids and enable alternative energy sources (e.g., solar and wind), AI models can be helpful. They optimize carbon capture and storage technologies; therefore, reduction of carbon dioxide emissions is effectively implemented (Marsh et al., 2021). In addition, AI can be employed to create green transport systems that enable emission reduction. This can be done through traffic optimization and encouraging electric vehicles. Nevertheless, the use of AI does not end with energy efficiency (Arfanuzzaman, 2021). It can also enable post-disaster relief by assessing damage, forecasting recovery requirements and maximizing the utilization of key resources such as food, medication and relief supplies. (Varshney & Tuberosa, 2007). Additionally, AI systems use data analysis to gain insights into how communities respond when confronted with climate catastrophes (Usige et al., 2023). Such information enables policy-makers to come up with better resilience-building measures. While AI models can maximize economic systems that support adaptive climate change response, they are especially useful to vulnerable communities residing in climate-risk-prone regions (Thudi et al., 2021). Additionally, AI algorithms can allow governments and institutions to make smart climate policy decisions, plan for disasters and climate change adaptation policies (Thudi et al., 2021).

3. THE NEXUS OF CLIMATE RESILIENCE AND SUSTAINABILITY

Climate resilience and the sustainability nexus is typically considered to be one of the pillars of sustainability (Bhat et al., 2023). It is included in the overall goal of sustainable development, which seeks to integrate environmental, social and economic considerations in a manner that meets present needs without compromising the ability of future generations to meet their own needs. The connection between climate resilience and sustainability is that they both have a common focus on long-term success, reduction of risk and the capacity of communities, economies and ecosystems to thrive in the face of changing environmental conditions (Sirmacek et al., 2023). Sustainability relies on resilience since it ensures that natural systems can continue to function even after disruption caused by climate change. For example, resilient ecosystems like forests or wetlands are very important to biodiversity, hydrological cycle regulation and flood alleviation. But the complexities of creating these resilient systems are daunting, although they are required to help create a sustainable world (Fan et al., 2021).

Climate resilience however, promotes social sustainability by reducing people's exposure, particularly those from marginalized communities, to climate impacts. This ensures their access to essential services and enhances social cohesion (Satornino et al., 2024). Climate resilience incorporation into practice of sustainable development involves prioritization of risk reduction and prevention (Alsamhi et al., 2024). Strategies of sustainable development target reduction of long-term climate change risks through proactive measures, including infrastructure development, disaster preparedness and adaptation through ecosystems. Adaptive capacity matters; resilience and sustainability both highlight the imperative of developing the capacity of communities to respond to changing situations (Qiao et al., 2022). This is through leverage of innovations in technology, governance and local knowledge. There are synergies between adaptation and mitigation; sustainability emphasizes dealing with the causes of climate change (e.g., carbon emissions) An integrated approach that merges climate adaptation (resilience) with mitigation strategies (sustainability) can yield co-benefits including reduced vulnerability and improved long-term environmental health (Raihan et al., 2024Nabwire et al., 2021). Although challenges exist, the benefits of such integration cannot be overstated. Effective governance requires not just coordination, but also a commitment to shared goals. Inclusive decision-making involves engaging communities in decisions regarding resilience-building to ensure that their needs and local knowledge are integrated (Kole et al., 2015Mushtaq et al., 2024).

Figure 1. Sustainable environmental technology

Source *(Sahoo et al., 2024)*

4. SUSTAINABLE DEVELOPMENT THROUGH AI INNOVATIONS

Technological innovations in climate resilience, particularly artificial intelligence (AI), are having far-reaching effects on how societies and ecosystems are learning to cope with, and mitigate, climate change. AI is being used across industries to promote climate resilience it does so by strengthening predictive functions, rationalizing resources and enabling decision-making through information (Rane et al., 2024). Precise weather prediction (AI models), especially those based on deep learning and neural networks, are enormously enhancing the precision of weather forecasts. Such systems are able to analyze large amounts of past data and current conditions in order to predict weather patterns, storms, flooding and drought more precisely. Additionally, AI is also used to tune global climate models, and this simplifies simulating long-term climate change scenarios. Such models can forecast effects like sea-level rise, extreme heat events and changes to ecosystems (González-Rodríguez et al., 2024). In smart farming and precision agriculture, AI-powered image recognition systems use drones and satellites to track crops in real-time. This makes one capable of detecting problems like infestation by pests, nutrient deficiencies or water stress. However, the use of these technologies is not without issues, as stakeholders must balance various factors to achieve maximum efficacy and sustainability (Srivastava & Maity, 2023). Machine learning algorithms can process this data which is very important to streamline irrigation schedules and maximize crop yield forecasts. This ability helps farmers to adjust to the constantly changing climate patterns. AI helps farmers to choose those crops that are more tolerant of these changes, especially those tolerant of drought. Examining factors like soil health, weather predictions and past patterns, AI provides actionable recommendations on those crops that are most suitable for specific regions (Mittal et al., 2025; Ueda et al., 2024). There are, nonetheless, challenges that need to be addressed and these are important to resolve in order to optimize the potential of AI applications in agricultural and disaster relief situations. AI systems can optimize water supply networks by analyzing real-time data on water consumption, rainfall and soil moisture. AI can, however, predict drought conditions, optimizing the utilization of water resources across different sectors (e.g., agriculture, urbanization and industry). This avoids wastage, although it also provides a sustainable means of utilizing water resources (Gangurde et al., 2019). Examples are China's investments in AI-based smart city projects to help lower emissions and increase urban sustainability.

- For instance, Alibaba's "ET City Brain" optimizes metropolitan infrastructure such as energy use and traffic, and this reduces congestion and improves air quality. In the realms of finance and publications, China leads the world when it comes to climate research and AI. The use of AI approaches to all facets of climate research ranging from climate system models to operations of renewable energy has been greatly aided by the Chinese Academy of Sciences.
- AI is being implemented in the US in several innovative applications to support climate resilience. AI-enabled early warning systems, such as those used by NOAA, allow communities to prepare in advance by predicting extreme weather events such as hurricanes and floods. AI-enabled monitoring systems, such as Austin Energy's 360-degree monitoring, provide early warning to prevent hazards from wildfires. The Biden administration is supporting AI-enabled clean energy projects to maximize power grid efficiency and minimize carbon footprints, again reflecting how AI is being put to use within energy infrastructure. These initiatives demonstrate how artificial intelligence

(AI) is being strategically deployed to support environmental monitoring, sustainable energy management, and catastrophic planning in the United States.

Figure 2. Trends of major issues of climate change

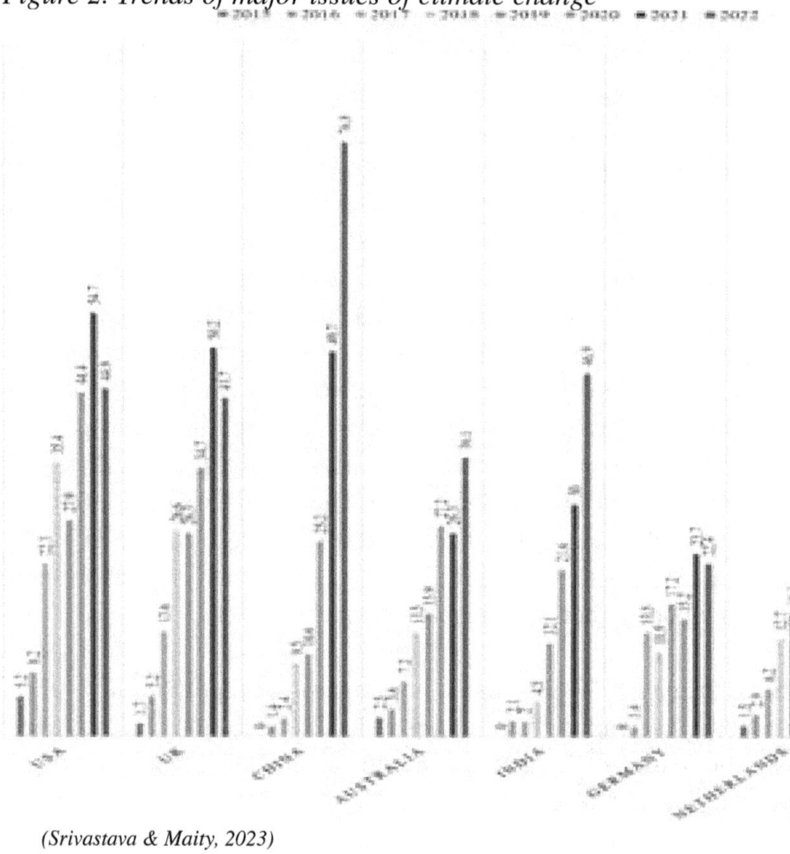

(Srivastava & Maity, 2023)

5. AI-DRIVEN INNOVATIONS IN CLIMATE RESILIENCE AND SUSTAINABLE DEVELOPMENT

Artificial intelligence (AI) is increasingly being leveraged to enhance climate resilience and promote sustainable development across various sectors. Here are several case studies illustrating the effective application of AI in these areas:

5.1 AI Climate Resilience in the Agricultural Sector

- AI is contributing significantly to climate resilience and sustainable development. AI-enabled precision farming maximizes crop yields, reduces pesticide use, and minimizes water use through live feedback on plants, soil health, and weather predictions. Drought does not strike as severely when AI-based irrigation systems are installed. (Al-Raeei, 2024).
- Machine learning algorithms prevent environmental degradation by predicting infestations and prescribing eco-friendly means of pest control (Argyroudis et al., 2022) AI-based climate forecasting also allows farmers to react to changing weather conditions and maintain food security while staying away from losses. AI is rendering the agriculture sector climate resilient while promoting sustainability through the use of resources and environmentally friendly practices (Srivastava & Maity, 2023).
- Artificial intelligence-based weather forecasting tools lowered debt among small farmers by 50% and saved them 10% annually in India (Reuters, 2024).
- Artificial intelligence-assisted precision agriculture can help improve global yields by 25% while decreasing water consumption by 20% (Liu et al., 2022).

5.2 AI Climate Resilience in the Manufacturing Sector

- AI is being deployed in the manufacturing sector to facilitate sustainable growth and climate resilience. Through optimizing machinery performance and avoiding breakdowns, AI-based predictive maintenance reduces energy consumption and waste production. (Leal Filho et al., 2022)
- AI is being used to analyse supply chain emissions, enabling companies to select greener production and sourcing processes. Moreover, through optimizing production processes and reducing material waste and carbon footprints, AI-based automation maximizes the use of resources. (Yadav, 2024) In the automotive and electronics sectors, AI is being leveraged to optimize recycling processes and design energy-efficient products.
- AI-based flood forecasting system, spans 80 countries and issues alert messages to more than 360 million people (Sahil et al., 2023)
- Early warning systems improved by AI lowered fatalities from disasters by 40% in areas with high exposure to extreme weather (UNEP, 2023).

5.3 AI in Disaster Prediction and Risk Management

- The implementation of sustainable practices will significantly augment employees' capacity to contribute to sustainability objectives and to proficiently manage artificial intelligence (AI) systems. (Miailhe et al., 2019). However, managers need to incorporate these AI tools into their enterprise risk management strategies (Kaack et al., 2022).

- By integrating AI-based climate risk models into business continuity and disaster response planning, managers can simulate various scenarios and prepare adaptive strategies for potential climate impacts on their operations. Businesses are increasingly held accountable for their environmental footprint (Rayhan, 2023).
- AI can streamline sustainability reporting, automating data collection, analysis and reporting processes. Managers must ensure that AI technologies support compliance with regulations and sustainability goals (Khan et al., 2022). Use AI to automate the tracking of emissions, resource usage and waste management; this ensures that sustainability reports are accurate and meet both internal and external standards. (Falk & Van Wynsberghe, 2024) allow companies to minimize waste, improve efficiency and achieve sustainability targets (Rane, 2023). Consumers are increasingly concerned about sustainability and climate resilience; however, they often seek transparency regarding the practices of businesses.
- Managers must utilize AI to cultivate products and services that resonate with consumer preferences regarding sustainable practices (Al-Raeei, 2024). They must also maintain transparency about the environmental impact.
- For instance, employing AI can enhance product lifecycle tracking, ensure supply chain sustainability and create consumer-facing platforms that disclose real-time data on the environmental effects of products. This, in turn, boosts brand loyalty and fosters consumer confidence (Falk & Van Wynsberghe, 2024).
- Innovation in business models is essential; AI can unlock new avenues focused on sustainable products and services, such as renewable energy solutions, smart cities and climate-resilient infrastructure. Although managers need to spearhead the transformation toward these innovative business models, they should also encourage innovation in AI-driven solutions that generate new revenue streams. These might include energy-efficient products or climate-adaptive technologies. Managers ought to lead pilot programs, test novel models and scale those that exhibit both financial and environmental benefits (Bolte & Van Wynsberghe, 2024).
- Microsoft has spent $10 billion on 10.5 gigawatts of renewable energy to compensate for AI/cloud computing emissions (Time, 2024).
- Solar and wind energy forecasting with AI increases efficiency by 10-20%, lowering fossil fuel back-up dependency (IEA, 2023).

- Integration of AI in energy grids has the potential to cut CO_2 emissions by 2.6 gigatons annually by 2030 (WEF, 2023).

Figure 3. Climate resilience practices

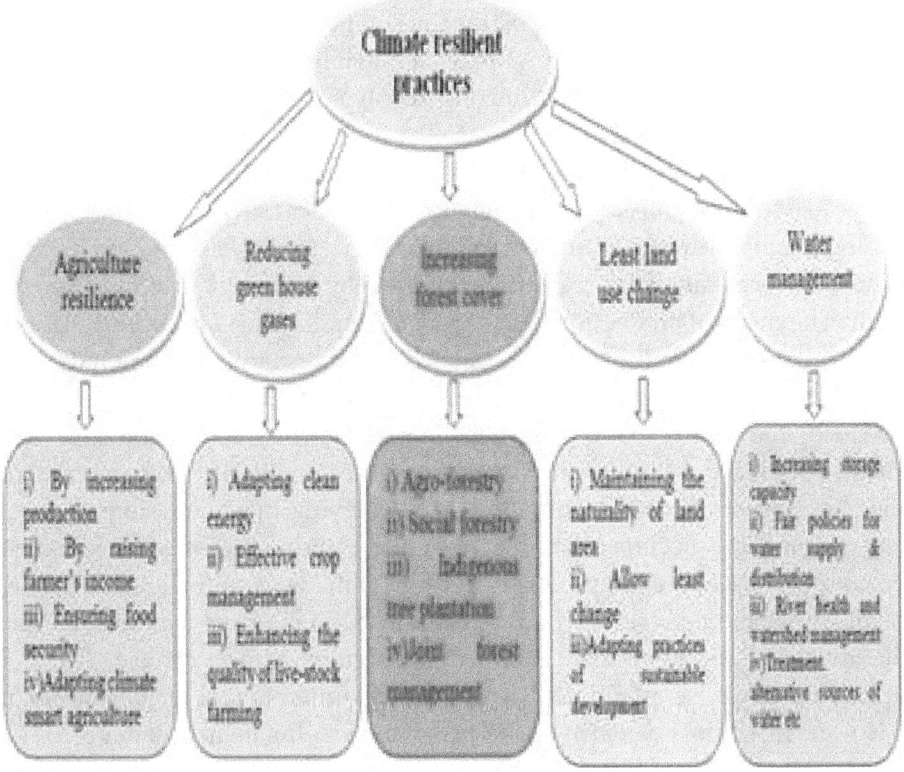

6. AI POWERED CLIMATE RESILIENCE: A NEW PARADIGM

AI technologies significantly enhance flood forecasting by examining river levels, rainfall data and topography to anticipate floods before they occur. These systems can also aid in the design of improved flood defenses and more resilient infrastructure. Climate-Smart Urban Planning and Infrastructure however is another domain where AI is being utilized (Leal Filho et al., 2022). Algorithms (notably) can analyze camera trap images to monitor animal populations and detect behavioral changes resulting from shifts in climate. AI-Enhanced Climate Policy and Decision Making involves policy simulations and scenario analysis; this is how AI is utilized to simulate the potential effects of various climate policies. Governments and organizations are able to design the most effective interventions, however, this process relies on the analysis of vast datasets (Sinha et al., 2023). AI provides insights into the economic, environmental and social impacts of climate change policies. AI tools assist policymakers in making real-time decisions during climate emergencies (using up-to-date data on weather, resource availability and infrastructure status), which helps guide appropriate responses. AI for Carbon Footprint Reduction in Business Operations is another area where AI plays a crucial role. AI helps businesses reduce their carbon footprints by optimizing supply chains, improving logistics and reducing energy consumption.

For instance, machine learning algorithms can analyze transport routes and inventories to minimize emissions and enhance resource efficiency (Yadav, 2024). AI technologies empower various industries (for instance, manufacturing) to recognize opportunities for enhancing energy efficiency, thereby reducing waste and decreasing the carbon intensity of production lines. Artificial Intelligence for Carbon Offsetting and Reforestation can optimize efforts related to reforestation by pinpointing ideal planting locations and species, based on climate and soil data. AI-driven drones, in this context, can be utilized to plant trees more efficiently and monitor the health of forests over time. AI is also employed to oversee and verify carbon offset programs; this ensures that such initiatives are achieving their emissions reduction targets and provides accurate data regarding the carbon sequestration capacity of diverse projects (Kang & Nam, 2024). AI for Public Awareness and Community Engagement is another critical area, as AI-powered platforms can analyze social media trends and discern public sentiment concerning climate change. This helps organizations to devise effective communication strategies that engage communities. However, AI tools can also create personalized climate action plans for both individuals and organizations, offering tailored recommendations on how to reduce carbon footprints and build resilience. Although this technology presents numerous advantages, it is essential to consider the ethical implications of its implementation (Batley & Edwards, 2016).

Figure 4. Sustainable AI in environmental economics and management: current trends and post-COVID perspective

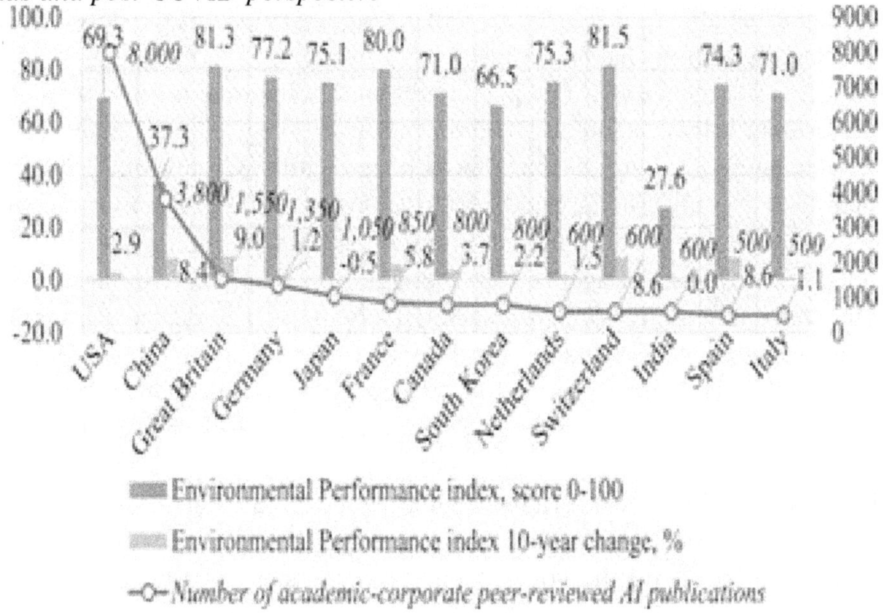

7. ETHICAL AND POLICY CONSIDERATION IN AI INTEGRATION

The transition to renewable energy sources is fundamental to sustainable development; however, these energy sources (e.g., solar, wind) tend to be intermittent and variable. AI has the capability of maximizing energy production, storage and distribution, thereby contributing to renewable energy system reliability and efficiency. Energy sector managers can apply AI algorithms for the purpose of optimizing energy grids, predicting energy generation by renewable resources and demand-response system management (Van Wynsberghe, 2021). In addition, AI can be used to advance energy storage devices (like batteries) to better utilize renewable energy when it is in abundance and more effectively distribute it when there is peak demand. Consumer engagement and behavioral information are also critical in order to realize sustainable behaviors. To have consumers' behaviors in sync with climate objectives, such dynamics must be taken into consideration since knowledge of consumers' motivations can enable more effective strategies (Floridi et al., 2018). Managers can be supported by Artificial Intelligence (AI) to craft more effective strategies to shape consumer choice towards environmentally friendly products and services. Personalization solutions offered by AI can learn consumer choice and preference in order to trigger environmentally friendly substitutes. For example, AI may suggest energy-saving products, low-carbon travel options, or sustainable food items; this will encourage consumers to develop more environmentally friendly habits.

Artificial Intelligence (AI) can potentially greatly improve the construction of climate-resilient infrastructure, or recommend specific agricultural practices that adjust to changing growing seasons. Partnerships with environmental NGOs and regulatory agencies are crucial; building AI for climate resilience and sustainability requires collaboration with numerous external partners, including environmental organizations, government agencies and foreign organizations (Heilinger, 2022). Managers have to interact with such stakeholders in order to design AI tools and frameworks that align with international climate goals. Through this interaction, AI-enhanced business strategies end up having a positive impact on the overall environmental ecosystem. Frequent interaction with regulators and NGOs can also aid in the ability to foresee upcoming policies or legislation that can shape the use of AI in sustainability efforts (Bélisle-Pipon et al., 2023). For instance, AI may suggest energy-efficient goods, low-carbon transportation, or eco-friendly foods; it can initiate the consumer to modify their habits to more sustainable orientations (Falk & Van Wynsberghe, 2024). Management, though, must implement AI in its marketing and interactions in order to offer consumer education around sustainability as well as build products to fulfill greener tendencies. Moreover, AI predictive analytics can assist companies to prepare and react to future climate conditions that are going to occur. This can be done by making changes in sea-level rise, extreme weather, or drought impact on operations. Managers can use AI solutions that enable the simulation of different climate conditions and their possible effects on business operations. This can include evaluating risks like flooding, wildfires, or extreme temperatures to create adaptive responses. There are many challenges, but the use of AI in these fields is a promising way of achieving sustainability in business activities (Ghallab, 2019).In addition, AI can have a critical function in climate finance and investment because it maximizes the allocation of resources to climate resilience initiatives. Investment can be guided by data, which can determine the most effective sustainability projects; however, this must be done with utmost care and scrutiny. Finance and investment managers possess the ability to utilize AI technologies to analyze and predict the financial returns of sustainability projects or to assess the environmental risks associated with investment portfolios (Stahl, 2021).

8. CHALLENGES AND LIMITATION OF AI IN CLIMATE RESILIENCE

Despite the huge potential of AI to support both climate resilience and sustainable development, harnessing AI for both poses significant challenges. AI models require high-quality, accurate and diverse datasets. Climate-related data including long-term weather trends, biodiversity indicators and carbon output often seems fragmented, inconsistent or narrow in focus. AI systems will simply not produce accurate predictions without quality and high-resolution data. This can, in turn, produce sub-optimal decisions on climate adaptation and sustainable development. Biases can be transferred from the data used to train AI systems. Such biases in the context of both climate change and sustainability, for example, may reflect disparities in access to data across regions, communities, or ecosystems (Falk & Van Wynsberghe, 2024). At the same time, solutions enabled by AI and the means to design and deploy them may favor more prosperous regions at the expense of the vulnerable populations already hit hardest by anthropogenic climate change. But with the tremendous potential AI presents comes the need to grapple with these challenges to deliver equitable outcomes (Borenstein & Howard, 2021).

AI solutions need significant computational power as well as technological and financial levels in order to be implemented on a global scale. This aspect severely restricts the opportunity of AI to bridge the gap between areas with high and low income, therefore limiting its contribution to global sustainable initiatives (Dodda et al., 2021). AI systems can unintentionally have social consequences for example, worker displacement or widening inequalities. For all its role in surveillance and decision-making, AI could also pose risks to privacy. AI solutions, and consequently, ethical issues might hinder their widespread adoption, especially in industries where social equity is key to achieving sustainability (Green, 2018). The complexity of Climate Systems is also a challenge. Earth's climate is very complex with numerous interdependent systems that are hard to model accurately. AI models can be tested to encompass the complete set of variables that are involved in climate processes and human-environment interactions (Shneiderman, 2021). Consequently, misdirected policies or interventions towards climate resilience and sustainability will be in vain because of incorrect projections or overly reduced models. (Stahl et al., 2023). Failing to seamlessly incorporate AI may delay the advancement toward climate resilience and sustainable development. This is the way AI can become a supporting resource without being used for transformation. (Dignum, 2020).

In addition, there are governance and policy concerns that test this landscape: there are no overarching global frameworks to regulate the application of AI for sustainable development and climate resilience. Various nations will have different priorities, ethical norms and regulations regarding AI applications. Varying governance and regulatory norms might result in disjointed efforts; however, this might cause inefficiencies that will prevent the mass adoption of AI-based solutions (Heilinger, 2022). Even though countries give precedence to innovation, they are not always successful, because their methods are quite distinct from one another. However, differences do not prove that AI potential is an impossibility, rather that more needs to be done in a uniform strategy (Floridi, 2023).

9. CONCLUSION

Green banks were established to address climate change and various environmental challenges. They play a crucial role in expediting the transition to clean energy; however, they must be incorporated into a more comprehensive climate policy framework. Banks need to do more than merely adhere to ESG (Environmental, Social and Governance) guidelines they ought to raise awareness and assist customers in taking action through their banking services. Green banking practices represent a constructive step towards establishing a fairer and more sustainable financial system. By emphasizing financial inclusion and responsible lending, green banks mitigate the adverse effects of traditional banking on marginalized communities.

But along with this integration, there will be a need for robust governance frameworks, ethical considerations and global collaboration for the prosperity of all but also vulnerable communities that are often left behind as AI technologies become omnipresent. In the end, the potential for AI to significantly improve climate resilience and sustainable development ultimately depends on the willingness of governments, the private sector and civil society to work together to ensure these technologies are deployed in a way that promotes sustainability, equity and long-term environmental stewardship. The potential role of AI in improving climate resilience (and) advancing sustainable development is enormous. Combining AI technologies across energy systems, agriculture, disaster management and urban planning, we could create a more robust and sustainable future.

10. MANAGERIAL IMPLICATIONS

The confluence of climate resilience, sustainability and technology innovations (AI, etc.) makes for a dynamic and interesting place for managers across all sectors. There are multiple important managerial implications of using AI to promote climate resilience and sustainability. Strategically invest in AI technologies: Managers need to invest funds, responsibly and wisely, in AI solutions that enable sustainability and climate resilience. This could mean the adoption of AI-powered tools used for environmental monitoring, predictive climate modelling, & resource optimization in business operations. Leaders must nevertheless allocate investment resources to AI research and development (particularly in sectors like energy use, water conservation, agriculture and disaster response systems) and ensure that those new investments are aligned with long-term sustainability goals. Cross-disciplinary collaboration is essential, climate resilience and sustainability necessitate cooperation among fields such as environmental science, economics and technology. AI serves as an enabler; however, managers must cultivate collaboration among teams to effectively integrate AI-driven insights with practical environmental strategies. It is crucial for managers to promote cross-functional teams that encompass AI specialists, climate scientists, sustainability experts and operations leaders, thereby developing comprehensive solutions to address climate-related challenges.

11. FUTURE RESEARCH

Future inquiries into enhancing climate resilience and sustainable development through the incorporation of Artificial Intelligence (AI) can concentrate on several vital domains, merging AI's potential with environmental science, policy-making and social frameworks. Predictive modelling for climate change impacts is one promising area: AI has the capacity to transform the manner in which we anticipate and prepare for climate-related phenomena, including extreme weather events, rising sea levels, or alterations in ecosystems. Future research ought to emphasize advanced climate modelling (however, it is essential to note) that integrates machine learning (ML) and deep learning (DL) to enhance the precision of climate forecasts, accommodating intricate interactions within the Earth system. Early warning systems are also crucial; developing AI-driven models for predicting climate-induced disasters (such as hurricanes, wildfires and floods) with greater accuracy will enable proactive adaptation strategies.

Future research may focus on smart grids and energy management: employing artificial intelligence (AI) to enhance the distribution of energy, minimize waste, while simultaneously integrating renewable sources into power grids more efficiently. Findings focused on energy storage optimization are also important exploring how AI can optimize the performance and scalability of energy storage technologies is critical since we need these technologies to integrate intermittent renewable energy sources like solar and wind. Moreover, decarbonizing industrial sectors is critical difficulty hence, AI can be used for real-time energy consumption monitoring and predictive maintenance, contributing to emission reduction in manufacturing, transportation, and mining industries. AI is also involved in ecosystem management and biodiversity Conservation. Artificial intelligence could potentially play a powerful role in the management of natural ecosystems and biodiversity conservation.

REFERENCES

Akter, M. S. (2024). AI for sustainability: Leveraging technology to address global environmental. *Journal of Artificial Intelligence General Science (JAIGS) ISSN: 3006-4023, 3*(1), 40–48.

Al-Raeei, M. (2024). The smart future for sustainable development: Artificial intelligence solutions for sustainable urbanization. *Sustainable Development*, sd.3131. https://doi.org/DOI: 10.1002/sd.3131

Al-Raeei, M. (2024). Artificial intelligence for climate resilience: Advancing sustainable goals in SDGs 11 and 13 and its relationship to pandemics. *Discover Sustainability*, *5*(1), 513. DOI: 10.1007/s43621-024-00775-5

Alsamhi, S. H., Rowan, N. J., Hawbani, A., Nashwan, F., & O'Brolchain, N. (2024). AI-empowered peatland management for climate-resilient and sustainable future: Challenges and opportunities. *2024 4th International Conference on Emerging Smart Technologies and Applications (eSmarTA)*, 1–8. https://ieeexplore.ieee.org/abstract/document/10638930/

Amiri, Z., Heidari, A., & Navimipour, N. J. (2024). Comprehensive survey of artificial intelligence techniques and strategies for climate change mitigation. *Energy*, *308*, 132827. DOI: 10.1016/j.energy.2024.132827

Arfanuzzaman, M. (2021). Harnessing artificial intelligence and big data for SDGs and prosperous urban future in South Asia. *Environmental and Sustainability Indicators*, *11*, 100127. DOI: 10.1016/j.indic.2021.100127

Argyroudis, S. A., Mitoulis, S. A., Chatzi, E., Baker, J. W., Brilakis, I., Gkoumas, K., Vousdoukas, M., Hynes, W., Carluccio, S., Keou, O., Frangopol, D. M., & Linkov, I. (2022). Digital technologies can enhance climate resilience of critical infrastructure. *Climate Risk Management*, *35*, 100387. DOI: 10.1016/j.crm.2021.100387

Balogun, A.-L., Marks, D., Sharma, R., Shekhar, H., Balmes, C., Maheng, D., Arshad, A., & Salehi, P. (2020). Assessing the potentials of digitalization as a tool for climate change adaptation and sustainable development in urban centres. *Sustainable Cities and Society*, *53*, 101888. DOI: 10.1016/j.scs.2019.101888

Batley, J., & Edwards, D. (2016). The application of genomics and bioinformatics to accelerate crop improvement in a changing climate. *Current Opinion in Plant Biology*, *30*, 78–81. DOI: 10.1016/j.pbi.2016.02.002 PMID: 26926905

Bélisle-Pipon, J.-C., Monteferrante, E., Roy, M.-C., & Couture, V. (2023). Artificial intelligence ethics has a black box problem. *AI & Society*, *38*(4), 1507–1522. DOI: 10.1007/s00146-021-01380-0

Bhat, J. A., Feng, X., Mir, Z. A., Raina, A., & Siddique, K. H. M. (2023). Recent advances in artificial intelligence, mechanistic models, and speed breeding offer exciting opportunities for precise and accelerated genomics-assisted breeding. *Physiologia Plantarum*, *175*(4), e13969. DOI: 10.1111/ppl.13969 PMID: 37401892

Bolte, L., & Van Wynsberghe, A. (2024). Sustainable AI and the third wave of AI ethics: A structural turn. *AI and Ethics*. Advance online publication. DOI: 10.1007/s43681-024-00522-6

Borenstein, J., & Howard, A. (2021). Emerging challenges in AI and the need for AI ethics education. *AI and Ethics*, *1*(1), 61–65. DOI: 10.1007/s43681-020-00002-7 PMID: 38624388

Brevini, B. (2023). Artificial intelligence, artificial solutions: Placing the climate emergency at the center of AI developments. In *Technology and Sustainable Development* (pp. 23–33). Routledge., https://www.taylorfrancis.com/chapters/oa-edit/10.1201/9781003325086-3/artificial-intelligence-artificial-solutions-benedetta-brevini DOI: 10.1201/9781003325086-3

Cembrowska-Lech, D., Krzemińska, A., Miller, T., Nowakowska, A., Adamski, C., Radaczyńska, M., Mikiciuk, G., & Mikiciuk, M. (2023). An integrated multi-omics and artificial intelligence framework for advance plant phenotyping in horticulture. *Biology (Basel)*, *12*(10), 1298. DOI: 10.3390/biology12101298 PMID: 37887008

Dignum, V. (2020). Responsibility and artificial intelligence. *The Oxford Handbook of Ethics of AI, 4698*, 215.

Dodda, S. B., Maruthi, S., Yellu, R. R., Thuniki, P., & Reddy, S. R. B. (2021). Ethical Deliberations in the Nexus of Artificial Intelligence and Moral Philosophy. *Journal of Artificial Intelligence Research and Applications*, *1*(1), 31–43.

Dubey, A. K., Kumar, A., Narang, S. K., Khan, M. A., & Srivastav, A. L. (2022). *Visualization Techniques for Climate Change with Machine Learning and Artificial Intelligence*. Elsevier.

Falk, S., & Van Wynsberghe, A. (2024). Challenging AI for Sustainability: What ought it mean? *AI and Ethics*, *4*(4), 1345–1355. DOI: 10.1007/s43681-023-00323-3

Fan, J., Zhang, Y., Wen, W., Gu, S., Lu, X., & Guo, X. (2021). The future of Internet of Things in agriculture: Plant high-throughput phenotypic platform. *Journal of Cleaner Production*, *280*, 123651. DOI: 10.1016/j.jclepro.2020.123651

Farooq, M. A., Gao, S., Hassan, M. A., Huang, Z., Rasheed, A., Hearne, S., Prasanna, B., Li, X., & Li, H. (2024). Artificial intelligence in plant breeding. *Trends in Genetics*. https://www.cell.com/trends/genetics/fulltext/S0168-9525(24)00167-7

Floridi, L. (2023). *The ethics of artificial intelligence: Principles, challenges, and opportunities*.

Floridi, L., Cowls, J., Beltrametti, M., Chatila, R., Chazerand, P., Dignum, V., Luetge, C., Madelin, R., Pagallo, U., Rossi, F., Schafer, B., Valcke, P., & Vayena, E. (2018). AI4People—An Ethical Framework for a Good AI Society: Opportunities, Risks, Principles, and Recommendations. *Minds and Machines*, *28*(4), 689–707. DOI: 10.1007/s11023-018-9482-5 PMID: 30930541

Gangurde, S. S., Kumar, R., Pandey, A. K., Burow, M., Laza, H. E., Nayak, S. N., Guo, B., Liao, B., Bhat, R. S., Madhuri, N., Hemalatha, S., Sudini, H. K., Janila, P., Latha, P., Khan, H., Motagi, B. N., Radhakrishnan, T., Puppala, N., Varshney, R. K., & Pandey, M. K. (2019). Climate-Smart Groundnuts for Achieving High Productivity and Improved Quality: Current Status, Challenges, and Opportunities. In Kole, C. (Ed.), *Genomic Designing of Climate-Smart Oilseed Crops* (pp. 133–172). Springer International Publishing., DOI: 10.1007/978-3-319-93536-2_3

Ghallab, M. (2019b). Responsible AI: Requirements and challenges. *AI Perspectives*, *1*(1), 3. DOI: 10.1186/s42467-019-0003-z

González-Rodríguez, V. E., Izquierdo-Bueno, I., Cantoral, J. M., Carbú, M., & Garrido, C. (2024). Artificial intelligence: A promising tool for application in phytopathology. *Horticulturae*, *10*(3), 197. DOI: 10.3390/horticulturae10030197

Green, B. P. (2018). *Ethical reflections on artificial intelligence*. https://dadun.unav.edu/handle/10171/58244

Habila, M. A., Ouladsmane, M., & Alothman, Z. A. (2023). Role of artificial intelligence in environmental sustainability. In *Visualization techniques for climate change with machine learning and artificial intelligence* (pp. 449–469). Elsevier., https://www.sciencedirect.com/science/article/pii/B9780323997140000091 DOI: 10.1016/B978-0-323-99714-0.00009-1

Haldorai, A., Babitha Lincy, R., Suriya, M., & Balakrishnan, M. (2024). Climate Resilience Via Smart Technologies Over Natural Disaster. In A. Haldorai, B. L. R, S. Murugan, & M. Balakrishnan, *Artificial Intelligence for Sustainable Development* (pp. 383–406). Springer Nature Switzerland. https://doi.org/DOI: 10.1007/978-3-031-53972-5_20

Heilinger, J.-C. (2022). The Ethics of AI Ethics. A Constructive Critique. *Philosophy & Technology*, *35*(3), 61. DOI: 10.1007/s13347-022-00557-9

Kaack, L. H., Donti, P. L., Strubell, E., Kamiya, G., Creutzig, F., & Rolnick, D. (2022). Aligning artificial intelligence with climate change mitigation. *Nature Climate Change*, *12*(6), 518–527. DOI: 10.1038/s41558-022-01377-7

Kang, B.-G., & Nam, Y. (2024). Responsible Artificial Intelligence for Climate Action: A Theoretical Framework for Sustainable Development. *Sustainable Machine Intelligence Journal*, *8*, 1–1. DOI: 10.61356/SMIJ.2024.88101

Khan, M. H. U., Wang, S., Wang, J., Ahmar, S., Saeed, S., Khan, S. U., Xu, X., Chen, H., Bhat, J. A., & Feng, X. (2022). Applications of artificial intelligence in climate-resilient smart-crop breeding. *International Journal of Molecular Sciences*, *23*(19), 11156. DOI: 10.3390/ijms231911156 PMID: 36232455

Khan, M. H. U., Wang, S., Wang, J., Ahmar, S., Saeed, S., Khan, S. U., Xu, X., Chen, H., Bhat, J. A., & Feng, X. (2022). Applications of artificial intelligence in climate-resilient smart-crop breeding. *International Journal of Molecular Sciences*, *23*(19), 11156. DOI: 10.3390/ijms231911156 PMID: 36232455

Kole, C., Muthamilarasan, M., Henry, R., Edwards, D., Sharma, R., Abberton, M., Batley, J., Bentley, A., Blakeney, M., Bryant, J., Cai, H., Cakir, M., Cseke, L. J., Cockram, J., de Oliveira, A. C., De Pace, C., Dempewolf, H., Ellison, S., Gepts, P., & Prasad, M. (2015). Application of genomics-assisted breeding for generation of climate resilient crops: Progress and prospects. *Frontiers in Plant Science*, *6*, 563. DOI: 10.3389/fpls.2015.00563 PMID: 26322050

Kumari, N., & Pandey, S. (2023). Application of artificial intelligence in environmental sustainability and climate change. In *Visualization techniques for climate change with machine learning and artificial intelligence* (pp. 293–316). Elsevier., https://www.sciencedirect.com/science/article/pii/B9780323997140000182 DOI: 10.1016/B978-0-323-99714-0.00018-2

Leal Filho, W., Wall, T., Mucova, S. A. R., Nagy, G. J., Balogun, A.-L., Luetz, J. M., Ng, A. W., Kovaleva, M., Azam, F. M. S., & Alves, F. (2022). Deploying artificial intelligence for climate change adaptation. *Technological Forecasting and Social Change*, *180*, 121662. DOI: 10.1016/j.techfore.2022.121662

Leal Filho, W., Wall, T., Mucova, S. A. R., Nagy, G. J., Balogun, A.-L., Luetz, J. M., Ng, A. W., Kovaleva, M., Azam, F. M. S., & Alves, F. (2022). Deploying artificial intelligence for climate change adaptation. *Technological Forecasting and Social Change*, *180*, 121662. DOI: 10.1016/j.techfore.2022.121662

Liu, Y., Tang, L., & Liu, G. (2022). Carbon Dioxide Emissions Reduction through Technological Innovation: Empirical Evidence from Chinese Provinces. *International Journal of Environmental Research and Public Health, 19*(15), 15. Advance online publication. DOI: 10.3390/ijerph19159543 PMID: 35954894

Marsh, J. I., Hu, H., Gill, M., Batley, J., & Edwards, D. (2021). Crop breeding for a changing climate: Integrating phenomics and genomics with bioinformatics. *Theoretical and Applied Genetics, 134*(6), 1677–1690. DOI: 10.1007/s00122-021-03820-3 PMID: 33852055

Miailhe, N., Hodes, C., Jain, A., Iliadis, N., Alanoca, S., & Png, J. (2019). AI for sustainable development goals. *Delphi, 2*, 207.

Mittal, R., Pathak, V., & Gaballa, M. (2025). Building Climate Resilience Through AI-Enabled Education Strategies for Community Empowerment. In *Community Resilience and Climate Change Challenges: Pursuit of Sustainable Development Goals (SDGs)* (pp. 29–40). IGI Global Scientific Publishing., https://www.igi-global.com/chapter/building-climate-resilience-through-ai-enabled-education-strategies-for-community-empowerment/363460

Mushtaq, M. A., Ahmed, H. G. M.-D., & Zeng, Y. (2024). Applications of artificial intelligence in wheat breeding for sustainable food security. *Sustainability (Basel), 16*(13), 5688. DOI: 10.3390/su16135688

Nabwire, S., Suh, H.-K., Kim, M. S., Baek, I., & Cho, B.-K. (2021). Application of artificial intelligence in phenomics. *Sensors (Basel), 21*(13), 4363. DOI: 10.3390/s21134363 PMID: 34202291

Orr, W., & Davis, J. L. (2020). Attributions of ethical responsibility by Artificial Intelligence practitioners. *Information Communication and Society, 23*(5), 719–735. DOI: 10.1080/1369118X.2020.1713842

Qiao, Y., Valente, J., Su, D., Zhang, Z., & He, D. (2022). AI, sensors and robotics in plant phenotyping and precision agriculture. In *Frontiers in Plant Science* (Vol. 13, p. 1064219). Frontiers Media SA. https://www.frontiersin.org/articles/10.3389/fpls.2022.1064219/full

Raihan, A., Ridwan, M., & Rahman, M. S. (2024). An exploration of the latest developments, obstacles, and potential future pathways for climate-smart agriculture. *Climate Smart Agriculture*, 100020.

Rane, N. (2023). Roles and challenges of ChatGPT and similar generative artificial intelligence for achieving the sustainable development goals (SDGs). *SSRN, 4603244*, •••. https://papers.ssrn.com/sol3/papers.cfm?abstract_id=4603244. DOI: 10.2139/ssrn.4603244

Rane, N., Choudhary, S., & Rane, J. (2024). Artificial intelligence for enhancing resilience. *Journal of Applied Artificial Intelligence*, *5*(2), 1–33. DOI: 10.48185/jaai.v5i2.1053

Rane, N., Choudhary, S., & Rane, J. (2024). Contribution of chatgpt and similar generative artificial intelligence for enhanced climate change mitigation strategies. *SSRN, 4681720*, •••. https://papers.ssrn.com/sol3/papers.cfm?abstract_id=4681720. DOI: 10.2139/ssrn.4681720

Rayhan, A. (2023). AI and the environment: Toward sustainable development and conservation. *Research Gate, July, 10.* https://www.researchgate.net/profile/Abu-Rayhan-

Roychowdhury, R., Das, S. P., Gupta, A., Parihar, P., Chandrasekhar, K., Sarker, U., Kumar, A., Ramrao, D. P., & Sudhakar, C. (2023). Multi-omics pipeline and omics-integration approach to decipher plant's abiotic stress tolerance responses. *Genes*, *14*(6), 1281. DOI: 10.3390/genes14061281 PMID: 37372461

Saad, N. S. M., Neik, T. X., Thomas, W. J., Amas, J. C., Cantila, A. Y., Craig, R. J., Edwards, D., & Batley, J. (2022). Advancing designer crops for climate resilience through an integrated genomics approach. *Current Opinion in Plant Biology*, *67*, 102220. DOI: 10.1016/j.pbi.2022.102220 PMID: 35489163

Sahil, K., Mehta, P., Bhardwaj, S. K., & Dhaliwal, L. K. (2023). Development of mitigation strategies for the climate change using artificial intelligence to attain sustainability. In *Visualization techniques for climate change with machine learning and artificial intelligence* (pp. 421–448). Elsevier., https://www.sciencedirect.com/science/article/pii/B9780323997140000212 DOI: 10.1016/B978-0-323-99714-0.00021-2

Satish, M., Babu, S. M., Kumar, P. P., Devi, S., & Reddy, K. P. (2023). Artificial Intelligence (AI) and the Prediction of Climate Change Impacts. *2023 IEEE 5th International Conference on Cybernetics, Cognition and Machine Learning Applications (ICCCMLA)*, 660–664. https://ieeexplore.ieee.org/abstract/document/10346636/

Satornino, C. B., Du, S., & Grewal, D. (2024). Using artificial intelligence to advance sustainable development in industrial markets: A complex adaptive systems perspective. *Industrial Marketing Management*, *116*, 145–157. DOI: 10.1016/j.indmarman.2023.11.011

Selvapriya, R., Gopinath, M., Velmurugan, D., & Tamilchelvan, P. (2024). Climate change mitigation through AI solutions. In *Global Challenges for the Environment and Climate Change* (pp. 252–277). IGI Global., https://www.igi-global.com/chapter/climate-change-mitigation-through-ai-solutions/351390

Shaamala, A., Yigitcanlar, T., Nili, A., & Nyandega, D. (2024). Algorithmic green infrastructure optimisation: Review of artificial intelligence driven approaches for tackling climate change. *Sustainable Cities and Society*, *101*, 105182. DOI: 10.1016/j.scs.2024.105182

Shneiderman, B. (2021). Responsible AI: Bridging from ethics to practice. *Communications of the ACM*, *64*(8), 32–35. DOI: 10.1145/3445973

Sinha, D., Maurya, A. K., Abdi, G., Majeed, M., Agarwal, R., Mukherjee, R., Ganguly, S., Aziz, R., Bhatia, M., Majgaonkar, A., Seal, S., Das, M., Banerjee, S., Chowdhury, S., Adeyemi, S. B., & Chen, J.-T. (2023). Integrated genomic selection for accelerating breeding programs of climate-smart cereals. *Genes*, *14*(7), 1484. DOI: 10.3390/genes14071484 PMID: 37510388

Sirmacek, B., Gupta, S., Mallor, F., Azizpour, H., Ban, Y., Eivazi, H., Fang, H., Golzar, F., Leite, I., Melsion, G. I., Smith, K., Fuso Nerini, F., & Vinuesa, R. (2023). The Potential of Artificial Intelligence for Achieving Healthy and Sustainable Societies. In Mazzi, F., & Floridi, L. (Eds.), *The Ethics of Artificial Intelligence for the Sustainable Development Goals* (Vol. 152, pp. 65–96). Springer International Publishing., DOI: 10.1007/978-3-031-21147-8_5

Srivastava, A., & Maity, R. (2023). Assessing the potential of AI–ML in urban climate change adaptation and sustainable development. *Sustainability (Basel)*, *15*(23), 16461. DOI: 10.3390/su152316461

Stahl, B. C. (2021). *Artificial intelligence for a better future: An ecosystem perspective on the ethics of AI and emerging digital technologies*. Springer Nature. https://library.oapen.org/handle/20.500.12657/48228

Stahl, B. C., Schroeder, D., & Rodrigues, R. (2023). *Ethics of artificial intelligence: Case studies and options for addressing ethical challenges*. Springer Nature. https://library.oapen.org/handle/20.500.12657/59315

Thudi, M., Palakurthi, R., Schnable, J. C., Chitikineni, A., Dreisigacker, S., Mace, E., Srivastava, R. K., Satyavathi, C. T., Odeny, D., Tiwari, V. K., Lam, H.-M., Hong, Y. B., Singh, V. K., Li, G., Xu, Y., Chen, X., Kaila, S., Nguyen, H., Sivasankar, S., & Varshney, R. K. (2021). Genomic resources in plant breeding for sustainable agriculture. *Journal of Plant Physiology*, *257*, 153351. DOI: 10.1016/j.jplph.2020.153351 PMID: 33412425

Ueda, D., Walston, S. L., Fujita, S., Fushimi, Y., Tsuboyama, T., Kamagata, K., Yamada, A., Yanagawa, M., Ito, R., Fujima, N., Kawamura, M., Nakaura, T., Matsui, Y., Tatsugami, F., Fujioka, T., Nozaki, T., Hirata, K., & Naganawa, S. (2024). Climate change and artificial intelligence in healthcare: Review and recommendations towards a sustainable future. *Diagnostic and Interventional Imaging*, *105*(11), 453–459. DOI: 10.1016/j.diii.2024.06.002 PMID: 38918123

Usigbe, M. J., Asem-Hiablie, S., Uyeh, D. D., Iyiola, O., Park, T., & Mallipeddi, R. (2023). Enhancing resilience in agricultural production systems with AI-based technologies. *Environment, Development and Sustainability*, *26*(9), 21955–21983. DOI: 10.1007/s10668-023-03588-0

Van Wynsberghe, A. (2021). Sustainable AI: AI for sustainability and the sustainability of AI. *AI and Ethics*, *1*(3), 213–218. DOI: 10.1007/s43681-021-00043-6

Varshney, R. K., & Tuberosa, R. (Eds.). (2007). *Genomics-Assisted Crop Improvement*. Springer Netherlands., DOI: 10.1007/978-1-4020-6297-1

Yadav, A. (2024). Leveraging Artificial Intelligence for Sustainable Development and Environmental Resilience. In *Exploring Ethical Dimensions of Environmental Sustainability and Use of AI* (pp. 140–165). IGI Global., https://www.igi-global.com/chapter/leveraging-artificial-intelligence-for-sustainable-development-and-environmental-resilience/334959

Chapter 10
Climate Neutrality-Based Sustainable Practices and AI for Developing Countries

Angel Paudel
https://orcid.org/0009-0003-1369-0952
Westcliff University, USA

Krishna Khanal
Westcliff University, USA

ABSTRACT

Climate change has become an emerging global issue in recent decades. As a result, the 'Climate Neutrality' concept has become a key framework for countries for balancing the carbon emissions generated. The chapter examines how AI technologies can enhance resource efficiency, optimize various human practices, and improve energy management, thereby contributing to climate neutrality. Case studies from developing countries illustrate successful implementations of AI-driven solutions that address local environmental challenges while contributing to community resilience. Additionally, we discuss the socio-economic implications of adopting these technologies, emphasizing the importance of inclusive policies that ensure equitable benefits. Ultimately, this chapter argues that integrating AI into sustainable development strategies in developing countries presents a transformative opportunity to achieve climate neutrality and foster a sustainable future.

DOI: 10.4018/979-8-3373-0680-3.ch010

1. INTRODUCTION

1.1 Importance of Climate Neutrality and Sustainable Practices in the Context of Developing Countries

The concept of climate neutrality has evolved over decades, rooted in the broader environmental movements of the 20th century. The 1972 Stockholm Conference marked the first global recognition of environmental sustainability, followed by the 1987 Brundtland Report, which introduced the idea of sustainable development (Mensah, 2019). The 1997 Kyoto Protocol and the 2015 Paris Agreement further cemented the global commitment to reducing carbon footprints (Mor et al., 2023).

Global attention has shifted from developed nations, which have made significant progress toward climate neutrality (CN), to developing countries, where there is an urgent need to borrow and adapt these strategies (Chen et al., 2024). The stakes are particularly high for developing countries as these countries have to bear both environmental degradation and socio-economic development challenges. CN is when greenhouse gas emissions are minimized as much as possible by transitioning to 100% renewable energy and implementing various other strategies. CN includes offsetting emissions that cannot be further reduced by purchasing credits, contributing to sustainable initiatives, or following other UN-recognized mechanisms to achieve a net-zero emissions balance. Achieving net zero emissions is crucial for securing a sustainable future and fulfilling the energy and environmental goals outlined in the Sustainable Development Goals (Vats & Mathur, 2024; Tiwari et al., 2024). Sustainable development focuses on using limited resources for the betterment of humans so that economic and social progress can be achieved (Pachar & Singh, 2021). For developing countries, reaching this goal is even more critical, as they are most vulnerable to the impacts of climate-related events (Adom, 2024).

For instance, according to a report from the European Commission (2024), Nepal contributes only 0.11% to global emissions. However, the country faces severe consequences, including disruptions in agriculture, erratic weather patterns characterized by heavy rainfall at times, drought-like situations at others, and glacial melt (Shrestha & Aryal, 2011). These events have seriously impacted people's livelihoods by reducing agricultural yields and creating powerful currents for certain months while experiencing low flow during others. This variability complicates electricity generation from specific projects, directly affecting the revenue of governmental organizations.

In Papua New Guinea (PNG), communities have been highly vulnerable to climate change-related issues, particularly rising sea levels (Kuir-Ayius, 2024). This has jeopardized traditional fishing practices, a cornerstone of their economy, leading to the displacement and forced relocation of communities (Jolly et al., 2024).

Consequently, PNG, despite contributing only 0.02% of global emissions according to the European Commission (2024), faces significant challenges that are largely beyond its control.

Similarly, the Maldives faces increasing risk due to rising sea levels. This rising sea level means saltwater intrusion jeopardizes freshwater supplies, which would affect 12% of the fish-catching rate by 2050 (UN ESCAP, 2024). Additionally, the country is projected to see a loss of USD 1.2 billion on a per-year basis, with a reduction of 30% in tourists visiting the nation by 2100 (Hosterman & Smith, 2025).

Likewise, in Sri Lanka, unpredictable weather patterns including drought have impacted approximately 10% of the population (Amarnath et al., 2017). These climate shocks have affected the livelihoods of the people with about a 40% decline in paddy production due to worse drought conditions in late 2016 (Gunaratne et al., 2021). But again, heavy rainfall in May of 2017 resulted in the deterioration of crops– heavily impacting over two hundred thousand families due to localized flooding in southwestern parts of the country (Zappacosta & Hollema, 2017).

These examples from four developing countries highlight their interconnected but diverse challenges. Together, they contribute to 0.21% of global emissions, as per the European Commission (2024), but they face several climate-related impacts that threaten their livelihoods, economies, and entire ecosystems.

The potential for solar and wind energy remains underexplored, highlighting a broader trend in developing nations. For example, Papua New Guinea, despite its significant solar potential, faces similar issues with limited investments and technical expertise (The World Bank, 2020a). In the Maldives, due to high initial cost and geographic dispersion – the integration of renewal energy is significantly hindered (The World Bank, 2020b). However, solar microgrids have shown some promise in addressing energy access issues (Asian Development Bank, 2020). Sri Lanka's renewable energy initiatives, such as wind power in Hambantota, demonstrate progress but remain limited by outdated infrastructure and financial constraints (Koswatte et al., 2024). All these examples exemplify various challenges but also opportunities within developing nations where, if those untapped renewable resources are well leveraged, they would significantly help advance sustainability goals.

1.2 Role AI Can Play in Addressing Climate-Related Challenges

Technological advancements have helped significantly impact our social life while also changing the ways we think and act (Fransisca & Ningsih, 2023). Artificial intelligence (AI) is part of that advancement in technology which has introduced a new angle for tackling complex environmental challenges through innovative solutions. These solutions range from enabling precision agriculture to optimizing renewable energy grids to monitoring biodiversity (Chisom et al., 2024; Akter,

2024). Developing countries often face the challenges of gathering resources – in cash or materials to undertake large-scale interventions. This is where AI comes to the rescue with an affordable solution through efficient and scalable solutions at low cost (Olawade et al., 2024).

Energy management for renewable sources like hydropower often can be simplified with the involvement of AI and Machine Learning (Entezari et al., 2023). With these energy sources – the common approach used is Peaking Run-of-River (PRoR), which means that the power is generated without having storage (Bhatt & Joshi, 2024). AI and Machine learning algorithms help optimize energy supply, minimize wastage, and maximize efficiency by predicting energy demand (Mersad Shoaei et al., 2024). This would be significant for a developing country like Nepal which experienced a 9.59% overall loss while some province offices like Karnali recorded this at 21.79% in the year 2023/24 records (Nepal Electricity Authority, 2024). In Kenya, the use of AI has helped analyze seismic activity, geological data, and temperature gradients to explore and identify geothermal energy resources (Ibegbulam et al., 2023).

These examples show the potential to act as a driving force towards climate resilience in developing countries. By addressing the critical gaps present in sectors including energy, agriculture, and disaster management, AI technologies can pave the way to maintain/limit the adverse effects of climate change while also achieving sustainability goals.

1.3 Challenges Faced by Developing Countries in Achieving Sustainability

Sustainability has been a main concern for different organizations, regulatory authorities, and business professionals in today's business environment (Singh et al., 2022). Developing countries when it comes to achieving sustainability face significant challenges due to a lack of fully developed technological infrastructure, a lack of qualified manpower to operate that advanced system, inadequate financial resources, and a lack of policies (Sarvajayakesavalu, 2015). The deployment of AI technologies to achieve sustainability requires a solid digital ecosystem with high internet speed, low latency, and high computational and data storage capacity (Rohde et al., 2024). However, many developing nations lack these elements. For example, Nepal does have disproportionate distribution of internet connectivity and lacks stable connection with poor infrastructure as we move away from the urban town (Pandey & Regmi, 2017). This results in difficulty in adopting AI-driven solutions for energy optimization and agriculture. In Sub-Saharan Africa, only 28.5% of individuals live in households with internet access (Seuyong et al., 2023). Such shortages for critical enablers for digital sustainability initiatives create a substantial barrier creating delays in implementing and scaling climate-resilient technology.

Sustainability-focused projects often demand large upfront investment in technology, training, and capacity building. Public debt in developing countries increases twice as fast and has reached 60% in 2021 (UN Trade and Development, 2024). This signals a massive challenge to these countries. In countries like Bangladesh, which is ranked as the seventh extreme disaster risk nation (United Nations Development Programme, 2022); the progress in AI infrastructure is often stalled due to a lack of funding for such initiatives.

Lack of or weaker frameworks and policies in place in developing countries compound the financial and technological issues further. These developing countries often set very ambitious climate targets; but with the absence of a robust regulatory framework and enforcement mechanism, the implementation of it is heavily impacted. For example, Nepal announced in COP26 to be cumulatively 'net zero carbon' from 2022-2045 and 'carbon negative' after that. However, there's no policy or law made about it to achieve the ambitious target set forth.

2. BACKGROUND AND CONTEXT

2.1 Global Frameworks

The Paris Agreement and United Nations Sustainable Development Goals (SDGs) have been the key frameworks for combatting climate change efforts around the world. Adopted in the year 2015, the Paris Agreement saw 196 parties come together to limit the increase in the global average temperature to well below 2°C above pre-industrial levels and continue further efforts to limit the temperature increase to 1.5°C above pre-industrial levels (UNFCCC, 2016). It is a landmark, binding treaty pulling all the nations to put up a fight against climate change effects. At the core of it, the agreement focuses on the nationally determined contributions (NDCs) allowing for communication of the actions by countries to reach the agreement's goals while reducing greenhouse gas emissions. The developing nations do not have the needed technical and financial strength to be able to fulfill these commitments. This is where collaboration and international support about the commitments is critical and it could be achieved through the effective implementation of the Green Climate Fund.

United Nations's SDG goal 13 calls to, "take urgent action to combat climate change and its impacts" – which provides a detailed roadmap for sustainable development. The framework encourages nations to incorporate into their policies climate measures to improve education and awareness and build resilience to deal against any climate-related threats. In Nepal, the nation has incorporated SDG 13 into the national adaptation plan including agriculture, livestock, irrigation, industry, tourism, health, forests, biodiversity, wildlife conservation, urban development, and disaster management (Government of Nepal, 2021). However, due to limited data-driven insights and institutional capacity – it remains a challenge to align local initiatives with a global framework to achieve the goals.

2.2 Disproportionate Impact of Climate Change on Developing Countries

Developing countries face a disproportionate impact due to climate change. This is even when these countries contribute very minimally to global emissions. Let's consider Nepal, which only contributes to 0.05% of global CO2 emissions (European Commission, 2024). However, the nation still sees dire impacts due to climate change effects. The snow-covered mountains have slowly lost their snow and many of those mountains appear like a bundle of rocks. The glacial retreat and erratic monsoons mean that the frequency of floods and drought is high in different parts of the nation – adding to the displacement of people in rural areas and increasing food insecurity. Nepal also heavily relies on hydropower – and a decrease in runoff water flow would impact electricity generation while the increase in it in the form of floods would result in the project needing to be halted – all those resulting in the energy generation efforts to see some disturbance (Pathak, 2010).

Climate change effects have also resulted in the rise of sea levels at an increasing rate (Griggs & Reguero, 2021). Throughout most of the 20th century, the rate at which the sea level rose was 1.4 millimeters but that has more than doubled to 3.6 millimeters per year from 2006 to 2015 (Lindsey, 2023). Maldives, which is comprised of nearly 1,200 small coral islands is also one of the flattest countries on Earth with over 80% of its land mass being 1 meter above mean sea level (Garrett, 2023). Due to this rise in sea level, it is expected that around 77% of the nation's land will be submerged by 2100. Similarly, Kiribati has an extremely low elevation of around 6 feet (World Bank Group, 2021). This means, that if the sea level is to rise by 3 feet from the current level – the nation will lose two-thirds of its land. Due to these events, several cities including Dhaka (22.4 million people); Lagos (15.3 million people); and Bangkok (9 million people) could be impacted as these cities are entirely drowned or have vast tracts of land underwater and unusable (Lakritz, 2019).

All these challenges highlight the urgent need for global strategies that address these unique challenges that developing nations face while also ensuring that sustainability and inclusive growth are achieved.

2.3 AI Applications in Environmental Science and Resource Management

Artificial intelligence (AI) can play a very important role through its scalable, data-driven solutions which can help in the optimal utilization of resources. When it comes to renewable energy management, most of the leakages are due to the inability to predict power demand with precision. AI-powered algorithms can optimize the energy grid by predicting demand to ensure the best utilization strategy. This ensures the supply of electricity is regulated with excess demand sold in case there's no means to store it. If deficient, the purchase of required demand can be pre-planned to ensure a smooth supply of energy. As such, with proper adaptation of AI – it can help design efficient energy grids to meet consumer demand, improving disaster preparedness while also generating revenue from it.

AI's contributions do also extend to biodiversity conversation. AI's object recognition system can help recognize particular objects in photos and videos (Bhatt & Sharma, 2022). The same tech, combined with deep learning models, can be used to evaluate imagery data from satellites or other means to automate the monitoring process for deforestation, track the population of wildlife, and detect illegal activities. One of the examples of this use case comes from the International Centre for Integrated Mountain Development (ICIMOD) headquartered in Nepal, which installed a camera trap at its Knowledge Park at Godavari, Nepal. The data that came from the camera was used for the study, training it against different models. The current system helped the organization identify animals with >93.8% accuracy while using it to describe the behaviors of 48 species in 3.2 million images from the Snapshot Serengeti dataset (Shrestha et al., 2022). The system, if it is to be altered to just classify images that it's confident for could help increase the accuracy to 99.3%. If the system is to only perform at the same 96.6% accuracy rate that the human volunteers have would save >8.4 years i.e. 17,000 hours at 40 hours/week work hours (Shrestha et al., 2022). This is significant and the rising data set would only help more time and energy while automating the overall process.

Another prominent example is from the second largest state in Brazil – Pará. The state alone contributes 38% of total deforestation from 2021 to 2023 for the Brazilian Legal Amazon (the largest socio-geographic division in Brazil). A model called "PrevisIA" is suggested to help predict the risk and help counter that. PrevisIA makes use of socioeconomic data, historical deforestation, topological data, and waterbodies to identify risks (Sales et al., 2017). The model also makes use of AI to identify unofficial roads – as most of these activities are carried out within a 5.5km radius, which would not be possible without road access at scale (Barber et al., 2014). The model successfully helped identify no risk to very high-risk areas, which would help consolidate efforts to reduce deforestation.

Further, AI is also used as a powerful tool in climate science, where it helps in climate modeling and disaster recovery. AI can help process huge amounts of data to identify patterns that are generally missed by traditional methods (Othman, 2024). This helps in creating more detailed insights and accurate forecasts. It also helps create predictions that are beyond the current range for days in the future that it can predict. This helps decision makers anticipate the event in the future, allowing them to take required preventive and protective actions to ensure the damage is reduced. This helps countries like Bangladesh, which experiences 5% of the global cyclones but more than 80% of global casualties, as per Debsarma (2009), be better prepared. The country can relocate its citizens while also having medical and other help available right away to ensure that the casualties are significantly reduced.

The use case of AI goes beyond like in transport sector where AI is used by tech platforms to calculate and recommend the shortest path to a destination – not just by distance but based on how busy the roads are (Sudheer Varma Surimalla, 2018). This helps reduce the emission of CO2 to a certain degree. A similar use case is also seen in the local context in Nepal, where a tech platform Pathao makes use of AI to compute the shortest path and also calculate the fare for each trip (Suman, 2022).

All these examples show the transformative potential that AI holds in addressing resource constraints while helping boost productivity and fostering climate resilience in developing nations.

3. CLIMATE NEUTRALITY-BASED SUSTAINABILITY PRACTICES

The impact of climate change is well established and there's no refuting the fact. This is where CN-based sustainability practices become vital to minimize the effect on developing nations. To ensure that social and economic development are fostered – the adoption of a circular economy can be key. This focuses on reducing wastage, reusing resources, and recycling resources to create a system that helps minimize harm to the environment (Arruda et al., 2021). Circular economy integrates environmental benefits with economic representing a paradigm shift towards sustainable practices (Geissdoerfer et al., 2017). Nepal and Rwanda demonstrate this model in action with strict bans on single-use plastic throughout the nation (Bharadwaj, 2016; Ogutu et al., 2023). Through this initiative nationwide, these countries aimed to reduce waste and promote sustainable use of resources.

Additionally, strategies to achieve zero carbon is critical to ensure that CN goals are achieved. This can be done through renewable energy sources, including hydropower, geothermal plants, wind, and solar. All these energy sources can provide a clean source of energy and help the nation transition to achieve CN. One such developing country that has been carbon-negative for a long time now is Bhutan. The nation reached carbon-negative status through the development and promotion of hydropower as its primary source of energy while drafting policies to prioritize vast forest cover as well (Yangka et al., 2018). This approach not only helps achieve economic growth but also increases energy security while mitigating greenhouse gas emissions.

Yet another, solid foundation of climate-neutral practices could be the promotion and implementation of sustainable urban development. In developing countries, if well implemented urban centers can minimize energy consumption and resultant emissions. Singapore is a perfect example of an extensive public transport system and green building without compromising on the standard of living and ensuring environmental sustainability (Diao, 2019). The governmental agency in Singapore's Building and Construction Authority (BCA) set up a policy to promote environmentally friendly and energy-efficient building practices with the Green Mark certification scheme in 2005 (Lam & Toan, 2006). Additionally, Singapore also has one of the most efficient public transport systems serving over 3 million people through the city's Mass Rapid Transit (MRT) system (Patel & Padhya, 2021). This helped people opt for public transit versus private vehicles helping encourage sustainable practice. All this establishes Singapore at the forefront of sustainable urban development.

All these practices show that developing countries, despite facing resource constraints can achieve climate-neutral objectives by aligning their developmental goals with sustainable practices. For example, Ethiopia through their works and adoption of frameworks like the Climate Resilient Green Economy (CRGE) strategy and the second Growth and Transformation Plan (GTP II), shows how countries can address climate challenges while developing sustainably (Federal Democratic Republic of Ethiopia, 2019). The government of Ethiopia, through reforestation programs and investments in renewable energy, plans to become a middle-income economy and a carbon-neutral nation by 2030 (Kassa et al., 2022). In the year 2016, Ethiopia pledged to restore 15 million hectares of degraded land as part of the African Forest Landscape Restoration Initiative (Kassa et al., 2022). The country also planned for significant investment towards renewal sources with projects like the Grand Ethiopian Renaissance Dam – which, however, due to politics, was stalled (Matthews & Vlado Vivoda, 2023). Regardless of the current slowdown, this ambitious project not only aims to meet the local demands for energy locally but also aims to export electricity to the neighboring countries. This would eventually help Ethiopia generate revenue from it while also helping cut down on other non-clean energy source usage.

All these policies, practices, and actions highlight that regardless of the developing countries facing constraints when it comes to resources, they can align their development goals with climate-neutral objectives and lead the way from the front – setting benchmarks for others to follow.

4. PRECISION AGRICULTURE, ORGANIC AGRICULTURE, PREDICTIVE ANALYTICS FOR DISASTER MANAGEMENT, SMART GRIDS, AND AI

Artificial intelligence is improving the efficiency, resilience, and sustainability of traditional agriculture and resource management. Most developing countries continue to face issues related to resource limitations, climate vulnerability, and population pressures. However, with the use of AI-driven solutions, these nations are trying to tackle those challenges sustainably.

4.1 Precision Agriculture

Precision agriculture makes use of AI-driven technologies like drones, the Internet of Things (IoT), and deep learning algorithms to help reduce the impact on the environment, optimize resource usage, and increase the yield rate of crops. A study by Pajany et al. (2023) shows that the use of AI in irrigation management has helped reduce the wastage of water in India by 53% compared to traditional methods. Considering for nation like India where agriculture is a major water consumer – this significant reduction in water wastage helps a lot. Similarly, AI tools can also be used to analyze the health of the soil and predict weather patterns to create a recommendation for the type of crop to be planted for the best result (Kumar Apat et al., 2023).

Similarly, AI-powered application has the potential to help farmers identify crop diseases through smartphone images and provide instant solutions for pest control and management of plant disease (Mrisho et al., 2020). Solutions like the one by PlantVillage with Nuru, an AI assistant that works inside a standard smartphone and does not require any additional device or setup, give ease and control to the farmers. The app is capable of accurately diagnosing cassava diseases offline, without an internet connection. According to the research done by L. Mrisho et al. (2020), symptom recognition when using six leaves was 74 to 88%, depending on the condition similar to that of experts, 1.5 times higher than agricultural extension agents and two times higher than farmers. Cassava mosaic disease (CMD) stands out as a major threat to cassava cultivation in many African nations (Fauquet & Fargette, 1990). This disease reduces yield from 20 to 95% resulting in losses between USD 1.9 to 2.7 billion a year (Ahoya et al., 2024). This simple yet powerful application alone, it shows that AI-based tools can empower farmers to address agricultural challenges while contributing towards food security and increasing productivity.

4.2 Organic Agriculture

Organic agriculture is critical to ensure farmers and nations as a whole contribute to adopting sustainable farming practices. AI technologies can help towards the goal. This is by helping farmers monitor soil conditions, and crop growth and identify pest infestation without synthetic chemical usage. One of the research projects around it from Akkas Ali et al. (2024) suggests the DenseNet model which is considered to be an AI-enabled IoT-based pest prevention and controlling system using sound analytics. The model researched achieved pest detection accuracy of up to 99.65%, which is superior to all the current models in use (Akkas Ali et al., 2024). One successful implementation of AI-driven recommendation comes from Brazil's 5,000-acre soybean farm where they implemented Climate FieldView's AI-driven recommendation system for seed selection, planting dates, and management of the nutrient, which helped the farm reduce its cost by 11% and increased yield by 9.5% (World Economic Forum, 2018).

Additionally, AI solutions can guide the use of fertilizers and pesticides only when and where they are required. Such, AI-powered solutions have helped reduce fertilizer use by up to 35% and herbicide by up to 70% with no impact on crop yield (European Union, 2016). This demonstrates that natural control measures can not only prevent negative impacts on soils but also ensure that the farmers can ensure they use less of those chemicals while retaining the same or higher yield rate.

4.3 Predictive Analytics for Disaster Management

Able to predict what might happen would not only help reduce the damage but also save lives – this is what AI could assist in disaster management via predictive analytics. In Bangladesh, a country that experiences a higher number of floods and casualties, being able to forecast floods accurately would help prevent economic losses and reduce human casualties through an early warning system (Rajab et al., 2023).

AI can also be used to predict droughts, especially in dry regions like in Africa. Abebe and Endalie (2023), used AI-based models to predict monthly rainfall at 92 Ethiopian meteorological stations, helping monitor the pattern, and soil moisture and enabling timely interventions to prevent crop failure. As the nation sees a higher number of floods and droughts due to unpredictable rainfall – this can help form strategies at the nation level to limits its damage. These innovations are vital for developing nations to ensure a climate-resilient society.

4.4 Smart Grids

AI-powered smart grid solutions can help reduce carbon footprint and enhance energy efficiency. The smart grid is considered as one of the most promising technologies for optimizing energy distribution and consumption (Silva et al., 2014). With over 10% of the distributed energy being wasted – this results in high inefficiency, corresponding to thousands of tons of avoidable CO2 emitted (Daví-Arderius et al., 2017). The use of smart grids can help reduce transmission losses while helping monitor and ensure the regular supply of energy is uninterrupted. The grid-enabled system can save up to 29% of energy while also helping optimize supply and demand (Swapnil Mahadik et al., 2024). According to Swapnil Mahadik et al. (2024), the use of a smart grid can help in reducing greenhouse gas emissions by up to 20%.

4.5 AI for Climate-Neutral Development

The transformative potential of achieving climate-neutral development is indeed possible to achieve with the integration of AI across all those domains including precision agriculture, organic agriculture, smart grids, and predictive analysis. AI helps developing countries deal with climate-related challenges while promoting economic growth. This is by promoting sustainable practices while reducing wastage and enhancing efficiency. All these benefits start from the very short term but also form a very strong foundation for sustainability in the long term.

With the integration of AI in those different areas, developing countries can take a leap forward towards achieving climate neutrality. And, become a model for other nations to replicate and achieve the same success.

5. IMPORTANCE OF PUBLIC-PRIVATE PARTNERSHIPS

Public-private partnerships (PPPs) are being used globally at a higher rate as ways to provide climate mitigation and adaptation interventions, including low-carbon, climate-resilient infrastructure (Casady et al., 2024). PPP refers to joint agreements between public and private sector enterprises by bringing together expertise, resources, and risk-sharing mechanisms to achieve mutual objectives. In this case, this might involve combining the regulatory framework from governments with the innovations and efficiency from the private sector to achieve climate-neutral goals. By sharing the risks with PPP and coming to the same page of the shared vision – this helps address systemic challenges to achieve climate resilience and sustainability.

In Nepal, the government offered a PPP scheme for the construction of new hydropower projects along with the provision of a guaranteed 10% share to the affected general public of the reason for all such projects (Acharya et al., 2013). The first such project was completed under this scheme in the year 2003 with Chilime Hydropower which produced 22MW of energy (Acharya et al., 2013). This success meant further bigger power projects were initiated under the same PPP model accounting for 570MW (Acharya et al., 2013). Similarly, Olando et al. (2024) through the study of 48 geothermal energy development projects in Kenya found that the PPP financing structure meant that the higher the level of PPP involvement, the faster the project completion time as well.

Regardless of all the advantages that PPP offers, it also faces its fair share of challenges. These challenges come in the form of unequal risk distribution, such as the private sector being at a loss more so than the government, inefficient bureaucratic processes, and misaligned objectives. To ensure these challenges do not derail future collaborations, it is important to promote transparency and improve regulatory frameworks. While also ensuring that such initiatives prioritize just benefit-sharing for both the parties involved to efficiently address climate-related challenges.

6. AI FOR BIODIVERSITY MONITORING, CARBON CAPTURE, AND CLIMATE MODELING

AI-based tools are emerging as significant contributors to addressing complex environmental challenges especially when it comes to climate modeling, biodiversity monitoring, and carbon capture. This advancement gives hope to developing countries to ensure that CN does not come at the cost of stalled economic growth. But with these tools, the nations can balance economic growth with CN efforts.

AI in biodiversity monitoring offers hope and something different as the current traditional system has mostly failed to achieve its goals. This is as not even one out of 20 agreed-upon goals (also called Aichi Biodiversity Targets) for the period of 2011 to 2020 has been fulfilled (Secretariat of the Convention on Biological Diversity, 2020). This calls for a more urgent need to design solutions that are more realistic and relevant for the current time (Sterner et al., 2019). The power of AI offers the opportunity to optimize the protection of biodiversity in a rapidly evolving and complex setup. Deep learning algorithms can be used to analyze a vast array of data collected from different sensors, camera traps, and satellite imagery – all of which can help identify patterns and understand changes in habitat and distribution of species.

In citizen science initiatives as well AI-driven solutions can provide significant contributions. An example of this can be eBird which pulls the contributions of human observers and processes their contributed data with AI aiding to democratize biodiversity monitoring while engaging the local community towards conservation efforts (Kelling et al., 2013). This is critical in developing countries where resources are very limited when it comes to large-scale biodiversity assessments.

Carbon capture is critical to negate climate change impacts resulting from the release of carbon into the atmosphere. In today's world, the use of digital technologies to address different issues has risen, and AI algorithms play a vital role by enabling real-time monitoring of carbon storage sites – ensuring safe and effective capture (Priya et al., 2023). One such example comes from Iceland, a developed country, and the project Carbfix 2. The later version, with the inclusion of AI, offers a faster way to convert the captured CO2 (Ratouis et al., 2022).

Climate predictions have improved with the use of advanced algorithms with AI, which has largely benefitted climate modeling. The ability to analyze historical climate data to predict/simulate future events, including extreme weather conditions, is now possible with deep learning techniques (Yang et al., 2024). The use of technology and interconnectedness can help improve sustainability initiatives, reduce carbon emissions, and help in economic growth (Nwokolo et al., 2023). In developing countries, AI-powered climate models enable the country to run risk assessments and design mitigation strategies. For example, this can help sub-Saharan Africa predict drought and flood patterns to enable people to adopt climate-resilient practices. In South Asia countries like Bangladesh – this can offer early warning for cyclones, reducing the impact of climate-related events on vulnerable communities.

7. ETHICAL CONSIDERATIONS

Ethical considerations in climate neutrality-based sustainable practices and AI for developing countries revolve around fairness, transparency, and inclusivity. AI-driven solutions must be designed to prevent biases that could disproportionately affect vulnerable populations, ensuring equitable access to clean energy, sustainable agriculture, and climate adaptation technologies (Hanna et al., 2025; Ukoba et al., 2024; Mana et al., 2024). Data privacy and security also present significant challenges, as the collection and use of environmental and socio-economic data must align with ethical standards and local regulations. Additionally, the reliance on AI should not exacerbate existing digital divides; instead, it should empower communities with accessible and contextually relevant solutions. The environmental footprint of AI itself, including energy-intensive computing, must also be addressed to prevent technological advancements from undermining sustainability goals. Ethical

AI governance, stakeholder engagement, and policies that balance innovation with social responsibility will be key to ensuring that AI-driven sustainability initiatives truly benefit developing nations.

8. CONCLUSION

Achieving CN is an environmental target and a holistic approach where various sectors like energy, agriculture, and waste management come together. The interconnectedness of sustainable practices, climate neutrality, and AI presents a pathway for developing countries to combat the challenges created by climate change. By using artificial intelligence, developing countries can harness the potential of AI and adopt solutions that protect the environment and reduce pollution. However, it is crucial to acknowledge the potential risks and challenges associated with AI-driven climate solutions regarding data privacy, algorithmic bias, and the digital divide, which must be carefully addressed to ensure equitable and inclusive development. AI has a big role in this situation by providing solutions to data analysis, resource optimization, and predictive modeling that can help increase the effectiveness of sustainability initiatives. AI can help developing countries make informed decisions and implement sustainable and economically viable strategies. Developing nations need to invest in digital literacy, digital capabilities, AI, and sustainable technologies to achieve climate neutrality. Various actors like governments, NGOs, and businesses should collaborate to create synergies that drive AI-based sustainable development initiatives. Collaboration can help fill the gaps in knowledge, technology, and funding, helping developing countries to use these technologies for a sustainable future. By prioritizing these partnerships, we can create a more resilient, equitable, and climate-neutral future.

9. FUTURE RESEARCH DIRECTIONS

Future research on climate neutrality-based sustainable practices and AI in developing countries should focus on enhancing AI-driven climate modeling, optimizing renewable energy integration, and improving climate-smart agriculture. Investigating the role of AI in circular economy strategies, waste reduction, and carbon capture technologies could yield significant insights. Additionally, ethical concerns related to AI deployment, such as data accessibility, bias, and governance, warrant deeper exploration. Research should also examine the socio-economic impact of AI adoption in sustainability, ensuring that technological advancements align with local contexts and promote inclusive development. Interdisciplinary collaboration between policymakers, technologists, and environmental scientists will be crucial in designing scalable, cost-effective AI solutions that drive sustainability while addressing the unique challenges faced by developing nations.

REFERENCES

Abebe, W. T., & Endalie, D. (2023). Artificial intelligence models for prediction of monthly rainfall without climatic data for meteorological stations in Ethiopia. *Journal of Big Data*, *10*(1), 2. Advance online publication. DOI: 10.1186/s40537-022-00683-3

Acharya, B., Akil, A. M., Ridhanda, P., Na, Y. J., Hong, C. Y., & Park, K. W. (2013, June). Public Private Partnership in Hydropower Development: Prospects for Nepal's Development. *Conference: Energy Transition and Policy Challenges, 36th IAEE International Conference*. https://www.researchgate.net/publication/353209903_Public_Private_Partnership_in_Hydropower_Development_Prospects_for_Nepal's_Development

Adom, P. K. (2024). The socioeconomic impact of climate change in developing countries over the next decades: A literature survey. *Heliyon*, *10*(15), e35134. DOI: 10.1016/j.heliyon.2024.e35134 PMID: 39170312

Ahoya, D. K. D., Sawadogo-Compaore, E. M. F. W., Yabi, J. A., Zandjanakou-Tachin, M., Houngue, J. A., Houedjissin, S. S., Pita, J. S., & Ahanhanzo, C. (2024). Factors Influencing Adoption of the PlantVillage Nuru Application for Cassava Mosaic Disease Diagnosis Among Farmers in Benin. *Agriculture*, *14*(11), 2001. DOI: 10.3390/agriculture14112001

Akter, M. S. (2024). AI for Sustainability: Leveraging Technology to Address Global Environmental. *Journal of Artificial Intelligence General Science (JAIGS) ISSN:3006-4023, 3*(1), 40–48. DOI: 10.60087/jaigs.v3i1.64

Amarnath, G., Alahacoon, N., Smakhtin, V., & Aggarwal, P. (2017). *Mapping multiple climate-related hazards in South Asia*. International Water Management Institute (IWMI).

Arruda, E. H., Melatto, R. A. P. B., Neto, W. L. B. da S., & Conti, D. de M. (2021). Circular Economy: A Brief Literature Review (2015-2020). *Sustainable Operations and Computers*, *2*, 79–86. DOI: 10.1016/j.susoc.2021.05.001

Asian Development Bank. (2020). *A Brighter Future for Maldives Powered by Renewables*. Asian Development Bank.

Barber, C. P., Cochrane, M. A., Souza, C. M.Jr, & Laurance, W. F. (2014). Roads, deforestation, and the mitigating effect of protected areas in the Amazon. *Biological Conservation*, *177*, 203–209. DOI: 10.1016/j.biocon.2014.07.004

Bharadwaj, B. (2016). *Plastic Bag Ban in Nepal: Enforcement and Effectiveness*. South Asian Network for Development and Environmental Economics (SANDEE).

Bhatt, M. W., & Sharma, S. (2022). An Object Recognition-Based Neuroscience Engineering: A Study for Future Implementations. *Istanbul University -. Journal of Electrical and Electronics Engineering (Oradea), 23*(2). Advance online publication. DOI: 10.5152/electrica.2022.22125

Bhatt, P., & Joshi, K. R. (2024). Hydropower Development in Nepal: Status, Opportunities and Challenges. *Journal of UTEC Engineering Management, 2*(01), 125–135. DOI: 10.36344/utecem.2024.v02i01.011

Casady, C. B., Cepparulo, A., & Giuriato, L. (2024). Public-private partnerships for low-carbon, climate-resilient infrastructure: Insights from the literature. *Journal of Cleaner Production, 470*, 143338–143338. DOI: 10.1016/j.jclepro.2024.143338

Chisom, O. N., Biu, P. W., Umoh, A. A., Obaedo, B. O., Adegbite, A. O., & Abatan, A. (2024). Reviewing the Role of AI in Environmental Monitoring and conservation: A data-driven Revolution for Our Planet. *World Journal of Advanced Research and Reviews, 21*(1), 161–171. DOI: 10.30574/wjarr.2024.21.1.2720

da Silva, R. M., Schettino, S., de Lima, P. M. R., & Adriano, F. C. (2014). SMART GRID: Evaluation and Trend in Brazil. *Independent Journal of Management & Production, 5*(3), 752–776. Advance online publication. DOI: 10.14807/ijmp.v5i3.122

Daví-Arderius, D., Sanin, M.-E., & Trujillo-Baute, E. (2017). CO2 content of electricity losses. *Energy Policy, 104*, 439–445. DOI: 10.1016/j.enpol.2017.01.011

Debsarma, S. K. (2009). Simulations of Storm Surges in the Bay of Bengal. *Marine Geodesy, 32*(2), 178–198. DOI: 10.1080/01490410902869458

Diao, M. (2019). Towards sustainable urban transport in Singapore: Policy instruments and mobility trends. *Transport Policy, 81*, 320–330. DOI: 10.1016/j.tranpol.2018.05.005

Entezari, A., Aslani, A., Zahedi, R., & Noorollahi, Y. (2023). *Artificial intelligence and machine learning in energy systems: A bibliographic perspective. Energy Strategy Reviews, 45(101017), 101017*. Elsevier., DOI: 10.1016/j.esr.2022.101017

European Commission. (2024). *EDGAR - The Emissions Database for Global Atmospheric Research*. Europa.eu. https://edgar.jrc.ec.europa.eu/report_2024

European Union. (2016). Precision agriculture and the future of farming in Europe Scientific Foresight Study. In *europa.eu*. https://www.europarl.europa.eu/RegData/etudes/STUD/2016/581892/EPRS_STU(2016)581892_EN.pdf

Fauquet, C., & Fargette, D. (1990). African Cassava Mosaic Virus: Etiology, Epidemiology, and Control. *Plant Disease*, *74*(6), 404. DOI: 10.1094/PD-74-0404

Federal Democratic Republic of Ethiopia. (2019). *Ethiopia's National Adaptation Plan*. https://www.preventionweb.net/media/97388/download?startDownload=20250104

Fransisca, V., & Ningsih, W. (2023). The Advancement of Technology and its Impact on Social Life in Indonesia. *Devotion Journal of Community Service*, *4*(3), 860–864. DOI: 10.36418/devotion.v4i3.445

Garrett, K. (2023). *The Adaptive Capacity to implement Climate Migration Policy in response to Sea-Level Rise in the Maldives*. https://digitalcollections.sit.edu/capstones/3275?utm_source=digitalcollections.sit.edu%2Fcapstones%2F3275&utm_medium=PDF&utm_campaign=PDFCoverPages

Geissdoerfer, M., Savaget, P., Bocken, N. M. P., & Hultink, E. J. (2017). The Circular Economy - a New Sustainability Paradigm? *Journal of Cleaner Production*, *143*(1), 757–768. DOI: 10.1016/j.jclepro.2016.12.048

Government of Nepal. (2021). *National Adaptation Plan (NAP) 2021-2050*. https://unfccc.int/sites/default/files/resource/NAP_Nepal_2021.pdf

Griggs, G., & Reguero, B. G. (2021). Coastal Adaptation to Climate Change and Sea-Level Rise. *Water (Basel)*, *13*(16), 2151. DOI: 10.3390/w13162151

Gunaratne, M. S., Radin Firdaus, R. B., & Rathnasooriya, S. I. (2021). Climate change and food security in Sri Lanka: Towards food sovereignty. *Humanities & Social Sciences Communications*, *8*(1), 229. Advance online publication. DOI: 10.1057/s41599-021-00917-4

Hanna, M. G., Pantanowitz, L., Jackson, B., Palmer, O., Visweswaran, S., Pantanowitz, J., Deebajah, M., & Rashidi, H. H. (2025). Ethical and bias considerations in artificial intelligence/machine learning. *Modern Pathology*, *38*(3), 100686. DOI: 10.1016/j.modpat.2024.100686 PMID: 39694331

Hosterman, H., & Smith, J. (2025). *Economic Costs and Benefits of Climate Change Impacts and Adaptation to the Maldives Tourism Industry*. Tourism.gov.mv;Ministry of Tourism. Male', Republic of Maldives. https://resources.tourism.gov.mv/wp-content/uploads/2024/09/Economic-Costs-and-benefits-of-Climate-Change-Impacts-and-Adaptation-to-the-Maldives-Tourism-Industry.pdf

Ibegbulam, C., & Ibegbulam, A.Ibegbulam C.MAigbovbiosa, O.JOlowonubi, J.AFatounde, S.A. (2023). Role of Artificial Intelligence in Electrification of Africa. *Engineering Science & Technology Journal*, *4*(6), 456–472. DOI: 10.51594/estj.v4i6.667

Jolly, S., Ahmad, N., & Scott, M. (Eds.). (2024). *Climate-Related Human Mobility in Asia and the Pacific* (1st ed.). Springer Nature. DOI: 10.1007/978-981-97-3234-0

Kassa, H., Abiyu, A., Hagazi, N., Mokria, M., Kassawmar, T., & Gitz, V. (2022). Forest landscape restoration in Ethiopia: Progress and challenges. *Frontiers in Forests and Global Change*, *5*, 796106. Advance online publication. DOI: 10.3389/ffgc.2022.796106

Kelling, S., Gerbracht, J., Fink, D., Lagoze, C., Wong, W.-K., Yu, J., Damoulas, T., & Gomes, C. (2013). A Human/Computer Learning Network to Improve Biodiversity Conservation and Research. *AI Magazine*, *34*(1), 10–10. DOI: 10.1609/aimag.v34i1.2431

Koswatte, I., Iddawala, J., Kulasekara, R., Ranaweera, P., Dasanayaka, C. H., & Abeykoon, C. (2024). Can Sri Lanka be a Net-Zero Nation by 2050? – Current Renewable Energy Profile, Opportunities, Challenges, and Recommendations. *Cleaner Energy Systems*, *8*, 100126. Advance online publication. DOI: 10.1016/j.cles.2024.100126

Kuir-Ayius, D. (2024). Impact of climate change and accessing services in Papua New Guinea. In Jolly, S., Ahmad, N., & Scott, M. (Eds.), *Climate-related human mobility in Asia and the Pacific*. Sustainable Development Goals Series. Springer., DOI: 10.1007/978-981-97-3234-0_11

Kumar Apat, S., & Mishra, J. (2023). An Artificial Intelligence-based Crop Recommendation System using Machine Learning. *Journal of Scientific and Industrial Research*, *82*(05), 558–567. DOI: 10.56042/jsir.v82i05.1092

Lakritz, T. (2019, September 10). *These 11 sinking cities are in danger of disappearing*. World Economic Forum. https://www.weforum.org/stories/2019/09/11-sinking-cities-that-could-soon-be-underwater/

Lam, S. H., & Toan, T. D. (2006). Land Transport Policy and Public Transport in Singapore. *Transportation*, *33*(2), 171–188. DOI: 10.1007/s11116-005-3049-z

Lindsey, R. (2023, August 22). *Climate change: Global sea level* (R. Lumpkin, G. Johnson, P. Thompson, & W. Sweet, Eds.). Climate.gov; National Oceanic and Atmospheric Administration. https://www.climate.gov/news-features/understanding-climate/climate-change-global-sea-level

Mana, A. A., Allouhi, A., Hamrani, A., Rehman, S., el Jamaoui, I., & Jayachandran, K. (2024). Sustainable AI-based production agriculture: Exploring AI applications and implications in agricultural practices. *Smart Agricultural Technology*, *7*, 100416. DOI: 10.1016/j.atech.2024.100416

Matthews, R., & Vivoda, V. (2023). "Water Wars": Strategic implications of the grand Ethiopian Renaissance Dam. *Conflict Security and Development, 23*(4), 1–34. DOI: 10.1080/14678802.2023.2257137

Mensah, J. (2019). Sustainable development: Meaning, history, principles, pillars, and implications for human action: Literature review. *Cogent Social Sciences, 5*(1), 1653531. Advance online publication. DOI: 10.1080/23311886.2019.1653531

Molden, D., Oweis, T., Steduto, P., Bindraban, P., Hanjra, M. A., & Kijne, J. (2010). Improving agricultural water productivity: Between optimism and caution. *Agricultural Water Management, 97*(4), 528–535. DOI: 10.1016/j.agwat.2009.03.023

Mor, S., Aneja, R., Madan, S., & Ghimire, M. (2023). Kyoto Protocol and Paris Agreement: Transition from Bindings to Pledges – A Review. *Millennial Asia, 15*(4), 690–711. DOI: 10.1177/09763996221141546

Mrisho, L. M., Mbilinyi, N. A., Ndalahwa, M., Ramcharan, A. M., Kehs, A. K., McCloskey, P. C., Murithi, H., Hughes, D. P., & Legg, J. P. (2020). Accuracy of a Smartphone-Based Object Detection Model, PlantVillage Nuru, in Identifying the Foliar Symptoms of the Viral Diseases of Cassava-CMD and CBSD. *Frontiers in Plant Science, 11*, 590889. DOI: 10.3389/fpls.2020.590889 PMID: 33391304

Nepal Electricity Authority. (2024). *A year in review - Fiscal Year - 2023/2024*. Nepal Electricity Authority.

Nwokolo, S. C., Singh, R., Khan, S., Kumar, A., & Luthra, S. (2023). Technological pathways to net-zero goals in Africa. In *Africa's Path to Net-Zero: Exploring Scenarios for a Sustainable Energy Transition* (pp. 93–210). Springer., DOI: 10.1007/978-3-031-44514-9_5

Ogutu, M. O., Akor, J., Mulindwa, M. S., Heshima, O., & Nsengimana, C. (2023). Implementing circular economy and sustainability policies in Rwanda: Experiences of Rwandan manufacturers with the plastic ban policy. *Frontiers in Sustainability, 4*, 1092107. Advance online publication. DOI: 10.3389/frsus.2023.1092107

Olando, S., Mwenda, M. N., Kikwatha, R. W., & Rambo, C. M. (2024). Effect Of Public Private Partnership Financing Structure On Completion Of Geothermal Energy Development Projects In Kenya. *International Journal of Professional Business Review, 9*(11), e05048–e05048. DOI: 10.26668/businessreview/2024.v9i11.5048

Olawade, D. B., Wada, O. Z., Ige, A. O., Egbewole, B. I., Olojo, A., & Oladapo, B. I. (2024). Artificial Intelligence in Environmental Monitoring: Advancements, Challenges, and Future Directions. *Hygiene and Environmental Health Advances, 12*, 100114–100114. DOI: 10.1016/j.heha.2024.100114

Othman, A. (2024, November). *AI in Climate Change Modeling and Prediction.* ResearchGate. https://www.researchgate.net/publication/385985206_AI_in_Climate_Change_Modeling_and_Prediction

Pachar, S., & Singh, R. (2021). Role of Sustainable Development Goals and Corporate Social Responsibility in India's Growth: Opportunities and Challenges. *The Empirical Economics Letters*, 20(2).

Pandey, S., & Regmi, N. (2017). Changing connectivities and renewed priorities: Status and challenges facing Nepali Internet. *First Monday*, *23*(1). Advance online publication. DOI: 10.5210/fm.v23i1.8071

Pathak, M. (2010). Climate Change: Uncertainty for Hydropower Development in Nepal. *Hyrdo Nepal*, *6*, 31–34. DOI: 10.3126/hn.v6i0.4190

Priya, A. K., Devarajan, B., Alagumalai, A., & Song, H. (2023). Artificial intelligence enabled carbon capture: A review. *The Science of the Total Environment*, *886*, 163913. DOI: 10.1016/j.scitotenv.2023.163913 PMID: 37150463

Rajab, A., Farman, H., Islam, N., Syed, D., Elmagzoub, M. A., Shaikh, A., Akram, M., & Alrizq, M. (2023). Flood Forecasting by Using Machine Learning: A Study Leveraging Historic Climatic Records of Bangladesh. *Water (Basel)*, *15*(22), 3970. DOI: 10.3390/w15223970

Ratouis, T. M. P., Snæbjörnsdóttir, S. Ó., Voigt, M. J., Sigfússon, B., Gunnarsson, G., Aradóttir, E. S., & Hjörleifsdóttir, V. (2022). Carbfix 2: A transport model of long-term CO2 and H2S injection into basaltic rocks at Hellisheidi, SW-Iceland. *International Journal of Greenhouse Gas Control*, *114*, 103586. DOI: 10.1016/j.ijggc.2022.103586

Rohde, F., Wagner, J., Meyer, A., Reinhard, P., Voss, M., Petschow, U., & Mollen, A. (2024). Broadening the perspective for sustainable artificial intelligence: Sustainability criteria and indicators for Artificial Intelligence systems. *Current Opinion in Environmental Sustainability*, *66*, 101411. DOI: 10.1016/j.cosust.2023.101411

Sales, M., de Bruin, S., Herold, M., Kyriakidis, P., & Souza, C.Jr. (2017). A spatio-temporal geostatistical hurdle model approach for short-term deforestation prediction. *Spatial Statistics*, *21*, 304–318. DOI: 10.1016/j.spasta.2017.06.003

Sarvajayakesavalu, S. (2015). Addressing challenges of developing countries in implementing five priorities for sustainable development goals. *Ecosystem Health and Sustainability*, *1*(7), 1–4. DOI: 10.1890/EHS15-0028.1

Secretariat of the Convention on Biological Diversity. (2020) Global Biodiversity Outlook 5 – Summary for Policy Makers. Montréal.

Seuyong, F. T., Silwal, A. R., Begazo, T., Newhouse, D., & N'Ghauran, A. (2023). The Size and Distribution of Digital Connectivity Gaps in Sub-Saharan Africa. *The World Bank*. https://documents1.worldbank.org/curated/en/099241003142325200/pdf/IDU0cb2e42f3050260484d0b8370b84eee303ecf.pdf

Shoaei, M., Noorollahi, Y., Hajinezhad, A., & Moosavian, S. F. (2024). A review of the applications of artificial intelligence in renewable energy systems: An approach-based study. *Energy Conversion and Management*, *306*, 118207–118207. DOI: 10.1016/j.enconman.2024.118207

Shrestha, A. B., & Aryal, R. (2011). Climate change in Nepal and its impact on Himalayan glaciers. *Regional Environmental Change*, *11*(S1), 65–77. DOI: 10.1007/s10113-010-0174-9

Shrestha, T. A., Matin, M., Chitale, V., & Thomas, S. (2022). Exploring the potential of deep learning for classifying camera trap data: A case study from Nepal. In *ICIMOD.org*. International Centre for Integrated Mountain Development (ICIMOD). https://lib.icimod.org/record/35997/files/HimalDoc2022_WP_ExploringDeepLearning_ForApproval.pdf

Silva, R. F., Fava, M. C., Saraiva, A. M., Mendiondo, E. M., Cugnasca, C. E., & Delbem, A. C. B. (2021). A Theoretical Framework for Multi-Hazard Risk Mapping on Agricultural Areas Considering Artificial Intelligence, IoT, and Climate Change Scenarios. *Engineering Proceedings*, *9*(1), 39. DOI: 10.3390/engproc2021009039

Singh, R., Bhattarai, N., Prajapati, A., & Shakya, S. R. (2022). Impact of Variation in Climatic Parameters on Hydropower Generation: A Case of Hydropower Project in Nepal. *Heliyon, 8*(12), e12240. ScienceDirect. https://doi.org/DOI: 10.1016/j.heliyon.2022.e12240

Singh, R., Khan, S., & Centobelli, P. (2022). Investigating the Interplay between Social Performance and Organisational Factors Supporting Circular Economy Practices. *Sustainability (Basel)*, *14*(24), 16781. DOI: 10.3390/su142416781

Sterner, T., Barbier, E. B., Bateman, I., van den Bijgaart, I., Crépin, A.-S., Edenhofer, O., Fischer, C., Habla, W., Hassler, J., Johansson-Stenman, O., Lange, A., Polasky, S., Rockström, J., Smith, H. G., Steffen, W., Wagner, G., Wilen, J. E., Alpízar, F., Azar, C., & Robinson, A. (2019). Policy design for the Anthropocene. *Nature Sustainability*, *2*(1), 14–21. DOI: 10.1038/s41893-018-0194-x

Sudheer Varma Surimalla, S. (2018). Intelligent Traffic Management Solution(ITMS) through the implementation of Google Maps and Artificial Intelligence. *International Journal of Engineering, Science and Mathematics*. https://www.researchgate.net/publication/327020444_Intelligent_Traffic_Management_SolutionITMS_through_the_implementation_of_Google_Maps_and_Artificial_Intelligence

Suman, S. (2022). Augmented Intelligence for National Security and Development. *Unity Journal, 3*(01), 245–252. DOI: 10.3126/unityj.v3i01.43329

Swapnil Mahadik, S., Gedam, M., & Shah, D. (2024). Environment sustainability with smart grid sensor. *Frontiers in Artificial Intelligence, 7*, 1510410. Advance online publication. DOI: 10.3389/frai.2024.1510410 PMID: 40045982

The World Bank. (2020a, June 30). *Implementation Completion and Results Report on PNG Energy Sector Development Project*. WorldBank.org. https://documents1.worldbank.org/curated/en/412371594163838703/pdf/Papua-New-Guinea-Energy-Sector-Development-Project.pdf

The World Bank. (2020b, November 26). *Accelerating Renewable Energy Integration and Sustainable Energy*. WorldBank.org. https://documents1.worldbank.org/curated/en/554641608001729081/pdf/Maldives-Accelerating-Renewable-Energy-Integration-and-Sustainable-Energy-Project.pdf

Tiwari, S., Bashir, S., Sarker, T., & Shahzad, U. (2024). Sustainable pathways for attaining net zero emissions in selected South Asian countries: Role of green energy market and pricing. *Humanities & Social Sciences Communications, 11*(1), 93. DOI: 10.1057/s41599-023-02552-7

UN Trade and Development. (2024). A world of debt: A growing burden to global prosperity. *UN Trade and Development*.

Ukoba, K., Olatunji, K. O., Adeoye, E., Jen, T.-C., & Madyira, D. M. (2024). Optimizing renewable energy systems through artificial intelligence: Review and future prospects. *Energy & Environment, 35*(7), 3833–3879. DOI: 10.1177/0958305X241256293

UN. ESCAP. (2024). Strengthening national and subnational capacity for sustainable disaster risk reduction, climate change adaptation and mitigation in Maldives. *Handle.net*. https://hdl.handle.net/20.500.12870/7370

UNFCCC. (2016). *The Paris Agreement*. Unfccc.int; UNFCCC. https://unfccc.int/process-and-meetings/the-paris-agreement

United Nations Development Programme. (2022). *Local Government Initiative on Climate Change (LoGIC)*. https://www.undp.org/sites/g/files/zskgke326/files/2023-03/Climate%20Vulnerability%20Index%20Book.pdf

Vats, G., & Mathur, R. (2022). A net-zero emissions energy system in India by 2050: An exploration. *Journal of Cleaner Production, 352*, 131417. DOI: 10.1016/j.jclepro.2022.131417

World Bank Group. (2021). *KIRIBATI CLIMATE RISK COUNTRY PROFILE.* https://climateknowledgeportal.worldbank.org/sites/default/files/country-profiles/15816-WB_Kiribati%20Country%20Profile-WEB.pdf

World Economic Forum. (2018). Innovation with a Purpose: The role of technology innovation in accelerating food systems transformation. In *weforum.org*. https://www3.weforum.org/docs/WEF_Innovation_with_a_Purpose_VF-reduced.pdf

Yang, R., Hu, J., Li, Z., Mu, J., Yu, T., Xia, J., Li, X., Dasgupta, A., & Xiong, H. (2024). Interpretable machine learning for weather and climate prediction: A review. *Atmospheric Environment, 338*, 120797–120797. DOI: 10.1016/j.atmosenv.2024.120797

Yangka, D., Rauland, V., & Newman, P. (2018). Carbon neutral policy in action: The case of Bhutan. *Climate Policy, 19*(6), 672–687. DOI: 10.1080/14693062.2018.1551187

Zappacosta, M., & Hollema, S. (2017). *FAO/WFP crop and food security assessment mission to Sri Lanka*. Food and Agriculture Organization of the United Nations (FAO). https://openknowledge.fao.org/server/api/core/bitstreams/a7fbafb5-41db-48e2-97e7-e8f415b8df0e/content

Zerssa, G., Feyssa, D., Kim, D.-G., & Eichler-Löbermann, B. (2021). Challenges of Smallholder Farming in Ethiopia and Opportunities by Adopting Climate-Smart Agriculture. *Agriculture, 11*(3), 192. DOI: 10.3390/agriculture11030192

Chapter 11
Harnessing Solar Energy for a Sustainable Society 5.0:
Legal and Regulatory Challenges in India

Anwesha Ghosh
https://orcid.org/0000-0003-4167-2121
Amity University, Kolkata, India

Naresh Prajapati
https://orcid.org/0000-0002-0752-3651
Central University of Punjab, Bathinda, India

Sachin Singh Parihar
https://orcid.org/0009-0007-5965-012X
Central University of Punjab, Bathinda, India

Deepak Kumar Chauhan
https://orcid.org/0000-0003-2344-1215
Central University of Punjab, Bathinda, India

ABSTRACT

The increasing scarcity and depletion of traditional energy sources and the environmental harm they inflict need a transition to renewable energy. India has significant geographical advantages for harnessing solar energy, although this potential has only been partially exploited. The proposed chapter seeks to examine the legal and regulatory framework for renewable energy, specifically focusing on solar energy. It will further analyze how the existing regulatory framework contributes to sustainable

DOI: 10.4018/979-8-3373-0680-3.ch011

development and the idea of Society 5.0. Society 5.0 envisions a future world where sophisticated technology such as the Internet of Things (IoT), Artificial Intelligence (AI), and robots converge with humans to create a smarter, more sustainable society. The proposed chapter thus aims to comprehensively study India's renewable energy policy to identify shortcomings and recommend improvements that might enhance the country's transition to a more sustainable energy system.

INTRODUCTION

Today, the world, with a particular emphasis on India, face a significant predicament: the deterioration of the environment and unsustainable development practices that put lives at risk. Environmental degradation has become a pervasive global challenge, casting a shadow over both the present and the future. Safeguarding the environment has now ascended to the highest priority, as it directly impacts the fundamental right to life. Human progress has historically been reliant on the environment due to the intricate interdependence among the physical, biological, and social surroundings, which collectively sustain all life forms (United Nations, 2022).

The destruction of the environment triggers a domino effect, ultimately leading to the demise of all living organisms, including humanity. Factors contributing to environmental degradation include excessive production, resource overexploitation, the adverse effects of nuclear radiation, industrial waste disposal, industrial accidents, ruthless exploitation of forests, unchecked quarrying activities, river, and water resource pollution, and the rapid escalation of air and noise pollution (Lampert, 2019). These elements collectively contribute to the deterioration of the environment. Furthermore, global warming, ozone layer depletion, and pollution are adverse outcomes resulting from the current development paradigm, directly impacting human well-being (Cirit & Aydemir, 2021). A heavy dependence on traditional energy sources such as coal, oil, and natural gas primarily characterises India's energy environment. Conventional energy sources such as these are responsible for causing environmental harm, including the release of air pollutants and greenhouse gases, which negatively impact human health and the environment. An immediate transition to renewable energy is desirable to guarantee a sustainable future (Jacob et al., 2023). The geographical location of India on the tropic of cancer offers abundant solar radiation, making it an excellent choice for harnessing solar energy. With an estimated annual average of 300 days of sunshine, the nation can produce substantial quantities of solar energy (Allen et al., 2024). Nevertheless, despite the promise, solar energy's contribution to India's energy combination is still restricted. The primary factors contributing to this underutilisation include inadequate regulatory frameworks, absence of incentives, and insufficient infrastructure. India has implemented many laws and legislations to foster the adoption and growth of renewable energy sources (Govind Bhutada, 2022). The National Solar Mission (NSM), a component of the National Action Plan on Climate Change, seeks to place India as a prominent solar energy player worldwide by developing a supportive policy structure. Additional notable legislations include the Electricity Act 2003, which promotes the production of renewable energy, and the Tariff Policy 2006, which offers directives for establishing rates for renewable energy. India has several obstacles and difficulties that inhibit the successful execution of its renewable energy policy despite its ongoing efforts. These factors incorporate variations in policy implementation among different states, financial obstacles such as increase in expenses and absence of financial incentives, regulatory challenges involving intricate procedures and administrative delays, and technological barriers such as restricted availability of cutting-edge technologies and technical proficiency (Kerr, 2024).

India's favourable geography offers substantial possibilities for solar energy development. Nevertheless, the absence of a comprehensive legislative framework and financial incentives obstructs achieving this potential. The current laws and policies have had a limited influence on the acceptance of solar energy in India. The lack of uniformity in policy execution among states of India and the absence of a cohesive regulatory framework are significant obstacles. In order to promote the use of solar energy and accomplish the Sustainable Development Goals (SDGs), the chapter shall propose the following suggestions: creating a standardised regulatory framework to guarantee consistent policy implementation in all states, offering financial incentives like subsidies, tax advantages, and low-interest loans to stimulate investment in solar energy initiatives, allocating resources to research and development to foster technological progress in solar energy, and raising public awareness about the advantages of solar energy and the significance of sustainable development (Bjelland & Mehta, 2021).

Efficient execution of renewable energy policies may greatly aid in the attainment of the SDGs, precisely Goal 7 (Ensuring access to affordable, reliable, sustainable, and modern energy for everyone) and Goal 13 (Taking urgent action to combat climate change and its impacts). India can diminish its dependence on traditional energy sources, lower greenhouse gas emissions, and foster environmental sustainability by using its solar energy potential effectively (Maka & Alabid, 2022). By harmonising renewable energy regulations with the principles of Society 5.0, it is possible to establish a more intelligent and environmentally conscious society. Incorporating innovative technologies like IoT and AI into renewable energy systems may increase efficiency, lower expenses, and enhance overall quality of life (Ramírez-Márquez et al., 2024).

The shift from traditional to renewable energy is essential for achieving sustainable development. This chapter aims to explore the intersection of renewable energy policies and the transformative vision of Society 5.0, focusing on how India can harness both to drive sustainable development. Society 5.0 envisions a future where advanced technologies—such as AI, the IoT, and robotics—are integrated into daily life, creating smarter, more efficient systems that cater to human needs while addressing environmental challenges. By examining renewable energy policies within this context, the chapter seeks to highlight how India can adopt innovative, technology-driven approaches to accelerate its transition to renewable energy. This integration of policy with technological advancement can enhance policy effectiveness, streamline resource management, and promote a sustainable society. The findings will offer valuable insights for policymakers, researchers, and industry leaders into the role of renewable energy in shaping a Society 5.0-driven future, where economic growth, social well-being, and environmental sustainability are harmoniously balanced.

RESEARCH METHODOLOGY

This chapter employs a doctrinal approach as its primary methodology, aiming to analyze renewable energy policies within the transformative framework of Society 5.0. The research entails a comprehensive examination of provisions related to sustainable development, including specific environmental laws and other statutes indirectly promoting sustainability. Emphasis has been placed on how renewable energy policies can align with Society 5.0's vision—a future society where technologies like AI, the IoT, and robotics merge with human life to foster a smarter, sustainable society. Additionally, this study probes into the international legal landscape governing sustainable development and renewable energy, considering both the direct and indirect legal statutes and the broader regulatory implications of global environmental policies. Secondary sources of data, including specialized books, articles from national and international journals, relevant case law, judicial decisions, and law journals, have been extensively consulted to ensure a robust, well-informed analysis that bridges renewable energy policies and the Society 5.0 vision for a balanced, technologically integrated, and environmentally sustainable future.

SIGNIFICANCE OF THE PROBLEM

The degradation of our environment has grown into a matter of global concern, posing a significant challenge for both current and future generations of humanity. Prioritizing environmental protection is not only essential for the preservation of life but also for safeguarding the fundamental right to life itself.

Environmental law, as a relatively recent branch of legal practice, has primarily taken shape in the last three decades. It is still in its formative stages and is experiencing rapid development, driven by our ever-growing understanding of the environmental crisis. The present generation bears a profound responsibility, given its unique opportunity to address environmental issues, as the consequences of inaction will be borne by future generations who may not even have a chance to do so. Human life is intricately linked not only to the environment but also to ecological factors. Hence, every country must undertake the imperative task of protecting the environment and fostering sustainable development.

SCOPE OF THE STUDY

The realization of sustainable development hinges on the unwavering dedication of all branches of government, including the legislative, executive, and judicial arms. Within this context, the judiciary assumes a critical role in advancing environmental governance, upholding the rule of law, and ensuring a just balance among environmental, societal, and developmental considerations. The fusion of environmental preservation and development objectives should be seamlessly woven into the policies and practices of each nation.

The primary objective of this study is to explore the imperative for a sustainable environment and to assess its profound significance concerning the fundamental Right to Life. Furthermore, the enactment and rigorous enforcement of laws and regulations, whether at the regional, national, state/provincial, or local/municipal level, are of paramount importance for translating international agreements in the fields of environment and development into tangible action. This is highlighted by the frequent inclusion of treaty obligations that mandate the reporting of legislative measures as part of international commitments.

SOCIETY 5.0: AN ANALYSIS

Society 5.0, put forward by the Japanese Government, is a straightforward concept (Narvaez Rojas et al., 2021). It was drafted in the 5th Science and Technology Basic Plan by the Council for Science, Technology and Innovation and approved by a Cabinet decision in January 2016 (Society 5.0, 2016).

Upon reflection on human history, several phases of society may be delineated. Society 1.0 is characterised by groups of individuals engaging in hunting and gathering while coexisting harmoniously with nature; Society 2.0 emerged with the establishment of agricultural communities, fostering enhanced organisation and nation-building; Society 3.0 is marked by the promotion of industrialisation through the industrial revolution, facilitating mass production; and Society 4.0 represents an information society that enhances value by integrating intangible assets into information networks. Society 5.0 is an information society that evolves from Society 4.0, intending to create a prosperous, human-centred society (Fukuyama, M., 2018).

Society 5.0 aims to establish a human-centric society that simultaneously fosters economic progress and addresses social concerns, enabling individuals to have a high quality of life that is dynamic and pleasant. It is a society that will meticulously address the diverse requirements of individuals, irrespective of area, age, gender, language, etc., by supplying essential commodities and services. The essential factor for its realisation is the integration of cyberspace and the physical realm to produce high-quality data, which subsequently facilitates the creation of novel values and solutions to address difficulties. Japan's national vision is cultivating a new, human-centred society while addressing various social challenges. Although Society 5.0 is Japan's development plan, it extends beyond Japan, aligning its objectives with the SDGs. Japan has problems, including an ageing population, a dropping birthrate, population reduction, and deteriorating infrastructure, which are issues several other nations will encounter (Nakatani, 2019). Japan is among the first countries to confront these difficulties. Through the prompt settlement of these concerns via Society 5.0 and the dissemination of these answers globally, Japan may aid in addressing analogous issues internationally and advancing the SDGs (Deguchi A. et al., 2020).

Merging of Cyberspace and Physical Space

The essential element for realising Society 5.0 is the integration of cyberspace and physical space with the principles of a human-centred society. In Society 5.0, every component of society will be constructed as a digital twin in cyberspace, reorganised regarding systems, business design, urban and regional development, and then manifested in the physical realm to revolutionise society. By integrating human-centred principles into these new processes, we will empower individual

Japanese citizens and global citizens to assume a pivotal role in decision-making, facilitating a flexible and rapid transformation of society into a more favourable state (Marr, B. 2018).

Creation of Knowledge as a Source of Value Creation

To transition to Society 5.0 and address the Ethical, Legal, and Social Implications (ELSI) associated with the implementation of new technologies, it is essential to adopt a comprehensive perspective and establish a system that facilitates the application of "Convergence Knowledge," encompassing both natural sciences and the humanities and social sciences (Report on the 5th Science and Technology Basic Plan, 2016).

Developing Human Resources

In the age of Society 5.0, it is essential to cultivate the skills and attributes attainable via exploratory activities to identify issues independently and devise solutions. It is essential to cultivate human resources capable of generating new value globally and to establish an educational and human resource development system that facilitates this objective (Ellen MacArthur Foundation, 2017).

INDIA AND SOLAR ENERGY

Ministry of Statistics and Programme Implementation released the Energy Statistics India, 2024 (Government of India, 2024); this report highlighted India's potential in the energy sector. In response to escalating concerns over the impact of manmade greenhouse gases on global climate, a worldwide consensus has been reached to mitigate emissions. Renewable energy is being investigated with increased commitments as a viable way to tackle poverty and global warming issues. For sustainable global development, it is essential to provide access to cheap, dependable, sustainable, and contemporary energy services while decreasing greenhouse gas emissions and the energy sector's carbon footprint. This report mentioned that India has a high potential for generating renewable energy from various sources such as Solar, wind, etc. Solar power potential is estimated at around 7 48 990 MV (35.50%) of the total renewable energy in India (Government of India, 2024).

Figure 1. Estimated renewable power potential by source in India

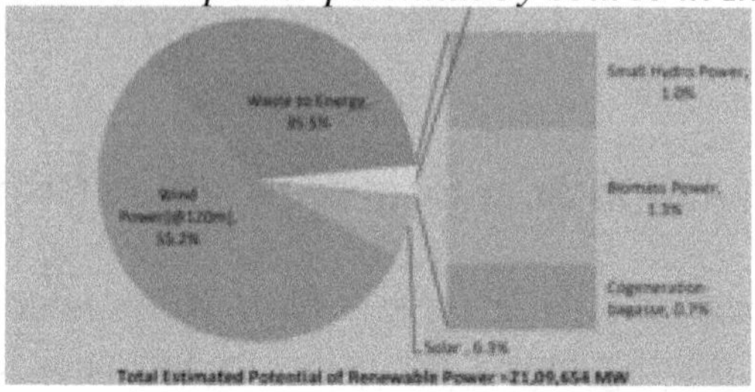

Source: *Ministry of Statistics and Programme Implementation. (2024). Energy Statistics India 2024. Government of India. (Government of India, 2024)*

Figure 1 shows India's sources-wise estimated renewable power potential as of 31.03.2023 (Government of India, 2024). Solar energy contribution is 6.3.%, the third highest contribution on renewable energy sources (Government of India, 2024).

The table 1 shows the state-wise solar energy production in India's estimated unit on 31st March 2023. Rajasthan is leading in solar energy, followed by Maharashtra, J&K, and Ladakh.

Table 1. State-wise solar energy production in India

Sl. No.	States/UTs	Wind Power @ 150m (MW)	Small Hydro Power (MW)	Biomass Power (MW)	Cogeneration Bagasse (MW)	Solar Energy (MW)	Large Hydro (MW)	Total (MW)	Distribution (%)
1	Andhra Pradesh	123336	409	1999	280	38440	2596	167060	7.92
2	Arunachal Pradesh	246	2065	18	0	8650	50394	61373	2.91
3	Assam	459	202	322	0	13760	643	15386	0.73
4	Bihar	4023	527	964	347	11200	130	17191	0.81
5	Chhattisgarh	2749	1098	354	0	18270	1311	23782	1.13
6	Goa	14	5	33	0	880	0	932	0.04
7	Gujarat	180790	202	2638	555	35770	550	220505	10.45
8	Haryana	593	107	1353	362	4560	0	6976	0.33
9	Himachal Pradesh	239	3460	70	0	33840	18305	55914	2.65
10	Jammu & Kashmir (Ladakh)	1707	83	0	0	111050	12972	125812	5.96
11	Jharkhand	16	223	146	0	18180	300	18870	0.89
12	Karnataka	169251	3726	1794	1762	24700	4414	205648	9.75
13	Kerala	2621	647	778	0	6110	2473	12629	0.6
14	Madhya Pradesh	55423	820	2516	0	61660	2819	123239	5.84
15	Maharashtra	173868	786	2630	3917	64320	3144	248665	11.79
16	Manipur	0	100	62	0	10630	615	11407	0.54
17	Meghalaya	55	230	69	0	5860	2026	8240	0.39
18	Mizoram	0	169	3	0	9090	1927	11189	0.53
19	Nagaland	0	132	54	0	7290	325	7851	0.37
20	Odisha	12129	236	299	0	25780	2825	41318	1.96

continued on following page

Table 1. Continued

Sl. No.	States/UTs	Wind Power @ 150m (MW)	Small Hydro Power (MW)	Biomass Power (MW)	Cogeneration Bagasse (MW)	Solar Energy (MW)	Large Hydro (MW)	Total (MW)	Distribution (%)
21	Punjab	428	578	3022	414	2810	1301	8554	0.41
22	Rajasthan	284250	52	1300	0	142310	411	428322	20.3
23	Sikkim	0	267	5	0	4940	6051	11262	0.53
24	Tamil Nadu	95107	604	1560	639	17670	1785	117366	5.56
25	Telangana	54717	102	1678	117	20410	1302	78327	3.71
26	Tripura	0	47	34	0	2080	0	2161	0.1
27	Uttar Pradesh	510	461	2800	4926	22830	502	32028	1.52
28	Uttarakhand	49	1664	93	215	16800	13481	32303	1.53
29	West Bengal	1281	392	1742	0	6260	809	10484	0.5
30	Andaman & Nicobar	1245	7	18	0	0	0	1270	0.06
31	Chandigarh	0	0	0	0	0	0	0	0
32	Dadar & Nagar Haveli, Daman & Diu	17	0	2	0	0	0	19	0
33	Delhi	0	0	0	0	2050	0	2050	0.1
34	Lakshadweep	31	0	1	0	0	0	32	0
35	Puducherry	408	0	5	0	0	0	413	0.02
36	Others$	0	0	0	284	790	0	1074	0.05

Source: Ministry of Statistics and Programme Implementation. (2024). *Energy Statistics India 2024*. (Government of India, 2024)

SCHEMES RELATED TO SOLAR ENERGY IN INDIA

The Union Minister for New and Renewable Energy and Power has informed that the Government has set a target of achieving 500 GW of installed capacity from non-fossil fuels by 2030, in line with the Prime Minister's announcement at COP-26 (*Major Ongoing Renewable Energy Schemes and Programmes*, 2023).

Initiative for the Establishment of Solar Parks and Ultra-Mega Solar Power Projects aiming for a capacity of 40,000 MW. Under the program, infrastructure, including land, roads, power evacuation systems, and water facilities, is created with the necessary regulatory permissions and approvals. Consequently, the initiative facilitates the rapid development of utility-scale solar projects nationwide.

The Central Public Sector Undertaking (CPSU) Scheme Phase-II (Government Producer Scheme) facilitates the establishment of grid-connected Solar Photovoltaic (PV) Power Projects by Government Producers, utilising domestically produced solar PV cells and modules, with Viability Gap Funding (VGF) assistance, for self-consumption or utilisation by Government/Government entities, either directly or via Distribution Companies (DISCOMS).

Production Linked Incentive program 'National Programme on High-Efficiency Solar PV Modules' aimed at attaining giga-watt (GW) scale manufacturing capability in High-Efficiency Solar PV modules (Tranche I & II).

Pradhan Mantri Kisan Urja Suraksha evam Utthaan Mahabhiyan Initiative (Ministry of New and Renewable Energy, 2019) to advance tiny grid-connected solar power plants, autonomous solar-powered agricultural pumps, and the solarisation of current grid-connected agricultural pumps. The initiative is advantageous not just for farmers but also for states and DISCOMs. States will reduce expenditures on subsidies for energy supplied to agricultural users, while DISCOMs will acquire more affordable solar power, hence minimising transmission and distribution losses.

Phase II of the Rooftop Solar Programme for grid-connected solar rooftop electricity installations. This programme offers subsidies for the residential sector and performance-linked incentives to DISCOMs to exceed baseline capacity additions in rooftop solar installations.

Pm-Surya Ghar: Muft Bijli Yojana, 2024

The Union Cabinet, under the leadership of Prime Minister Shri Narendra Modi, has sanctioned the PM-Surya Ghar: Muft Bijli Yojana, a 75,021 crore initiative aimed at installing rooftop solar systems and supplying free power to one crore families (Press Information Bureau, 2024). The program provides Central Financial Assistance (CFA) up to 60% of the system cost for 2 kW systems and 40% of the extra system cost for systems with a capacity ranging from 2 to 3 kW (Press Information Bureau,

2024). The program offers collateral-free, low-interest financing solutions at about 7% for residential rooftop solar systems up to 3 kW (Press Information Bureau, 2024). A Model Solar Village will be established in each district to encourage the use of rooftop solar energy in rural regions. Urban Local Bodies and Panchayati Raj Institutions would get incentives to advance rooftop solar installations. The initiative seeks to produce 30 GW of solar capacity in the residential sector, yielding 1000 BUs of energy and decreasing 720 million tonnes of CO2 equivalent emissions throughout the 25-year lifespan of rooftop systems (Press Information Bureau, 2024). It is projected to provide over 1.7 million direct employment opportunities in manufacturing, logistics, supply chain, sales, installation, operations and maintenance, and many other services (Press Information Bureau, 2024)

These renewable energy schemes reflect the principles of Society 5.0 by integrating advanced technologies with societal needs to promote a sustainable and efficient future. Initiatives like the PM-Surya Ghar: Muft Bijli Yojana and rooftop solar programs not only drive environmental sustainability but also foster economic development, energy independence, and social empowerment. By utilizing solar energy and smart technologies, these programs contribute to a low-carbon, resilient society, aligning with Society 5.0's goal of creating a more intelligent, automated, and sustainable world.

SUSTAINABLE DEVELOPMENT: A COMPREHENSIVE UNDERSTANDING

The idea of Sustainable Development (SD) has its roots in various environmental movements from earlier decades but gained significant prominence following the Earth Summit in Rio de Janeiro, Brazil, in 1992. The concept of using Earth's resources sustainably has ancient origins, but the specific term SD came into common usage during the Cocoyoc Declaration on Environment and Development in the early 1970s. Since then, it has become a hallmark of international organizations dedicated to achieving environmentally responsible or beneficial development (Redclift, 2003). SD represents a delicate balance between the objectives of environmental preservation and economic progress, with a focus on meeting the needs of both current and future generations. The underlying premise is that natural resources, such as the ocean floor, are not solely the result of the labor of present generations, and therefore, their exploitation must be conducted with due consideration for the rights of future generations (Krämer, 2021).

The most effective approach to addressing climate change involves adopting a SD path, which entails transitioning to environmentally friendly technologies and promoting energy efficiency, renewable energy, and similar measures (Fülöp, 2021).

SD serves as a mechanism through which society can engage with the environment without jeopardizing the planet's resources for future generations. It is both a development paradigm and a guiding principle that seeks to enhance living standards without causing harm to Earth's ecosystems or giving rise to environmental issues like deforestation, water and air pollution, which can lead to problems such as climate change and species extinction.

Over the past three decades, SD has garnered widespread support in non-binding international legal documents, manifesting itself in numerous state declarations, international organization resolutions, and action programs. Nevertheless, these efforts have not translated into legally binding rules pertaining to SD. SD is featured in 112 multilateral treaties, with 30 of them aiming for universal participation (Sustainable Development, Future Generations, and Public Trust, 2014). This indicates a certain level of consensus within the international community regarding the importance of SD within international law. SD serves as a reconciling force among conflicting norms related to environmental protection, economic development, social progress, and human rights. It is a central theme in many international treaties and conventions and features prominently in the decisions of international courts of justice. Importantly, environmental law and developmental law are not opposing concepts but rather integral elements that must coexist. Development should not come at the expense of the environment, and environmental protection should not hinder development.

Fundamental Principles of Sustainable Development

The core principles of SD are encapsulated in several key international agreements, including the Rio Declaration on Environment and Development, Agenda 21, the Climate Change Convention, the Biodiversity Convention, and the Non-binding Principles on Forestry from 1992. These principles serve as the foundation for the practice of SD (United Nations Conference on Environment and Development, *Rio Declaration on Environment and Development; Agenda 21; United Nations Framework Convention on Climate Change; Convention on Biological Diversity; Non-binding Principles on Forestry*, 1992), Principle of Conservation underscores the importance of conserving and wisely managing natural resources to ensure their availability for present and future generations (Rio Declaration, Principle 7). Principle of the Right to Development recognizes that all people have the right to economic, social, and cultural development while emphasizing that such development should be sustainable and not harm the environment (Rio Declaration, Principle 25). Principle of Key Elements of Sustainable Development and Interconnections Precaution highlights the interconnectedness of various elements of SD and encourages precautionary measures to prevent harm to the environment and society (Art. 3, UNFCCC & Bio-diversity Convention, 1992). Principle of Integration emphasizes the need to integrate economic, social, and environmental considerations into decision-making processes and policies to achieve a balanced and sustainable approach to development (Rio Declaration, Principle 2). Principle of Polluter Pays places the responsibility on those who cause pollution or environmental damage to bear the associated costs and encourages the internalization of environmental costs in economic activities (Rio Declaration, Principle 16). Principle of Inter-generational Equity stresses the importance of ensuring that the needs and interests of present generations do not compromise the ability of future generations to meet their own needs, thereby safeguarding the rights and well-being of future inhabitants (Rio Declaration, Principle 3); and Principle of Common but Differential Responsibility acknowledging that not all countries bear equal historical responsibility for environmental degradation, this principle recognizes that nations have a shared responsibility for addressing global environmental challenges but allows for differentiated actions based on varying levels of development and capacity (Rio Declaration, Principle 3). These principles collectively provide a framework for guiding policies and actions that promote SD, taking into account the complex and interrelated nature of environmental, social, and economic factors.

International Law and Sustainable Development

Environmental law has emerged as one of the most prominent legal developments in the 21st century, gaining widespread popularity worldwide. Virtually all nations have enacted environmental protection laws to address pollution and ecological concerns. Consequently, both international and national perspectives on environmental law have evolved. Courts have gradually developed pockets of environmental jurisprudence, particularly in cases involving judicial activism and public interest litigation. Over the past 30 years, SD has become an integral part of international law due to United Nations conventions, gaining global recognition and acceptance (Reimer, 2021). While there is a growing demand for the documentation of SD within the international community, it is important to note that SD is already embedded in numerous international declarations, resolutions, and treaties.

International law serves as a mechanism to make the global community's objectives more effective, robust, and clear. As Decleris notes, "Like all social institutions created by humans, the institution of SD will also develop its specific form through the science of law and its application in court decisions" (Directorate-General for Environment (European Commission) & Decleris, 2000).

Although international legal norms in the realm of environmental protection were first prominently initiated during the Stockholm Conference in 1972, it's essential to recognize that there was not a complete absence of International Environmental Law before the Stockholm Conference. This historical period can be divided as follows:

Pre-Stockholm International Environmental Law

The growing public concern over international crises prompted efforts to establish International Environmental Regulation as signs of irreparable damage to the environment became increasingly evident. The roots of international environmental law can be traced back to a few bilateral and multilateral agreements that were largely concluded in the first half of the 20th century and contained some provisions related to the preservation of natural resources. For instance, the 1909 Boundary Waters Treaty between Great Britain (on behalf of Canada) and the United States included provisions stating that boundary waters and waters flowing across the boundary should not be polluted on either side, causing harm to the health or property of the other party (Sand, 2021). Similarly, the 1929 Agreement between Germany and Belgium concerning their frontier included clauses prohibiting the discharge of substances into streams and ditches that could harm public health. Although these early treaties and agreements had limited scopes and were primarily intended to resolve disputes related to the sharing of water resources, they had the potential to influence the development of international norms for environmental protection. The mid-20th century marked a turning point with the convening of an International Convention in 1954 specifically addressing the issue of sea pollution by oil (Laursen, 2016). This convention explicitly prohibited the discharge of oil or oily mixtures from tankers that could pollute the surface of the sea, and it outlined penalties in accordance with the laws of the relevant territory.

The adoption of the Antarctic Treaty further accelerated the development of international environmental law and strengthened environmental regulation in areas shared internationally. The treaty emphasized the peaceful use of Antarctica and imposed bans on nuclear explosions and the disposal of radioactive wastes, serving as a notable example of international efforts to protect the environment (Jariwala, 1993).

Post-Stockholm International Environmental Law

Stockholm Declaration 1972

The principle of SD gained significant momentum with the Stockholm Declaration in 1972, which introduced several key principles. For instance, Principle 3 of the Stockholm Declaration emphasizes the importance of preserving the Earth's capacity to produce vital renewable resources and, whenever feasible, restoring these resources to their optimal state. Principle 5 highlights the responsible use of non-renewable resources to protect them from the risk of future depletion. It underscores the need for sustainable management. Furthermore, Principle 11 of the Declaration emphasizes that the environmental protection policies of all countries should align with and support the development efforts, particularly the energy-related development, of developing nations. It calls for a balance between environmental preservation and the economic development aspirations of less developed countries (Person et al., 1987).

These principles laid out in the Stockholm Declaration serve as foundational guidelines for integrating environmental considerations into policymaking and development practices, fostering a more balanced and sustainable approach to resource utilization and global development.

Brundtland Commission Report: Our Common Future 1987

The concept of SD was first defined by the Brundtland Commission in its 1987 report 'Our Common Future.' According to this report, SD is described as "development that meets the needs of the present without compromising the ability of future generations to meet their own needs" (Report of the World Commission on Environment and Development: 1987).

The Brundtland Report introduces two fundamental concepts within the framework of SD. First, it emphasizes the importance of addressing the essential needs of the world's poor as a top priority. Second, it recognizes that technology and social organization impose limitations on the environment's ability to meet both current and future needs (Report of the World Commission on Environment and Development: 1987: P. 13).

Furthermore, the Brundtland Report stresses that SD requires the integration of economic and ecological considerations into decision-making processes at all levels. It is viewed as a strategy that seeks to reconcile the often-conflicting objectives of economic development and ecological preservation, with the goal of ensuring the well-being of both present and future generations.

Earth Summit

The United Nations Conference on Environment and Development (UNCED), also known as the Earth Summit, took place in June 1992 in Rio de Janeiro and saw the participation of over 150 governments. This historic event marked the largest UN Conference ever held and was driven by UNCED's mission to set the world on a path toward SD, with a primary goal of preserving resources for future generations.

The momentum for the Earth Summit stemmed from the 1987 Brundtland Report, which had a profound impact by compelling people worldwide to consider how their lives impact natural resources. The Earth Summit yielded five significant documents:

Rio Declaration on Environment and Development: This declaration encapsulates 27 principles that nations pledged to uphold in their pursuit of more environmentally SD patterns.

Agenda 21: Adopted at UNCED, Agenda 21 serves as a comprehensive plan of action for environmental and SD. It offers guidance for actions at local, national, and global levels and encompasses a wide array of recommended measures to address environmental challenges and promote SD.

United Nations Commission on Sustainable Development 1993: Originating from Chapter 38 of Agenda 21, the UN Commission on Sustainable Development (UNCSD) was established to follow up on the Rio Conference's outcomes. It was tasked with promoting international cooperation, streamlining intergovernmental decision-making processes, and overseeing the integration of environmental and developmental considerations. Its role included assessing the progress in implementing Agenda 21.

Earth Summit plus Five 1997: Held in June 1997 in New York, this event aimed to review and reinforce commitments made at the Earth Summit. Participants renewed their dedication to global partnership and collaborative efforts in addressing the needs of both present and future generations.

These documents collectively represent a significant global effort to address environmental challenges and advance SD, underscoring the importance of international cooperation and shared commitment to a more sustainable and equitable future (Take action for the sustainable development goals - United Nations Sustainable Development, 2004).

Kyoto Conference and Pact on Global Warming

On December 11, 1997, a momentous summit convened to address the pressing issue of global warming. Delegates representing 159 nations gathered at the World Climate Conference and achieved a historic accord. This landmark agreement called for mandatory reductions in the greenhouse gas emissions of industrialized nations in the coming millennium, signifying a significant stride in the collective effort to combat the potentially devastating consequences of global warming (UNFCCC, 2019).

This accord served to reinforce and bolster the commitments originally laid out in the 1992 Convention on Climate Change. It demonstrated the international community's commitment to combating climate change and reducing the environmental damage associated with industrial activity by putting legally binding limitations on greenhouse gas emissions from industrialised countries. The summit marked a growing global consensus regarding the urgency of addressing this critical environmental challenge (Poppick, 2017).

United Nations Climate Change Conference 2005

The conference took place from November 28 to December 10, 2005, in Montreal, Canada. During this gathering, many participating countries committed to specific and legally binding emission reduction targets in accordance with the Convention on Climate Change. This marked a significant step forward in international efforts to address and combat climate change (*Summary Report 28 November – 10 December 2005*, 2005).

Statute of the International Renewable Energy Agency, 2009

The International Renewable Energy Agency attains full status as an International Organisation. Established in January 2009 with 75 Member States, IRENA has rapidly evolved within a year to become one of the largest international organisations. One hundred forty-seven nations and the European Union have already ratified the legislation. The 25th instrument of ratification was submitted in Berlin on June 8th, and pursuant to Article XIX, the treaty shall take effect 30 days thereafter. Twenty-nine Member States have approved the Treaty. In June 2009, Abu Dhabi was designated as the headquarters of IRENA. The inaugural international organisation of the 21st century is also the first to be based in the Gulf Region. Enhancing the proportion of renewable energy in the energy mix is a direct method for addressing climate change and reducing greenhouse gas emissions. Simultaneously promoting energy security and autonomy. Moreover, it provides robust support for both economic and social development. The International Renewable Energy Agency seeks to establish itself as the preeminent advocate for renewable energy worldwide. The Agency's mission is to assist its Member States in formulating their strategies for all renewable energy sectors: bioenergy, geothermal, hydropower, ocean, solar, and wind (*IRENAs Statute Enters into Force,* 2010).

The Paris Agreement, 2015

The Paris Agreement is a legally enforceable international convention concerning climate change. It was ratified by 196 Parties at the UN Climate Change Conference (COP21) in Paris, France, on December 12, 2015. It becomes effective on 4 November 2016. The primary objective is to restrict the rise in global average temperature to well below 2°C over pre-industrial levels and to strive to limit the increase to 1.5°C above pre-industrial levels. In recent years, global leaders have emphasised the need of restricting global warming to 1.5°C by the century's conclusion. The UN's Intergovernmental Panel on Climate Change asserts that beyond the 1.5°C barrier poses a significant danger of triggering much more severe climate change consequences, such as increased frequency and intensity of droughts, heatwaves, and precipitation events. To restrict global warming to 1.5°C, greenhouse gas emissions must reach their zenith no later than 2025 and thereafter decrease by 43% by 2030. Recognising the need to advance universal access to sustainable energy in developing nations, especially in Africa, via the increased implementation of renewable energy sources (UNFCCC, 2024).

International Energy Charter, 2015

The signatories seek sustainable energy development, enhanced energy security, and maximised efficiency in the production, conversion, transport, distribution, and utilisation of energy, aiming to improve safety in a socially acceptable, economically viable, and environmentally sound manner. Acknowledging the sovereignty of each State over its energy resources and its authority to regulate energy transmission and transportation within its territory, while adhering to pertinent international obligations, and fostering political and economic cooperation, they resolve to advance the establishment of efficient, stable, and transparent energy markets at both regional and global levels. This initiative will be grounded in the principles of non-discrimination and market-oriented price form. They are resolute in establishing an environment conducive to business operations and the influx of capital and technology to attain the goals (Mundi, 2024).

Conference of Parties 26: Together for our Planet (2021)

During the recent climate conference, several key outcomes and commitments were reached (United Nations, 2021):

Recognizing the Emergency: Participating countries reaffirmed their commitment to the Paris Agreement's objective of limiting the global average temperature increase to well below 2°C above pre-industrial levels, with a strong emphasis on striving to limit it to 1.5°C. This recognition highlights the urgent need for decisive action to address climate change.

Accelerating Action: Countries emphasized the critical nature of the current decade, underlining that carbon dioxide emissions must be reduced by 45 percent during this period to achieve the goal of reaching net-zero emissions around the mid-century mark. This underscores the imperative for immediate and substantial action to combat climate change.

Moving Away from Fossil Fuels: One of the most contentious decisions in Glasgow revolved around a commitment to phase down coal power and phase out "inefficient" fossil fuel subsidies. Although debated, this commitment signifies a collective effort to transition away from fossil fuels and reduce their environmental impact.

Delivering on Climate Finance: Developed countries entered the conference with a shortfall in their pledge to provide US$100 billion annually to support climate action in developing countries. The Glasgow outcome expressed "regret" over this shortfall and reaffirmed the commitment to delivering the promised US$100 billion per year urgently, underlining the importance of financial support in addressing climate challenges.

Stepping up Support for Adaptation: The Glasgow Pact called for a significant increase in financial support to assist developing countries in adapting to the impacts of climate change and enhancing their resilience. This commitment acknowledges the growing importance of adaptation efforts to address the consequences of a changing climate.

These outcomes collectively reflect a commitment by nations to take meaningful steps toward addressing the climate crisis and achieving the goals outlined in the Paris Agreement, with a strong emphasis on urgent action, transitioning to cleaner energy sources, and supporting both mitigation and adaptation efforts.

UN Climate Change Conference - United Arab Emirates, 2023

COP28 was significant as it signified the completion of the first 'global stocktake' of global initiatives to combat climate change under the Paris Agreement. Having demonstrated that advancements in climate action were insufficiently rapid in all domains – including the reduction of greenhouse gas emissions, enhancement of resilience to climate change, and provision of financial and technological assistance to vulnerable nations – countries reached a consensus on measures to expedite action across all sectors by 2030. This entails urging nations to accelerate the shift from fossil fuels to renewable energy sources, such as wind and solar power, in their forthcoming climate pledges (United Nations, 2024).

INDIAN LAW AND SUSTAINABLE DEVELOPMENT

The challenges of industrialization, urbanization, rapid population growth, and poverty have given rise to significant environmental issues on a formidable scale. In response to these environmental challenges, numerous legislative and policy measures have been adopted at various levels, both nationally and internationally. To ensure the effective implementation of these measures, multiple authorities and regulatory agencies have been established. However, despite their efforts, various reports have pointed to shortcomings in addressing the complex issues at the intersection of development and the environment. This has raised questions about whether the legal aspects of SD are effectively integrated into environmental governance in India.

For a country like India, embarking on a path of industrial development required addressing a myriad of environmental concerns. The legal framework had to adapt to the diverse socioeconomic demands while safeguarding existing resources. Consequently, the development of environmental law in India needed to strike a delicate balance between fostering economic development and ensuring sustainability. This dynamic process is characteristic of a society governed by the rule of law, where the law serves as a dynamic instrument for positive change toward a better environment.

Programme and Policies

Several significant policies and initiatives have been introduced in India to address environmental concerns and promote sustainability. Here are some of the notable ones:

National Environment Awareness Campaign (NEAC), 1986: Launched in 1986, NEAC aims to create environmental awareness at the national level by providing nominal financial assistance to NGOs, educational institutions, government departments, and other organizations. It encourages voluntary actions to address environmental issues.

National Water Policy (NWP), 1987: This policy emphasizes the need for multipurpose water development projects that address priorities like providing drinking water and mitigating floods. It sets a goal of improving water use efficiency by 20 percent through measures such as pricing. Integrated water resources management is promoted to conserve water, reduce wastage, and ensure equitable distribution.

National Environment Policy (NEP), 2006: NEP, 2006 responds to India's commitment to a clean environment, as mandated in the Indian Constitution. It recognizes that maintaining a healthy environment is the responsibility of both the state and every citizen. NEP places human beings at the center of concerns for SD.

National Action Plan on Climate Change (NAPCC), 2008: Launched in 2008, NAPCC emphasizes the importance of maintaining a high growth rate to improve living standards and address climate change impacts. It identifies measures that align with India's SD goals while also contributing to climate change mitigation. NAPCC is coordinated by the Ministry of Environment and Forests and implemented through various nodal ministries.

National Ganga River Basin Authority (NGRBA), 2009: Established in 2009, NGRBA is responsible for the conservation of the Ganga River. It has been empowered to plan, finance, monitor, and coordinate activities aimed at conserving the river holistically under the Environment (Protection) Act, 1986.

Renewable Energy Certificates (RECs), 2010: The renewable energy certificate (REC) incentivises producers of renewable energy (RE). India started the development of the renewable energy sector over a decade ago. The Pan-India market-based Renewable Energy Certificate (REC) Mechanism was established in 2010 under the Electricity Act 2003 ("EA 2003") and the National Tariff Policy 2006 ("Tariff Policy 2006") to rectify the disparity between resource availability and the obligations of entities to meet their renewable purchase obligation (RPO) (Sawhney, 2022).

National Solar Mission, 2015: The National Solar Mission is a significant project by the Government of India and State Governments aimed at fostering environmentally sustainable development while tackling India's energy security issues. It would represent a significant contribution by India to the global endeavour to address the issues of climate change.

Constitutional Environmentalism of India: In 1976, the 42nd Amendment Act introduced environmental protection into the Indian Constitution through various provisions, Article 48-A Directive Principles of State Policy (*The Constitution of India, Article 48-A*) was added, focusing on the protection and improvement of the environment. Courts have drawn guidance from this article in environmental cases (*Virender Gaur v. State of Haryana*, 1995(4) SCC 571, 580; *Indian Council for Enviro-Legal Action v. Union of India*, AIR 1996 SC 1446-9; M.C. *Mehta v. Union of India,* AIR 1988 SC 1037). Fundamental Duties, article 51-A (g) introduced fundamental duties with respect to the environment (*The Constitution of India, Article 51-A(g)*). Fundamental Rights, Article 21 guarantees the fundamental right to life with dignity, livelihood, health, and medical care. The judiciary has interpreted this right to encompass the right to a healthy environment, emphasizing the interconnection between a clean environment, health, and a meaningful life (*The Constitution of India, Article 21*). These policies and constitutional provisions underscore India's commitment to environmental protection, SD, and the well-being of its citizens.

Legislative Framework of India

Several key environmental laws and programs have been enacted in India to address various aspects of environmental protection and sustainability. Here are some notable ones:

Water (Prevention and Control of Pollution) Act, 1974: This act aims to prevent and control water pollution and restore the wholesomeness of water in the country. It requires industrial units likely to discharge effluents into water bodies to obtain consent for effluent discharge. The act is implemented through State Pollution Control Boards (SPCBs) and monitored by the Central Pollution Control Board (CPCB) through a network of water quality monitoring stations.

Air (Prevention and Control of Pollution) Act, 1981: Enacted in 1981 and amended in 1987, this law addresses the prevention, control, and abatement of air pollution in India. It assigns various functions to the CPCB and SPCBs at the state level to regulate and control air pollution.

Biosphere Reserves Programme, 1986: India designates certain areas as biosphere reserves, apart from protected areas under the Wildlife (Protection) Act, 1972. Biosphere reserves typically incorporate significant portions of sanctuaries or national parks, subject to applicable restrictions.

The National Environment Tribunal Act, 1995: This act established a National Environment Tribunal to expedite the resolution of cases arising from accidents, damages, or strict liability related to environmental issues. Its primary objective is to provide relief and compensation for damages to individuals, property, and the environment.

Energy Conservation Act, 2001: Enacted in 2001, this act promotes energy conservation in India. It mandates large energy consumers to adhere to energy consumption norms, requires new buildings to follow Energy Conservation Building Codes (ECBCs), and sets energy performance standards for appliances, along with energy consumption labeling. The act provides a legal framework for energy efficiency initiatives in the country.

National Green Tribunal Act (NGTA), 2010: The National Green Tribunal, established in 2010, is dedicated exclusively to environmental issues. It consists of a chairman and members with expertise in environmental sciences. The NGT has the authority to issue directions for compensation and restitution of damage caused by environmental negligence. It operates based on principles such as the polluter pays principle and SD.

These laws and programs reflect India's commitment to addressing environmental challenges, conserving natural resources, and promoting sustainability through legal and regulatory frameworks.

The Role of Judiciary in India

The judiciary in India plays a pivotal role in advancing the cause of SD. It serves as an essential partner in promoting effective environmental governance, upholding the rule of law, and ensuring a delicate balance between environmental, social, and developmental considerations through its judicial decisions and declarations (Global judges symposium from 18-20 August 2002, members of the ... - eufje 2002).

SD encompassing principles at international, national, provincial, and local levels, presents a complex legal landscape. Therefore, the judiciary faces the significant responsibility of interpreting and applying SD principles on a case-by-case basis (UNEP Global Judges programme - united nations environment programme

2005). In fulfilling this role, national judiciaries can benefit from the exchange of judicial decisions, information sharing, and the collective experience of different jurisdictions. Collaboration and cooperation among judicial bodies can contribute to a more coherent and effective approach to SD within the legal framework.

In essence, the judiciary is an imperative partner in advancing environmental governance, upholding the principles of the rule of law, and ensuring a harmonious equilibrium among environmental, social, and developmental factors through its judgments and declarations. This role emphasizes the judiciary's vital contribution to the pursuit of sustainability in India and beyond.

Several landmark judicial decisions in India have contributed to the advancement of environmental protection and SD. Here are some noteworthy cases and the principles they established, *MC Mehta* v. *Union of India & Ors* (Writ Petition Civil No 860 of 1991), this case emphasized the importance of judicial reasoning in environmental matters and the promotion of public awareness and environmental education at secondary and tertiary levels. It underscored the judiciary's role in raising awareness and fostering environmental education to address environmental challenges effectively. *Rural Litigation and Entitlement Kendra* v. *State of U.P.* (AIR 1988 SC 2187), this case, the Supreme Court of India prohibited limestone mining operations in the Himalayan hills. The court took into account the interests of future generations and recognized the unique legacy of the Himalayan ecosystem. This decision highlighted the importance of preserving ecologically sensitive regions for the benefit of present and future generations. *Vellore Citizens Welfare Forum* v. *Union of India* (AIR 1996 SC 2715), this case addressed the leather industry's impact on the environment. While acknowledging the industry's economic significance, the Supreme Court held that it had no right to destroy the ecology, degrade the environment, or pose health hazards. This decision emphasized the need for industries to operate in an environmentally responsible manner. *Indian Council for Enviro Legal Action* v. *Union of India*, (1996 AIR SCW 1069), this case reaffirmed the 'Polluter Pays' principle and stressed the importance of simplicity and practicality in applying this principle in India. The court ruled that activities considered hazardous or inherently dangerous make those responsible liable for compensating any loss caused to others by their activities. This decision highlighted the need for accountability in environmental matters. *Hindustan Zinc Limited* v. *Rajasthan Electricity Regulatory Commission* (2015), India's Supreme Court recently delivered a verdict that conveys a strong message: Industries must comply with renewable energy objectives or face penalties. high court rejected a case submitted by Hindustan Zinc Limited, a prominent zinc manufacturer in the nation, which contested the Rajasthan Electricity Regulatory Commission's (RERC) laws requiring certain users to get a percentage of their electricity from renewable sources. This significant Supreme Court ruling mandates that industries with captive power plants—where the consumer holds a minimum 26 percent ownership and utilises at least 51 percent of the generated energy—must acquire a portion of their energy from renewable sources. Additionally, it authorises state electricity regulators and agencies to impose penalties on entities that do not fulfil these requirements. Rajasthan is a landlocked state located in the northwest of India, next to Pakistan to the west. The state has an installed capacity of 15,917 megawatts (MW), with about one-fourth (3,916 MW) derived from renewable sources such as wind, solar, and biomass energy plants. The Rajasthan energy Regulatory Commission (RERC) oversees the regulation of energy within the state, including the advancement of power production from renewable sources. In 2007, RERC promulgated a set of mandates, delineating three categories of companies required to begin procuring a fraction of their energy from renewable sources. The "Renewable Purchase Obligation" pertains to distribution utilities and certain major electrical users (Krishnan, 2015).

Solar Energy Corporation of India Limited (Seci) Appellant(S) v. *Wind Four Renergy Private Limited & Ors,* (2024 INSC 160) in this case the Solar Energy Corporation of India Limited (SECI) contested the judgement and ruling of the Appellate Tribunal for Electricity, which permitted a delay of 132 days. The court determined that SECI served as the implementing agency for a renewable energy initiative, but the inter-state transmission licensee, Power Grid Corporation of India Limited, failed to operationalise the Long-Term Access (LTA) necessary for the project. The court agreed with WFRPL's assertion, allowing an extension of 9 months beyond the original date, accompanied by liquidated damages and a tariff decrease. WFRPL appealed to the APTEL, which permitted the postponement. The court determined that the judgement and its operational directives were untenable and inconsistent with the plan and PPA. The Supreme Court has annulled a judgement dated January 11, 2022, which was predicated on the order issued by the CERC. The court determined that the order issued on 08.03.2021 by the CERC would be effective and obligatory for the parties involved. The court further observed that the appellant, SECI, had reimbursed Rs.10 crores to WFRPL after encashing the performance bank guarantee. The appellant is entitled to recover the sum together with simple interest at a rate of 12% per year from the date of payment until the money is reimbursed by WFRPL.

M.K. Ranjitsinh v. *Union of India*, (2024 INSC 280), The Indian Supreme Court affirmed that the right to be free from the detrimental impacts of climate change is a fundamental right under Articles 21 and 14 of the Constitution of India. The lawsuit pertained to the conservation of the Great Indian Bustard and Lesser Florican bird species, which were threatened with extinction owing to overhead electricity wires. The court-imposed limitations on the installation of overhead transmission lines in vital habitats and ordered the Union to provide a detailed status report recommending a strategy that balances the preservation of the Bustard with the advancement of solar energy. The court underscored India's commitment to renewable energy and the effects of climate change on public health. The Court underscored the obligation of the State and the essential responsibility of people under Article 48A and 51A to safeguard the environment. The Indian Supreme Court has instructed the Union and relevant ministries to adopt strategies to address climate change, such as establishing predator-proof cages for avian species, fostering a conducive environment, and partnering with research institutions. The Court's finding is based on prior judgements in Entertainment Network (India) Ltd. v. Super Cassette Industries Ltd. and Apparel Export Promotion Council v. A.K. Chopra. Climate change is a genuine and pressing hazard, characterised by abbreviated winters, prolonged heatwaves, intensified precipitation, and forecasts indicating the potential submersion of coastal communities soon. The Court's determination to include the right against the detrimental impacts of climate change within the framework of basic rights is commendable; nonetheless, it prompts enquiries over its practical enforcement. The expansion of Article 21 is a favourable development for the domestic implementation of commitments outlined in international treaties and conventions.

These judicial decisions demonstrate the Indian judiciary's commitment to protecting the environment, promoting SD, and holding those responsible for environmental harm accountable for their actions. They serve as significant legal precedents in the field of environmental law and governance in India.

CONCLUSION

India stands at a pivotal juncture in its journey toward sustainable energy and environmental stewardship. The nation's substantial potential in solar energy, supported by favorable geography and abundant sunlight, positions it as a leader in renewable energy transformation. However, realizing this potential requires a robust, cohesive regulatory framework that can uniformly guide implementation across states, address financial and technological challenges, and incentivize investment in solar initiatives.

The legal and policy landscape governing solar energy in India has made significant strides, with initiatives like the National Solar Mission, PM-Surya Ghar: Muft Bijli Yojana, and the Rooftop Solar Programme. These programs are essential steps toward meeting India's climate commitments under the Paris Agreement and achieving Sustainable Development Goals (SDGs) 7 and 13, which focus on affordable, reliable, sustainable energy access and urgent climate action. Nonetheless, further reforms are needed to streamline policies, provide consistent incentives, and facilitate a more seamless integration of technological advancements such as IoT and AI, aligning with the vision of Society 5.0.

Incorporating the principles of Society 5.0 in renewable energy policies could transform India into a super-smart society where technological innovation coexists with environmental preservation, promoting an intelligent, sustainable quality of life. By fostering a legal and regulatory environment that supports sustainable energy, India not only advances its energy independence but also sets an example for integrating sustainable development with technological growth. Moving forward, the synergy between legal frameworks, policy incentives, and technological adoption will be crucial for India's renewable energy ambitions and its contribution to global environmental sustainability.

REFERENCES

Allen, N. (2024, May 28). *How much power does a solar panel produce?* Forbes. https://www.forbes.com/home-improvement/solar/how-much-power-does-a-solar-panel-produce/

Bjelland, O., & Mehta, S. (2021, July 19). Here's how India can lead the world in solar-based growth. World Economic Forum. https://www.weforum.org/stories/2021/07/india-can-lead-the-world-in-solar-based-growth/

Cirit, D. K., & Aydemir, S. (2021). *Exploring Levels of Secondary School Students' Knowledge: Global Warming, Acid Rain, and Ozone Layer Depletion*. DOI: 10.31235/osf.io/ehq2s

Deguchi, A., Hirai, C., Matsuoka, H., Nakano, T., Oshima, K., Tai, M., & Tani, S. (2020). What is society 5.0. *Society*, *5*(0), 1–24.

Publications Office of the European Union. (2000). Directorate-General for Environment (European Commission) & Decleris. "The law of sustainable development: General principles." Photo of Publications Office of the European Union. Retrieved August 1, 2022, from https://op.europa.eu/en/publication-detail/-/publication/28d5c957-eb81-4ef2-8cdc-31aa28436206/language-en

Fukuyama, M. (2018). Society 5.0: Aiming for a new human-centred society. *Japan Spotlight*, *27*(5), 47–50.

Fülöp, S. 2021. "Future generations institutions to implement international obligations towards Future Generations." In *Intergenerational Justice in Sustainable Development Treaty Implementation*, 137–162. DOI: 10.1017/9781108768511.010

*Global judges symposium from 18-20 August 2002, members of the ... - eufje.2002. Retrieved August 1, 2022*https://www.eufje.org/images/DocDivers/Johannesburg%20Principles.pdf

Government of India. (2024). (rep.). *Energy Statistics India 2024*. Retrieved from https://mospi.gov.in/publication/energy-statistics-india-2024-1

Govind Bhutada. (2022, April 13). The world's energy transitions: a history told in infographics. World Economic Forum. https://www.weforum.org/stories/2022/04/visualizing-the-history-of-energy-transitions/https://www8.cao.go.jp/cstp/english//society5_0/index.html

IRENAs statute enters into force. (2010, July 8). Irena.org. https://www.irena.org/News/articles/2010/Jul/IRENAs-statute-enters-into-force

Jacob, R., Hoffmann, M., Weinand, J. M., Linßen, J., Stolten, D., & Müller, M. (2023). The future role of thermal energy storage in 100% Renewable Electricity Systems. *Renewable and Sustainable Energy Transition*, *4*, 100059. DOI: 10.1016/j.rset.2023.100059

Jariwala, C. M. (1993). Direction Of Environmental Justice in India: Critical Appraisal Of 1987 Case Law. *Journal of the Indian Law Institute*, *35*(1/2), 92–114. https://www.jstor.org/stable/43952326

Kerr, T. (2024, February 5). India is making strides on climate policy that others could follow. World Economic Forum. https://www.weforum.org/stories/2024/02/climate-policy-india/

Krämer, L. 2021. "Time to think: Sustainable Development, future generations and the individual." In *Intergenerational Justice in Sustainable Development Treaty Implementation*, 211–224. DOI: 10.1017/9781108768511.014

Krishnan, D. S. (2015). India's Supreme Court Reinforces Renewable Energy Targets for Industry. World Resources Institute. https://www.wri.org/insights/indias-supreme-court-reinforces-renewable-energy-targets-industry

Laursen, F. (2016). The Founding Treaties of the European Union and Their Reform. In *Oxford Research Encyclopedia of Politics*. Oxford University Press., DOI: 10.1093/acrefore/9780190228637.013.151

MacArthur Foundation. (2017). *Let's Build a Circular Economy*. Ellenmacarthurfoundation.org. https://ellenmacarthurfoundation.org/

Major ongoing Renewable Energy Schemes and Programmes. (2023). Pib.gov.in. https://pib.gov.in/Pressreleaseshare.aspx?PRID=1983766

Maka, A. O. M., & Alabid, J. M. (2022). Solar energy technology and its roles in sustainable development. *Clean Energy*, *6*(3), 476–483. DOI: 10.1093/ce/zkac023

Marr, B. (2018). *Data-Driven HR: How to Use Analytics and Metrics to Drive Performance*. Kogan Page.

Ministry of New and Renewable Energy, Government of India. (2019). *Pradhan Mantri Kisan Urja Suraksha evam Utthaan Mahabhiyaan (PM-KUSUM)*. Ministry of New and Renewable Energy. https://mnre.gov.in/pradhan-mantri-kisan-urja-suraksha-evam-utthaan-mahabhiyaan-pm-kusum/

Mundi, J. (2024). The Energy Charter Treaty. Jusmundi.com. https://jusmundi.com/en/document/treaty/en-the-energy-charter-treaty-the-energy-charter-treaty-1994-saturday-17th-december-1994

Nakatani, H. (2019, October 31). *Population aging in Japan: Policy transformation, sustainable development goals, Universal Health Coverage, and social determinates of health.* Global health & medicine. https://pmc.ncbi.nlm.nih.gov/articles/PMC7731274/

Narvaez Rojas, C., Alomia Peñafiel, G. A., Loaiza Buitrago, D. F., & Tavera Romero, C. A. (2021). Society 5.0: A Japanese concept for a superintelligent society. *Sustainability (Basel), 13*(12), 6567. DOI: 10.3390/su13126567

Person, Michael, & Redclift. 1987. *Sustainable development: Exploring the contradictions.* Taylor & Francis. Retrieved August 1, 2022, from https://www.taylorfrancis.com/books/mono/10.4324/9780203408889/sustainable-development-michael-redclift

Poppick, L. (2017, February 17). Twelve Years Ago, the Kyoto Protocol Set the Stage for Global Climate Change Policy. Smithsonian Magazine. https://www.smithsonianmag.com/science-nature/twelve-years-ago-kyoto-protocol-set-stage-global-climate-change-policy-180962229/

Press Information Bureau. (2024, February 29). *Cabinet approves PM-Surya Ghar: Muft Bijli Yojana for installing rooftop solar in one crore households.* Ministry of Information and Broadcasting, Government of India. Retrieved November 10, 2024, from https://pib.gov.in/PressReleasePage.aspx?PRID=2010133

Ramírez-Márquez, C., Posadas-Paredes, T., Raya-Tapia, A. Y., & Ponce-Ortega, J. M. (2024). Natural Resource Optimization and Sustainability in Society 5.0: A Comprehensive Review. *Resources, 13*(2), 19. DOI: 10.3390/resources13020019

Redclift, M. (2003). *Sustainable development: Exploring the contradictions.* Routledge.

Reimer, F. 2021. "Institutions for a sustainable future: The German Parliamentary Advisory Council on Sustainable Development." In *Intergenerational Justice in Sustainable Development Treaty Implementation*, 374–394. DOI. DOI: 10.1017/9781108768511.025

Sand, P. H. (2021). Origin and History. Oxford University Press EBooks, 50-C3.S15. DOI: 10.1093/law/9780198849155.003.0003

Sawhney, A. (2022). Renewable Energy Certificates Trading in India: A Decade in Review. In www.adb.org. Asian Development Bank. https://www.adb.org/publications/renewable-energy-certificates-trading-in-india-a-decade-in-review

Society 5.0 (2016). Cabinet Office Home Page.

Summary report 28 November – 10 December 2005. (2005). IISD Earth Negotiations Bulletin. https://enb.iisd.org/events/unfccc-cop-11/summary-report-28-november-10-december-2005

Sustainable development, future generations, and Public Trust. 2014. *Global Environmental Constitutionalism*, 329–342. DOI: 10.1017/CBO9781139135559.019

UNEP Global Judges programme - united nations environment programme. 2005. Retrieved August 1, 2022, *from* https://stg-wedocs.unep.org/bitstream/handle/20.500.11822/8406/-UNEP%20Global%20Judges%20Programme-20053747.pdf

UNFCCC. (2019). What Is the Kyoto Protocol? UNFCCC. https://unfccc.int/kyoto_protocol

UNFCCC. (2024). The Paris Agreement. United Nations Climate Change; United Nations. https://unfccc.int/process-and-meetings/the-paris-agreement

United Nations. 1987. *Report of the World Commission on Environment and Development*. Retrieved August 1, 2022, from https://digitallibrary.un.org/record/139811

United Nations. 2004. "Take action for the sustainable development goals - united nations sustainable development." United Nations. Retrieved August 1, 2022, https://www.un.org/sustainabledevelopment/sustainable-development-goals/

United Nations. (2021). COP26: Together for our planet. United Nations; United Nations. https://www.un.org/en/climatechange/cop26

United Nations. (2022). As Humanity's Environment Footprint Becomes Increasingly Unsustainable, Global Leaders Recommit to Joint Climate Action, at Opening of Stockholm Summit | UN Press. Press.un.org. https://press.un.org/en/2022/envdev2046.doc.htm

United Nations. (2024, January 12). UN Climate Change Conference - United Arab Emirates. Unfccc.int. https://unfccc.int/cop28

Chapter 12
Rule-Based Energy Management Optimized With Artificial Intelligence and a Case Study Analysis

Rampelli Manojkumar
https://orcid.org/0000-0002-8141-0321
BVRIT HYDERABAD College of Engineering for Women, India

Gumpu Srinivasulu
Maulana Azad National Institute of Technology, India

Pulipati Anusha
BVRIT HYDERABAD College of Engineering for Women, India

K. Rathnamma
BVRIT HYDERABAD College of Engineering for Women, India

Ankitha Bodupalli
BVRIT HYDERABAD College of Engineering for Women, India

Deekshitha Kanumuri
BVRIT HYDERABAD College of Engineering for Women, India

ABSTRACT

This chapter discusses rule-based control for energy management optimized with artificial intelligence in AC/DC hybrid systems. A case study at an educational institution is considered with the photovoltaic-battery energy storage system in order to show the significance of optimized demand response method in reducing energy consumption costs. The energy cost is minimized while considering state of charge of battery thresholds and other constraints of battery. The study demonstrates particle swarm optimization's efficiency over genetic algorithm with respect to run

DOI: 10.4018/979-8-3373-0680-3.ch012

time. It is observed that autumn season achieves 39.44% cost savings by prioritizing PV energy and grid injection, while winter achieves 76.47% savings by aligning discharge with peak demand.

INTRODUCTION

The development of new methods of storing energy along with the increasing popularity of renewable energy sources in the past years has led to changes in power distribution systems. This shift to more sustainable and effective solutions resulted in the use of AC/DC hybrid systems. They serve as the base for the integration of numerous energy resources, load management, and grid stability. The adoption of AC/DC systems necessitates the development of smart control techniques that are capable of improving the performance of these hybrid systems. Guided by this idea, the chapter outlines the optimized rule-based control approaches with particular attention paid to their application in price-driven demand response (DR) strategies in AC/DC hybrid systems.

This chapter however focuses on the importance of energy management techniques in improving the robust functionality of AC/DC hybrid systems. It investigates the possibility of their full exploitation with regards to RES, energy storage, and load management. By implementing AI, these techniques allow the re-alteration of energy consumption according to the electricity pricing signals, which makes them respond on real time (Häseler & Wulf, 2024; Manojkumar, Kumar, & Ganguly, 2020). This chapter offers insights into the principles of smart control methods used in modern power distribution networks (Kumar, Manojkumar, Ganguly, & Liserre, 2020; Das, Manojkumar, Kumar, & Ganguly, 2021). The primary motivation behind this chapter is the increasing significance of energy management programs aimed at reducing energy costs and optimizing resource utilization across various sectors (Mariano-Hernández et al., 2021; Mishra & Singh, 2023). It also discusses the role of AI in implementing these energy management strategies. By employing smart rule-based methodologies, AC/DC hybrid systems can actively participate in demand response (DR) initiatives and improve grid stability (Abdelkader et al., 2024). To illustrate the advantages of these methods, a detailed case study within an educational institution is presented. This case study highlights improvements in energy usage and the assurance of system reliability (Alghassab, 2024; Sharma, Gupta, & Kumar, 2023).

In summary, the objectives of this chapter are to examine the role of power electronics in facilitating smart control of AC/DC hybrid systems, as well as to discuss key features such as efficient AC-DC conversion, flexible power flow control, and energy usage optimization. It aims to provide a comprehensive understanding of energy management in hybrid systems while emphasizing the importance of peak shaving methods and DR mechanisms. The chapter explores various DR strategies, including time-based, incentive-based, and automated DR, along with the integration of energy-efficient loads, on-site generators, and renewable energy sources (RES). Additionally, it delves into the application of AI in energy management, addressing topics like load forecasting, genetic algorithms (GA), and particle swarm optimization (PSO). The focus is given to optimized rule-based DR methods with a thorough methodology. The effectiveness of the proposed approach is validated through a case study using PSO. Further, the system performance across different seasons to reduce energy consumption cost is analysed.

BACKGROUND OF STUDY AND LITERATURE SURVEY

Role of Power Electronics for Smart Control of AC/DC Hybrid Systems

Power electronics play an important role in the smart control of AC/DC hybrid systems by providing an efficient and adaptive management of energy flow between various sources and loads as discussed follows.

Conversion Between AC and DC

Power electronics devices such as rectifiers and inverters are used for the conversion between AC and DC. In hybrid systems combining AC and DC sources, power electronics manage these conversions efficiently (Luo & Zhang, 2022; Yazdkhasti & Diduch, 2020). For example, solar photovoltaic panels generate DC power, which needs to be converted to AC for use in conventional AC power grids. Similarly, AC power from the grid can be converted to DC for battery charging (Khojasteh, Faria, & Vale, 2020).

Flexible Power Flow Control

Power electronic converters provide the ability to control the flow of power in hybrid systems. They can manage power distribution based on real-time demand and availability of energy sources. This flexibility provides smart energy management i.e., power can be routed optimally to meet load demands, charge storage systems like batteries, export surplus power back to the grid, etc (Ali et al., 2022; Rezaei-mozafar et al., 2022).

Efficiency Optimization

Power electronics help enhance the overall efficiency of hybrid systems. By using efficient conversion techniques and controlling power flows dynamically, losses can be minimized. This is particularly required with RES where energy availability is intermittent in nature (Wang, Yuan, Sun, & Wennersten, 2022; Gernaat et al., 2021).

Integration of Energy Storage

In hybrid systems including energy storage devices such as batteries, supercapacitors, the power electronics are essential. They ensure that energy from different sources is stored and retrieved as needed for smoothing out fluctuations associated with renewable energy generation and load (Khalid, 2024; Barra et al., 2021).

Adaptive Control Methods

Advanced control algorithms can be implemented in power converters for dynamically adjusting the operation based on parameters like load demand and energy availability (Imani, Yousefpour, Andani, & Ghadi, 2019; Manojkumar et al., 2022).

Grid Interaction and Stability

Power electronics play an important role in maintaining stability and reliability. Converters can help manage voltage fluctuations, power quality, and grid synchronization. It is useful for ensuring seamless integration and operation within the larger grid infrastructure (Kabalci, 2024; Taha, Abdeltawab, & Mohamed, 2018).

Smart Grid Integration

As part of smart control systems, power electronics contribute to the execution of smart grid functionalities. They enable bidirectional power flow, grid monitoring, and communication capabilities which are required for grid interaction, DR, and other smart grid applications (Manojkumar, Kumar, Ganguly, & Catalão, 2021; Anteneh et al., 2021).

In summary, power electronics technology forms the backbone of smart control in AC/DC hybrid systems and enable efficient energy conversion, flexible power management, use of RES, etc. This integration leads to more sustainable, resilient, and efficient energy systems.

Energy Management in AC/DC Hybrid Systems

The Energy Management System (EMS) is required for the optimal operation of distributed energy resources. For example, Taha, Abdeltawab, and Mohamed (2018) propose a model that takes into account regulated loads, renewable energy sources (RES) like wind and solar, and hybrid energy sources while using a diesel generator and energy storage system to reduce environmental emissions. Based on the demand requirements, especially during peak hours, it can enhance the proportion of energy generation for the generators and avoid the costlier energy sources (Manojkumar, Kumar, Ganguly, & Catalão, 2021). Moreover, in a distribution system, network reconfiguration planning lowers system losses and reduces the system's operational expenses (Anteneh et al., 2021; Aziz et al., 2021; Shi et al., 2021; Rasheed & Abhyankar, 2023).

Out of many applications of the energy management system, peak shaving and demand response (DR) are the most important applications.

Peak Shaving

Peak shaving entails proactively managing overall demand to exclude demand spikes, which result in a higher peak. This process reduces peak loads, lowering the overall cost of demand charges. For both the grid and end users, high peak demand drives up electricity costs. Peak shaving is able to reduce voltage drop and voltage rise issues, reduce ratings of the different components, lower energy consumption costs, and reduce power losses. Peak shaving also provides the possibility of increasing the penetration levels of RES and satisfying the increased load demand.

Demand Response

DR represents a significant move in the electricity sector, responding to the inherent challenges of traditional power systems. In conventional setups, grid operators face constraints in immediately aligning electricity demand with supply fluctuations. This limitation becomes particularly critical during peak demand periods when the grid is under stress. DR is a strategic solution to this challenge and offers a flexible and interactive framework (Stanelyte, Radziukyniene, & Radziukynas, 2022).

DR involves the strategic change of electricity consumption patterns among end-users. In addition to relying only on adjusting the supply side of the grid, DR enables the modification of demand. This is achieved through a combination of influencing consumer behavior and introducing economic incentives (Xue et al., 2021).

The inputs for DR are the signals from grid operators, such as variations in electricity prices. These signals help consumers adapt their electricity usage either by reducing consumption during periods of high demand or by shifting their usage to times when electricity is more readily available. This dynamic adjustment helps create a more balanced and resilient grid while addressing peak demand. One of the main benefits of DR is its capacity to alleviate grid congestion. Congestion typically occurs during peak hours when electricity demand exceeds the available transmission capacity, leading to reliability issues. By strategically changing demand during these critical times, DR mitigates congestion and enhances the overall stability of the grid (Abdelkader et al., 2024).

Moreover, DR aligns with the broader goals of sustainability. The intermittency of renewable sources like solar and wind power poses challenges for grid operators, as they may not always align with periods of high demand. Therefore, DR steps in as a complementary strategy, allowing consumers to adjust their electricity usage to coincide with the availability of renewable energy. This enhances the use of clean energy and supports the transition towards a more sustainable power grid (Khalid, 2024). DR programs are implemented through time-of-use (ToU) pricing structures, encouraging consumers to shift their electricity use to off-peak hours when prices are lower. Automated demand response refers to the automatic adjustment of appliances and devices based on real-time signals from the grid. The advantages of DR are explored in detail below.

Grid Reliability

Demand Response (DR) helps in maintaining grid stability and reduces the risk of blackouts by matching supply and demand more effectively (Mishra & Singh, 2023).

Cost Savings

Consumers who participate in DR programs benefit from lower electricity prices during off-peak hours, thereby reducing costs (Alghassab, 2024).

Environmental Impact

DR can contribute to lower carbon emissions by promoting the use of renewable energy and reducing reliance on conventional power plants during peak periods (Wang, Yuan, Sun, & Wennersten, 2022).

Overall, DR is a dynamic and growing aspect of modern electricity systems, contributing to the development of smarter, more resilient, and sustainable grids (Stanelyte, Radziukyniene, & Radziukynas, 2022).

DR has various types and methods that enable the dynamic adjustment of electricity consumption patterns in response to signals from grid operators. These types of DR are categorized based on their mechanisms and objectives, as discussed below.

Time-Based Demand Response

Electricity prices vary based on the time of day, encouraging consumers to shift their usage to off-peak hours (Xue et al., 2021).

Incentive-Based Demand Response

Consumers voluntarily reduce their electricity consumption during specified periods in exchange for incentives such as bill credits or discounts (Rezaeimozafar et al., 2022). Moreover, consumers bid to receive payments for committing to reduce their electricity usage during peak times (Liu & Cui, 2020).

Automated Demand Response (ADR)

ADR utilizes advanced metering systems and smart grid technologies to automatically adjust or reduce electricity usage in response to signals from the grid operator. This often involves smart appliances, thermostats, and other devices (Manojkumar et al., 2022). There is an open standard that provides communication between utilities and end-use customers to enable ADR (Imani, Yousefpour, Andani, & Ghadi, 2019). Moreover, it is possible for consumers to bid to provide a predetermined amount of demand reduction capacity to the grid during periods of high demand (Sharma, Gupta, & Kumar, 2023).

Energy Efficient Loads, Load Shifting, and On-Site Generators for Energy Management

Energy management is achievable through the use of energy-efficient loads, on-site generators, renewable energy sources (RES), and energy storage devices as discussed below.

Usage of Energy Efficient Loads and Shifting of Loads

By switching to more energy-efficient loads during specified hours of the day, industrial and commercial clients can lower the amount of power they consume from the grid (Avordeh, Gyamfi, & Opoku, 2022). For instance, commercial buildings can switch to LED lighting, which uses less energy, or turn off heating elements. Similarly, variable-frequency drives (VFDs) can reduce grid energy consumption in motor-driven applications. During peak demand hours, a VFD assists in adjusting the speed of the motor. Further, load shifting is also used for peak shaving purposes (Wang et al., 2023). Load shifting is a strategy used to reduce electricity consumption during periods of peak power demand by shifting the usage of electrical devices to off-peak hours. This helps reduce strain on the power grid, increase its efficiency, and potentially reduce the need for additional power generation capacity.

Installing On-Site Generators

One of the most common ways to reduce grid power use in commercial and industrial buildings is to install on-site generators. Diesel or gas-powered generators can provide auxiliary power during times of high demand to sustain certain loads (Mosbah, Guerra, & Barrera, 2022; Mohamed et al., 2021).

For example, an automatic changeover circuit can be used to transfer chillers to backup generators if a plant's peak demand hours are between 3 and 5 p.m. Although employing generators is an effective strategy to avoid peak energy prices, the long-term cost of running them could be high due to fuel and maintenance expenses.

Renewable Energy Sources and Battery Energy Storage

Recently, the use of battery energy storage systems (BESS) along with RES has increased for implementing energy management applications due to the continuous reduction in the cost of installing BESS and RES. The ESS's appealing qualities, such as its quick response and declining cost, make it suitable for use in a range of electrical system scenarios. Energy storage integration is projected to be significant in power distribution networks in the future to manage power variations from decentralized renewable energy resources and reliably satisfy load demand. Industrial or commercial loads may benefit from extra power supplied by power storage units like batteries and supercapacitor modules during periods of high demand. BESS has several potential applications in low-voltage unbalanced distribution networks, including peak load shaving and load leveling (Zupančič, Lakić, Medved, & Gubina, 2017).

Consumers who reside in warmer climates or whose peak hours are in the evening may find this strategy extremely helpful. Batteries or supercapacitors are charged by solar panels using electricity produced by the sun during the day. Tesla's Powerpack is an energy storage device that can be used alongside a solar array to reduce peak demand.

The growing low-voltage network operation issues are a result of the rising use of PV systems. With the help of grid-based energy storage, the EU project STORY is addressing these issues (Karthikeyan et al., 2018). Using a peak demand control technique, a novel control concept is proposed in this research, which is applied to storage units in LV distribution networks. In Karthikeyan et al. (2018), the management of network congestion, lowering network power loss, and peak shaving for a low-voltage active distribution network are discussed with the usage of battery storage owned by the distribution system operator. It aims to decrease the peak PV power input to the grid while taking network power loss reduction and battery degradation into account.

Electricity prices increase as a result of high peak demand, impacting end consumers as well as utilities. In Chua, Lih Bong, Lim, Wong, and Wang (2020), the problems caused by high peak demand are addressed. Four categories that can be used for peak shaving applications are discussed. In Boyouk, Munzke, and Hiller (2018), a novel peak-shaving method with a PV-battery storage system is proposed. When energy production fluctuates in response to changing power demand, a BESS is an efficient approach to reduce peaks and smooth the load. Peak shaving applies a threshold based on load and battery State of Charge (SOC). In Chua, Lim, and Morris (2013), a method for employing a BESS to reduce electrical network voltage imbalance and peak demand is discussed. The BESS is developed to reduce peak demand and consequently the electricity bill of customers. With the foreseeable large use of BESS, the stress on utility companies can be reduced during high peak power demands. In Wang, Zarghami, and Vaziri (2016), the use of BESS in improving the performance of distribution grids heavily infused with PV is discussed. The paper features a state-space-based model for BESS and implements a simple and effective approach for peak load shaving by considering device constraints. This work provides an energy management strategy with the goal of peak shaving, which also considers degradation and aging limits of the batteries.

The paper by Lehmann et al. (2016) deals with production facilities that require improvements in the use of renewable energies and energy efficiency due to growing electricity prices, rising environmental consciousness, and stricter legal requirements. Here, a techno-economic evaluation of photovoltaic battery storage systems for uses like industrial peak shaving and self-consumption is conducted. The paper by Weng et al. (2022) shows that an essential flexibility resource of the power grid has traditionally been the peak regulation capability of gas-fired power units. Coal power facilities are gradually being shut down to reduce carbon emissions, which emphasizes the importance of gas-fired power plants. However, in reality, gas-fired power plants frequently lack an acceptable level of flexibility. Through the research of this study, it is found that the main reason for the limited flexibility of gas-fired power plants is the lack of coordination among multiple entities belonging to different energy systems, such as electricity and gas. The proposed mechanism for cooperation is an attempt to actualize the cooperation between the gas and electric power systems, which serves as a model for deeper cooperation among various energy systems in the future.

Various approaches are used in the literature to control the charge/discharge schedules of the battery, such as genetic algorithms (GA), dynamic programming (DP), rule-based algorithms, etc. (Reddy & Meikandasivam, 2020; Urooj & Nasir, 2024; Hafiz et al., 2020). Out of these, rule-based methods are simple to implement and develop compared to other methods. Further, rule-based approaches can be enhanced using AI, as discussed in the following sections. Table 2 summarizes energy management and the role of DR and BESS.

Table 2. Energy management and the role of DR and BESS

References	Broad Topic	Key Points
Taha, Abdeltawab, & Mohamed (2018); Manojkumar et al. (2021)	EMS	EMS models include regulated loads, RES, diesel generators, and storage to reduce emissions.
Stanelyte, Radziukyniene, & Radziukynas (2022); Xue et al. (2021); Abdelkader et al. (2024)	DR	DR addresses grid stress during peak demand periods, enhances flexibility, and alleviates grid congestion during peak demand.
Khalid (2024)	RES	DR complements RES integration, optimizing renewable energy use during peak demand.
Imani, Yousefpour, Andani, & Ghadi (2019); Manojkumar et al. (2022)	Automated DR (ADR)	ADR uses smart appliances and devices for real-time consumption adjustments.
Avordeh, Gyamfi, & Opoku (2022)	Energy-Efficient Loads	Energy-efficient loads like LED lighting reduce grid energy consumption.
Wang et al. (2023)	Load Shifting	Load shifting helps manage peak shaving and reduces grid strain during high demand.
Mosbah, Guerra, & Barrera (2022); Mohamed et al. (2021)	On-Site Generators	On-site generators reduce grid dependency during peak hours but incur operational costs.
Zupančič et al. (2017)	BESS	BESS applications include peak shaving, load leveling, and managing decentralized RES power variations.
Karthikeyan et al. (2018)	Grid-Based Storage Systems	EU project STORY uses grid-based storage to address low-voltage network operation issues.
Karthikeyan et al. (2018)	Battery Storage in LV Networks	Battery storage in LV distribution networks helps reduce peak PV power input and power losses.
Chua et al. (2020)	BESS and Electricity Prices	High peak demand impacts electricity prices; BESS helps mitigate this.
Boyouk, Munzke, & Hiller (2018)	PV-BESS Integration	PV-BESS integration supports load smoothing and peak shaving based on load and SOC thresholds.
Wang, Zarghami, & Vaziri (2016)	Large-Scale BESS Applications	Large-scale BESS usage alleviates utility stress during peak power demands.

Role of Artificial Intelligence for Energy Management

The rapid advancement of artificial intelligence (AI) and machine learning (ML) is transforming EMSs across industries, including load forecasting, fault detection, electricity markets, energy management in buildings, electric vehicles (EVs), and more. Energy management has become a critical focus globally due to increasing energy demands, climate change, and the need for sustainable energy solutions (Perera & Kamalaruban, 2021). AI has emerged as a transformative technology in addressing these challenges by enhancing efficiency, optimizing energy systems, and enabling smart decision-making. The rapid advancement of AI and ML is altering EMSs across electricity markets, load forecasting, fault detection, buildings, and EVs. AI is revolutionizing the energy industry by making energy management and optimization effective and sustainable (Mosavi et al., 2019). The applications of AI are diverse, spanning from industrial optimization and consumption predictions to the enhancement of storage solutions and building performance. With the use of AI, energy management systems can predict consumption patterns, optimize resource allocation, and facilitate the transition to renewable energy sources. AI's incorporation into energy systems is predicted to increase as technology develops, leading to additional breakthroughs and a more sustainable energy future (Entezari, Aslani, Zahedi, & Noorollahi, 2023).

The future is expected to experience an increased role for AI in energy management and optimization, driven by the development of more complex algorithms and enhanced data accessibility. Smarter, more resilient, and more efficient energy systems result from the use of AI with other cutting-edge technologies like blockchain and the Internet of Things (IoT), which could further transform the energy industry (Ghoddusi, Creamer, & Rafizadeh, 2019, Rosero, Díaz, & Trujillo, 2021). As we move towards a more sustainable future, AI is going to play an important role in shaping the energy landscape.

Load Forecasting Using AI for Energy Management

Load forecasting is significant in power system energy planning and management. In the past, it relied more on conventional statistical techniques, time-series analysis, and simple prediction models. Such methods lacked the ability to capture the intricacies and dynamic behavior of energy usage patterns. However, as energy demand grows, advanced technology must address the evolving challenges of the power system (Das, Samal, & Luhach, 2023; Kermani et al., 2023; Xin et al., 2022).

The methodology for load forecasting varies for traditional and smart grid systems due to their nature and functionality. Traditional power grids rely on simple statistical models and past data to forecast future energy consumption. These approaches tend to lack the ability to factor in the dynamic patterns of energy usage and have a limited ability to adapt to shifting circumstances. Furthermore, traditional grids lack real-time monitoring and feedback mechanisms, which might not permit the implementation of demand-response programs.

Load forecasting in smart grids employs technologies such as sensors, smart meters, and communication networks to obtain real-time information regarding power demand, generation, and system status. This forecasting method employs sophisticated AI algorithms to investigate vast amounts of data, including historical consumption data, weather, and socio-economic factors. This innovation provides more precise predictions of electricity use and offers better electrical grid management. These methods enhance accuracy by incorporating historical consumption patterns, climate data, and other external parameters. ML brought a data-driven approach to forecasting future results, empowering utility companies to make strategic decisions about energy resource allocation and grid management.

In the 21st century, the introduction of deep learning (DL) techniques significantly changed how load forecasting works in energy management. Neural networks, in particular, have shown capabilities to discover complex patterns and relationships within large datasets related to energy resources and loads. Their ability to autonomously extract characteristics and adjust to fluctuating circumstances makes them suitable for the dynamic features of energy management (Yao, Zhang, Wang, & Zhao, 2022; Khan et al., 2022; Islam, Rasheed, & Ahmed, 2022). The use of AI, reinforcement learning, and hybrid models, which combine the strengths of various techniques, has the potential to improve the accuracy and applicability of load forecasting in smart grids. This evolution will contribute to a sustainable approach to energy management.

Genetic Algorithms in Energy Management Applications

Core Principles of GAs

GA operates on a population of candidate solutions, where each individual represents a potential solution to the problem. In energy management, these individuals might model energy allocation methods, load scheduling plans of distributed energy resources. The algorithm evaluates the quality of each solution using a fitness function tailored to the problem's objectives such as minimizing energy consumption. Through iterations called as generations, GA refines the population by applying genetic operators as discussed follows.

Selection

Solutions with higher fitness are more likely to be chosen for reproduction while simulating the survival of the fittest.

Crossover

Pairs of selected solutions exchange parts of their genetic information to produce offspring, fostering diversity and innovation in the solution space.

Mutation

Small random changes are introduced to some solutions to explore new areas of the search space and avoid premature convergence.

Applications in Energy Management

GAs have been widely applied in energy management due to their flexibility and ability to handle non-linear, constrained, and multi-objective problems. Key applications include:

Demand Response and Load Scheduling

GA is used to optimize load shifting and scheduling, balancing consumer demand with energy supply while minimizing costs and peak loads.

Renewable Energy Integration

In microgrids and distributed energy systems, GAs help determine optimal dispatch methods for renewable sources like solar and wind, alongside storage systems.

Power Flow Optimization

GAs can optimize power flow in distribution networks to lower losses, maintain voltage stability, and improve overall system efficiency.

Hybrid Energy Systems Design

For systems combining multiple energy sources, GA is used to identify the optimal configuration and control methods for maximizing performance and reduced costs.

The adaptability of GA allows them to use for diverse energy systems and makes them well suited for modern energy management applications.

Particle Swarm Optimization in Energy Management Applications

PSO is also another nature-inspired optimization technique, modelled based on the social behaviour of birds and fish. PSO is superior in solving continuous and non-linear optimization problems. It is an effective tool for energy management applications where dynamic and real-time solutions are required.

Core Principles of PSO

In PSO, a group of particles represents potential solutions to the optimization problem. Each particle has a position (representing a solution) and a velocity (defining its movement through the solution space). The algorithm tracks two key pieces of information:

Personal Best (pBest)

The best solution a particle has discovered so far.

Global Best (gBest)

The best solution discovered by the entire swarm.

Particles adjust their velocity and position based on their pBest, the gBest, and a stochastic component to balance exploration and exploitation of the solution space. This collaborative approach allows PSO to converge efficiently toward optimal solutions.

Applications in Energy Management

PSO has got significant attention in energy management for its simplicity, computational efficiency, and effectiveness in handling multi-objective optimization problems. Common applications are discussed as follows.

Economic Dispatch

PSO is used to allocate generation resources in a power system to reduce costs while satisfying demand and operational constraints.

Renewable Energy Forecasting

By optimizing parameters of forecasting models, PSO improves the accuracy of solar and wind power predictions.

Energy Storage Management

PSO optimizes charging and discharging schedules for batteries, ensuring optimal energy usage and cost savings.

Microgrid Energy Management

PSO helps in determining optimal power-sharing methods between distributed generators, storage systems, and loads to maintain system stability and reduce costs.

Grid Stability and Control

PSO optimizes voltage regulation and frequency stabilization and ensures reliable and efficient operation.

Rule-Based Approaches Optimized with Artificial Intelligence

Optimized rule-based approaches in EMS play a key role in enhancing the efficiency and reliability of various power systems like hybrid vehicles, microgrids, and residential PV storage systems. These methods combine predefined rules with optimization techniques to manage energy resources.

Hybrid Vehicles

In hybrid vehicles, rule-based energy management methods are used to balance the power flow between the internal combustion engine and the electric motor. In Jalil et al. (1997), a rule-based energy management strategy for a series hybrid vehicle is presented. The minimization of emissions is considered as the objective.

Microgrid Energy Management

Microgrids significantly benefit from rule-based inferential systems. In Kurukuru et al. (2021), a rule-based approach is used for microgrid EMS which considers real-time data and predefined rules to manage energy distribution efficiently. This approach ensures a reliable power supply and reduces operational costs. In Farhadi Gharibeh & Farrokhifar (2021), an online multi-level energy management strategy combining rule-based and optimization-based approaches is presented. The method dynamically adjusts the power split between fuel cells and batteries for optimizing the overall system performance and extending the vehicle's range.

Residential PV Storage Systems

In residential PV storage systems, rule-based approaches are employed to manage energy storage and consumption for ensuring optimal usage of PV power. In Manojkumar et al. (2023) and Hossain et al. (2024), the various rule-based peak shaving techniques, including dynamic demand and feed-in limits, and master-slave level optimization, are discussed. The objective is to decrease the peak demand on the grid, lower energy bills, and improve power supply reliability. Further, in Manojkumar et al. (2020), an optimal peak-shaving technique for dynamic DR is suggested. A rule-based mechanism is employed to efficiently control the energy flow and maximize overall energy consumption during peak hours. For residential PV storage systems, the application of rule-based DR techniques in Manojkumar et al. (2020) utilizes smart transformers with dynamic optimum setpoints along with limit tolerance.

Thus, by integrating predefined rules with dynamic optimization techniques, these methods ensure effective energy management.

Potential Limitations, Operational Challenges, and Practical Constraints

The integration of power electronics, EMSs, DR, and AI-driven solutions into AC/DC hybrid systems offers several technical, economic, regulatory, social, and environmental challenges as discussed follows.

Technical Limitations

Power electronics challenges are mainly of thermal stress, switching losses, and the need for additional filters to maintain power quality. BESS suffer from aging, temperature sensitivity, and degradation over time, which increases long-term costs. RESs like solar and wind are intermittent, requiring backup solutions to maintain grid stability. High penetration of renewables can also cause voltage fluctuations and frequency instability.

Economic Constraints

The high initial costs of power electronics, BESS, and smart grid infrastructure make large-scale adoption challenging. Even though costs are gradually reducing, the long payback periods are challenging. Regular maintenance, battery replacements, and software updates further add to operational expenses. Market structures and pricing mechanisms, such as ToU and DR incentives, often fail to align with consumer behaviour.

Regulatory and Infrastructure Challenges

Many existing power grids were not designed for bidirectional energy flow and smart control. A lack of standardization across different manufacturers results in interoperability issues, making system integration complex. Moreover, slow regulatory updates create uncertainty regarding grid-code compliance, renewable energy incentives, and DR program designs.

Social and Behavioural Barriers

Consumer participation in DR programs remains low due to a lack of awareness, privacy concerns, and reluctance to change energy usage habits. Automated DR systems face resistance as users fear losing control over their energy consumption. Further, energy access disparities mean that low-income households may struggle to afford energy-efficient appliances or rooftop solar, exacerbating energy inequality.

Environmental and Spatial Constraints

BESS depend on materials like lithium and cobalt, raising concerns about resource scarcity and environmental impact. Recycling infrastructure for these materials is still underdeveloped. Moreover, large-scale renewable energy projects and battery storage installations require significant land space, which can lead to land-use conflicts.

Ethical and Societal Challenges in Energy Systems

The transition to AI-driven energy management and smart grid technologies presents ethical and societal challenges as discussed follows.

Privacy and Data Security

Smart grids collect huge amounts of consumer energy data, raising privacy concerns and the risk of unauthorized access. Cybersecurity threats, including potential attacks on IoT devices and communication networks, could compromise grid stability and consumer safety.

Bias and Fairness in AI Systems

AI-based EMSs may introduce biases, unintentionally favouring certain user groups or regions in resource distribution. DR programs often benefit households with smart appliances, leaving low-income populations at a disadvantage. Additionally, lack of transparency in AI decision-making can reduce trust in automated energy management.

Ensuring Fairness in Demand-Response Programs

Time-of-use pricing and rebates may unintentionally disadvantage households with inflexible schedules or those lacking smart devices. Automated DR systems prioritizing grid stability might favor commercial users over residential consumers, impacting fairness.

METHODOLOGY OF OPTIMIZED RULE-BASED DEMAND RESPONSE METHOD

In this chapter, the optimized rule-based DR method is discussed. Firstly, the operating modes pf battery are decided as discussed follows in presence of PV source.

Operating Modes of the Battery

There are three operational modes to manage electricity price (EP(t)) and maintain it within electricity price limit (EP_{lim})using a battery alongside a PV source:

Discharging Mode

This mode is active during time when the electricity price exceeds the electricity price limit and the PV power is less than the load demand.

Charging Mode-1

This mode is active during time when the electricity price is at or below the electricity price limit.

Charging Mode-2

This mode is active during time when the electricity price exceeds the electricity price limit and the PV power exceeds the load demand.

Method of Determination of Inputs

The rule-based DR method relies on several key inputs derived from day-ahead electricity prices, predicted load demand, and PV powers.

Electricity Price Limit (EP_{lim})

EP_{lim} is selected within the range of minimum (EP_{min}) and maximum (EP_{max}) day-ahead electricity prices over a day.

$$EP_{min} <= EP_{lim} <= EP_{max} \quad (1)$$

Energy to be Discharged from Battery ($E_{b-disch}$)

It is calculated based on the difference between the required load demand and the power supplied by the PV source.

$$P_{b-disch}(t) = P_d(t) - P_{pv}(t), \forall t \in t_{disch} \quad (2)$$

$= 0$, otherwise

$P_{b-disch}$ represents the power intended to be discharged from the battery. The total energy to be discharged from the battery over a day is determined by aggregating the power values over the predictive horizon of 24 hours.

$$E_{b-disch} = {}^{T}\Sigma_{t=1} P_{b-disch}(t) \quad (3)$$

Required Energy for Charging Battery (E_{b-ch})

For day-to-day flexibility, the energy required to charge the battery should be equal to the energy to be discharged.

$$E_{b-ch} = E_{b-disch} \quad (4)$$

Available PV Energy to Charge Battery (E_{pv-ch})

The system calculates available PV energy for charging the battery, considering self-consumption.

$$P_{pv-ch} = P_{pv}(t), \forall t \in t_{ch1} \quad (5)$$

$$= P_{pv}(t) - P_{d}(t), \forall t \in t_{ch2}$$

$$= 0, \text{ otherwise.}$$

The storage system storing excess energy generated during the day is calculated as follows:

$$E_{pv\text{-}ch} = {}^{T}\Sigma_{t=1} P_{pv-ch}(t) \quad (6)$$

Coefficient of PV Energy to Charge Battery (C_{pv})

If available PV energy is more than required, a fraction of PV energy is considered, and the coefficient C_{pv} is calculated accordingly.

$$C_{pv} E_{pv-ch} = E_{b-ch} \quad (7)$$

$$C_{pv} = E_{b-ch}/E_{pv-ch}$$

Available Grid Energy to Charge Battery (E_{g-ch})

When available PV energy is insufficient, the system draws deficit energy from the grid during specific time periods.

$$P_{g-ch}(t) = P_{b-ch-max}, \forall t \in t_{ch1} \quad (8)$$

$$= 0, \text{ otherwise.}$$

The available grid energy for charging the battery over a day is calculated as follows:

$$Eg-ch = {}^{T}\Sigma_{t=1} \, Pg-ch(t) \quad (9)$$

Coefficient of Grid Energy to Charge Battery (C_g)

If grid energy is needed, a fraction of the grid energy is considered and accordingly the coefficient Cg is calculated as follows.

$$C_g E_{g-ch} = E_{b-ch} - E_{pv-ch} \tag{10}$$

$$C_g = E_{b-ch} - E_{pv-ch} / E_{g-ch}$$

This rule-based approach make sure effective use of renewable sources like PV and optimizes the DR by dynamically adjusting battery charging methods based on electricity prices.

Rules for Demand Response

In formulating rules to determine day-ahead charge/discharge schedules of the battery, a flexible day-to-day management strategy is developed. The rules with respect to discharging and charging modes are provided below.

Discharging Mode (During t_{disch})

Rule 1

The battery discharges by the amount of $P_d(t) - P_{pv}(t)$, as specified in equation (4).

Charging Mode-1 (During t_{ch1})

Rule 2

If E_{pv-ch} is less than or equal to E_{b-ch}, both the PV source and the grid are used to charge the battery by an amount of $P_{pv}(t) + Cg\,(P_{b-ch-max})$.

Rule 3

If E_{pv-ch} is greater than E_{b-ch}, the PV source alone is used to charge the battery by an amount of $C_{pv}(P_{pv}(t))$.

Charging Mode-2 (During t_{ch2})

Rule 4

In case of E_{pv-ch} less than or equal to E_{b-ch}, the PV source is used to charge the battery by $P_{pv}(t) - P_d(t)$.

Rule 5

In case if E_{pv-ch} is greater than E_{b-ch}, the PV source is used to charge the battery by the amount of C_{pv} (P_{pv} (t) − P_d (t)).

Optimized Demand Response Method

The price of electricity is selected as a control variable. The problem in consideration is an offline optimization problem represented by a nonlinear fitness function. The problem is solved by both GA solver and a PSO solver in MATLAB. A population size of 20 is taken for the optimization process. The approach for finding the optimal inputs by using the GA and PSO are explained as follows.

Genetic Algorithm

The GA maximizes electricity price limit with energy consumption cost as the fitness function. This maximization is crucial in EMSs, where pricing determines consumption patterns and costs. The process begins with initializing the GA framework by specifying parameters like population size, generations, crossover probability, mutation rate, and termination criteria. In the population initialization step, the solutions that represent candidate electricity price limits are created. These price limits dynamically affect energy consumption patterns and ensure that varied pricing methods are investigated.

Finally, the fitness assessment verifies every solution by total energy consumption cost as the fitness function in order to reduce overall cost without compromising the stability of the grid. Then algorithm verifies stop conditions to halt if specified requirements are fulfilled. Otherwise, move on to Selection, Crossover, and Mutation. Selection selects the best solutions to reproduce according to energy consumption cost by methods such as roulette wheel. Crossover offers solutions to share information for diversity. The Mutation applies small changes to avoid premature convergence.

These iterations progressively optimize price limits for cost savings. The Optimal Electricity Price Limit is identified once complete. Achieved fitness value and convergence rate are also reported as performance metrics. This methodology is scalable for uses such as demand response optimization, energy storage scheduling, and integration of renewable energy.

Particle Swarm Optimization

The PSO reduces energy cost consumption while the price limit of electricity is used as a control variable. Optimization in EMSs is very important, and it begins with the parameter initialization such as swarm size, maximum iterations, inertia weight, and acceleration coefficients that strike a balance between exploration and exploitation. Then problem parameters are initialized i.e., integration of total energy cost computation with constraints such as load limits, energy budget, and electricity price limits from DR programs.

Particles are then initialized with random positions and velocities which represent energy consumption schedules. The electricity price limit influences search space evaluation. Fitness evaluation follows to determine optimal solutions through cost function while ensuring adherence to pricing constraints. Each particle updates its Personal Best (pBest), storing the lowest energy cost found so far. The Global Best (gBest), the lowest cost across the swarm, guides all particles toward the optimal solution.

Further, the position and velocity will be updated. The particles adjust based on their pBest, gBest, and stochastic factors. The process repeats until the stop criteria is reached (e.g. reaching the maximum iterations or achieving a satisfactory cost reduction). The algorithm then outputs the global best solution which represent the optimal energy consumption schedule.

Genetic Algorithm vs. Particle Swarm Optimization: Comparative Analysis

Both GA and PSO are widely used for EMS optimization, but their effectiveness varies based on specific conditions. GA is highly effective when searching for optimal electricity price limits in dynamic pricing environments, as it explores a diverse set of candidate solutions through crossover and mutation. This method is beneficial in scenarios requiring a wide search space. However, GA can sometimes converge slowly due to its stochastic nature and reliance on mutation to introduce variability.

On the other hand, PSO is particularly effective in scenarios requiring fast convergence and smooth adaptation to constraints like load limits and electricity price caps. Since PSO updates solutions based on pBest and gBest, it tends to reach optimal solutions more quickly than GA. This makes PSO well-suited for real-time energy cost optimization where fast response is crucial. However, PSO may suffer from premature convergence if particles get trapped in local optima, particularly in highly complex optimization landscapes.

CASE STUDY SYSTEM AND RESULT ANALYSIS

The optimized rule-based DR control algorithm is tested on the considered system i.e., BVRIT HYDERABD College of Engineering for Women for different seasons through GA and PSO. The system is shown in Fig. 1. The ratings of the components are given in Table 3 and described as follows.

Load Demand

The residential end-user load, which is considered as ac load demand, is connected at the ac bus of the system. Two types of load demand profiles are considered.

Interfacing Converter

The ac and dc buses are connected by a bidirectional converter known as interfacing converter (IC). The IC acts as a rectifier and an inverter, while transferring power from ac bus to dc bus and dc bus to ac bus, respectively. The IC controls the power balance and maintains constant dc-link voltage.

Photovoltaic System

The PV source is connected at the dc bus of the system through a dc–dc converter. An installed PV power rating of 150 kW is considered. The application of dc-dc converter is discussed as follows.

DC-DC converters enable efficient power transfer and ensure maximum power point tracking (MPPT) to optimize energy harvest. The boost converter is the most commonly used which steps up the low voltage from a PV array to match the higher voltage of a DC bus (e.g., 400V). This converter uses an inductor, switch, diode, and capacitor. They are highly efficient (90–98%) and widely favoured in grid-tied systems. They integrate seamlessly with MPPT algorithms like Perturb and Observe to dynamically change the duty cycle. However, they cannot step down voltage. Moreover, they require robust design to manage fluctuating input voltages caused by varying irradiance/temperature.

If the PV voltage becomes more than the DC bus voltage a buck converter will be used. This topology steps down the input voltage using a switch, diode, inductor, and capacitor. Even though they are simple, they are less common in grid-tied systems as they cannot handle high-voltage DC buses. They are mainly suited for cases like charging low-voltage battery banks (e.g., 24 V or 48 V).

In case if PV systems experience much voltage variability due to shading, temperature changes, a buck-boost converter can be used. It offers flexibility with both step-up and step-down operations. This makes it suitable for dynamic conditions even though it has less efficiency compared to only boost or buck converters.

Battery Energy Storage System

The battery is linked to the dc bus of the system via a dc–dc converter. A 120 V, 5000 Ah rated battery is selected. The usage of dc-dc converter is explained as follows.

DC-DC converters are important in the connection of a battery to a DC bus because they provide bidirectional power flow like charging and discharging of the battery while maintaining voltage, efficiency, and safety. The converters should also handle variable battery voltages, high current, and incorporation into active energy management. Some of the significant topologies and design considerations for battery to DC bus applications are discussed in this section.

The bidirectional buck-boost converter is a special device that will operate in situations where the battery voltage is greater than or less than the DC bus voltage. For instance, in boost mode, a 48V battery can be discharged to a 400V DC bus, and in buck mode, it can be charged from the bus. This technique employs switches, inductors, and capacitors; the new designs employ a combination of Silicon Carbide and Gallium Nitrate as they are capable of over 90% efficiency. This topology is also very flexible, which makes it perfect for remote energy storage, electric vehicles, and microgrids, where battery state of charge (SoC) dictates battery voltage. But there is a downside: seamless transitions between buck and boost are not fully fleshed out today and require advances in control algorithms alongside protection circuitry to stop overcurrent switching.

For safety critical applications such as EVs and grid storage, equipment and - user protection necessitates galvanic isolation. In contrast, the Dual Active Bridge (DAB) raises the standard for performance with its integrated isolation. Using two H-bridges in combination with a high-frequency transformer allows for bidirectional power transfer, making DAB more efficient than its competitors. DAB implements soft-switching techniques to reduce loss and achieve efficiency rates of 97%. This power density serves as a perfect complement of DAB's intended application: EV fast chargers. Other easier to assemble isolated topologies such as flyback converters are more inefficient and uncompetitive, making them suitable for low power systems such as small backup systems. Furthermore, isolation improves safety and lowers the chance of electromagnet interference, which raises the overall cost incurred by the transformers and complicated control systems.

In systems where the voltage relationships are fixed, single or double direction buck or boost converters serve as a cost efficiency option. For instance, a buck voltage converter lowers a higher DC bus voltage from 400V to charge a 48V battery. The battery can also discharge into the bus through a bidirectional converter; however, additional circuitry is needed for mode switching.

The two-way operation is important for grid energy storage systems, enabling the charging of a grid to a battery and the discharging of the battery to a load or a grid. With these specifications, the converters also have to track the changing battery voltages. For instance: 40–60V for a 48V Li-ion bank. At the same time, the DC bus has to be steady. This is important in all situations, but critical in EV utilization where there are many discharge/charge events. Also, isolation, overcurrent and over-voltage protection, along with BMS ensure sufficient safety. In as well, the response to transients has to be quite fast for effective load change handling, coupled with thermal management of the components at power limits to maximize their lifetime.

SiC and GaN are some of the wide bandgap semiconductors that are changing the design of converters by permitting higher frequencies. This adds efficiency while reducing the size of passive components. AI can provide advanced methods for control, especially in predictive scenarios like MPPT, fault detection, and in scalable architectures for large storage systems, while enhancing modularity.

Figure 1. System configuration

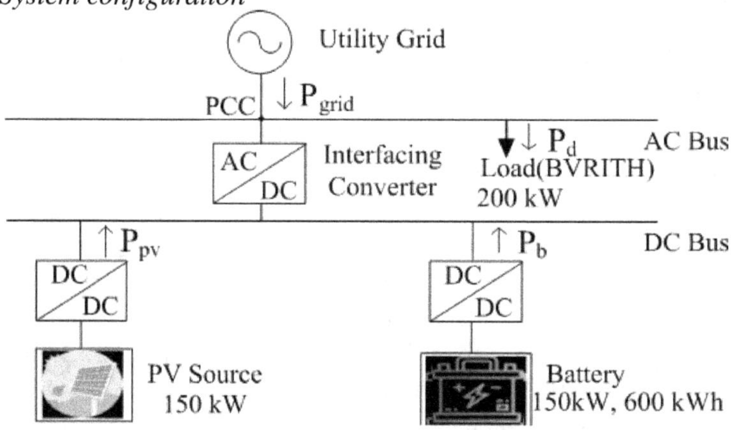

Table 3. Ratings of the components

S.No.	Parameter	Value
1	Load rating	200 kW
2	PV installed power	150 kW
3	Battery rated capacity	5000 Ah
4	Battery voltage rating	120 V
5	Battery charge/discharge power limit	150 kW

The time taken to obtain the results of PSO is less than that of GA for the same optimal fitness value. Therefore, the obtained results with the PSO are discussed as follows.

Autumn Season

In the Autumn season, the optimal buying price limit is 7.3166 INR/kWh (Fig. 2). The daily buying price is shown in Fig. 3(a), with load demand in black and PV energy in grey (Fig. 3(b)). The battery charge/discharge schedules are depicted in Fig. 3(c). The SOC and grid power are shown in Fig. 3(d) and (e). The discharging mode is from 12 to 21 hours, with the remaining hours for charging. PV energy is solely used to charge the battery since PV energy exceeds the required charging energy. The SoC remains at 50% at the end of the day, as seen in Fig. 3(d). When $EP(t) > EP_{olim}$, no power is drawn from the grid, and during charging mode, power is injected into the grid (Fig. 3(e)). This results in an energy cost of 363,786.4 INR, compared to 600,723.2 INR without PV-BESS, achieving a 39.44% cost saving.

Figure 2. Graphical representations of fitness value for autumn season

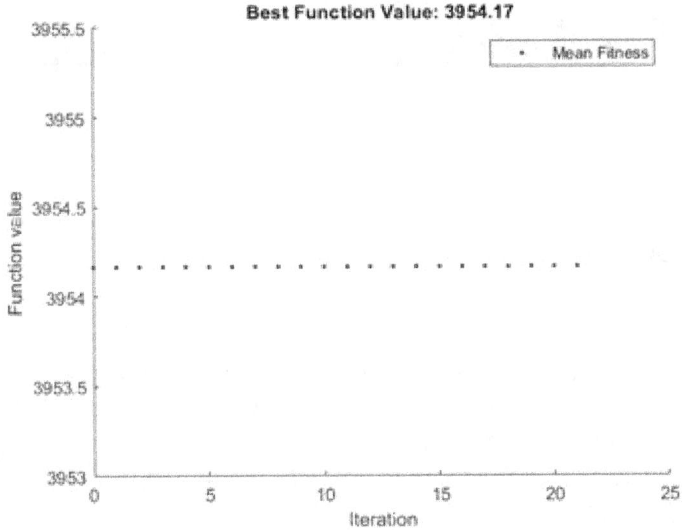

Figure 3. Graphical representation of autumn season (a) Buying price, (b) Load demand and PV power, (c) Battery power, (d) SOC, (e) Grid power.

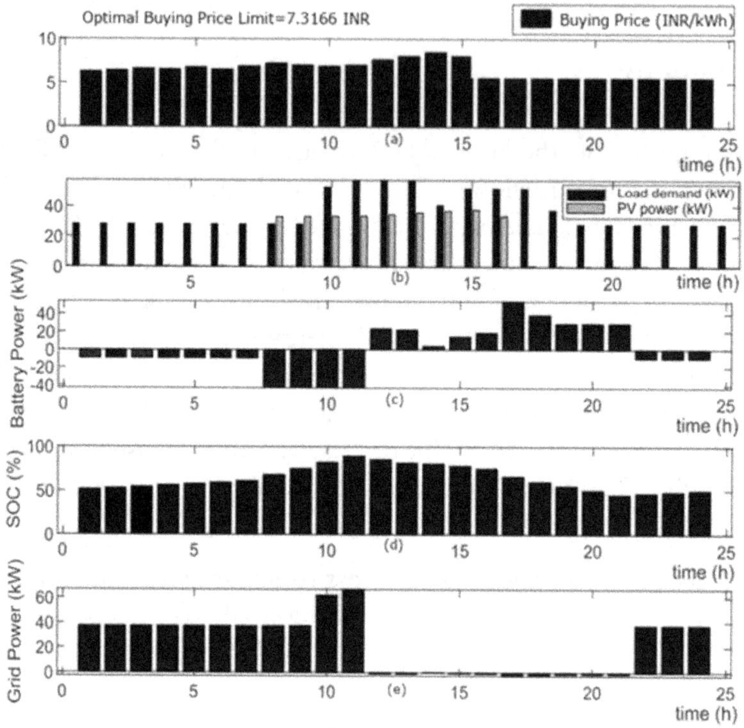

Winter Season

For the Winter season, the optimal buying price limit is shown in Fig. 4. The daily buying price is depicted in Fig. 5(a), with the load demand profile in black and PV energy availability in grey (Fig. 5(b)). The battery charge and discharge schedules are displayed in Fig. 5(c). The SoC and $P_{grid}(t)$ are illustrated in Fig. 5(d) and 5(e), respectively. The determined EP_{olim} is 7.0423 INR/kWh. The discharging mode is active from 17 to 22 hours, with the remaining hours for charging. Fig. 5(c) indicates that only the PV source is used for battery charging, as the available PV energy exceeds the required charging energy. Fig. 5(e) shows no grid demand during the discharging mode and power injection to the grid during the charging mode. This results in an energy cost of 113,826 INR, compared to 483,870 INR without the PV-BESS, achieving 76.47% cost savings.

Figure 4. Graphical representation of fitness value for winter season

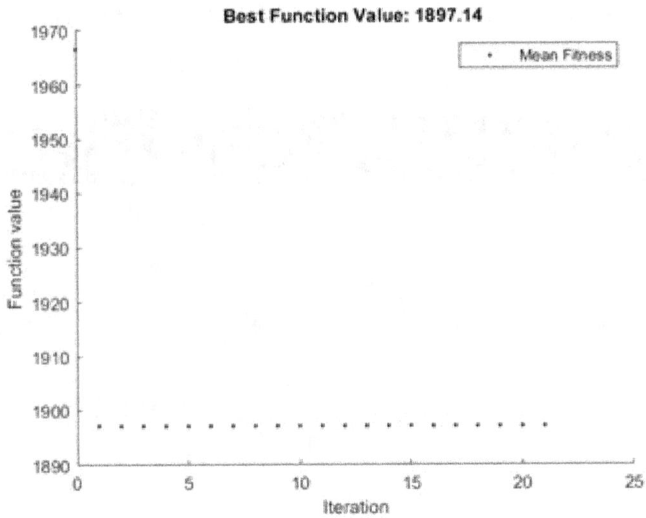

Figure 5. Graphical representation of winter season (a) Buying price, (b) Load demand and PV power, (c) Battery power, (d) SOC, (e) Grid power.

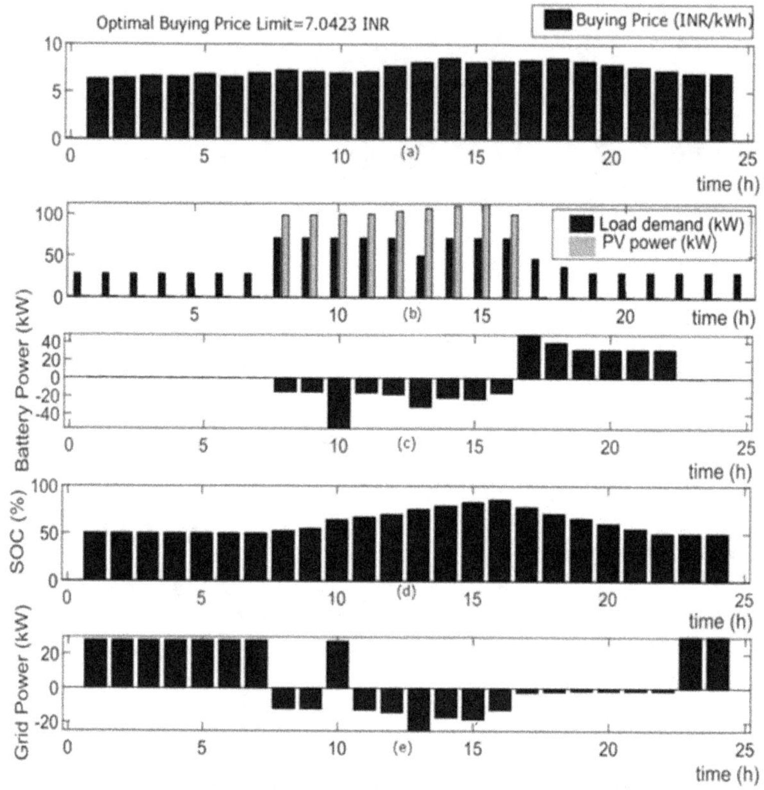

Stakeholder Feedback and Benefits

Stakeholders observed significant cost savings and improved energy efficiency, with energy costs reducing by 39.44% in autumn and 76.47% in winter. The algorithm effectively manages battery charging and discharging based on PV availability and pricing limits, reducing reliance on the grid. Users suggested incorporating ML techniques to enhance forecasting accuracy and decision-making.

Implementation Challenges

Ensuring smooth transitions between converter modes and managing bidirectional power flow for batteries posed challenges. The variability in PV generation due to shading and temperature fluctuations required robust control strategies. Cybersecurity concerns and the need for additional infrastructure to integrate IoT and blockchain remain key areas for improvement.

FURTHER RESEARCH

This chapter opens opportunities for further research in rule-based approaches for hybrid AC/DC systems. One crucial direction is the use of ML techniques, such as predictive analytics and reinforcement learning, to refine energy management methods. These methods improve forecasting accuracy for energy demand and renewable generation. This will lead to more informed decision-making. Moreover, cybersecurity remains a critical concern in DR and EMS operations. Safeguarding hybrid systems against cyber-attacks is necessary to ensure stability and resilience.

Integration of new technologies such as blockchain and IoT will improve the efficiency of DR and energy management practices is another area for the future. Blockchain will increase transparency and security in energy transactions. IoT can be utilized for real-time monitoring and control. Investigating these technologies' application in hybrid AC/DC systems will establish a more efficient, secure, and dynamic energy infrastructure.

MANAGERIAL IMPLICATIONS

Rule-based methods optimized for peak demand reduction allow organizations to balance energy usage against real-time pricing. They are able to cut operational expenses in addition to minimizing peak demand fees. Additionally, these techniques have a significant function in attaining sustainability objectives through enabling the application of RES. Through optimizing the use of energy, organizations are able to be actively involved in sustainability programs and carbon reduction targets.

Beyond cost reduction and sustainability, these intelligent strategies improve system reliability through dynamic response to variations in energy demand and supply. Real-time data analysis and intelligent control systems further enhance decision-making through the provision of insights. Additionally, customer participation is enhanced through incentive-based DR programs that can foster improved relationships with consumers. The insights can also inform policy-making and enable policymakers to craft regulations and incentives. This will encourage the use of EMSs and result in a resilient and sustainable energy industry.

CONCLUSIONS

This chapter discusses an optimized DR control algorithm developed for a grid-connected hybrid ac/dc microgrid while using the BESSs and RESs. The study has established day-ahead charge/discharge schedules for the battery while minimizing the energy costs based on dynamic electricity prices, PV power predictions, and load demands. The energy cost with PV-BESS has been reduced significantly. This substantial decrease proves the effectiveness of integrating PV-BESS in reducing energy expenses. The buying price is limited to optimal buying price limit. SoC of battery at the end of the day maintained equal to SOC of start of the day for flexible day-to-day management. Moreover, PSO algorithm runtime is less than the GA. Reduced runtime leads to faster task completion, enhancing productivity and efficiency. It lowers resource consumption, optimizing cost and energy usage. Additionally, it enhances user experience by delivering quicker results and responsiveness.

REFERENCES

Abdelkader, S., Amissah, J., Kinga, S., Mugerwa, G., Emmanuel, E., Mansour, D. E. A., Bajaj, M., Blazek, V., & Prokop, L. (2024). Securing modern power systems: Implementing comprehensive strategies to enhance resilience and reliability against cyber-attacks. *Results in Engineering*, *23*, 102647. DOI: 10.1016/j.rineng.2024.102647

Alghassab, M. A. (2024). Fuzzy-based smart energy management system for residential buildings in Saudi Arabia: A comparative study. *Energy Reports*, *11*, 1212–1224. DOI: 10.1016/j.egyr.2023.12.039

Ali, A. O., Elmarghany, M. R., Abdelsalam, M. M., Sabry, M. N., & Hamed, A. M. (2022). Closed-loop home energy management system with renewable energy sources in a smart grid: A comprehensive review. *Journal of Energy Storage*, *50*, 104609. DOI: 10.1016/j.est.2022.104609

Anteneh, D., Khan, B., Mahela, O. P., Alhelou, H. H., & Guerrero, J. M. (2021). Distribution network reliability enhancement and power loss reduction by optimal network reconfiguration. *Computers & Electrical Engineering*, *96*, 107518. DOI: 10.1016/j.compeleceng.2021.107518

Asghar, Z., Hafeez, K., Sabir, D., Ijaz, B., Bukhari, S. S. H., & Ro, J. (2023). RECLAIM: Renewable energy based demand-side management using machine learning models. *IEEE Access : Practical Innovations, Open Solutions*, *11*, 3846–3857. DOI: 10.1109/ACCESS.2023.3235209

Avordeh, T. K., Gyamfi, S., & Opoku, A. A. (2022). The role of demand response in residential electricity load reduction using appliance shifting techniques. *International Journal of Energy Sector Management*, *16*(4), 605–635. DOI: 10.1108/IJESM-05-2020-0014

Aziz, T., Lin, Z., Waseem, M., & Liu, S. (2021). Review on optimization methodologies in transmission network reconfiguration of power systems for grid resilience. *International Transactions on Electrical Energy Systems*, *31*(3), e12704. DOI: 10.1002/2050-7038.12704

Barra, P. H., de Carvalho, W. C., Menezes, T. S., Fernandes, R. A. S., & Coury, D. V. (2021). A review on wind power smoothing using high-power energy storage systems. *Renewable & Sustainable Energy Reviews*, *137*, 110455. DOI: 10.1016/j.rser.2020.110455

Boyouk, N., Munzke, N., & Hiller, M. (2018). Peak shaving of a grid-connected photovoltaic battery system at Helmholtz Institute Ulm (HIU). In *2018 IEEE PES Innovative Smart Grid Technologies Europe (ISGT-Europe)* (pp. 1–5). IEEE.

Chua, K. H., Lih Bong, H., Lim, Y. S., Wong, J., & Wang, L. (2020). The state-of-the-arts of peak shaving technologies: A review. In *2020 International Conference on Smart Grid and Clean Energy Technologies (ICSGCE)* (pp. 162–166). IEEE. DOI: 10.1109/ICSGCE49177.2020.9275637

Chua, K. H., Lim, Y. S., & Morris, S. (2013). Battery energy storage system for peak shaving and voltage unbalance mitigation. *International Journal of Smart Grid and Clean Energy*, *2*(3), 357–363. DOI: 10.12720/sgce.2.3.357-363

Das, D., Manojkumar, R., Kumar, C., & Ganguly, S. (2021). Optimal power management for islanded operation of ST-based meshed hybrid LV microgrid. In *2021 IEEE 12th Energy Conversion Congress & Exposition-Asia (ECCE-Asia)* (pp. 183–188). IEEE. DOI: 10.1109/ECCE-Asia49820.2021.9479110

Das, R. P., Samal, T. K., & Luhach, A. K. (2023). An energy efficient evolutionary approach for smart city-based IoT applications. *Mathematical Problems in Engineering*, *2023*(1), 2023. DOI: 10.1155/2023/9937949

Entezari, A., Aslani, A., Zahedi, R., & Noorollahi, Y. (2023). Artificial intelligence and machine learning in energy systems: A bibliographic perspective. *Energy Strategy Reviews*, *45*, 101017. DOI: 10.1016/j.esr.2022.101017

Farhadi Gharibeh, H., & Farrokhifar, M. (2021). Online multi-level energy management strategy based on rule-based and optimization-based approaches for fuel cell hybrid electric vehicles. *Applied Sciences (Basel, Switzerland)*, *11*(9), 3849. DOI: 10.3390/app11093849

Gernaat, D. E., de Boer, H. S., Daioglou, V., Yalew, S. G., Müller, C., & van Vuuren, D. P. (2021). Climate change impacts on renewable energy supply. *Nature Climate Change*, *11*(2), 119–125. DOI: 10.1038/s41558-020-00949-9

Ghoddusi, H., Creamer, G. G., & Rafizadeh, N. (2019). Machine learning in energy economics and finance: A review. *Energy Economics*, *81*, 709–727. DOI: 10.1016/j.eneco.2019.05.006

Hafiz, F., Awal, M. A., de Queiroz, A. R., & Husain, I. (2020). Real-time stochastic optimization of energy storage management using deep learning-based forecasts for residential PV applications. *IEEE Transactions on Industry Applications*, *56*(3), 2216–2226. DOI: 10.1109/TIA.2020.2968534

Häseler, S., & Wulf, A. J. (2024). Promoting real-time electricity tariffs for more demand response from German households: A review of four policy options. *Energy, Sustainability and Society*, *14*(1), 59. DOI: 10.1186/s13705-024-00490-z

Hossain, J., Saeed, N., Manojkumar, R., Marzband, M., Sedraoui, K., & Al-Turki, Y. (2024). Optimal peak-shaving for dynamic demand response in smart Malaysian commercial buildings utilizing an efficient PV-BES system. *Sustainable Cities and Society*, *101*, 105107. DOI: 10.1016/j.scs.2023.105107

Imani, M. H., Yousefpour, K., Andani, M. T., & Ghadi, M. J. (2019). Effect of changes in incentives and penalties on interruptible/curtailable demand response program in microgrid operation. In *2019 IEEE Texas Power and Energy Conference (TPEC)* (pp. 1–6). IEEE.

Islam, B., Rasheed, M., & Ahmed, S. F. (2022). Review of short-term load forecasting for smart grids using deep neural networks and metaheuristic methods. *Mathematical Problems in Engineering*, *2022*, 2022. DOI: 10.1155/2022/4049685

Jalil, N., Kheir, N. A., & Salman, M. (1997). A rule-based energy management strategy for a series hybrid vehicle. In *Proceedings of the 1997 American Control Conference (ACC)* (Vol. 1, pp. 689–693). IEEE. DOI: 10.1109/ACC.1997.611889

Kabalci, E. (2024). Power electronics applications in smart grid. In *Power Electronics Handbook* (pp. 993–1013). Butterworth-Heinemann. DOI: 10.1016/B978-0-323-99216-9.00006-8

Karthikeyan, N., Pokhrel, B. R., Pillai, J. R., & Bak-Jensen, B. (2018). Utilization of battery storage for flexible power management in active distribution networks. In *2018 IEEE Power & Energy Society General Meeting (PESGM)* (pp. 1–5). IEEE. DOI: 10.1109/PESGM.2018.8586215

Khalid, M. (2024). Smart grids and renewable energy systems: Perspectives and grid integration challenges. *Energy Strategy Reviews*, *51*, 101299. DOI: 10.1016/j.esr.2024.101299

Khan, N., Shahid, Z., Alam, M. M., Bakar Sajak, A. A., Mazliham, M. S., Khan, T. A., & Ali Rizvi, S. S. (2022). Energy management systems using smart grids: An exhaustive parametric comprehensive analysis of existing trends, significance, opportunities, and challenges. *International Transactions on Electrical Energy Systems*, *2022*, 2022. DOI: 10.1155/2022/3358795

Khojasteh, M., Faria, P., & Vale, Z. (2020). Scheduling of battery energy storage and demand response resource in balancing ancillary service. In *2020 IEEE PES Innovative Smart Grid Technologies Europe (ISGT-Europe)* (pp. 299–303). IEEE. DOI: 10.1109/ISGT-Europe47291.2020.9248798

Kumar, C., Manojkumar, R., Ganguly, S., & Liserre, M. (2020). Power loss minimization in smart transformer based meshed hybrid distribution network. In *IECON 2020 The 46th Annual Conference of the IEEE Industrial Electronics Society* (pp. 1670–1675). IEEE. DOI: 10.1109/IECON43393.2020.9254324

Kurukuru, V. S. B., Haque, A., Padmanaban, S., & Khan, M. A. (2021). Rule-based inferential system for microgrid energy management system. *IEEE Systems Journal, 16*(1), 1582–1591. DOI: 10.1109/JSYST.2021.3094403

Lehmann, C., Weeber, M., Böhner, J., & Steinhilper, R. (2016). Techno-economical analysis of photovoltaic-battery storage systems for peak-shaving applications and self-consumption optimization in existing production plants. *Procedia CIRP, 48*, 313–318. DOI: 10.1016/j.procir.2016.03.017

Liu, X., & Cui, W. (2020). *Economic analysis and optimal design for a grid-connected microgrid system. In 2020 IEEE Sustainable Power and Energy Conference (iSPEC)*. IEEE.

Luo, Q., & Zhang, S. (2022). Two-stage optimal scheduling of virtual power plant considering demand response and forecast errors. In *2022 4th International Conference on Electrical Engineering and Control Technologies (CEECT)* (pp. 852–856). IEEE. DOI: 10.1109/CEECT55960.2022.10030666

Manojkumar, R., Kumar, C., & Ganguly, S. (2020). Optimal demand response using dynamic electricity price limit in a hybrid AC/DC system. In *2020 21st National Power Systems Conference (NPSC)* (pp. 1–6). IEEE.

Manojkumar, R., Kumar, C., Ganguly, S., & Catalão, J. P. S. (2021). Optimal peak shaving control using dynamic demand and feed-in limits for grid-connected PV sources with batteries. *IEEE Systems Journal, 15*(4), 5560–5570. DOI: 10.1109/JSYST.2020.3045020

Manojkumar, R., Kumar, C., Ganguly, S., & Catalão, J. P. S. (2023). Net load forecast error compensation for peak shaving in a grid-connected PV storage system. *IEEE Transactions on Power Systems*.

Manojkumar, R., Kumar, C., Ganguly, S., Gooi, H. B., & Mekhilef, S. (2020). Voltage control using smart transformer via dynamic optimal setpoints and limit tolerance in a residential distribution network with PV sources. *IET Generation, Transmission & Distribution*, *14*(22), 5143–5151. DOI: 10.1049/iet-gtd.2020.0213

Manojkumar, R., Kumar, C., Ganguly, S., Gooi, H. B., Mekhilef, S., & Catalão, J. P. S. (2022). Rule-based peak shaving using master-slave level optimization in a diesel generator supplied microgrid. *IEEE Transactions on Power Systems*, *38*(3), 2177–2188. DOI: 10.1109/TPWRS.2022.3187069

Mariano-Hernández, D., Hernández-Callejo, L., Zorita-Lamadrid, A., Duque-Pérez, O., & García, F. S. (2021). A review of strategies for building energy management system: Model predictive control, demand side management, optimization, and fault detect & diagnosis. *Journal of Building Engineering*, *33*, 101692. DOI: 10.1016/j.jobe.2020.101692

Mishra, P., & Singh, G. (2023). Energy management systems in sustainable smart cities based on the internet of energy: A technical review. *Energies*, *16*(19), 6903. DOI: 10.3390/en16196903

Mohamed, A. A. R., Best, R. J., Morrow, D. J., Cupples, A., & Bailie, I. (2021). Impact of the deployment of solar photovoltaic and electrical vehicle on the low voltage unbalanced networks and the role of battery energy storage systems. *Journal of Energy Storage*, *42*, 102975. DOI: 10.1016/j.est.2021.102975

Mosavi, A., Salimi, M., Ardabili, S. F., Rabczuk, T., Shamshirband, S., & Varkonyi-Koczy, A. R. (2019). State of the art of machine learning models in energy systems, a systematic review. *Energies*, *12*(7), 1301. DOI: 10.3390/en12071301

Mosbah, H., Guerra, E. C., & Barrera, J. L. C. (2022). Maximizing the electricity cost-savings for local distribution system using a new peak-shaving approach based on mixed integer linear programming. *Electronics (Basel)*, *11*(21), 3610. DOI: 10.3390/electronics11213610

Perera, A., & Kamalaruban, P. (2021). Applications of reinforcement learning in energy systems. *Renewable & Sustainable Energy Reviews*, *137*, 110618. DOI: 10.1016/j.rser.2020.110618

Rasheed, S., & Abhyankar, A. R. (2023). Efficient operational planning of active distribution network by embedding uncertainties and network reconfiguration. *Electric Power Systems Research*, *216*, 109036. DOI: 10.1016/j.epsr.2022.109036

Reddy, K. R., & Meikandasivam, S. (2020). Load flattening and voltage regulation using plug-in electric vehicle's storage capacity with vehicle prioritization using ANFIS. *IEEE Transactions on Sustainable Energy*, *11*(1), 260–270. DOI: 10.1109/TSTE.2018.2890145

Refaai, M. R. A., Vonteddu, S. N. R., Nunna, P. K., Kumar, P. S., Anbu, C., & Markos, M. (2022). Energy management prediction in hybrid PV-battery systems using deep learning architecture. *International Journal of Photoenergy*, *2022*, 6844855. DOI: 10.1155/2022/6844853

Rezaeimozafar, M., Monaghan, R. F., Barrett, E., & Duffy, M. (2022). A review of behind-the-meter energy storage systems in smart grids. *Renewable & Sustainable Energy Reviews*, *164*, 112573. DOI: 10.1016/j.rser.2022.112573

Rosero, D., Díaz, N., & Trujillo, C. (2021). Cloud and machine learning experiments applied to the energy management in a microgrid cluster. *Applied Energy*, *304*, 117770. DOI: 10.1016/j.apenergy.2021.117770

Sharma, R., Gupta, S., & Kumar, A. (2023). Integration of rule-based demand response and reinforcement learning for enhanced energy efficiency. In *2023 International Conference on Sustainable Energy Technologies and Smart Grid (SESG)* (pp. 1–8). IEEE.

Shi, Q., Li, F., Olama, M., Dong, J., Xue, Y., Starke, M., Winstead, C., & Kuruganti, T. (2021). Network reconfiguration and distributed energy resource scheduling for improved distribution system resilience. *International Journal of Electrical Power & Energy Systems*, *124*, 106355. DOI: 10.1016/j.ijepes.2020.106355

Stanelyte, D., Radziukyniene, N., & Radziukynas, V. (2022). Overview of demand-response services: A review. *Energies*, *15*(5), 1659. DOI: 10.3390/en15051659

Taha, M. S., Abdeltawab, H. H., & Mohamed, Y. A. I. (2018). An online energy management system for a grid-connected hybrid energy source. *IEEE Journal of Emerging and Selected Topics in Power Electronics*, *6*(4), 2015–2030. DOI: 10.1109/JESTPE.2018.2828803

Urooj, A., & Nasir, A. (2024). Review of intelligent energy management techniques for hybrid electric vehicles. *Journal of Energy Storage*, *92*, 112132. DOI: 10.1016/j.est.2024.112132

Wang, B., Zarghami, M., & Vaziri, M. (2016). Energy management and peak-shaving in grid-connected photovoltaic systems integrated with battery storage. In *2016 North American Power Symposium (NAPS)* (pp. 1–5). IEEE. DOI: 10.1109/NAPS.2016.7747844

Wang, C., Wu, Z., Lin, Z., & Liu, J. (2023). Multi-agent interaction of source, load and storage to realize peak shaving and valley filling under the guidance of the market mechanism. *Frontiers in Energy Research*, *11*, 1192587. DOI: 10.3389/fenrg.2023.1192587

Wang, W., Yuan, B., Sun, Q., & Wennersten, R. (2022). Application of energy storage in integrated energy systems—A solution to fluctuation and uncertainty of renewable energy. *Journal of Energy Storage*, *52*, 104812. DOI: 10.1016/j.est.2022.104812

Weng, G., Bi, M., Fan, H., Fan, Y., Huang, C., Shi, Y., & Hua, H. (2022). Key problems of gas-fired power plants participating in peak load regulation: A review. *IET Cyber-Physical Systems: Theory & Applications*, *7*(4), 1–10.

Xin, Q., Alazab, M., Díaz, V. G., Montenegro-Marin, C. E., & Crespo, R. G. (2022). A deep learning architecture for power management in smart cities. *Energy Reports*, *8*, 1568–1577. DOI: 10.1016/j.egyr.2021.12.053

Xue, C., Zhou, H., Wu, Q., Wu, X., & Xu, X. (2021). Impact of incentive policies and other socio-economic factors on electric vehicle market share: A panel data analysis from the 20 countries. *Sustainability (Basel)*, *13*(5), 2928. DOI: 10.3390/su13052928

Yao, Z., Zhang, T., Wang, Q., & Zhao, Y. (2022). Short-term power load forecasting of integrated energy system based on attention-CNN-DBILSTM. *Mathematical Problems in Engineering*, *2022*, 2022. DOI: 10.1155/2022/1075698

Yazdkhasti, P., & Diduch, C. P. (2020). Demand side management using model-free fuzzy controller in a direct load control program. In *2020 IEEE Electric Power and Energy Conference (EPEC)* (pp. 1–5). IEEE. DOI: 10.1109/EPEC48502.2020.9320090

Zupančič, J., Lakić, E., Medved, T., & Gubina, A. F. (2017). *Advanced peak shaving control strategies for battery storage operation in low voltage distribution network. In 2017 IEEE Manchester PowerTech*. IEEE.

Chapter 13
AI for Community Engagement and Environmental Awareness:
Harnessing Artificial Intelligence to Promote Sustainable Practices in the Global South

Rumbidzai Matindike
https://orcid.org/0009-0003-1804-2858
University of Johannesburg, South Africa

Danmore Ngoni Matutu
https://orcid.org/0000-0002-2113-1187
University of Zimbabwe, Zimbabwe

ABSTRACT

The use of artificial intelligence (AI) is becoming more powerful for transformative community engagement and raising environmental awareness, particularly in the Global South. This chapter examines how Ai-driven technologies can promote sustainable practices in resource-limited environments. To enable local communities make knowledgeable decisions about waste management, energy conservation and sustainable agriculture, the study makes use of case studies and illustrations of Ai applications, including chatbots, mobile apps and predictive analytics. Findings demonstrate that Ai systems effectively disseminate environmental knowledge, support participatory decision-making and drive behavioral change towards

DOI: 10.4018/979-8-3373-0680-3.ch013

sustainability. Ai-enabled platforms improved access to localized environmental information, increasing community participation in conservation efforts. While AI offers immense potential to address ecological challenges, the discussion highlights barriers, including limited technological infrastructure, lack of digital literacy, and the need for culturally relevant solutions.

1. INTRODUCTION

Artificial Intelligence (AI) introduced a new paradigm progressively transforming from a niche technological innovation to a powerful tool capable of solving multidimensional and complex problems across various sectors (Cazzaniga et al., 2024). The ability of AI to predict potential outcomes and perform analytics associated with given data in an automated way has revolutionized various sectors in terms of productivity, innovation, and sustainability (Chatterjee et al., 2023). The application of AI in sustainable development is promising, offering state-of-the-art changes that are crucial for economic growth, social equity, and environmental sustainability (Triguero et al. 2024). The use of artificial intelligence (AI) to solve urgent problems in Africa, a continent marked by fast industrialization, urbanization, and environmental deterioration, has gained popularity in recent years.

AI can be defined within computer science as the pursuit of developing intelligent agents. These agents possess the capacity to perceive their surroundings and execute actions strategically to optimize the likelihood of achieving desired objectives (Tiwari, 2024). This endeavor necessitates emulating cognitive functions characteristic of human intelligence, encompassing learning, problem-solving, perceptual abilities, and language comprehension. The systems use algorithms and dynamic modelling to process data, make decisions, and learn from outcomes, often surpassing the speed and accuracy of human capabilities in various tasks (Mienye & Jere, 2024). Natural language processing (NLP), image recognition, and speech recognition are examples of specific AI applications. On the other side, community engagement is the process of addressing challenges impacting the well-being of groups of people who are connected by shared circumstances, physical closeness, or unique interests.

The applications of AI in promoting sustainable practices span areas such as climate change mitigation, agriculture optimization, ocean health monitoring, water resource management, weather forecasting, and disaster resilience amongst others (Rautela et al., 2020). Current advancements in AI necessitate robust regulatory oversight to ensure transparency, community engagement, and ethical practices to promote sustainable practices (Ansari et al., 2022). To effectively utilize AI's potential to promote sustainable practices, more study on environmental awareness and community engagement is required. Addressing these gaps will bridge the divide between AI's potential and its sustainable, practical, and ethical applications to promote sustainable practices (Schiff et al., 2024).

The study seeks to reconcile harnessing AI to promote sustainable practices in the Global South through community engagement and environmental awareness. While we remain optimistic about AI's potential, there remains a concern about its limitations and the importance of harnessing AI's potential in participatory sustainable practices (Chisika & Yeom, 2024). Despite the potential, there is still a significant gap in the literature on AI's application to sustainable practices since most of the previous research applies results from other areas, frequently ignoring the unique circumstances of the Global South, particularly in Zimbabwe. The study also discusses future research paths and possible opportunities, which will help ensure that AI is successfully included in participatory environmental awareness processes.

2. BACKGROUND OF STUDY

The global environmental crisis marked by climate change, deforestation and biodiversity loss, presents significant challenges for communities worldwide including Zimbabwe. Climate change has disrupted traditional weather patterns, affecting ecosystems and livelihoods, particularly in rural areas (Sango & Godwell, 2015). As a country heavily dependent on natural resources and agriculture, Zimbabwe is especially vulnerable to these environmental pressures. The unpredictability of weather has undermined the reliability of traditional farming practices, resulting in decreased agricultural output. In regions such as the Makonde Communal Lands climate change has degraded ecosystem productivity and biodiversity, further endangering human livelihoods (Sango & Godwell, 2015). Despite the pressing environmental challenges, community engagement and environmental awareness in rural areas remain inadequate. Artificial Intelligence (AI) has emerged as a transformative technology for addressing these challenges by offering advanced tools to predict both short and long-term climate impacts including weather patterns, floods and drought (Rutenberg et al., 2021). Such predictive capabilities are essential for informing adaptation strategies. However, local populations often lack access to timely information regarding climate adaptation measures, sustainable practices, and environmental conservation, which limits their ability to respond effectively to environmental threats.

AI enhances the accuracy and reliability of weather forecasts, playing a crucial role in managing natural resources and preparing for extreme weather events (Chen et al., 2023). In the context of Zimbabwe, AI significantly benefits farmers through optimizing crop yields and improving resource management, which is vital given the country's dependence on agriculture. However, the widespread adoption of AI introduces theoretical, factual and ethical considerations that must be addressed to ensure equitable and sustainable implementation (Ciaralli, 2023). While traditional knowledge remains valuable, it is insufficient to tackle the complex and dynamic issues brought on by environmental degradation and climate change. This underscores the importance of innovative approaches that integrate AI technologies to strengthen community resilience and promote sustainable practices.

AI tools have proven highly effective in improving community awareness and participation within educational settings. For example, AI-assisted educational platforms have been shown to enhance student engagement, participation, and satisfaction in national community education programs (Yan & Fengyi, 2024). This demonstrates the potential of adapting similar AI tools to enhance community engagement in Zimbabwe by making educational activities more engaging and participatory. AI technologies also support participatory approaches in environmental management by involving diverse stakeholders in decision-making processes. These tools facilitate adaptive management strategies and strengthen collaboration among stakeholders (R. C. Santos & Cagica Carvalho, 2025). Applying such AI-driven participatory frameworks in Zimbabwe could significantly enhance community involvement in environmental initiatives, fostering greater engagement and collective action.

AI technologies are increasingly being utilized to enhance agricultural practices and monitor environmental impacts, offering transformative potential for countries like Zimbabwe, where agriculture is a critical component of the economy (Pelet et al., 2022). Through earth monitoring techniques and time-series satellite imagery, AI enables the optimization of agricultural practices, thereby improving resource management and promoting sustainability. For example, AI-driven tools provide real-time data on soil health, weather conditions, and crop health, facilitating more efficient resource use and increasing crop yields (Kumar Tyagi et al., 2023). Additionally, AI-powered weather forecasting tools deliver accurate and timely weather predictions, allowing farmers to better plan planting and harvesting schedules, thus mitigating risks associated with climate-induced weather variability.

In the Umzingwane Catchment, AI has been applied to map illegal mining activities and assess their impact on water quality (Dube et al., 2024). Using satellite data with machine learning methods, researchers identified the spatial distribution of illegal mining operations and their influence on river nutrient and metal concentrations. This data is critical for informing regulatory measures to protect riverine ecosystems. Furthermore, organizations such as the Zimbabwe Farmers Union (ZFU) are exploring partnerships to incorporate AI technologies into agricultural advisory services, aiming to enhance decision-making and improve outcomes for farmers. By analysing large datasets, AI not only optimises resource management but also reduces environmental impacts, fostering sustainable agricultural practices critical for solving the problems caused by climate change.

2.1 Environmental Challenges and AI's Possible Contribution to Their Resolution

Zimbabwe faces a multitude of environmental challenges, driven by both natural and human-induced factors. The impacts of climate change are becoming increasingly severe, with more frequent droughts and floods affecting both rural and urban areas (Munanga & Mafuku, 2021). These extreme weather events have caused agricultural production to drop productivity, posing a significant threat to food security and livelihoods, particularly for rural communities that depend on rain-fed farming. AI offers a promising solution by optimizing climate prediction models, enabling better planning and mitigation strategies to address the effects of extreme weather events on infrastructure and communities (Raghav et al., 2024).

Forests cover approximately 45% of Zimbabwe's land area, yet deforestation is occurring at an alarming rate due to the heavy reliance on natural resources for livelihoods, especially in rural regions (Nhapi, 2023). Agricultural expansion, tobacco curing, and using wood as a primary energy source are major drivers of deforestation. This has resulted in biodiversity loss, soil erosion and the degradation of forest ecosystems. AI technologies can play a crucial role in addressing these issues by enhancing the monitoring of deforestation, species distribution, and pollution levels through tools such as remote sensing and machine learning (Jain & Mitra, 2024). These technologies enable real-time tracking of environmental changes and provide valuable predictions on the impacts of climate change (Todorova et al., 2024). By exploiting AI, Zimbabwe can improve its capacity to monitor, manage and mitigate environmental challenges more effectively.

The Fast Track Land Reform Programme in Zimbabwe has significantly contributed to land degradation, marked by the loss of plantations, small farms and forests, alongside an increase in bare land and settlements (Jombo et al., 2017). Overgrazing, poor farming practices and unsustainable land use have led to severe soil erosion and declining soil fertility. Desertification is also becoming a pressing concern in certain regions. While Zimbabwe is home to rich biodiversity, environmental degradation poses a serious threat to its ecosystems (Nhapi, 2023). Habitat destruction and poaching have caused a decline in wildlife populations, with protected areas facing increasing pressure from human encroachment and illegal activities. By enhancing natural resource management, artificial intelligence (AI) technologies present a substantial opportunity to tackle these issues. AI can optimize agricultural practices, enhance water management, and promote sustainable land use (Putra et al., 2024). These tools could support conservation efforts by monitoring biodiversity, predicting habitat changes, and mitigating the impact of human activities on ecosystems.

Air and water pollution are ongoing issues in Zimbabwe, adversely affecting environmental and public health (Ndumeya & Maluleka, 2024). Prolonged droughts and mismanagement of water resources have caused widespread water shortages. Additionally, pollution from industrial waste, mining activities and urban runoff has further exacerbated the issue. AI can assist in developing smart waste management systems, reducing pollution, and improving recycling processes (Putra et al., 2024). Furthermore, AI-driven optimization of energy systems can reduce carbon emissions and improve the efficiency of renewable energy sources (Raghav et al., 2024). However, despite the promise of AI, several challenges impede its effective implementation. The handling of large datasets raises concerns about data privacy and security (Mogha & Hasteer, 2023). Additionally, AI systems can be energy-intensive, potentially contributing to environmental problems if not managed sustainably (Gundeti et al., 2023). Moreover, the successful deployment of AI requires specialized technical expertise, which remains limited in Zimbabwe (Jain & Mitra, 2024). To fully realize AI's promise in advancing environmental sustainability, several issues must be resolved.

Harnessing artificial intelligence (AI) to promote sustainable practices in Zimbabwe offers numerous opportunities, as highlighted by various studies. AI can play a transformative role in enhancing agricultural productivity by utilizing satellite imagery and chlorophyll data to monitor soil aridity and vegetation trends (Pelet et al., 2022). This approach can optimize farming practices, increase crop yields, and promote sustainable land use.

AI technologies can also support biodiversity conservation, climate modelling, and waste management (Yadav, 2023). These applications enable the monitoring of environmental changes, prediction of climate impacts, and more efficient management of natural resources (Singh et al., 2024). By integrating AI into these areas, Zimbabwe can address critical environmental challenges and improve resource management. AI applications will support more general sustainability goals if AI activities are in line with the Sustainable Development Goals (SDGs). These include reducing poverty, enhancing public health, and fostering environmental sustainability, creating a comprehensive framework for progress and resilience.

3. RESEARCH OBJECTIVES

This study looks into the aspects that affect Zimbabwe's adoption of AI for sustainable development. Specifically, the study seeks to:

- To explore how AI technologies can enhance community engagement to increase participation and awareness of environmental sustainability.

- Identify and analyse socio-economic, cultural, and technological barriers to adopting AI-driven solutions for environmental awareness and sustainable practices.

4. RESEARCH METHODOLOGY

This paper aims to comprehensively analyse how AI can be harnessed for community engagement and environmental awareness to promote sustainable practices in Zimbabwe. To provide a comprehensive yet targeted analysis of the current state of research in AI-driven sustainable practices, especially in the context of the Global South, the study used an organized method of examining relevant information.

4.1 Search Criteria and Literature Search

The paper used a systematic process to ensure that the most relevant, recent, and impactful literature is included. The following criteria guided our literature selection:

a. Time Frame: Considering the last ten years (2010–2024) have seen the greatest developments in artificial intelligence (AI) technologies and their applications in sustainable practices, the paper mainly concentrated on literature produced during this time.
b. Database Search: The writers employed scholarly databases such as IEEE Xplore, Scopus, and Google Scholar to ensure thorough coverage. The databases were chosen because they promote sustainable practices in a variety of fields and include a comprehensive index of artificial intelligence.
c. Keywords: Key phrases were used in conjunction as part of the search strategy and their synonyms such as ''artificial intelligence OR smart technology OR automated reasoning OR machine intelligence" AND ''sustainable practices OR conservation practices OR sustainable methods OR sustainable management" AND ''community engagement OR community involvement OR stakeholder engagement OR participatory*'' AND ''AI in Zimbabwe'' AND ''AI in Global South'', AND ''environmental awareness OR ecological awareness OR ecological sensitivity". This enabled it possible to incorporate a large number of papers that were pertinent to AI and encouraging sustainable practices in a variety of academic fields.
d. Inclusion and Exclusion Criteria: Peer-reviewed journal publications, conference proceedings, and high-impact white papers were given priority in the study. Studies that did not contain a focus on the Global South or that just addressed

technological developments in AI without any connection to sustainable practices were excluded.

4.2 Review Process and Analysis

The study used a two-phase review procedure:

a. Initial Screening: To meet the inclusion requirements for additional analysis, each paper was evaluated for relevance using its title, abstract, and keywords.
b. In-Depth Review and Thematic Categorization: Following this, a thorough review of the chosen papers' full texts was conducted. The articles' contributions to AI applications in environmental awareness and community involvement to support sustainable practices were the focus of their subject category. Additionally, we focused on research publications that discussed the prospects, risks, and difficulties of adopting AI in the Global South, particularly in Zimbabwe, and offered insights into strategies that can support sustainable development.

5. LITERATURE REVIEW

Several strategies and insights can be explored to influence Artificial Intelligence (AI) for enhancing community engagement and promoting environmental awareness in Zimbabwe. This literature review examines how AI can be explored to achieve these objectives in Zimbabwe, drawing on insights from a range of studies. AI technologies, particularly machine learning, are increasingly being applied to climate modelling and environmental change prediction. These models provide valuable insights into the potential impacts of climate change, enabling communities to understand these challenges and adopt proactive measures (Al-Raeei, 2024).

5.1 Enhancing Community Engagement and Environmental Awareness

Several strategies and insights can be considered to enhance community engagement and environmental awareness through artificial intelligence (AI) in Zimbabwe. AI tools have proven effective in increasing community awareness and participation in educational initiatives. For example, a study comparing traditional teaching methods with AI-assisted tools found that the latter significantly improved students' awareness, engagement, and satisfaction (Yan & Fengyi, 2024). This indicates that

AI can be used to create more engaging and participatory educational activities in Zimbabwe, fostering greater community involvement.

AI-powered technologies such as chatbots, mobile applications, and voice assistants are widely used worldwide to disseminate information on health, agriculture, and environmental conservation (Schiavo, 2021). These tools help address challenges related to literacy and language barriers, promoting broader participation in sustainability efforts. In Zimbabwe, platforms like Eco Farmer deliver personalized weather forecasts and market updates, helping rural communities adapt to environmental and economic challenges. Additionally, AI-enabled chatbots capable of providing instant weather information are being introduced, offering accessible and timely assistance. AI-based educational platforms also provide interactive learning opportunities that encourage behavioural change at the community level. AI is essential to climate action and environmental monitoring. Its ability to collect and analyse extensive datasets enhances the assessment of air and water quality, land use changes, and biodiversity (Al-Raeei, 2024). In Zimbabwe, Artificial intelligence (AI) tools like remote sensing and predictive analytics are used to track environmental metrics and encourage sustainable practices.

AI can enhance participatory approaches in environmental management by improving data acquisition and management. For instance, in Kenya, AI has been shown to increase the efficiency and transparency of data processes in participatory forest management (Chisika & Yeom, 2024). A similar approach could be implemented in Zimbabwe to actively involve local communities in environmental decision-making processes. Agriculture, the backbone of Zimbabwe's economy, has been significantly affected by climate change. AI technologies, such as precision farming and pest detection systems, have demonstrated their ability to increase productivity while promoting sustainable practices. For example, AI-powered tools that analyse soil health and recommend optimal inputs have been piloted in Zimbabwe, showing potential to reduce environmental degradation and improve livelihoods (Parwada & Marufu, 2023). However, the usefulness of these instruments has been limited by the lack of experience, which has made the implementation of AI-powered machinery difficult, especially in rural areas.

AI-driven technologies can also be utilized for early warning systems, risk assessments, and adaptive response strategies, which are essential for building community resilience against natural disasters (Zolkafli et al., 2024). Implementing such systems in Zimbabwe can help communities better prepare for and respond to environmental challenges, ultimately promoting sustainable development and reducing vulnerability.

5.2 AI Applications for Environmental Sustainability

AI technologies have been widely applied across various domains to address environmental sustainability challenges, particularly in the Global South and Zimbabwe. By optimizing natural resource management, AI enables the monitoring of environmental changes and the prediction of resource needs. This includes tracking deforestation, analysing species distribution, and assessing pollution levels (Jain & Mitra, 2024). AI plays a vital role in managing environmental resources and promoting sustainability. In water resource management, AI-powered models analyse satellite and sensor data to predict water availability, detect leaks, and optimize irrigation systems for more efficient and sustainable water use. Additionally, AI facilitates conflict resolution and enhances communication among diverse stakeholders, contributing to improved environmental resource governance (Naim et al., 2024).

AI plays a crucial role in enhancing climate resilience by improving predictive climate models, optimizing energy consumption, and refining waste management strategies (Chaudhary, 2023). As a powerful tool in climate monitoring, AI offers innovative solutions to address the complex challenges posed by climate change. One of AI's key contributions is enhancing the accuracy of climate models by processing vast amounts of data from multiple sources, such as satellite imagery and sensor networks. This capability enables more precise weather forecasting and climate impact predictions (Jones et al., 2023). Additionally, AI helps quantify uncertainties in climate projections, providing more reliable data for informed decision-making (Jones et al., 2023). Machine learning algorithms analyse satellite imagery and meteorological data to identify climate trends and detect changes in environmental patterns. AI plays a vital role in biodiversity research by predicting environmental services and enhancing conservation efforts (Zehouani et al., 2024). Advanced AI technologies, particularly machine learning and computer vision, are used to detect and classify species, including those at risk of extinction. Platforms such as Conservation AI utilize visual spectrum and thermal infrared cameras to monitor wildlife populations and prevent poaching (Fergus et al., 2024). AI systems also outperform traditional methods in identifying fine-grained morphological features of fish, significantly improving the classification and taxonomy of aquatic species. This capability supports conservationists in protecting endangered species and advancing aquatic biodiversity research (Wasik & Pattinson, 2024). By integrating AI-driven tools into conservation strategies, researchers can develop more effective measures for preserving biodiversity and mitigating threats to wildlife.

AI enhances crop management and pest control, leading to increased food productivity while minimizing environmental impact (Al-Raeei, 2024). When integrated with the Internet of Things (IoT), AI enables real-time monitoring of crop health through sensors, drones, and remote sensing technologies. These tools help assess soil conditions, plant water status, and various environmental factors, providing valuable insights for precision agriculture (Eli-Chukwu, 2019).

Machine learning algorithms analyse both historical and real-time data to predict crop yields, determine optimal planting times, and select the most suitable seed varieties based on specific weather conditions (Alhalwachi et al., 2024). Additionally, AI-powered image recognition and machine learning techniques allow for the early detection of diseases and pests, enabling timely interventions and reducing crop losses (Fulkar et al., 2024). By leveraging AI-driven solutions, farmers can optimize resource use, enhance agricultural productivity, and promote sustainable farming practices.

6. RESULTS AND DISCUSSION

6.1 Community Engagement in Environmental Sustainability

The systematic review shows that communities nationwide actively engage in environmental sustainability programs. The Forestry Commission of Zimbabwe (FCZ) launched a National Tree-Planting Day flagship campaign commemorated every first Saturday of December every year. The goal of the campaign is to raise awareness of the value of forests, biodiversity, and household food security while encouraging and strengthening community involvement in tree planting and conservation. For the past few years, the initiative has attracted interest from several organizations, local government departments, local authorities, civil society organizations, and traditional local leadership lobbying for sustainable management of forests in Zimbabwe.

Significantly, monthly campaigns in Zimbabwe have improved the cleanliness of local areas by reducing litter and waste, fostering community engagement across the country for a common goal and raising awareness about environmental sustainability. In 2023, during an interview with the Herald newspaper, the Provincial Manager for the Environmental Management Agency (EMA) highlighted that they engaged community-based organisations to find households ready to separate and classify waste to protect the environment and raise awareness nationwide. Government and corporate sectors have embraced the campaigns with notable participation from local authorities and community members. For the past years, EMA has played a significant role in coordinating clean-up campaigns nationwide. These initiatives have positively improved the state of the ecosystems and raised momentum in recognition of environmental stewardship.

Advocacy groups in Zimbabwe, such as the Environmental Management Agency, Friends of the Earth Zimbabwe, and the Zimbabwe Environmental Law Association, have significantly impacted policy changes by promoting effective waste management policies, strengthening legal frameworks for environmental protection, and increasing public awareness about climate change and biodiversity conservation. Initiatives like Zero Waste Zimbabwe and Green Governance Zimbabwe have encouraged sustainable practices at both community and governmental levels, Youth for Conservation has successfully engaged the younger generation in environmental advocacy. These collective actions have led to tangible improvements in environmental governance and a heightened commitment to sustainability nationwide.

In addition, scholarly evidence shows that government departments have actively engaged with local farmers and schools to promote environmental institutions and education through various programs and initiatives. EMA has partnered with educational institutions to implement workshops and training sessions focused on waste management, sustainable practices and conservation. Several schools and local farmers often participate in tree-planting campaigns for students to learn about climate change mitigation and the importance of biodiversity.

The systematic review shows that the Communal Areas Management Programme for Indigenous Resources (CAMPFIRE) project is the most acknowledged community engagement initiative covering 58 districts in Zimbabwe. The CAMPFIRE program has been under the Zimbabwe Parks and Wildlife Management Authority since the late 1980s as a community-based approach to managing Indigenous resources (see Table 1). Scholarly evidence shows that the program was built on the assumption that engaging local people in managing wildlife and economic benefits will help ensure the long-term sustainability of local resources and enhance rural livelihoods and development (Harrison & carr, 2015) .

Table 1. Scope of the CAMPFIRE program

Land area under CAMPFIRE	**50,000 km2 – 12% of Zimbabwe**
Number of CAMPFIRE Districts	58
Number of wildlife districts	28 (15 active in hunting)
Number of Safari Operators	33
Number of Photographic safaris	10
Number of leased Parks Safari Areas	4

Source: Campfire Association (2025)

Despite the challenges facing the CAMPFIRE program over the past decade, it has remained robust in several communities. An analysis of Sengwe communal land around Gonarezhou National Park shows that more robust communities joined the CAMPFIRE program earlier and had more time to develop robust institutions such as Malipati Development Trust characterised by higher levels of trust and cooperation (Ntuli & Muchapondwa, 2018).

Non-governmental organisations have collaborated in community engagement activities across the country to address environmental issues and sustainable practices. For example, the United Nations Development Program (UNDP) Equator initiative implemented the Chibememe Earth Healing Association (CHIEHA), located in Zimbabwe's Masvingo district in the small settlement of Chibememe. To capitalize on the region's rich natural and cultural history, CHIEHA is a community-based program that aims to increase awareness of the preservation and processing of non-timber forest products as well as the promotion of ecotourism. The 17 houses that comprise the Chibememe community have spearheaded the CHIEHA initiative in reaction to habitat destruction and deforestation in the Zivambava Island Forest and the Chibememe mainland forest, according to scholarly data. With further assistance from the growth of environmental outreach and sustainable livelihoods initiatives that seek to manage natural resources in the Great Limpopo Transfrontier Park Conservation Area (GLTFCA) and the larger Sengwe communal land area in a sustainable manner, this initiative has expanded to neighbouring communities.

6.2 Artificial Intelligence Tools in Zimbabwe

An exhaustive literature review shows a significant increase in AI use across various sectors in the country (see Figure 1) showing a structured progression of AI adoption in Zimbabwe since 2010.

Figure 1. AI adoption in Zimbabwe

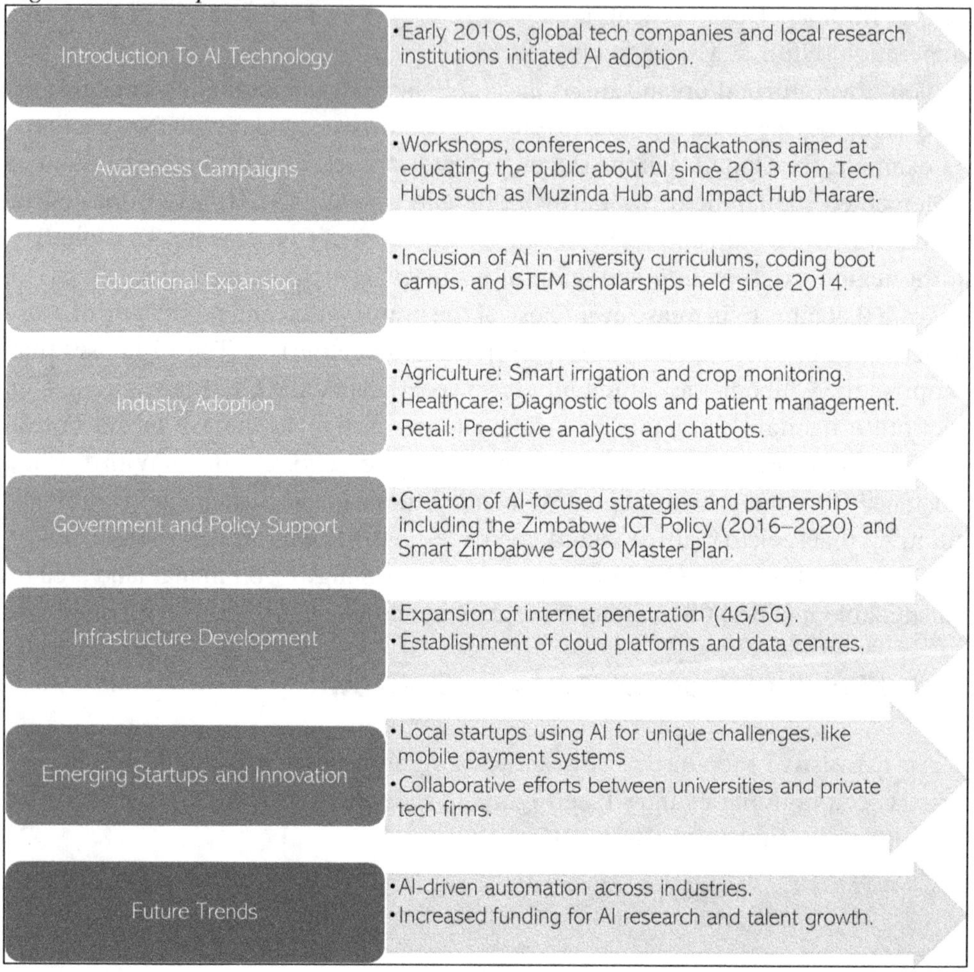

This structured approach reflects Zimbabwe's evolving engagement with AI technology and highlights its potential to influence various aspects of society and the economy. A notable case study is that machine learning algorithms can analyse historical climate data with agricultural yield data to identify patterns and correlations to forecast future climate scenarios and the likelihood of impacts on crop production (Mlambo et al. 2021). The International Crops Research Institute for the Semi-Arid Tropics (ICRISAT) developed and implemented AI models in Masvingo, Chivi and Beitbridge to assess the climate variability impact of sorghum yields. These models are being used to analyse historical climate data and predict future scenarios to help farmers understand the best time frame for seeding and suitable sorghum varieties for their specific locations. Local farmers in Masvingo province have adopted early maturing drought-tolerant sorghum varieties as a way of improving food security across the regions. The table below summarises the positive impacts of AI, challenges and examples of applications.

Zimbabwe's agricultural sector is undergoing a digital revolution, embracing innovative technologies to boost productivity and efficiency. Integrating AI with the agriculture sector has presented smart farming opportunities that facilitate sustainable agricultural practices. (Nyoni et al., 2024) highlights that the usage of drones and remote sensing technology is gaining traction to enhance the efficiency of farm management practices, particularly for large-scale farmers, to reduce labour. Findings show that local farmers have been using drone-equipped with AI to capture high-resolution images to identify pest infestations and nutrient deficiencies in real time to minimise environmental pollution and health risks to farm workers. The Drone and Research Group at the University of Zimbabwe deployed drones to monitor forest health and assess the impact of human activities to create maps highlighting areas at risk of deforestation to inform targeted conservation efforts with assistance from the Ministry of Environment, Climate and Wildlife and Hospitality.

Local farmers have been leveraging the use of AI in pest and disease management (Chikozho, 2019). Pest and disease outbreaks are critical challenges for local farmers across the country. Scholarly evidence shows that there is potential for AI applications to diagnose crop diseases. For instance, the PlantVillage application uses machine learning to identify diseases from images uploaded by local farmers and recommend mitigating strategies. The application has been beneficial for small-scale farmers with a lack of resources to agricultural extension officers from the Agricultural and Rural Development Advisory Services Directorate (AGRITEX), Chikozho (2019).

For the past years, Zimbabwe has experienced notable deforestation due to illegal logging, agricultural expansion and infrastructure development. In this regard, Munyati et al. (2021) highlight that the country has been leveraging advanced AI technology for stakeholders to gain insights critical for effective environmental management. For instance, Zimbabwe's Eastern Highlands employed an AI-driven remote sensing technique to monitor forest cover. The project employs high-resolution satellite imagery through machine-learning models for stakeholders to detect subtle changes and strategies for timely interventions for conservationists and policymakers.

Scholarly evidence shows that AI tools integration in the management of water resources in these past few years has offered innovative solutions optimizing water usage enhancing agricultural productivity and promoting sustainable resource use. For example, the application of AI in water resource management through the WaterNet initiative employed an algorithm that monitors and manages water resources effectively (Kahanda et al. 2022). A case study conducted by Zimbabwe Agricultural Research Trust (ZART) in collaboration with local agricultural extension services in the Hwedza district shows the effectiveness of AI-driven precision irrigation systems (Mlambo et al. 2021). During the project, local farmers utilised AI tools to monitor soil moisture levels in real-time. In addition, AI tools have also been deployed to engage local communities to sustainably manage water resources. The WaterNet initiative developed a WaterSmart application that has allowed local farmers to report water usage and receive tailored irrigation advice based on AI analysis from contextual data enabling the community to make informed decisions about water use, contributing to more sustainable practices (WaterNet, 2021; Kahanda, 2022; Mlambo and Mhlanga, 2021).

Table 2.

Category	Positive Impacts	Challenges	Examples/Applications
Community Engagement	- AI-powered chatbots provide real-time environmental information.	- Limited internet access in rural areas.	- WhatsApp chatbots for climate alerts e.g. Mhondoro Ngezi Irrigation Community.
	- AI-driven social media analysis tracks public opinion on sustainability.	- Low digital literacy in some communities.	- AI tools for community sentiment analysis.
	- AI-enhanced translation tools facilitate communication in local languages.	- High cost of AI implementation.	- AI-powered translation for Indigenous languages e.g. The 'Hurudza AI Contact Centre for Farmers' in Murehwa
	- AI-driven citizen science platforms encourage community participation.	- Mistrust of AI technologies.	- Mobile apps for community-driven environmental data collection.
Environmental Awareness	- AI-powered early warning systems predict extreme weather events.	- Limited data availability for accurate AI modelling.	- AI-driven weather forecasting systems.
	- AI detects deforestation and land degradation via satellite imagery.	- Need for skilled personnel to interpret AI data.	- AI-based monitoring of forest cover changes.
	- AI-powered agriculture tools optimize water and fertilizer use.	- High initial costs for AI adoption.	- Smart irrigation systems using AI sensors.
	- AI-driven waste management systems improve recycling efficiency.	- Resistance to new technologies.	- AI-based waste sorting and recycling programs.

6.3 Platforms of Community Engagement and Knowledge Gap with Artificial Intelligence

In Zimbabwe, community workshops have raised awareness and education through interactive sessions to learn about waste management, agroforestry, water conversation and biodiversity preservation, often organised by government agencies, environmental groups and NGOs. From an exhaustive literature review, despite the efforts, there remains a significant gap in environmental sustainability knowledge among marginalised communities, often due to limited access to resources and education, resulting in the exploitation of local resources. AI can significantly provide tailored education resources and tools for communities nationwide. AI-driven applications offer real-time information on the local environmental conditions, suggesting sustainable practices in sustainability practices based on specific community needs and analysing data to predict outcomes on various environmental sustainability practices (Bidari & Chickerur, 2024). Enhancing access to knowledge and resources empowers communities to adopt environmentally sustainable practices with better outcomes.

For the past decade, local radio programs have played a crucial role in broadcasting environmental news and tips to reach diverse audiences, mostly in rural areas where access to information is limited. The Zimbabwe Broadcasting Corporation (ZBC), Star FM, and Power FM, amongst others, often feature interviews with environmental experts, discussions on sustainable practices and segments with upcoming community engagement events focused on sustainability and conservation. However, low literacy rates, limited access to formal education, exploitation of local resources and environmental change have contributed to massive unsustainable practices. With the use of AI, local radio stations can analyse audience data to tailor content to meet community needs, preferences and knowledge levels. The use of AI technology can create an interactive platform where listeners can ask questions and receive real-time data feedback. AI-driven tools assist in developing educational materials and campaign strategies that resonate with audiences to bridge the knowledge gap, promote sustainable practices more effectively, and foster greater community engagement in environmental stewardship.

Environmental education is increasingly being integrated into school curricula by the Ministry of Primary and Secondary Education in Zimbabwe, emphasising conservation and sustainability across all subjects such as environmental science and geography. The ministry initiatives include hands-on activities like tree planting and waste management, fostering a sense of responsibility toward the environment. However, significant gaps remain among the teachers and students due to resource limitations, insufficient teacher training, and a lack of localised content. The use of AI presents an opportunity to address these challenges by providing personalised learning experiences, developing interactive educational materials, analysing student performance for targeted feedback, facilitating partnerships with environmental organisations nationwide and leveraging AI in schools to enhance environmental education, empower students to become more informed stewards and custodians of their communities.

An exhaustive literature review shows that community leaders play a crucial role in hosting discussions on environmental sustainability during local gatherings in the Ward Development Committees (WADCO) and Village Development Committees (VIDCO) in Zimbabwe. The discussions are often focused on pressing environmental issues in the community. Although traditional leaders and other relevant stakeholders play a critical role in interactive, participatory engagement to raise awareness, many community members lack an understanding of the long-term impacts of environmental degradation echoing with observations of Zhou et al. (2020), which highlight the critical need for accessible environmental education to combat unsustainable practices. In this regard, there is a need to enhance community discussions by providing data-driven insights and resources tailored to the local context. There is a need to implement AI-driven tools that analyze community-specific environmental data, identify key issues, and develop educational materials that resonate with the WADCOs and VIDCOs. AI-driven platforms facilitate real-time feedback and engagement, allowing community members to ask questions and access information instantly during the engagement process. In addition, the integration of AI technology into community discussions fosters a more informed and proactive approach to environmental sustainability and empowers the community's implementation of sustainable practices.

6.4 AI-Driven Solutions Barriers to Environmental Awareness and Sustainable Practices

Using AI-powered solutions in Zimbabwe to promote sustainability and environmental awareness presents several challenges, categorized into technological, financial, ethical, policy, and infrastructural barriers. Developing countries face a significant challenge in trusting the value of investing in AI technology, hence the prevalent skepticism regarding AI's potential benefits, resulting in hesitance in adopting AI solutions due to perceived high acquisition costs and implementation. Adopting AI in Zimbabwe is limited to the availability of funding and resources. For the past decade, the country has been facing economic instability, often characterised by hyperinflation, which has diverted attention and resources away from technological advancements as it limits the ability to make sustainable, long-term investments in AI (Shahedi et al., 2023). The costs associated with acquiring and maintaining AI solutions have hindered the progress of several organisations, particularly Small and Medium Enterprises (SMEs) and NGOs with limited funds.

In addition, the lack of capacity-building programs across various sectors poses a significant gap in training programs that equip stakeholders with crucial skills to implement and manage AI technology. Several institutions are lagging in integrating technology-focused curricula that prepare the workforce to meet the demands of AI applications. The gap has limited local engagement with AI initiatives and restricted overall potential for technological advancements in environmental sustainability. Data quality and availability is a critical challenge for AI implementation. The implementation of effective AI systems requires access to accurate and comprehensive datasets across various sectors. Data collection methods in Zimbabwe are often inconsistent, and there is a lack of standardized metrics for environmental monitoring and sustainability.

A significant obstacle lies in Zimbabwe's limited technological infrastructure. The populace has struggled with inadequate infrastructure, including unreliable and stable internet access and access to modern computing equipment. Marginalised communities in remote areas lack the necessary technological infrastructure to support AI applications exacerbating existing inequalities between the urban and rural populations. Insufficient access to high-speed internet and inadequate technological resources hinders the effective deployment of AI solutions (Wang et al., 2021) . Frequent power outages and unreliable electricity supply across most regions of Zimbabwe further complicate the consistent operation of AI technologies. Another issue is the absence of renewable energy sources to run environmentally friendly AI systems. While AI has the potential to optimize energy consumption, reduce greenhouse gas emissions, and promote renewable energy usage (Kaur et al., 2024) implementing such solutions is particularly difficult given Zimbabwe's current energy management struggles.

The rural-urban divide worsens these challenges, as advanced technologies are primarily concentrated in urban areas leaving rural communities at a disadvantage. Remote areas, such as Gokwe, often experience a complete lack of AI implementation due to their isolation. Limited outreach programs to educate rural populations about AI's potential further impede the technology's adoption in these regions (Tingzon et al., 2023). The quality and accessibility of data is also essential to the efficacy of AI applications. In many developing regions including Zimbabwe, gaps in data collection and management weaken the impact of AI-driven environmental solutions (Wanyama et al., 2024). Moreover, the high initial costs of implementing AI technologies present a major financial barrier. Economic instability, including high inflation rates, complicates the financial landscape and limits the ability to make sustainable, long-term investments in AI (Shahedi et al., 2023). For households reliant on agriculture, climate change-induced declines in production further reduce cash flow and diminish the financial capacity to invest in advanced technologies. Addressing these barriers will require targeted investments in infrastructure, capacity building, and policy frameworks to support the equitable and sustainable adoption of AI-driven solutions in Zimbabwe.

Human capital and expertise are significant barriers to adopting AI in Zimbabwe, primarily due to the lack of technical proficiency. There is often a shortage of skilled professionals capable of developing, implementing, and maintaining AI systems (Jain & Mitra, 2024). This scarcity of expertise is particularly pronounced in remote and rural areas such as Mount Darwin and Masvingo, slowing the adoption of AI technologies. Additionally, technological literacy gaps among the general population and decision-makers further hinder AI implementation. Effective use of AI requires a foundational understanding of the technology, which is typically acquired through training conducted by experts (Tingzon et al., 2023).

Ethical concerns also pose challenges to AI integration. Issues such as data privacy, algorithmic bias, and the potential for exacerbating existing inequalities must be addressed to build public trust and ensure equitable implementation (Jain & Mitra, 2024). Language barriers further complicate adoption, as most AI solutions are developed in English, making them less accessible to Zimbabwe's linguistically diverse population. For instance, communities in Beitbridge, where Venda and Tonga are predominantly spoken, rely heavily on Indigenous Knowledge Systems and may resist adopting AI technologies perceived as foreign or lacking clear local benefits. Cultural resistance to AI highlights the need for solutions tailored to align with local communities' social and cultural contexts. Western-centric AI frameworks often fail to address African communities' values and unique needs, creating an additional barrier to adoption (Odero et al., 2024).

Policy and regulatory challenges are another significant hurdle. The absence of legal and regulatory frameworks limits the structured and effective implementation of AI technologies (Hernandez et al., 2023). Clear guidelines and policies ensure that AI aligns with national development goals and enhances productivity, particularly in critical sectors like agriculture. However, the regulatory system in Zimbabwe also presents a major challenge in adopting AI-driven solutions. Discrepancies and a lack of clear frameworks and policies supporting AI integration in environmental sustainability practices have created uncertainties for organisations seeking to innovate and implement AI technology. Administrative hurdles may also present another major challenge in the approval process for AI-related projects deters potential investors and stakeholders from pursuing AI initiatives. Existing policies do not adequately address the complexities of integrating AI across several sectors including environmental management and sustainability. Strong involvement with local stakeholders, including communities, lawmakers, and industry actors, is essential for overcoming these obstacles since comprehensive rules are required in this area to encourage the responsible use of AI while protecting the interests and ecosystems of local populations.

7. FURTHER RESEARCH

AI has the potential to play a transformative role in empowering marginalized communities, particularly in the Global South. By focusing on the development and application of AI systems tailored to underserved populations, significant progress can be achieved in critical areas such as environmental sustainability and healthcare. However, current AI systems often fail to adequately address the needs of these communities, largely due to biases inherent in their design and a lack of localized, tailored solutions (Okolo, 2022). Investigating the creation of human-centered AI systems that have an emphasis on inclusion and adaptability is crucial to closing this gap. These systems should consider the unique requirements, cultural contexts, and environmental challenges faced by different groups. Such research can lead to AI solutions that are more equitable, effective, and impactful, ultimately driving positive change in marginalized communities.

AI can completely change how we handle urgent ecological issues and is rapidly developing into an effective instrument in environmental monitoring and management. Real-time monitoring of vital environmental elements including biodiversity, deforestation, and air and water quality can be greatly enhanced by concentrating on the creation and improvement of AI applications. Technologies like remote sensing and predictive analytics enable more accurate and efficient data collection and analysis, providing actionable insights for conservation efforts and resource optimization (Li & Li, 2024). This advancement can drive better decision making and foster sustainable environmental practices, ultimately supporting global efforts to mitigate environmental degradation (Putra et al., 2024).

Research on the application of AI to improve climate resilience and catastrophe management has become crucial, with an emphasis on how well AI can forecast and control climate-related disasters. This includes using AI to create and conduct advanced early warning systems and flexible response plans. There are several possibilities to increase the precision and promptness of climatic event forecasts using AI technologies like machine learning, data analytics and predictive modelling (McGovern et al., 2024). AI can offer practical insights that help communities and authorities prepare for and lessen the effects of disasters by evaluating enormous information on weather patterns, environmental changes, and past disaster occurrences. AI-powered communication solutions can also improve the distribution of real-time notifications, guaranteeing that susceptible groups get important information quickly. To improve resilience and lessen the socioeconomic and environmental effects of climate-related disasters, this area of study will attempt to investigate how AI might close the gap between scientific data and community readiness.

Future research in the use of AI to sustainable urban development is promising, with an emphasis on how AI technologies might revolutionize urban planning to maximize mobility, waste management, and energy consumption. AI provides creative ways to improve urban infrastructure while reducing environmental effects as cities deal with mounting pressures from population expansion and environmental issues. Through advanced data analysis and predictive modeling, AI can improve energy efficiency by identifying patterns in consumption and suggesting optimal usage strategies (Leal Filho et al., 2024). By enabling intelligent garbage collection, sorting, and recycling procedures, AI-driven solutions can also transform waste management by lowering landfill contributions and advancing a circular economy. In transportation, AI can optimize traffic flow, enhance public transport efficiency, and support the transition to sustainable mobility through electric and autonomous vehicles (Al-Raeei, 2024). This area of study will investigate how AI can be incorporated into urban planning frameworks to build smart, flexible, and sustainable cities, supporting international initiatives to address the problems of urbanization and climate change.

The creation of AI solutions that are not only technologically sophisticated but also socially and environmentally responsible will be made possible by more study in these fields. Such research can support more equitable and sustainable development outcomes by addressing the opportunities and difficulties encountered by the Global South, promoting resilience and inclusive progress.

8. MANAGERIAL IMPLICATIONS

Managers should prioritise exploiting AI to analyse environmental data, identify trends, and generate actionable insights for addressing local environmental challenges. This requires investing in AI infrastructure and developing expertise within organizations. By providing predictive analytics and data-driven insights (Sulich et al., 2023), AI empowers managers to make informed decisions. This approach is particularly valuable in areas such as resource management, waste reduction, and environmental monitoring, where precise and timely interventions can have significant impacts.

The successful integration of AI relies on collaboration among key stakeholders, including governments, businesses, and communities, to ensure that AI-driven solutions are inclusive and address the needs of marginalized populations. Managers should actively promote partnerships and encourage open data sharing to support the implementation of AI-powered sustainability initiatives (Gulyamov, 2024). The efficacy and reach of environmental awareness initiatives could be greatly increased by such cooperative efforts.

Managers must fund capacity-building programs that equip local organizations and communities with the know-how to use AI technologies efficiently if they want to fully reap the benefits of this technology. This involves putting in place training initiatives that prioritize environmentally friendly procedures and the moral use of technology. AI can also enhance citizen engagement by creating platforms that encourage community involvement in sustainability initiatives (Tan et al., 2023). For instance, AI-powered applications for waste management and food donation can promote community responsibility and active participation, fostering a collective commitment to sustainability (Kumar Tyagi et al., 2023).

Challenges related to trust in AI technology in most developing countries are prevalent in both local and national contexts. Trusting AI encompasses a range of factors but more importantly hinges on establishing social and technical trust within organisations, among business stakeholders, AI experts, and the broader public. The cultural attitude towards technology plays a role in the adoption of AI technology in Zimbabwe however, several communities value traditional knowledge and practices, causing scepticism about the effectiveness of AI-driven solutions. The perception that technology undermines local customs and ways of sustenance has created resistance to its adoption fueled by fears of job displacement and concerns that AI could exacerbate existing inequalities.

Managers must ensure that AI applications adhere to ethical standards and avoid worsening existing inequalities. Transparent governance frameworks that put equity, inclusivity, and data privacy first should direct the use of AI. Managers need to address the ethical implications of AI, including concerns related to data privacy, transparency, and potential biases (Sulich et al., 2023). Upholding ethical AI practices is vital for maintaining public trust and achieving long-term sustainability goals (Popescu & Yu, 2024). The managerial implications of exploiting AI for community engagement and environmental awareness in the Global South are significant. Managers may significantly contribute to the advancement of sustainable practices and the accomplishment of global sustainability goals by integrating AI strategically, addressing ethical issues, assisting in the formulation of policies and encouraging active community involvement.

9. CONCLUSION

In conclusion, using artificial intelligence to raise environmental awareness and engage communities offers a revolutionary chance to advance sustainable practices in the Global South. Communities, governments, and companies may use AI to solve urgent environmental issues by implementing creative solutions including improved communication tactics, predictive analytics, and intelligent resource management. However, a cooperative strategy that places a high priority on inclusion, ethical issues, and capacity building is necessary for the successful execution of AI-driven initiatives. Stakeholders may promote environmental stewardship and help create more resilient and sustainable communities by incorporating AI ethically and responsibly, which will ultimately aid in the accomplishment of global sustainability goals.

REFERENCES

Al-Raeei, M. (2024). Artificial intelligence for climate resilience: Advancing sustainable goals in SDGs 11 and 13 and its relationship to pandemics [Article]. *Discover Sustainability*, *5*(1), 513. Advance online publication. DOI: 10.1007/s43621-024-00775-5

Alhalwachi, L., Alkhater, N., Albalawi, N., Alharbi, M., Zainaldeen, Z., Aldhamin, M., & Alwarsh, M. (2024). Exploring the Potential Applications of Artificial Intelligence in the Agriculture Sector: Case Study Tabuk District. In *Studies in Big Data* (Vol. 163, pp. 151-158). https://doi.org/DOI: 10.1007/978-3-031-73632-2_13

Ansari, S. P. M. A., Odongo, J. O., Nomani, M. Z. M., Salahuddin, G., Hasan, M. F., & Pallathadka, L. K. (2022). Evaluating the role of environment management system based on artificial intelligence. *Materials Today: Proceedings*, *56*, 2240–2244. https://doi.org/https://doi.org/10.1016/j.matpr.2021.11.571. DOI: 10.1016/j.matpr.2021.11.571

Bidari, I., & Chickerur, S. (2024). Deep Recurrent Residual U-Net with Semi-Supervised Learning for Deforestation Change Detection. *SN Computer Science*, *5*(7), 893. DOI: 10.1007/s42979-024-03127-2

Cazzaniga, M., Jaumotte, F., Li, L., Melina, G., Panton, A. J., Pizzinelli, C., Rockall, E. J., & Mendes Tavares, M. (2024). *Gen-AI: Artificial Intelligence and the Future of Work* (Vol. 2024). International Monetary Fund.

Chatterjee, S., Chaudhuri, R., Kamble, S., Gupta, S., & Sivarajah, U. (2023). Adoption of Artificial Intelligence and Cutting-Edge Technologies for Production System Sustainability: A Moderator-Mediation Analysis. *Information Systems Frontiers*, *25*(5), 1779–1794. DOI: 10.1007/s10796-022-10317-x

Chaudhary, G. (2023). Environmental Sustainability: Can Artificial Intelligence be an Enabler for SDGs? [Article]. *Nature Environment and Pollution Technology*, *22*(3), 1411–1420. DOI: 10.46488/NEPT.2023.v22i03.027

Chen, L., Chen, Z., Zhang, Y., Liu, Y., Osman, A. I., Farghali, M., Hua, J., Al-Fatesh, A., Ihara, I., Rooney, D. W., & Yap, P. S. (2023). Artificial intelligence-based solutions for climate change: A review [Review]. *Environmental Chemistry Letters*, *21*(5), 2525–2557. DOI: 10.1007/s10311-023-01617-y

Chisika, S., & Yeom, C. (2024). Improving the role of communities in participatory forest management through artificial intelligence: The case of Nairobi city park community forest association [Article]. *International Forestry Review*, *26*(1), 1–15. DOI: 10.1505/146554824838457916

Ciaralli, C. A. (2023). Artificial intelligence, political decision-making and environmental transition: challenges and perspectives for constitutionalism [Article]. *Federalismi.it, 2023*(15), 40-87. https://www.scopus.com/inward/record.uri?eid=2-s2.0-85165143688&partnerID=40&md5=2a63020fb009c2cdfac5de37a10da951

Dube, T., Dube, T., Dalu, T., Gxokwe, S., & Marambanyika, T. (2024). Assessment of land use and land cover, water nutrient and metal concentration related to illegal mining activities in an Austral semi–arid river system: A remote sensing and multivariate analysis approach [Article]. *The Science of the Total Environment, 907*, 167919. Advance online publication. DOI: 10.1016/j.scitotenv.2023.167919 PMID: 37879464

Eli-Chukwu, N. C. (2019). Applications of Artificial Intelligence in Agriculture: A Review [Article]. *Engineering. Technology and Applied Science Research, 9*(4), 4377–4383. DOI: 10.48084/etasr.2756

Fergus, P., Chalmers, C., Longmore, S., & Wich, S. (2024). Harnessing Artificial Intelligence for Wildlife Conservation [Article]. *Conservation, 4*(4), 685–702. DOI: 10.3390/conservation4040041

Fulkar, B., Mendhe, S., & Patil, P. (2024). Artificial Intelligence Cultivation: Transforming Agriculture for a Smart and Sustainable Future. *2nd International Conference on Intelligent Data Communication Technologies and Internet of Things, IDCIoT 2024* DOI: 10.1109/IDCIoT59759.2024.10467857

Gulyamov, S. (2024). Intelligent waste management using IoT, blockchain technology and data analytics. E3S Web of Conferences

Gundeti, R., Vuppala, K., & Kasireddy, V. (2023). The future of AI and environmental sustainability: Challenges and opportunities. In *Exploring Ethical Dimensions of Environmental Sustainability and Use of AI* (pp. 346-371). https://doi.org/DOI: 10.4018/9798369308929.ch017

Carr, D., & Harrison, T. (2015). *Educating character through stories*. Andrews UK Limited.

Hernandez, A. A., Caballero, A. R., Albina, E. M., Balmes, I. L., & Niguidula, J. D. (2023). Artificial Intelligence for Sustainability: Evidence from select Small and Medium Enterprises in the Philippines. 2023 8th International Conference on Business and Industrial Research, ICBIR 2023 – Proceedings

Jain, V., & Mitra, A. (2024). Integrative innovations: Harnessing artificial intelligence for enhanced environmental sustainability. In *Maintaining a Sustainable World in the Nexus of Environmental Science and AI* (pp. 365-382). https://doi.org/DOI: 10.4018/979-8-3693-6336-2.ch014

Jombo, S., Adam, E., & Odindi, J. (2017). Quantification of landscape transformation due to the Fast Track Land Reform Programme (FTLRP) in Zimbabwe using remotely sensed data [Article]. *Land Use Policy, 68*, 287–294. DOI: 10.1016/j.landusepol.2017.07.023

Jones, A., Kuehnert, J., Fraccaro, P., Meuriot, O., Ishikawa, T., Edwards, B., Stoyanov, N., Remy, S. L., Weldemariam, K., & Assefa, S. (2023). AI for climate impacts: applications in flood risk [Article]. *npj Climate and Atmospheric Science, 6*(1), Article 63. https://doi.org/DOI: 10.1038/s41612-023-00388-1

Kaur, M., Aron, R., Tandon, R., Wadhwa, H., Kaur, G., Sandhu, R., & Ghai, D. (2024). Impact of Artificial Intelligence Techniques on Green Applications. In *Artificial Intelligence Techniques for Sustainable Development* (pp. 64-85). https://doi.org/DOI: 10.1201/9781003546382-4

Kumar Tyagi, P., Sharma, P., Kumar Singh, A., Tyagi, P., Sharma, P., & Jit Singh, V. (2023). The Role of Artificial Intelligence in Promoting Sustainable Business Operations and Autonomy. 2023 10th IEEE Uttar Pradesh Section International Conference on Electrical, Electronics and Computer Engineering, UPCON 2023

Leal Filho, W., Mbah, M. F., Dinis, M. A. P., Trevisan, L. V., de Lange, D., Mishra, A., Rebelatto, B., Ben Hassen, T., & Aina, Y. A. (2024). The role of artificial intelligence in the implementation of the UN Sustainable Development Goal 11: Fostering sustainable cities and communities [Article]. *Cities (London, England), 150*, 105021. Advance online publication. DOI: 10.1016/j.cities.2024.105021

Li, Y., & Li, X. Z. (2024). Utilizing unmanned aerial vehicle technology for environmental monitoring: Future trends, methods, and applications [Article]. *Internet Technology Letters*. Advance online publication. DOI: 10.1002/itl2.526

McGovern, A., Tissot, P., & Bostrom, A. (2024). Developing trustworthy AI for weather and climate [Article]. *Physics Today, 77*(1), 26–31. DOI: 10.1063/PT.3.5379

Mienye, I. D., & Jere, N. (2024). Deep Learning for Credit Card Fraud Detection: A Review of Algorithms, Challenges, and Solutions. *IEEE Access : Practical Innovations, Open Solutions, 12*, 96893–96910. DOI: 10.1109/ACCESS.2024.3426955

Mogha, H., & Hasteer, N. (2023). Challenges to Implement Artificial Intelligence for Environmental Sustainability. Lecture Notes in Civil Engineering, Munanga, Y., & Mafuku, S. H. (2021). Climate-Resilient Infrastructure for Water and Energy in Greater Harare. In *Advances in 21st Century Human Settlements* (pp. 69-85). https://doi.org/DOI: 10.1007/978-981-99-1388-6_31

Naim, A., Mohammed, A., Fatima, N., Khan, S. A., Alnfiai, M. M., & Malik, P. K. (2024). Applications of artificial intelligence in environmental resource business management and sustainability. In *AI Applications for Business* (pp. 1–22). Medical, and Agricultural Sustainability., DOI: 10.4018/979-8-3693-5266-3.ch001

Ndumeya, N., & Maluleka, P. (2024). Can History Teaching Contribute Towards Saving the Planet? Reflections on the Value of Including Environmental History in the Zimbabwean Advanced Level History Curriculum. In *History Education in Africa: Research, Perspectives and Practices* (pp. 47-68). https://doi.org/DOI: 10.1007/978-3-031-61388-3_3

Nhapi, T. (2023). Reflections on integrating environmental social work methods of intervention in Zimbabwean social work curriculum [Article]. *African Journal of Social Work, 13*(2), 118–126. DOI: 10.4314/ajsw.v13i2.8

Ntuli, H., & Muchapondwa, E. (2018). The role of institutions in community wildlife conservation in Zimbabwe. *The International Journal of the Commons, 12*(1), 134. DOI: 10.18352/ijc.803

Nyoni, R., Bruelle, G., Chikowo, R., & Andrieu, N. (2024). Targeting smallholder farmers for climate information services adoption in Africa: A systematic literature review. *Climate Services, 34*, 100450. DOI: 10.1016/j.cliser.2024.100450

Odero, B., Nderitu, D., & Samuel, G. (2024). The Ubuntu Way: Ensuring Ethical AI Integration in Health Research [Review]. *Wellcome Open Research, 9*, Article 625. https://doi.org/DOI: 10.12688/wellcomeopenres.23021.1

Okolo, C. T. (2022). Optimizing human-centered AI for healthcare in the Global South [Review]. *Patterns (New York, N.Y.), 3*(2), 100421. Advance online publication. DOI: 10.1016/j.patter.2021.100421 PMID: 35199066

Parwada, C., & Marufu, H. (2023). Digitalisation of agriculture in Zimbabwe: Challenges and opportunities Keywords Agricultural management Agricultural production Artificial intelligent Big data Increased efficiency. *International Journal of Sustainable Agricultural Research*. https://doi.org/DOI: 10.18488/ijsar.v10i1.3280

Pelet, J. É., Belda, S., & Arezki, D. (2022). Use of AI in Predicting Trends in Vegetation Dynamics in Africa. In *Encyclopedia of Data Science and Machine Learning* (pp. 31-56). https://doi.org/DOI: 10.4018/978-1-7998-9220-5.ch003

Popescu, C. R. G., & Yu, P. (2024). *Intersecting environmental social governance and AI for business sustainability* [Book]. https://doi.org/DOI: 10.4018/979-8-3693-1151-6

Putra, G. A. A., Wiraatmaja, P. P., & Adhiswari Wedagama, D. A. T. (2024). The Application of Artificial Intelligence (AI) in Addressing Environmental Issues. 2024 10th International Conference on Smart Computing and Communication, ICSCC 2024

Raghav, A., Singh, B., Jermsittiparsert, K., Raghav, R., & Yadav, U. (2024). Artificial intelligence in environmental and climate changes: A sustainable future. In *Maintaining a Sustainable World in the Nexus of Environmental Science and AI* (pp. 485-506). https://doi.org/DOI: 10.4018/979-8-3693-6336-2.ch019

Rautela, K., Kumawat, K., Rahman, F., & Patidar, H. O. (2020). Artificial Intelligence In Environmental Monitoring And Conservation. *International Journal of Psychosocial Rehabilitation*, 10031–10034. DOI: 10.61841/V24I2/400257

Rutenberg, I., Gwagwa, A., & Omino, M. (2021). Use and Impact of Artificial Intelligence on Climate Change Adaptation in Africa. In *African Handbook of Climate Change Adaptation: With 610 Figures and 361 Tables* (pp. 1107-1126). https://doi.org/DOI: 10.1007/978-3-030-45106-6_80

Sango, I., & Godwell, N. (2015). Climate change trends and environmental impacts in the Makonde Communal lands, Zimbabwe [Review]. *South African Journal of Science*, *111*(7-8), 6. Advance online publication. DOI: 10.17159/sajs.2015/20140266

Santos, R. C., M., & Cagica Carvalho, L. (. (2025). AI-driven participatory environmental management: Innovations, applications, and future prospects [Article]. *Journal of Environmental Management*, *373*, 123864. Advance online publication. DOI: 10.1016/j.jenvman.2024.123864 PMID: 39752951

Schiavo, R. (2021). What is true community engagement and why it matters (now more than ever). *Journal of Communication in Healthcare*, *14*(2), 91–92. DOI: 10.1080/17538068.2021.1935569

Schiff, D. S., Kelley, S., & Camacho Ibáñez, J. (2024). The emergence of artificial intelligence ethics auditing. *Big Data & Society*, *11*(4), 20539517241299732. DOI: 10.1177/20539517241299732

Shahedi, A., Dadashpour, I., & Rezaei, M. (2023). Barriers to the sustainable adoption of autonomous vehicles in developing countries: A multi-criteria decision-making approach [Article]. *Heliyon*, *9*(5), e15975. Advance online publication. DOI: 10.1016/j.heliyon.2023.e15975 PMID: 37229167

Singh, M., Arora, V., & Kulshreshta, K. (2024). AI and the environment: Innovative approaches to climate change. In *Maintaining a Sustainable World in the Nexus of Environmental Science and AI* (pp. 1-22). https://doi.org/DOI: 10.4018/979-8-3693-6336-2.ch001

Sulich, A., Soloducho-Pelc, L., & Grzesiak, S. (2023). Artificial Intelligence and Sustainable Development in Business Management Context - Bibliometric Review. *Procedia Computer Science*, *225*, 3727–3735. DOI: 10.1016/j.procs.2023.10.368

Tan, T., Dong, X., Xu, Q., & Zou, N. (2023). AI-Powered Waste Management: A Case Study on the Application of YOLOv5 and K-Means Clustering in Urban Communities. *Proceedings of SPIE - the International Society for Optical Engineering*, 176. DOI: 10.1117/12.3012490

Tingzon, I., Miraflor, J. M., Xiang Zhu, X., & Kochupillai, M. (2023). Towards Impactful Applications of AI4EO in the Global South. 2023 Joint Urban Remote Sensing Event, JURSE 2023

Tiwari, S. (2024). The Rise of Intelligent Machines. In *Artificial Intelligence and Machine Learning in Drug Design and Development* (pp. 1-22). https://doi.org/ https://doi.org/10.1002/9781394234196.ch1

Todorova, A., Kostadinova, I., Antonova, D., & Ruskova, S. (2024). Artificial Intelligence Against Climate Change Challenges: From Research to Action Towards a More Sustainable Future. ISMSIT 2024 - 8th International Symposium on Multidisciplinary Studies and Innovative Technologies, Proceedings

Wang, Y., Ksienzyk, A. K., Liu, M., & Brönner, M. (2021). Multigeophysical data integration using cluster analysis: Assisting geological mapping in Trøndelag, Mid-Norway [Article]. *Geophysical Journal International*, *225*(2), 1142–1157. DOI: 10.1093/gji/ggaa571

Wanyama, J., Bwambale, E., Kiraga, S., Katimbo, A., Nakawuka, P., Kabenge, I., & Oluk, I. (2024). A systematic review of fourth industrial revolution technologies in smart irrigation: Constraints, opportunities, and future prospects for sub-Saharan Africa [Review]. *Smart Agricultural Technology*, *7*, 100412. Advance online publication. DOI: 10.1016/j.atech.2024.100412

Wasik, S., & Pattinson, R. (2024). Artificial Intelligence Applications in Fish Classification and Taxonomy: Advancing Our Understanding of Aquatic Biodiversity [Article]. *Journal of Fish Taxonomy*, *31*, 11–21. https://www.scopus.com/inward/record.uri?eid=2-s2.0-85181698869&partnerID=40&md5=88b8729c2cbde458180a641c2777fcbf

Yadav, A. (2023). Leveraging artificial intelligence for sustainable development and environmental resilience. In *Exploring Ethical Dimensions of Environmental Sustainability and Use of AI* (pp. 140-165). https://doi.org/DOI: 10.4018/979-8-3693-0892-9.ch008

Yan, L., & Fengyi, C. (2024). Research on the Evaluation Model of National Community Consciousness Education Effect Based on Artificial Intelligence. In *Learning and Analytics in Intelligent Systems* (Vol. 41, pp. 49-58). https://doi.org/DOI: 10.1007/978-3-031-69457-8_5

Zehouani, N., Ababou, M., & Faquir, S. (2024). *Harnessing Artificial Intelligence for Environmental Sustainability: A Comprehensive Review of Its Applications in Biodiversity, Energy, Transportation, and Water Management*. Lecture Notes in Networks and Systems.

Zolkafli, A., Mansor, N. S., Omar, M., Ahmad, M., Ibrahim, H., & Yasin, A. (2024). AI for Smart Disaster Resilience among Communities. In *Studies in Systems* (Vol. 553, pp. 369–395). Decision and Control., DOI: 10.1007/978-3-031-67317-7_22

Chapter 14
AI in Health and Environmental Sustainability

Surendra Prakash Gupta
https://orcid.org/0000-0002-4873-6346
Shri Vaishnav Vidhyapeeth Viswavidhayalaya, India

Ankur Bhardwaj
Shri Vaishnav Vidyapeeth Vishwavidyalaya, India

ABSTRACT

The integration of artificial intelligence into the areas of health and environmental sustainability has become a topic of increasing interest and importance. As the world grapples with the pressing challenges of climate change and the need for sustainable development, the potential of AI-powered technologies to contribute to these efforts is widely recognized. AI-driven systems can actually be designed to optimize resource allocation for environmental sustainability in healthcare systems. These systems use advanced algorithms to increase efficiency, reduce waste and support sustainable practices, ultimately contributing to better healthcare while minimizing environmental impact. Review of the available literature shows that future studies on AI for sustainability should incorporate multi-level views. This chapter addresses AI-related significant opportunities to optimize resource allocation and improve sustainability, challenges such as data bias, ethical considerations, and the initial investments required for AI infrastructure to realize its potential.

DOI: 10.4018/979-8-3373-0680-3.ch014

1. INTRODUCTION

1.1 Definition of Artificial Intelligence (AI)

AI in healthcare and environmental sustainability represents a transformative intersection that enhances diagnostic accuracy, optimizes resource use, and addresses climate challenges. AI's capabilities in analyzing vast datasets enable improved patient outcomes and efficient healthcare delivery while promoting sustainable practices within the industry. Healthcare is a key setting where expectations are rising for the potentially life-enhancing benefits of AI "the science and engineering of creating intelligent machines that can achieve goals like humans via a constellation of technologies" (Topol, 2019).

1.2 Importance of AI in Modern Challenges

AI-powered monitoring systems significantly improve the tracking and assessment of the Sustainable Development Goals (SDGs) by leveraging advanced data analysis and predictive modeling capabilities. These systems facilitate informed decision-making, optimize resource allocation and improve overall efficiency across various sectors, accelerating progress toward sustainability.

AI systems can analyze large data sets to identify trends and patterns relevant to SDGs such as poverty reduction and climate change (Farahani & Ghasemi, 2024). For example, AI can predict crop yields for food security (SDG 2) and predict extreme weather events to strengthen disaster preparedness (SDG 13) (Yadav et al., 2024). In agriculture, AI-powered intelligent systems can automate labor-intensive processes, improve farm management, and facilitate transparent transactions through blockchain-based smart contracts, leading to greater efficiency and cost savings (Sirmacek et al., 2023). Additionally, AI can play an important role in improving energy efficiency, a crucial aspect of sustainable development. Smart grids using AI can optimize the balance between regional electricity production and local demand, thereby reducing overall electricity consumption (Bolón-Canedo et al., 2024). Furthermore, AI has the potential to enable the achievement of 134 of the 169 SDG goals, but can also hinder 59 of them, highlighting the need for responsible and ethical AI integration. As the development of AI accelerates, it is important to ensure that the necessary regulatory insights and oversight are in place to realize the technology's potential for sustainable development while mitigating potential risks or unintended consequences (Sirmacek et al., 2023).

AI technologies optimize resource management in sectors such as renewable energy, improving energy efficiency and reducing waste (Raman et al., 2024). Applications include smart energy systems that improve grid management and renewable energy forecasting, which is critical to achieving SDG 7 (Raman et al., 2024). AI empowers developing countries by providing scalable, cost-effective solutions, promoting inclusive growth and reducing inequalities (Farahani & Ghasemi, 2024). Initiatives such as precision agriculture and telemedicine illustrate how AI can address local challenges while contributing to broader SDG goals (Baig et al., 2024). Conversely, while AI offers transformative potential, it also raises ethical concerns such as privacy and algorithmic bias that could undermine trust and exacerbate inequalities if not handled responsibly (Yadav et al., 2024).

1.3 Overview of Health and Environmental Sustainability

Machine learning algorithms (MLA) have shown significant potential for predicting and measuring the effectiveness of sustainable development initiatives in various contexts. By leveraging large data sets and advanced analytical techniques, these algorithms can uncover patterns and provide insights that support decision-making for Sustainable Development Goals (SDGs).

1.3.1 Predictive Modeling for Sustainability

Machine learning models such as Random Forest and Support Vector Machines (SVM) have been effectively used to classify countries based on sustainability scores, achieving high accuracy rates (Vijayanand, 2023). A systematic literature review shows that ML can facilitate the identification of good practices and improve the monitoring of SDG progress (Kagzi et al., 2023). ML methods have been successfully applied to various projects, for example in Latin America and the Caribbean, demonstrating adaptability to local contexts and improving sustainability ratings of projects (García Villena et al., 2022).

Automated machine learning (AutoML) approaches have demonstrated superior performance in predicting SDG achievements, outperforming traditional methods such as back-propagation neural networks (Singpai & Wu, 2020). Techniques such as Self-Organizing Maps (SOM) and Classification and Regression Trees (CART) have been used to assess sustainability performance and achieve high prediction accuracy (Nilashi et al., 2021). While ML offers promising tools to improve sustainable development initiatives, challenges remain around data quality and the need for context-specific models. Addressing these issues is critical to maximizing the impact of these technologies on sustainability efforts.

2. AI IN HEALTHCARE IN DISEASE AND DIAGNOSIS

AI in tranforming disease diagnosis and treatment in various medical fields and diagnosis by increasing accuracy, personalizing care, and accelerating drug discovery. The integration of AI technologies such as ML and DL enables the analysis of large data sets, leading to improved patient outcomes and innovative therapeutic strategies. Despite these advances, challenges such as data protection, algorithmic bias, and the need for regulatory frameworks remain important concerns that must be addressed to ensure responsible AI implementation in healthcare (Singh et al., 2024; Mohammadi et al., 2024). Below are the key applications of AI in healthcare.

2.1 Applications of AI in Diagnostics

AI algorithms excel at analyzing medical images and significantly improve the detection of diseases such as cancer and cardiovascular diseases (Chowdhury et al., 2025). ML models are used in the diagnosis of specific diseases such as Kawasaki disease, where they match the performance of experienced clinicians in identifying heart problems (Pan & Jiao, 2024).

2.2 Personalized Medicine and Treatment Plans

AI integrates genomic data and electronic health records to tailor treatment plans, improving the effectiveness of interventions (Rakibul Chowdhury, 2024). Predictive analytics enable healthcare providers to predict treatment outcomes, enabling more informed decision-making (Darwish et al., 2024).

AI-powered chatbots, particularly those using large language models (LLMs) like ChatGPT, significantly improve patient engagement and healthcare outcomes (Andrew, 2024)(REF). These technologies enable improved communication between patients and healthcare providers, streamline administrative tasks, and support clinical decision-making, ultimately leading to better patient satisfaction and quality of care. Chatbots improve patient-doctor interaction by providing timely responses to queries, thereby reducing communication barriers (Nalongo, 2024). They can automate routine tasks, allowing healthcare professionals to focus on more complex patient needs (FNU et al., 2024). AI chatbots can analyze patient data to offer tailored health recommendations, improving the personalization of care (Triplett, 2024). They engage patients in goal-oriented conversations, which can lead to better adherence to treatment plans (Wen et al., 2024). Chatbots promote health literacy by providing accessible information about medical conditions and treatments and empowering patients to make informed decisions (Neha et al., 2024).

The field of digital health is undergoing rapid change, driven by the convergence of advanced technologies and the increasing availability of health data. LLMs, deep learning models trained on large amounts of text data, have proven to be a particularly promising tool for unlocking insights and automating tasks in various healthcare domains. Recent advances in LLMs have enabled breakthroughs in multimodal learning, particularly in text-to-image and text-to-video generation. As LLMs continue to evolve and integrate more modalities, they may open up further opportunities for integrating physiological signals, medical images, wearable and IoMT signals into patient-provider communication personalization of health measures (Warner et al., 2024). However, to realize the full potential of healthcare LLMs, multiple challenges related to privacy, bias, transparency, and regulatory compliance must be addressed. Ensuring the responsible development and use of medical LLMs requires ongoing interdisciplinary collaboration between healthcare stakeholders, AI researchers, policymakers and patient advocates.

2.3 AI in Drug Discovery

The integration of AI into drug development is revolutionizing the pharmaceutical industry and offering innovative solutions that improve sustainability in drug development. AI technologies streamline processes such as target identification, lead optimization and predictive modeling, to more efficient use of resources and less waste in drug development cycles. AI accelerates drug target identification and optimizes molecular designs, enabling faster drug development processes (Chowdhury, 2024). In the context of bacterial infections, AI helps in the rapid identification of pathogens and antimicrobial susceptibility, thereby improving treatment strategies (Zhang et al., 2024).

2.3.1 AI-Driven Efficiency

(a) **Target Identification**: AI algorithms analyze large-scale biological data to identify potential drug targets, significantly reducing the time and resources required for initial research (Wenteler et al., 2024). AI algorithms analyze large biological data sets, enabling the identification of disease-related targets and interactions with drug candidates, streamlining the drug development process. ML techniques predict pharmacokinetics and toxicity, enable prioritization of lead compounds, and reduce the need for extensive animal testing (Ghule, 2024). By identifying materials suitable for recycling, AI supports circular economy initiatives, further minimizing the environmental footprint of drug development (Faiz et al., 2024).

(b) **Lead Optimization**: ML has been an integral part of generating small molecule targets for decades. ML models can predict the efficacy and safety of drug candidates, minimizing the need for extensive laboratory testing (Kokudeva et al., 2024).

(c) **Data Integration**: AI facilitates the integration of omics data and electronic health records, improves the understanding of disease mechanisms and improves drug repurposing strategies (Kokudeva et al., 2024; Khan et al., 2024). AI offers significant opportunities to improve sustainability in drug discovery. To realize their full potential, the challenges and ethical considerations must be addressed. The effectiveness of AI depends on the quality of the data used, which can introduce bias and affect model reliability (Wenteler et al., 2024; Abbas et al., 2024). The use of AI raises ethical concerns regarding the privacy and interpretability of AI-driven decisions in drug development (Rahate & Mondal, 2024).

AI plays a transformative role in identifying sustainable drug development pathways by optimizing processes, minimizing waste and reducing carbon footprint. Through advanced algorithms and data analysis, AI increases efficiency in drug discovery and development and leads to more sustainable practices in the pharmaceutical industry.

AI enables cleaner production by optimizing resource utilization and minimizing waste through predictive analytics and ML (Jebbor et al., 2024; Pratap & Venkatesh, 2024). The integration of AI with green chemistry principles promotes environmentally friendly production processes and thus contributes to sustainable drug development (Rebecca et al., 2023). AI-driven solutions improve energy efficiency and reduce the environmental footprint in drug manufacturing (Pratap & Venkatesh, 2024). By automating processes and improving decision-making, AI contributes to lower operational costs and reduced carbon emissions (Haleem et al., 2023).

2.4 Remote Patient Monitoring and Telemedicine

Remote Patient Monitoring (RPM) and telemedicine significantly reduce the environmental footprint of clinical trials by minimizing patient travel and associated emissions. The integration of these technologies not only improves healthcare but is also in line with sustainability goals in the healthcare sector. Telemedicine has been shown to dramatically shorten patient travel distances. For example, a study of cystic fibrosis patients reported a reduction in mean travel distance from 266 km to 43 km, resulting in an 81.7 kg decrease in CO2 emissions per patient (Green et al., 2023). of the emissions come from the inpatient hospital sector. A study of CO2 emissions from operating rooms found annual emissions of up to 5 million kg of CO_2. In another study, remote postoperative follow-up visits resulted in an average reduction of 41.2 kg CO_2 per patient, demonstrating significant savings in CO_2 emissions compared to traditional in-person consultations (Lathan et al., 2024).

RPM and telemedicine not only reduce emissions but also reduce costs. The estimated cost of car journeys for cystic fibrosis patients fell from £23.10 to £3.76 after the introduction of telemedicine. Remote monitoring in clinical trials also reduces sponsors' travel costs, thereby increasing the overall efficiency of trial management (Pachiappan et al., 2023). It is critical to adopt policies to support telehealth integration while raising awareness of potential disparities and barriers that may impact implementation. Although the environmental benefits of RPM and telemedicine are obvious, it is important to recognize that not all medical conditions can be effectively treated remotely. Face-to-face interactions remain critical for certain patient populations, highlighting the need for a balanced approach to healthcare.

2.5 Ethical Concerns and Data Privacy

Ethical considerations, particularly privacy and bias, play a critical role in shaping sustainable drug development pathways. As the pharmaceutical industry increasingly relies on large data sets and AI, the ethical use of these resources is becoming increasingly important. Integrating AI into drug development can perpetuate existing biases if not handled properly. This can lead to unequal access to treatment and exacerbate health inequalities (Pitel et al., 2024). Using diverse and inclusive datasets for AI training is critical to mitigating these biases and ensuring fair outcomes in drug development (Pitel et al., 2024). The use of large clinical and real world data raises significant privacy concerns. Patients often fear that their data will be misused or not adequately protected, leading to distrust of the healthcare system (Heidelberg et al., 2020). Therefore, ethical frameworks are essential to balance the utility of data with data protection rights, emphasizing informed consent and transparency in data use (Valli et al., 2024).

Ethical oversight by independent ethics committees is crucial to protect participants in clinical trials and ensure that all studies comply with established ethical principles (Arányi, 2018). Continuous adaptation of the regulatory framework is required to address emerging ethical challenges in drug development, particularly as technologies evolve (Pitel et al., 2024). Conversely, some argue that strict ethical considerations could slow innovation in drug development and potentially delay access to new therapies. Balancing ethical imperatives and the need for rapid progress remains a complex challenge in this area.

3. AI IN ENVIRONMENTAL SUSTAINABILITY

In the ever-evolving landscape of environmental sustainability, the integration of AI has emerged as a transformative force, revolutionizing the way we address the pressing issue of climate change. From automated discovery and distribution to optimized decision-making, AI has the potential to significantly reduce energy consumption and serve as a driving force in the energy sector (Chen et al., 2023).

Environmental monitoring plays a critical role in understanding and mitigating the impacts of both natural and human activities that contribute to climate change. Advances in AI and robotics have accelerated the development of autonomous environmental monitoring solutions, allowing scientists to collect more empirical data at each site of interest (Mansfield & Montazeri, 2024). AI-powered systems can analyze large amounts of information much faster than humans, helping to address challenges such as waste detection, space debris and sustainability (Sharma & Vaid, 2021).

The area of AI-based solutions for environmental protection has become increasingly important and numerous examples show its potential. AI is being used to create more sustainable production methods and model climate change, making it a valuable tool in the fight against environmental degradation (Chen et al., 2023; Pachot & Patissier, 2023).

3.1 AI for Climate Change Monitoring

AI is emerging as a critical tool in monitoring climate change and environmental sustainability, leveraging its capabilities in data processing and predictive modeling to improve mitigation strategies. Integrating AI and environmental science can make significant progress in various areas, including climate modeling, resource optimization, and real-time environmental monitoring. AI improves predictive climate modeling and provides more accurate forecasts and simulations that influence policy decisions (Adegbite et al., 2024). MLA analyze huge data sets from satellites and sensors, refining climate predictions and enabling better planning of climate action (Raghav et al., 2024). AI-driven solutions facilitate real-time tracking of carbon emissions and waste across the supply chain, enabling instant adjustments to improve sustainability (Blessings Ameh, 2024). Integrating IoT devices with AI enables comprehensive insights into operational impacts and identifies areas for emissions reduction. AI optimizes energy systems through smart grid management and demand forecasting, thereby reducing dependence on fossil fuels (Adegbite et al., 2024). Predictive analytics help companies model scenarios to improve resource allocation and operational efficiency, leading to significant sustainability improvements (Blessings Ameh, 2024). AI offers transformative opportunities to mitigate climate change but also presents challenges, such as high energy consumption in AI training, which can contribute to greenhouse gas emissions (Raghav et al., 2024). Balancing these aspects is crucial for the sustainable use of AI in an environmental context.

3.2 Precision Agriculture

AI-driven precision agriculture offers innovative solutions to optimize crop yields while minimizing environmental degradation. By using advanced technologies such as ML, IoT sensors, and robotics, farmers can make data-driven decisions that increase productivity and sustainability (Gupta & Bhardwaj, 2024). The adoption of AI-driven precision agriculture in developing countries is influenced by a variety of factors spanning technological, social and economic dimensions. Understanding these factors is crucial to improving the implementation of AI technologies in agriculture, which can significantly improve productivity and sustainability. AI systems can reduce water use by 25-40% while increasing crop yields by 5-15% through precise irrigation practices. Variable rate fertilization techniques enabled by AI can reduce fertilizer consumption by 30-40%, resulting in healthier soils and reducing runoff, as discussed by Hoque and Padhiary (2024). Using AI technologies can improve plant health assessments by 30-50%, enabling timely interventions to prevent losses. Additionally, AI models are capable of predicting crop yields and pest damage with up to 90% accuracy, enabling predictive management strategies. (Hoque & Padhiary, 2024).

AI-driven solutions provide real-time insights into soil nutrient levels, enabling tailored nutrient management strategies that improve soil health. Predictive analytics help detect nutrient deficiencies and enable timely interventions that improve soil health and crop productivity (Reddy et al., 2024). Smart irrigation systems use AI algorithms to ensure efficient water distribution, reduce waste and promote conservation. AI-driven analysis of soil conditions enables precise fertilization practices, improves nutrient utilization, and minimizes runoff (Adewusi et al., 2024).

The alliance of AI and precision agriculture represents a beneficial direction for addressing these challenges through better resource management, improved crop monitoring and the promotion of a more sustainable and effective agricultural landscape. The collaboration between ML and decision support systems (DSS) in precision agriculture has proved to be an important development, giving farmers access to data-based insights and forecasts (Karunathilake et al., 2023; Akhter & Sofi, 2022).

3.3 Renewable Energy Optimization

In recent years, the renewable energy landscape has experienced unprecedented growth, driven by increasing awareness of climate change and a shared commitment to reducing dependence on fossil fuels (Burke and Stephens, 2018). However, the intermittent nature of renewable energy sources such as solar and wind present operational challenges that require innovative solutions.

Integrating AI-driven renewable energy optimization into the energy sector offers numerous benefits and increases efficiency, reliability and sustainability. Through the use of advanced algorithms, AI optimizes energy production, predictive maintenance and grid management, resulting in significant operational improvements. AI techniques such as ML and deep neural networks can predict equipment failures with up to 92% accuracy, reducing unplanned downtime by 35% (Bello et al., 2024). Data from renewable assets is analyzed through AI-driven predictive maintenance tools, ensuring continuous energy production and reducing maintenance costs. AI in smart grids promotes energy-saving behavior, which reduces carbon footprint and supports sustainability goals (Manuel et al. in 2024). AI examines massive amounts of sensor data and past performance using MLA to find trends that indicate potential problems. By extending the life of renewable energy infrastructure and reducing downtime, this proactive approach improves reliability and saves significant money.

The health of the energy network is continuously assessed through dynamic optimization algorithms that adapt electricity distribution to demand while reducing losses. This is particularly important when including variable output renewable energy sources such as solar and wind. Dynamic optimization algorithms control the charging and discharging cycles in energy storage systems, taking into account supply and demand conditions in real-time. This reduces waste and guarantees effective use of the stored energy. Thanks to the ability of dynamic optimization algorithms to instantly change operating parameters, energy systems can respond quickly to fluctuations in demand or environmental conditions (Xu et al., 2020).

AI improves grid stability and reduces dependence on non-renewable energy sources by optimizing energy distribution and storage and balancing supply and demand (Darwish et al., 2024). When AI is integrated into photovoltaic and energy storage systems, overall efficiency can increase by up to 28% compared to traditional techniques. By 2030, energy costs for solar-plus-storage projects could be reduced by 25% using AI applications (Shen et al., 2024). Real-time monitoring and sophisticated data analysis enable AI algorithms to adapt to changing environmental conditions, predict patterns of energy production and allocate resources as efficiently as possible. This guarantees that renewable energy sources produce the most energy and increases their competitiveness compared to conventional energy sources.

3.4 Waste Management and Recycling

Integrating AI-driven renewable energy optimization into waste management and recycling has a significant impact on greenhouse gas emissions in the energy sector. By using ML and other AI technologies, these systems increase the efficiency of renewable energy production from waste, resulting in lower energy consumption and emissions. AI-driven systems in waste incineration plants can achieve a 36.3% reduction in greenhouse gas emissions while generating significant energy yields (Hai et al., 2022). AI algorithms can optimize the recycling process, resulting in a reduction in energy consumption of up to 30% (Fernandez et al., 2024). Improved energy forecasting and grid management through AI minimizes energy waste and improves the integration of renewable sources (Nnajiofor et al., 2024). Predictive maintenance and real-time monitoring in renewable energy plants help prevent equipment failures, maintaining high operational efficiency and reducing dependence on fossil fuels. Despite the benefits, challenges such as the initial cost of AI implementation and the need for skilled personnel remain. Addressing these issues is critical to maximizing the potential of AI in reducing emissions and improving sustainability in the energy sector.

4. GREEN BIOETHICS

Green bioethics, a rapidly evolving field, combines the principles of environmental sustainability and ethical considerations in the context of healthcare and the broader environment. As AI continues to revolutionize various industries, its application in the field of green bioethics is becoming increasingly important (Pachot & Patissier, 2023).

The use of AI in the health and environmental sectors has numerous advantages, as current studies show. AI-driven smart health products can increase transparency, mitigate bias, and protect privacy while improving security and promoting equity and autonomy (Pasricha, 2022). In addition, AI is used to develop more sustainable production methods and model climate change, making it a valuable tool in the fight against environmental degradation (Pachot & Patissier, 2023). The principles of green bioethics have a significant impact on the development of sustainable healthcare practices by integrating environmental considerations into clinical ethics and healthcare operations (Table 1). This approach emphasizes the importance of environmental sustainability, patient autonomy, and institutional responsibility, ultimately promoting a healthcare system that prioritizes both health and environmental stewardship. Green bioethics advocates for incorporating environmental impacts into healthcare decision-making and encourages providers to inform patients about the ecological consequences of treatments (Resnik & Pugh, 2023). This perspective expands the ethical framework beyond individual patient values to include institutional responsibilities and promote sustainability at a systemic level (Herlitz et al., 2022).

The healthcare sector's significant impact on the environment requires ethical responsibility in adopting environmentally friendly measures that can drive policy changes and improve sustainability (Chen-Xu, 2022). Implementing energy-efficient practices and waste management strategies can improve both environmental and patient care outcomes (Rana, 2024). In her hypothesis, Richie uses concepts from environmental ethics to introduce a framework for green bioethics that assesses the sustainability of medical practices, techniques, and advances. The model consists of four guiding principles: Distributive justice considers the moral community comprehensively and insists on providing essential medical resources before addressing the needs of specific stakeholders. The emphasis is on conserving resources to meet basic health needs before addressing additional needs; adopting a minimalist approach to reduce reliance on medical treatments; and supporting ethical business practices to improve patient-centered care rather than profit-driven motives (Olawade et al., 2024).

Table 1. Green bioethics in health sustainability

Key Term	Main Idea	Explanation	Examples
Environmental Justice	Ethical distribution of health resources	Advocates for fair access to health resources while considering environmental impacts.	Addressing health disparities in areas affected by pollution or climate change.
Sustainable Healthcare	Eco-friendly healthcare practices	Focus on reducing the environmental footprint of healthcare systems.	Green hospitals use renewable energy; reducing medical waste.
Climate-Health Nexus	The link between climate change and health	Examines how climate change affects public health and promotes preventive measures.	Policies addressing heatwaves, vector-borne diseases, and air pollution.
Biodiversity and Medicine	Ethical sourcing of medical resources	Ensures sustainable and ethical use of biodiversity for drug discovery and healthcare innovations.	Development of antibiotics from rainforest plants while conserving ecosystems.
Circular Economy in Health	Waste reduction and resource reuse in health systems	Promotes recycling and sustainable resource use in healthcare operations.	Recycling medical plastics; designing biodegradable medical devices.
Bioethics of Technology	Evaluating health technologies for sustainability	Focuses on the ethical implications of AI and biotechnology in achieving health and environmental sustainability.	Use of AI in optimizing resource use in telemedicine; ethical considerations in genetic editing.

continued on following page

Table 1. Continued

Key Term	Main Idea	Explanation	Examples
Public Health Advocacy	Promoting sustainable health policies	Encourages policies that address both human health and environmental well-being.	Campaigns for clean air regulations to prevent respiratory illnesses.
Green Pharmacy	Developing eco-friendly pharmaceuticals	Focuses on designing drugs with minimal environmental impact during production and disposal.	Biodegradable drugs reduce aquatic pollution; eco-friendly drug packaging.
Renewable Energy in Healthcare	Transitioning healthcare systems to renewables	Advocates the use of renewable energy sources in health facilities to reduce carbon footprints.	Solar-powered rural clinics; and hospitals powered by wind energy.
Global Ethics Framework	Collaborative global health sustainability	Proposes a global ethical framework to ensure equitable and sustainable health initiatives.	Paris Agreement addressing health impacts of climate change; global vaccine equity efforts.

However, integrating AI into these sensitive areas also presents unique ethical challenges. Stakeholders need to carefully consider the organizational factors that influence the implementation of AI, such as: the availability of specialist staff, data management and the alignment of AI systems with organizational goals (Lebcir et al., 2021). Towards a more sustainable future, the application of AI in green bioethics must be guided by a comprehensive understanding of environmental impacts, social impacts and ethical considerations.

This ethical framework serves as a guide for healthcare professionals and researchers in the responsible conduct of clinical trials, with an emphasis on the need to respect the autonomy and rights of individual participants, whether they are in a group or acting independently. It is critical to establish research methods that minimize risks to individuals while maximizing benefits to society and to ensure fairness when recruiting participants for research initiatives.

5. CHALLENGES AND ETHICAL CONSIDERATIONS

Implementing AI in the health and environmental sectors presents significant ethical challenges that require careful navigation. These dilemmas include issues such as privacy, algorithmic bias, and the need for transparency and accountability. Addressing these concerns is critical to increasing trust and ensuring equitable access to AI technologies.

The integration of AI raises concerns about protecting sensitive patient information and requires robust frameworks such as blockchain for secure data management. AI systems can also perpetuate existing algorithm biases, leading to unequal treatment outcomes. Patient confidentiality is a fundamental ethical principle in healthcare that protects the privacy of patient information. Ensuring fairness in AI algorithms is critical to protecting patient rights. There is an urgent need for clear guidelines for AI decision-making processes to preserve professional autonomy and patient trust (Gami et al., 2023). As AI systems become increasingly integrated into healthcare, it is essential to establish solid ethical frameworks to guide their development, deployment, and use. Ethical frameworks serve as guiding principles and standards that govern the ethical behavior and decision-making of individuals, organizations, and industries (Olorunsogo et al., 2024).

The role of AI in environmental management needs to be reviewed to prevent misuse and ensure that decisions are made transparently and ethically. Ethical frameworks should prioritize sustainable practices and equitable access to AI technologies in environmental applications (Gadani & Bhattacharya, 2024). Although the ethical implications of AI are complex, they also provide opportunities for innovation and improved outcomes in both healthcare and environmental management. Balancing these aspects is crucial for responsible AI use. The implementation of AI-driven healthcare solutions in clinical settings is preceded by the identification and mitigation of potential risks and ethical concerns through independent oversight, guidance, and ethical review by ethics committees (Murphy et al., 2021).

6. FUTURE TRENDS AND OPPORTUNITIES

6.1 AI for Sustainable Development Goals (SDGs)

Integrating AI into sustainable development, particularly in the areas of environment and health, poses several challenges and limitations that must be addressed to maximize its potential. These challenges include ethical concerns, privacy issues, and the need for interdisciplinary collaboration, which are critical to responsible AI use. As the world grapples with pressing environmental and health challenges, the potential of AI to contribute to sustainable development has attracted significant attention. While AI offers promising solutions, its implementation also presents a number of several unique challenges and limitations that must be carefully considered.

A key challenge is the energy-intensive nature of AI systems. The high computing power required to train and run AI models can result in significant energy consumption and carbon emissions, potentially negating the environmental benefits of AI-driven solutions. (Pachot & Patissier, 2023) This paradox highlights the need for more energy-efficient AI architectures and careful consideration of the carbon footprint of AI deployment. (Pachot & Patissier, 2023). Another limitation is the potential for data bias and inaccuracy in AI-driven decision-making. AI systems are only as good as the data they are trained on. If this data is incomplete, distorted, or inaccurate, the resulting decisions and recommendations may be flawed, leading to unintended environmental and health consequences. In addition, the long-term impact of AI on social sustainability and intergenerational equity must be carefully examined. AI-based decision support systems that prioritize short-term economic benefits over long-term environmental and health considerations can have far-reaching impacts on future generations (Rohde et al., 2023)

AI systems can perpetuate algorithmic biases and lead to unfair outcomes in resource allocation and health interventions. The ethical implications of AI decision-making processes require careful consideration to ensure equal access and outcomes. The collection and analysis of large amounts of data raises significant privacy issues, especially in healthcare applications that involve sensitive information (Farahani & Ghasemi, 2024; Nazer et al., 2023). Ensuring data security and compliance is critical to maintaining public trust. Successful implementation of AI technologies requires specialized technical skills that may not be readily available in all regions, especially in developing countries (Farahani & Ghasemi, 2024). Bridging the digital divide is of critical to ensuring that the benefits of AI are distributed fairly.

6.2 Innovations in AI for Health and Environment

AI-powered data analysis and visualization play a critical role in identifying and mitigating the environmental and health impacts of human activities on ecosystems. By using advanced technologies such as ML and data analysis, AI improves the accuracy and efficiency of environmental assessments and enables companies to make informed decisions to promote sustainability. AI automates data collection from various sources, including IoT sensors and satellite imagery, improving the accuracy of environmental impact assessments. MLA can detect patterns and predict trends, enabling proactive measures to mitigate negative impacts on ecosystems. AI systems facilitate real-time monitoring of environmental conditions and enable timely reporting and compliance (Popescu et al., 2024). Self-organizing AI models can analyze visual data to detect subtle changes in landscapes and correlate them with human influences (Dresp-Langley, 2024). AI plays a crucial role in detecting and managing environmental pollution such as: Heavy metal contamination by identifying sources and guiding remediation strategies. Integrating AI into sustainable practices across various sectors helps reduce future environmental impacts (Ashkanani et al., 2024). AI offers significant advantages in environmental management. Challenges such as data protection and algorithmic transparency must be addressed to ensure ethical implementation and maximize their potential for sustainable development (Zhuk, 2023).

7. CONCLUSION

AI-powered predictive analytics can significantly optimize healthcare outcomes while minimizing AI-powered predictive analytics can significantly optimize healthcare outcomes while minimizing environmental impact through various strategies to improve efficiency and sustainability. By integrating AI with green computing practices, healthcare organizations can improve diagnostic accuracy, reduce resource waste, and streamline operations. AI algorithms process large amounts of medical data, resulting in improved diagnostic accuracy and early detection of diseases (Alowais et al., 2023). Predictive models can estimate disease progression, enabling timely interventions and personalized treatment plans (Hossain et al., 2024). Adopting energy-efficient AI models and green computing practices can reduce the carbon footprint of healthcare. Implementing telemedicine and remote monitoring reduces the need for physical visits, thereby reducing transportation emissions. Continuous monitoring and improvement of AI systems can ensure responsible data management and minimize e-waste (Olawade et al., 2024). AI-driven decision-making systems can be designed to prioritize human well-being and environmental sustainability through a multi-faceted approach that integrates ethical considerations, stakeholder engagement and innovative technologies. The aim is to create framework conditions that not only increase efficiency but are also aligned with social and ecological goals. AI technologies optimize resource use and reduce waste in sectors such as agriculture, energy, and urban planning, exemplified by smart farming initiatives such as Iron Ox (R. C. Santos & Cagica Carvalho, 2025). These applications demonstrate the potential of AI to effectively address global environmental challenges. Ethical use of AI, guided by principles of transparency and accountability, is critical to ensuring that the technology serves the best interests of humanity. Integrating AI into healthcare offers numerous benefits, but it is important to consider the ethical implications and environmental concerns associated with its use. Balancing technological advancement and sustainability will be critical to the future of healthcare and environmental sustainability.

REFERENCES

Abbas, M. K. G., Rassam, A., Karamshahi, F., Abunora, R., & Abouseada, M. (2024). The Role of AI in Drug Discovery. *ChemBioChem*, *14*(14), e202300816. Advance online publication. DOI: 10.1002/cbic.202300816 PMID: 38735845

Adegbite, A. O. (2024). Artificial Intelligence in Climate Change Mitigation: A Review of Predictive Modeling and Data-Driven Solutions for Reducing Greenhouse Gas Emissions. *World Journal of Advanced Research and Reviews*, *1*(1), 408–414. DOI: 10.30574/wjarr.2024.24.1.3043

Adewusi, A. O. (2024). AI in precision agriculture: A review of technologies for sustainable farming practices. *World Journal of Advanced Research and Reviews*, *1*(1), 2276–2285. DOI: 10.30574/wjarr.2024.21.1.0314

Akhter, R., & Sofi, S. A. (2022). Precision agriculture using IoT data analytics and machine learning. *Journal of King Saud University. Computer and Information Sciences*, *8*(8), 5602–5618. DOI: 10.1016/j.jksuci.2021.05.013

Alowais, S. A., Alghamdi, S. S., Alsuhebany, N., Alqahtani, T., Alshaya, A. I., Almohareb, S. N., Aldairem, A., Alrashed, M., Bin Saleh, K., Badreldin, H. A., Al Yami, M. S., Al Harbi, S., & Albekairy, A. M. (2023). Revolutionizing healthcare: The role of artificial intelligence in clinical practice. *BMC Medical Education*, *1*(1), 689. Advance online publication. DOI: 10.1186/s12909-023-04698-z PMID: 37740191

Ameh, B. (2024). Digital tools and AI: Using technology to monitor carbon emissions and waste at each stage of the supply chain, enabling real-time adjustments for sustainability improvements. *International Journal of Science and Research Archive*, *1*(1), 2741–2757. DOI: 10.30574/ijsra.2024.13.1.1995

Andrew, A. (2024). Potential applications and implications of large language models in primary care. *Family Medicine and Community Health*, *e002602*(Suppl 1), e002602. DOI: 10.1136/fmch-2023-002602 PMID: 38290759

Arányi, P. (2018). Ethical aspects of drug development. *Microchemical Journal*, *244–246*, 244–246. Advance online publication. DOI: 10.1016/j.microc.2016.11.015

Ashkanani, Z., Mohtar, R., Al-Enezi, S., Smith, P. K., Calabrese, S., Ma, X., & Abdullah, M. (2024). AI-assisted systematic review on remediation of contaminated soils with PAHs and heavy metals. *Journal of Hazardous Materials*, *133813*, 133813. Advance online publication. DOI: 10.1016/j.jhazmat.2024.133813 PMID: 38402679

Baig, M. A. A., Huang, J., & Jingzehua, Xu. (2024). *AI for United Nations Sustainable Development Goals*. https://doi.org/DOI: 10.31224/3866

Bello, S. F.Sakiru Folarin BelloIfeoluwa Uchechukwu WadaOlukayode B IgeErnest C ChianumbaSamod Adetunji Adebayo. (2024). AI-driven predictive maintenance and optimization of renewable energy systems for enhanced operational efficiency and longevity. *International Journal of Science and Research Archive*, *1*(1), 2823–2837. DOI: 10.30574/ijsra.2024.13.1.1992

Bolón-Canedo, V., Morán-Fernández, L., Cancela, B., & Alonso-Betanzos, A. (2024). A review of green artificial intelligence: Towards a more sustainable future. *Neurocomputing*, *128096*, 128096. Advance online publication. DOI: 10.1016/j.neucom.2024.128096

Burke, M. J., & Stephens, J. C. (2018). Political power and renewable energy futures: A critical review. *Energy Research &amp. Social Science*, *35*, 78–93. DOI: 10.1016/j.erss.2017.10.018

Chen, L., Chen, Z., Zhang, Y., Liu, Y., Osman, A. I., Farghali, M., Hua, J., Al-Fatesh, A., Ihara, I., Rooney, D. W., & Yap, P.-S. (2023). Artificial intelligence-based solutions for climate change: A review. *Environmental Chemistry Letters*, *5*(5), 2525–2557. DOI: 10.1007/s10311-023-01617-y

Chen-Xu, J. (2022). *Ethical Responsibility in Environmentally Sustainable Health Care*. https://doi.org/DOI: 10.1590/SciELOPreprints.4691

Chowdhury, M. A., Rizk, R., Chiu, C., Zhang, J. J., Scholl, J. L., Bosch, T. J., Singh, A., Baugh, L. A., McGough, J. S., Santosh, K., & Chen, W. C. W. (2025). The Heart of Transformation: Exploring Artificial Intelligence in Cardiovascular Disease. *Biomedicines*, *2*(2), 427. DOI: 10.3390/biomedicines13020427 PMID: 40002840

Darwish, A. S. K., Abbas, M. K., Al-Salim, W. L., & Al-Tameemi, M. R. J. (2024). Artificial Intelligence for Sustainable Energy Transition: Optimising Renewable Energy Integration and Management. *ARID International Journal for Science and Technology*, 55–79. https://doi.org/DOI: 10.36772/arid.aijst.2024.7134

Dresp-Langley, B. (2024). Explainable Self-Organizing Artificial Intelligence Captures Landscape Changes Correlated with Human Impact Data. *JCSSD*, *1*(1). Advance online publication. DOI: 10.17303/jcssd.2024.3.103

Faiz, F., Ninduwezuor-Ehiobu, N., Adanma, U. M., & Solomon, N. O. (2024). AI-Powered waste management: Predictive modeling for sustainable landfill operations. Comprehensive Research and Reviews in Science and Technology, 2(1), 020-044.

Familoni, B. T. (2024). Ethical frameworks for AI in healthcare entrepreneurship: A theoretical examination of challenges and approaches. *International Journal of Frontiers in Biology and Pharmacy Research, 1*, 057–065. DOI: 10.53294/ijfbpr.2024.5.1.0032

Farahani, M. S., & Ghasemi, G. (2024). How artificial intelligence plays a role in achieving sustainable development goals? *Sustainable Economies*, *3*(3), 66. DOI: 10.62617/se.v2i3.66

Fernandez, M., Faturahman, A., & Santoso, N. A. (2024). Harnessing Machine Learning to Optimize Renewable Energy Utilization in Waste Recycling. [ITEE]. *International Transactions on Education Technology*, *2*(2), 173–182. DOI: 10.33050/itee.v2i2.595

Fnu, N., Zeb, S., Abbasi, N., Umer Qayyum, M., & Fahad, M.FNU. (2024). AI in Healthcare: ChatGPT's Significance in Transforming Patient-Physician Communication and Clinical Assistance. *Asian Journal of Engineering. Socialism and Health*, *9*(9), 2058–2075. DOI: 10.46799/ajesh.v3i9.434

Gadani, N. N., & Bhattacharya, P. (2024). Ethical Considerations in AI Development for Cloud Computing and Data-Driven Software Solutions. In *Advances in Human and Social Aspects of Technology* (pp. 23–58). IGI Global. http://dx.doi.org/DOI: 10.4018/979-8-3693-4147-6.ch002

Gami, B., Agrawal, M., Mishra, D. K., Quasim, D., & Mehra, P. S. (2023). Artificial intelligence-based blockchain solutions for intelligent healthcare: A comprehensive review on privacy preserving techniques. *Transactions on Emerging Telecommunications Technologies*, *9*(9), e4824. Advance online publication. DOI: 10.1002/ett.4824

García Villena, E., Pascual Barrera, A., Álvarez, R. M., Dzul López, L. A., Tutusaus Pifarré, K., Vidal Mazón, J. L., Miró Vera, Y. A., Brie, S., & López Flores, M. A. (2022). Evaluation of the Sustainable Development Goals in the Diagnosis and Prediction of the Sustainability of Projects Aimed at Local Communities in Latin America and the Caribbean. *Applied Sciences (Basel, Switzerland)*, *21*(21), 11188. DOI: 10.3390/app122111188

Ghule, P. J., Kale, P. A., & Kale, S. P. (2024). Ai-enhanced drug discovery and pharmaceutical development. Futuristic Trends in Pharmacy & Nursing. *IIP: Chikkamagaluru, India*, *3*, 254–267.

Green, D., Fitzmaurice, T., Eason, M., Frost, F., Nazareth, D., & Walshaw, M. (2023). Environmental impact of a CF virtual clinic. *M-Health/e-Health*, PA2905. https://doi.org/DOI: 10.1183/13993003.congress-2023.PA2905

Gupta, S. P., & Bhardwaj, A. (2024). Technology of Sensors. In *Advances in Environmental Engineering and Green Technologies* (pp. 1–9). IGI Global. DOI: 10.4018/979-8-3693-1930-7.ch001

Hai, T., Alsharif, S., Dhahad, H. A., Attia, E.-A., Shamseldin, M. A., & Najat Ahmed, A. (2022). The evolutionary artificial intelligence-based algorithm to find the minimum GHG emission via the integrated energy system using the MSW as fuel in a waste heat recovery plant. *Sustainable Energy Technologies and Assessments, 102531*, 102531. Advance online publication. DOI: 10.1016/j.seta.2022.102531

Haleem, A., Javaid, M., Khan, I. H., & Mohan, S. (2023). Significant Applications of Artificial Intelligence Towards Attaining Sustainability. *Journal of Industrial Integration and Management, 04*(4), 489–520. DOI: 10.1142/S2424862223500331

Heidelberg, M., Kelman, A., Hopkins, J., & Allen, M. E. (2020). The evolution of data ethics in clinical research and drug development. *Ethics, Medicine, and Public Health, 100517*, 100517. Advance online publication. DOI: 10.1016/j.jemep.2020.100517

Herlitz, A., Malmqvist, E., & Munthe, C. (2022). 'Green' bioethics widens the scope of eligible values and overrides patient demand: Comment on Parker. *Journal of Medical Ethics, 2*(2), 100–101. DOI: 10.1136/jme-2022-108849 PMID: 36564173

Hoque, A., & Padhiary, M. (2024). Automation and AI in Precision Agriculture: Innovations for Enhanced Crop Management and Sustainability. *Asian Journal of Research in Computer Science, 10*(10), 95–109. DOI: 10.9734/ajrcos/2024/v17i10512

Hossain, S., Hossain, S., Khadka, U., Sarkar, S., & Khan, N. (2024). Ai-driven predictive Analytics, healthcare outcomes, cost reduction, machine learning, and patient monitoring. *Advanced International Journal of Multidisciplinary Research, 2*(5), 1104. Advance online publication. DOI: 10.62127/aijmr.2024.v02i05.1104

Jebbor, I., Benmamoun, Z., & Hachmi, H. (2024). Revolutionizing cleaner production: The role of artificial intelligence in enhancing sustainability across industries. *Journal of Infrastructure. Policy and Development, 10*(10), 7455. DOI: 10.24294/jipd.v8i10.7455

Kagzi, M., Khanra, S., & Paul, S. K. (2023). Machine learning for sustainable development: Leveraging technology for a greener future. *Journal of Systems and Information Technology, 4*(4), 440–479. DOI: 10.1108/JSIT-11-2022-0266

Karunathilake, E. M. B. M., Le, A. T., Heo, S., Chung, Y. S., & Mansoor, S. (2023). The Path to Smart Farming: Innovations and Opportunities in Precision Agriculture. *Agriculture, 8*(8), 1593. DOI: 10.3390/agriculture13081593

Khan, I. R., Grover, V., Fatima, I., Alam, M., & Ahmad, N. (2024). The Application of AI in Drug Discovery. In *Advances in Healthcare Information Systems and Administration* (pp. 181–198). IGI Global. http://dx.doi.org/DOI: 10.4018/979-8-3693-5468-1.ch010

Kokudeva, M., Vichev, M., Naseva, E., Miteva, D. G., & Velikova, T. (2024). Artificial intelligence as a tool in drug discovery and development. *World Journal of Experimental Medicine*, *3*(3). Advance online publication. DOI: 10.5493/wjem.v14.i3.96042 PMID: 39312699

Lathan, R., Hitchman, L., Walshaw, J., Ravindhran, B., Carradice, D., Smith, G., Chetter, I., & Yiasemidou, M. (2024). Telemedicine for sustainable postoperative follow-up: A prospective pilot study evaluating the hybrid life-cycle assessment approach to carbon footprint analysis. *Frontiers in Surgery*, *11*, 1300625. Advance online publication. DOI: 10.3389/fsurg.2024.1300625 PMID: 38562585

Lebcir, R., Hill, T., Atun, R., & Cubric, M. (2021). Stakeholders' views on the organisational factors affecting application of artificial intelligence in healthcare: A scoping review protocol. *BMJ Open*, *3*(3), e044074. DOI: 10.1136/bmjopen-2020-044074 PMID: 33753441

Mansfield, D., & Montazeri, A. (2024). A survey on autonomous environmental monitoring approaches: Towards unifying active sensing and reinforcement learning. *Frontiers in Robotics and AI*, *11*, 1336612. Advance online publication. DOI: 10.3389/frobt.2024.1336612 PMID: 38533524

Mohammadi, T., Özkal, A., & Özkal, A. S. (2024). Artificial Intelligence in Medicine: Opportunities and Challenges. *Black Sea Journal of Engineering and Science*, *5*(5), 1092–1099. DOI: 10.34248/bsengineering.1499831

Mohammadi, T. D., Özkal, A., & Özkal, A. S. (2024). Artificial Intelligence in Medicine: Opportunities and Challenges. *Black Sea Journal of Engineering and Science*, *5*(5), 1092–1099. DOI: 10.34248/bsengineering.1499831

Murphy, K., Di Ruggiero, E., Upshur, R., Willison, D. J., Malhotra, N., Cai, J. C., Malhotra, N., Lui, V., & Gibson, J. (2021). Artificial intelligence for good health: A scoping review of the ethics literature. *BMC Medical Ethics*, *22*(1), 14. DOI: 10.1186/s12910-021-00577-8 PMID: 33588803

Nalongo, J. A. (2024). The Use of AI in Enhancing Patient Communication. *Research Invention Journal of Scientific and Experimental Sciences*, *2*(2), 45–48. DOI: 10.59298/RIJSES/2024/424548

Neha, F., Bhati, D., Shukla, D. K., & Amiruzzaman, M. (2024). *ChatGPT: Transforming Healthcare with AI*. https://doi.org/DOI: 10.20944/preprints202410.0686.v1

Nilashi, M., Asadi, S., Abumalloh, R. A., Samad, S., Ghabban, F., Supriyanto, E., & Osman, R. (2021). Sustainability Performance Assessment Using Self-Organizing Maps (SOM) and Classification and Ensembles of Regression Trees (CART). *Sustainability (Basel)*, *7*(7), 3870. DOI: 10.3390/su13073870

Olawade, D. B., Fapohunda, O., Wada, O. Z., Usman, S. O., Ige, A. O., Ajisafe, O., & Oladapo, B. I. (2024). Smart waste management: A paradigm shift enabled by artificial intelligence. *Waste Management Bulletin*, *2*(2), 244–263. DOI: 10.1016/j.wmb.2024.05.001

Olorunsogo, T. O., Anyanwu, A., Abrahams, T. O., Olorunsogo, T., Ehimuan, B., & Reis, O. (2024). Emerging technologies in public health campaigns: Artificial intelligence and big data. *International Journal of Science and Research Archive*, *11*(1), 478–487. DOI: 10.30574/ijsra.2024.11.1.0060

Pachiappan, S., Gnanasambandam, A., Preethi Samundi, S., Parameshwaran, S. P., Hassan Ali, A. F., & Saravanan, P. P. (2023). Remote Monitoring In Clinical Trial – A Guide for Healthcare Professionals. *Asian Journal of Pharmaceutical Research and Development*, *4*(4), 139–142. DOI: 10.22270/ajprd.v11i4.1300

Pachot, A., & Patissier, C. (2023). *Towards Sustainable Artificial Intelligence: An Overview of Environmental Protection Uses and Issues*. Green and Low-Carbon Economy., DOI: 10.47852/bonviewGLCE3202608

Pan, Y., & Jiao, F.-Y. (2024). Application of artificial intelligence in the diagnosis and treatment of Kawasaki disease. *World Journal of Clinical Cases*, *23*(23), 5304–5307. DOI: 10.12998/wjcc.v12.i23.5304 PMID: 39156094

Pasricha, S. (2023). AI Ethics in Smart Healthcare. *IEEE Consumer Electronics Magazine*, *4*(4), 12–20. DOI: 10.1109/MCE.2022.3220001

Pitel, E., Lea u, F., Nicolau, A., & Rogozea, L. (2024). Ethical aspects in the use of artificial intelligence in the process of drug development I [Aspecte etice în utilizarea inteligen ei artificiale în procesul de dezvoltare a medicamentelor]. *Jurnal Medical Brasovean*, •••, 53–61. DOI: 10.31926/jmb.2023.2.8

Popescu, S. M., Mansoor, S., Wani, O. A., Kumar, S. S., Sharma, V., Sharma, A., Arya, V. M., Kirkham, M. B., Hou, D., Bolan, N., & Chung, Y. S. (2024). Artificial intelligence and IoT driven technologies for environmental pollution monitoring and management. *Frontiers in Environmental Science*, *12*, 1336088. Advance online publication. DOI: 10.3389/fenvs.2024.1336088

Pratap, S., & Venkatesh, K. S. (2024). The Role of AI in Enhancing Green Management and Advancing Digital Lean Practices for Sustainable Efficiency. *ShodhKosh: Journal of Visual and Performing Arts*, *6*(6). Advance online publication. DOI: 10.29121/shodhkosh.v5.i6.2024.1725

Raghav, A., Singh, B., Jermsittiparsert, K., Raghav, R., & Yadav, U. (2024). Artificial Intelligence in Environmental and Climate Changes: A Sustainable Future. In Maintaining a Sustainable World in the Nexus of Environmental Science and AI (pp. 485-506). IGI Global.

Rahate, K. P., & Mondal, R. (2024). Applications of AI in Drug Discovery. In *Advances in Medical Technologies and Clinical Practice* (pp. 86–120). IGI Global. http://dx.doi.org/DOI: 10.4018/979-8-3693-2238-3.ch005

Chowdhury, R. H. (2024). Intelligent systems for healthcare diagnostics and treatment. World Journal of Advanced Research and Reviews, 23(1), 007-015.

Raman, R., Gunasekar, S., Kaliyaperumal, D., & Nedungadi, P. (2024). Navigating the Nexus of Artificial Intelligence and Renewable Energy for the Advancement of Sustainable Development Goals. *Sustainability (Basel)*, *21*(21), 9144. DOI: 10.3390/su16219144

Rebecca, B., Kumar, K. P. M., Padmini, S., Srivastava, B. K., Halder, S., & Boopathi, S. (2023). Convergence of Data Science-AI-Green Chemistry-Affordable Medicine. In *Advances in Computational Intelligence and Robotics* (pp. 348–373). IGI Global. http://dx.doi.org/DOI: 10.4018/978-1-6684-9999-3.ch014

Reddy, K. P., Thanka, M. R., Edwin, E. B., Ebenezer, V., & Joy, P. (2024, April). Farm_era. Precision farming with GIS, AI pest management, smart irrigation, data analytics, and optimized crop planning. In *2024 International Conference on Inventive Computation Technologies (ICICT)* (pp. 1134-1140). IEEE.

Resnik, D. B., & Pugh, J. (2023). Green bioethics, patient autonomy and informed consent in healthcare. *Journal of Medical Ethics*, *7*(7), 489–493. DOI: 10.1136/jme-2023-109404 PMID: 37833040

Santos, R. C., M., & Cagica Carvalho, L. (. (2025). AI-driven participatory environmental management: Innovations, applications, and future prospects. *Journal of Environmental Management*, *123864*. Advance online publication. DOI: 10.1016/j.jenvman.2024.123864

Sharma, P., & Vaid, U. (2021). Emerging role of artificial intelligence in waste management practices. *IOP Conference Series. Earth and Environmental Science*, *1*(1), 012047. DOI: 10.1088/1755-1315/889/1/012047

Shen, Q., Wen, X., Xia, S., Zhou, S., & Zhang, H. (2024). AI-Based Analysis and Prediction of Synergistic Development Trends in U.S. Photovoltaic and Energy Storage Systems. *International Journal of Innovative Research in Computer Science and Technology*, *5*(5), 36–46. DOI: 10.55524/ijircst.2024.12.5.6

Singh, B., Chandra, A., Joshi, D., Semwal, N., Kukreti, G., & Saxena, U. (2024). Application of Artificial Intelligence Techniques in Healthcare. In *AI in the Social and Business World: A Comprehensive Approach* (pp. 67–101). Bentham Science Publishers., DOI: 10.2174/9789815256864124010005

Singpai, B., & Wu, D. (2020). Using a DEA–AutoML Approach to Track SDG Achievements. *Sustainability (Basel)*, *23*(23), 10124. DOI: 10.3390/su122310124

Sirmacek, B., Gupta, S., Mallor, F., Azizpour, H., Ban, Y., Eivazi, H., Fang, H., Golzar, F., Leite, I., Melsion, G. I., Smith, K., Fuso Nerini, F., & Vinuesa, R. (2023). The Potential of Artificial Intelligence for Achieving Healthy and Sustainable Societies. In *Philosophical Studies Series* (pp. 65–96). Springer International Publishing., DOI: 10.1007/978-3-031-21147-8_5

Topol, E. J. (2019). *Deep Medicine:How artificial intelligence can make healthcare human again*. Hachette Book Group.

Triplett, W. J. (2024). Artificial Intelligence and ChatGPT Models in Healthcare. In *Advances in Logistics, Operations, and Management Science* (pp. 35–60). IGI Global. http://dx.doi.org/DOI: 10.4018/979-8-3693-7327-9.ch003

Valli, L. N., Sujatha, N., Mech, M., & Lokesh, V. S. (2024). Ethical considerations in data science: Balancing privacy and utility. International Journal of Science and Research Archive, 11(1), 011-022.

Vijayanand, D. (2023). From data to impact: Machine learning models for sustainable development region classification- accelerating the achievement of Sustainable Development Goals through predictive analytics. *International Journal for Multidisciplinary Research*, *5*(5), 7180. Advance online publication. DOI: 10.36948/ijfmr.2023.v05i05.7180

Warner, E., Lee, J., Hsu, W., Syeda-Mahmood, T., Kahn, C. E., Gevaert, O., & Rao, A. (2024). Multimodal Machine Learning in Image-Based and Clinical Biomedicine: Survey and Prospects. *International Journal of Computer Vision*, *9*(9), 3753–3769. DOI: 10.1007/s11263-024-02032-8 PMID: 39211895

Wen, B., Norel, R., Liu, J., Stappenbeck, T., Zulkernine, F., & Chen, H. (2024). Leveraging Large Language Models for Patient Engagement: The Power of Conversational AI in Digital Health. *ArXiv*. https://doi.org/DOI: 10.1109/ICDH62654.2024.00027

Wenteler, A., Cabrera, C. P., Wei, W., Neduva, V., & Barnes, M. R. (2024). AI approaches for the discovery and validation of drug targets. *Cambridge Prisms. Precision Medicine*, *2*, e7. Advance online publication. DOI: 10.1017/pcm.2024.4 PMID: 39258224

Xu, X., Wen, H., Lin, H., Li, Z., & Huang, C. (2024). Online detection method for variable load conditions and anomalous sound of hydro turbines using correlation analysis and PCA-adaptive-K-means. *Measurement*, *113846*, 113846. Advance online publication. DOI: 10.1016/j.measurement.2023.113846

Yadav, P., Tudela, L. A. M., & Marco-Lajara, B. (2024). The role of AI in assessing and achieving the sustainable development goals (SDGs). In *Issues of Sustainability in AI and New-Age Thematic Investing* (pp. 1–17). IGI Global.

Zhang, X., Zhang, D., Zhang, X., & Zhang, X. (2024). Artificial intelligence applications in the diagnosis and treatment of bacterial infections. *Frontiers in Microbiology*, *15*, 1449844. Advance online publication. DOI: 10.3389/fmicb.2024.1449844 PMID: 39165576

Zhuk, A. (2023). Artificial Intelligence Impact on the Environment: Hidden Ecological Costs and Ethical-Legal Issues. *Journal of Digital Technologies and Law*, *4*(4), 932–954. DOI: 10.21202/jdtl.2023.40

Chapter 15
AI in Sustainable Farming Revolutionizing Precision Agriculture

Rehan Mehmood Sabir
https://orcid.org/0009-0007-4711-8304
University of Agriculture, Faisalabad, Pakistan

Samia Jawad
https://orcid.org/0009-0004-4106-1317
University of Agriculture, Faisalabad, Pakistan

Abdul Malik
University of Agriculture, Faisalabad, Pakistan

Fawad Ali
College of Chemistry and Chemical Engineering, Harbin Normal University, China

Muhammad Huzaifa Mahmood
https://orcid.org/0000-0001-7303-1500
University of Turin, Italy

Hassan Tanveer
https://orcid.org/0009-0003-1786-637X
Depaul University, Chicago, USA

Muhammad Shoaib
The Islamia University of Bahawalpur, Pakistan

Zeshan Quddos
MNS-University of Agriculture, Multan, Pakistan

Muhammad Soban Arif
Government College University, Faisalabad, Pakistan

Sadia Azam
University of Agriculture, Faisalabad, Pakistan

Naima Nawaz
University of Agriculture, Faisalabad, Pakistan

Nalain E. Muhammad
University of Agriculture, Faisalabad, Pakistan

DOI: 10.4018/979-8-3373-0680-3.ch015

ABSTRACT

Precision agriculture has emerged as a vital approach to enhance productivity, reduce waste, and promote environmental sustainability. Artificial intelligence (AI) is enhancing precision farming with innovative technology. Artificial intelligence facilitates precise and efficient agricultural practices by integrating sensors, drones, and IoT systems for data acquisition with machine learning and AI technologies for predictive analytics, resource optimization, and pest detection. Applications encompass sustainable land management, enhanced crop monitoring, and effective resource utilization, all of which contribute to reducing environmental impact and increasing productivity. Significant concerns include data security, accessibility for small-scale farms, and the necessity of human expertise to complement artificial intelligence solutions. AI revolutionizes agriculture through technologies such as robotic weed management and climate-resilient farming systems offers promising avenues for sustainable food production and global food security.

1. INTRODUCTION

The human population of the world grew slowly until 1700 due to significantly high infant mortality rates. The first billion was attained about 1800, succeeded by the second billion in 1928 and the third billion in 1960. In 2017, the global population exceeded seven billion. The rapid population increase in recent decades is primarily attributable to improved medical care. The United Nations predicts that the global population will reach 9.7 billion by 2050 and 10.9 billion by 2100 (Roser and Rodés-Guirao, 2013). The swift increase in population in recent decades has heightened the demand for agricultural products, thus resulting in significant expansion of farming (Fróna et al., 2019). To satisfy increasing population demands for food, biofuels, and animal products, crop yield productivity must double by 2050. To attain this objective, essential agricultural yields must enhance by 2.4% annually, while they are already rising by approximately 1.3% each year (Ray et al., 2013). Nonetheless, meeting these criteria will adversely affect the environment, resulting in diminished biodiversity and heightened greenhouse gas emissions.

Conventional agricultural practices are economically and environmentally unsustainable; therefore, it is essential to maximize resource utilization, including water and soil, to achieve high-yield crops (Fróna et al., 2019). Furthermore, illnesses and insect pests persistently jeopardize crop yields. It is estimated that 20% to 40% of annual agricultural production is lost globally owing to plant diseases and insect infestations, resulting in economic losses of $220 billion and $70 billion, respectively. The extent of these losses differs worldwide and frequently results from transboundary plant pests and diseases (FAO, 2017).

Pest damage and development are influenced by the increase in global temperatures due to climate change. As temperatures escalate, the metabolic rate of insects intensifies, prompting increased food consumption and greater damage. The growth rates of certain insect species are influenced by temperature. For every degree of average global warming of the Earth's surface, global agricultural losses attributable to insect pests are anticipated to rise by 10% to 25%. Farmers have historically employed pesticides and chemical treatments to deter pests. The utilization of pesticides for agricultural protection is increasing, resulting in adverse effects on human health and heightened environmental harm to soil and groundwater (Deutsch et al., 2018).

A system that can execute such duties is vital in mitigating the overuse of pesticides and chemicals, hence diminishing environmental harm and lowering production costs linked to their application (Kartikeyan and Shrivastava, 2021). The increasing accessibility of big data analysis techniques could catalyze more study and development in precision agriculture. In addition to fostering higher-yield crops sustainably, it also seeks to enhance event forecasting, disease detection, and the management of water and soil resources. Big data is infiltrating the agricultural sector through the aggregation of information from meteorological stations, remote sensors, historical records, and publicly accessible datasets (Kamilaris et al., 2017).

Machine learning methodologies have been effectively employed across diverse domains, such as disease identification from medical imaging, picture classification on extensive datasets, autonomous vehicles (Soni et al., 2021), and scholarly research disciplines. Disease categorization from photos can be accomplished via prevalent Convolutional Neural Network (CNN) architecture. Deep learning facilitates image-based plant disease detection across various plants and diseases (Mohanty et al., 2016); the correlation between meteorological data and pest prevalence can be analyzed using Long Short-Term Memory (LSTM) networks for predicting future pest invasions (Xiao et al., 2019); insect identification on foliage can be executed through object segmentation and deep learning methodologies (Gutierrez et al., 2019).

Farmers today have access to commercial technologies and services for smart farming that extensively utilize machine learning. Several instances are provided below. Plantix, developed by the German startup Progressive Environmental and Agricultural Technologies (PETA), is an Android-based agricultural assistant that offers crop health insights and aids in the identification of plant illnesses through computer vision and deep learning methodologies. Additional instances of such apps include Agrio and Crop Diagnosis. Gamaya is a Swiss firm that provides an array of smart agricultural services utilizing image analysis from drones integrated with IoT technologies. Asian iFarmer is a firm that provides IoT-driven soil analysis and satellite images for crop monitoring. See & Spray, created by Blue River Technology in California, is a substantial two-behind herbicide sprayer that employs computer vision and deep learning algorithms to autonomously detect and identify weeds in real time, applying herbicides precisely to identified locations rather than uniformly across the entire field (Chostner, 2017).

The implementation of precision agriculture (PA) methods, utilizing the extensive array of Precision Agriculture Technologies (PATs) in agricultural operations, might significantly aid in the reduction of greenhouse gas (GHG) emissions due to:

a) Improvement of soil capacity to function as a carbon reservoir through less tillage and decreased nitrogen fertilizer.
b) Decrease in fuel consumption by reduced in-field tractor activities, resulting in direct greenhouse gas reduction.
c) Minimization of inputs for agricultural field operations (indirect GHG reduction)

Conversely, these strategies enhance farm production by optimizing agricultural inputs, resulting in higher or equivalent yields at a reduced cost compared to conventional methods. Consequently, greenhouse gas mitigation strategies that incorporate innovative technologies and methodologies across all agricultural practices (such as precision and variable rate sowing, fertilization, spraying, and irrigation) can substantially diminish the inputs contributing to greenhouse gas emissions. These measures could assist in achieving minimal climate change impact from agriculture, while ensuring that crop production is sustained or even augmented to address the imperative of food security and safety for human consumption.

The implementation of automation in agriculture using information and communication technologies and robotic devices has garnered significant interest recently. Notable advancements in agricultural practices have occurred, especially with the implementation of ICT principles. Artificial intelligence (AI) is a fundamental technology in precision agriculture (PA) due to its capacity to evaluate extensive datasets, extract valuable insights, and assist agricultural professionals in comprehending farming strategies and making precise evaluations.

2. AI IN PRECISION AGRICULTURE: CORE TECHNOLOGIES

2.1 Data Acquisition

2.1.1 Sensors (Soil Moisture, Nutrient Levels)

Remote sensing systems utilized for precision agriculture and agricultural broadly can be categorized according to (i) sensor platform and (ii) sensor type. Sensors are generally installed on satellites, aerial platforms, and terrestrial systems (Figure 2.1). Since the 1970s, satellite goods have been widely utilized for precision agriculture. Aerial platforms, encompassing airplanes and unmanned aerial vehicles (UAVs), are being utilized in public administration. Ground-based platforms utilized for precision agriculture can be classified into three categories: (i) hand-held, (ii) free-standing in the field, and (iii) installed on tractors or agricultural machines. Ground-based systems, also known as proximal remote sensing systems, are situated close to the target surface (land or vegetation) in contrast to aerial or satellite-based platforms.

Figure 1. Layout for the remote, aircraft, and ground sensing systems for PA.

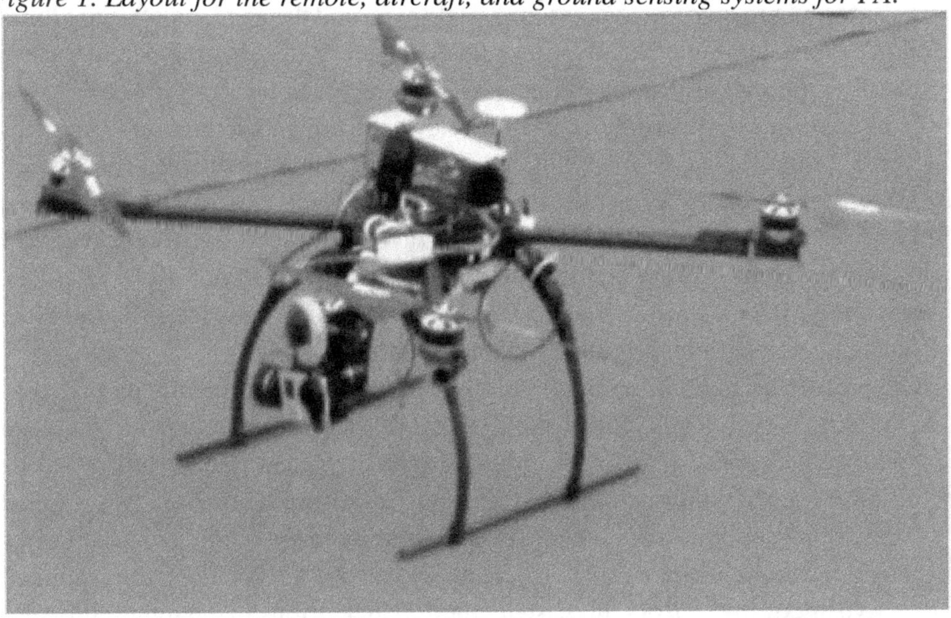

Sensors employed in remote sensing vary according to their spatial, spectral, radiometric, and temporal resolutions (Sundaresan et al., 2014). The spatial resolution of a sensor is determined by the dimensions of the pixel that corresponds to the ground area. Sensors with high spatial resolution typically possess tiny footprints, whereas sensors with large footprints generally exhibit low spatial resolution. Temporal resolution is more closely related to the sensor platform than to the sensor itself. Temporal resolution for a satellite refers to the duration required to complete an orbit and return to the identical observation area. The spectral resolution of a sensor is determined by the quantity of bands recorded within a specified region of the electromagnetic spectrum. Hyperspectral images comprise numerous contiguous bands of tiny width (<20 nm) spaced by minimal increments in wavelength (Teke et al., 2013). A variety of vegetation indices and statistical and machine learning techniques, including deep convolutional neural networks and random forests, have been utilized to diminish the dimensionality of hyperspectral data for the extraction of pertinent information regarding crop conditions. Recently, the estimation of solar-induced chlorophyll fluorescence (SIF) from hyperspectral images has been extensively utilized to estimate photosynthesis, plant nutrients, and both biotic and abiotic stressors, including disease and water stress (Chang et al., 2020).

Despite the availability of multiple contemporary satellites that offer high spatial (<5 m) and temporal (daily) resolution imagery, most publicly accessible satellite products are relatively coarse for various PA applications. The requisite spatio-temporal resolution for Precision Agriculture (PA) is contingent upon various elements, including management objectives, field size, and the capacity of agricultural machinery to adjust input application rates (irrigation, fertilizer, pesticide, etc.). Estimating crop biomass and yield generally necessitates better spatial resolution (1–3 m) than that required for variable rate fertilizer and irrigation treatments (5–10 m). Moreover, weed mapping and varied herbicide treatment necessitate a spatial resolution that exceeds the dimensions of the weed patches (e.g., 5–50 cm) (Fernández-Quintanilla et al., 2018). Aerial platforms, such as UAVs, typically yield photographs with superior spatial resolution (<5 m) in contrast to satellites. Consequently, UAVs and other terrestrial platforms have enhanced flexibility in delivering photos with high spatial and temporal resolution.

Sensors affixed to satellites, aircraft, and UAVs are predominantly passive sensors, meaning they lack an independent light source. Nevertheless, certain satellites are equipped with active sensors, including the active microwave instrument (AMI) on the ERS-1/2. A multitude of terrestrial remote sensing platforms are equipped with active proximity sensors. For instance, commercially available variable rate fertilizer application devices, such as Green Seeker and Crop Circle, are equipped with active proximity sensors. In these methods, fluctuations in daylight minimally impact recorded reflectance, hence yielding more precise and consistent normalized difference vegetation index (NDVI) or other vegetation indices (VI) utilized for evaluating crop nutritional status.

Recent satellites are progressively employing two additional sensors: thermal infrared and microwave, in agricultural applications. Thermal infrared sensors quantify the energy radiated from a target (e.g., crops) to ascertain its temperature, which can then be utilized to evaluate crop water stress, evapotranspiration, and irrigation needs (Khanal et al., 2017). Microwave sensors function analogously to thermal sensors by measuring the radiated energy, but at longer microwave wavelengths, from the land surface. Microwave sensors are primarily utilized to assess soil moisture levels and agricultural water consumption across extensive regions (Palazzi et al., 2019). Microwaves can penetrate clouds, providing an advantage over conventional sensors that utilize visible and near-infrared wavelengths.

The coarse spatial resolution (tens of kilometers) of microwave satellite sensors, particularly passive sensors, constrain their utility in precision agriculture. Recently, numerous techniques have been devised and employed to downscale passive microwave data to a higher resolution for application in PA (Babaeian et al., 2019). Active microwave sensors, such as synthetic aperture radar (SAR), typically provide superior spatial resolution and exhibit heightened sensitivity to surface roughness, such as vegetation, which may lead to inaccuracies in soil moisture measurement. A diverse array of remote sensors and platforms exists to produce high-resolution images (spatial, spectral, radiometric, and temporal) essential for the development and execution of site-specific management.

2.1.2 Drones and Satellite Imagery (Crop Health Monitoring)

Drones can operate autonomously using specialized software that enables the creation of flight plans and system deployment via GPS, incorporating numerous characteristics such as speed, altitude, ROI (Region of Interest), geo-fencing, and fail-safe protocols. Drones are favored over full-sized aircraft due to significant characteristics such as their high spatial resolution, rapid turnaround capabilities, low operational costs, and ease of activation. These attributes are essential in precision agriculture, where extensive areas are monitored, and analysis is conducted in minimal time. The utilization of aerial vehicles is facilitated by the shrinking of tiny cameras and other sensors, including infrared and sonar.

3. SYSTEM DESCRIPTION

3.1 Aerial Platform

Quad-rotor helicopter with a diameter of 700 mm (refer to Fig. 3.1). This vehicle can transport a 1kg cargo with a flight duration of 35 minutes at a maximum speed of 25 knots. The primary advantage of this equipment is its vertical takeoff and landing capabilities, allowing operation from highly restricted locations. This airplane is automated with an Arduino open-source development board, and the fly path is planned to use the open-source software PIX4D, which determines waypoints, altitude, heading direction, and speed. The user interface is streamlined, enabling individuals with two hours of training to effectively operate this instrument and provide valuable information.

Figure 2. Quadrotor helicopter and gyro-stabilized high-resolution camera.

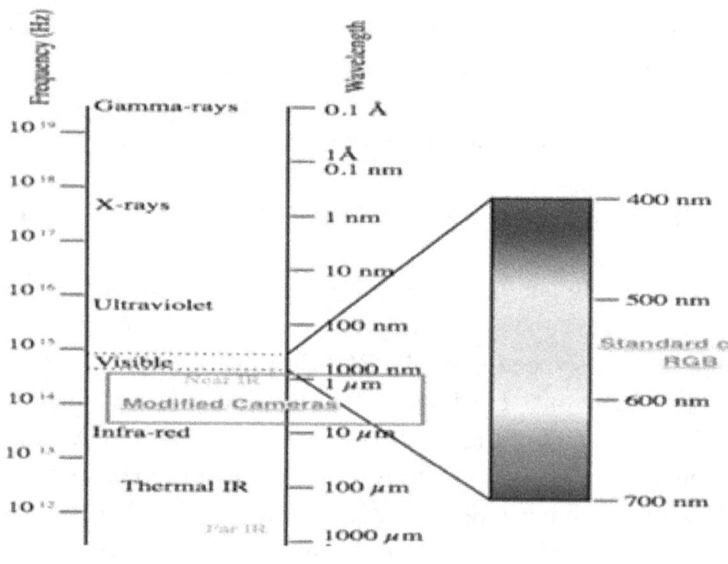

3.2 Image Sensor

Commercially available cameras suitable for drone platforms are compact enough to capture NIR footage; nevertheless, they are costly and exhibit limited resolution. This obstacle may be perceived as an affront, and the remedy involves transforming a Canon SX220 camera into an NDVI camera with minimum alterations. The output was a cost-effective high-resolution camera capable of observing near-infrared (NIR) wavelengths.

Figure 3. (a.) Modified Canon SX260 camera (without the infrared filter)

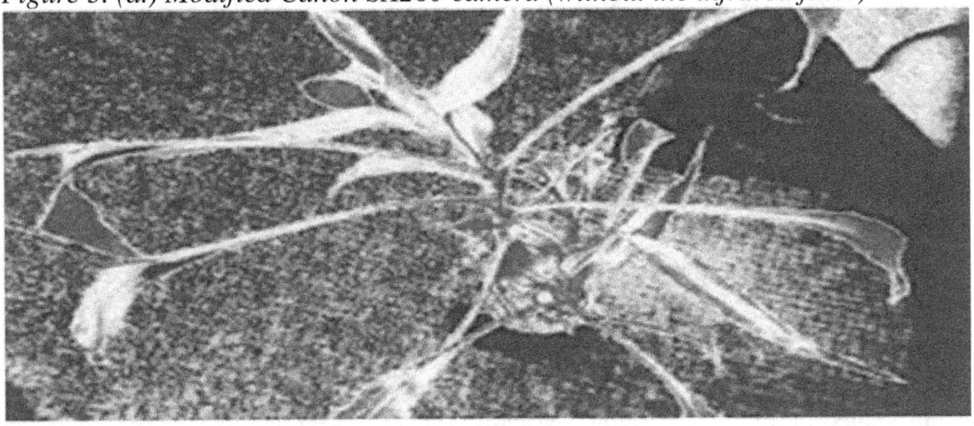

Figure 4. (b.) Output of the modified camera (NDVI image)

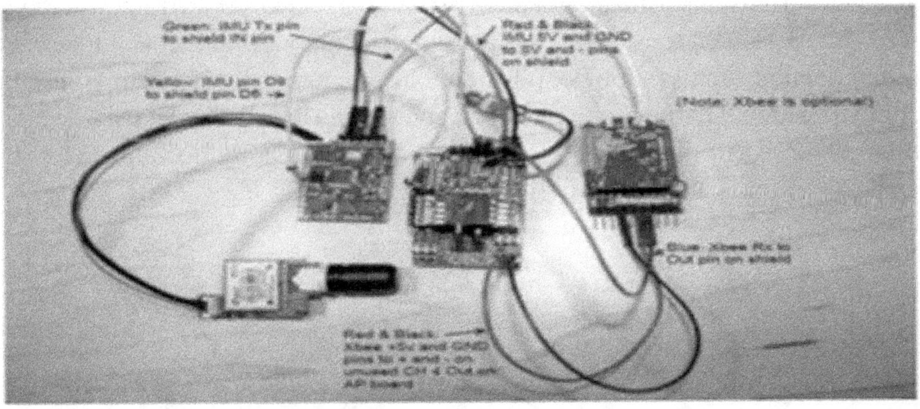

Disparity in normalization of the vegetation index is a straightforward statistic that signifies the vitality of green vegetation. The fundamental principle is that chlorophyll significantly reflects near-infrared light (NIR, approximately 750 nm) while absorbing red and blue wavelengths. Chlorophyll exhibits substantial reflection, which is why plants appear green; nevertheless, the reflection in the near-infrared spectrum is even more pronounced, significantly contributing to the accuracy of data for analysis. The NDVI principle is based on the observation that leaves, due to the spongy layers on their undersides, reflect significant amounts of near-infrared light, in sharp contrast to most non-plant materials. Upon dehydration or stress, the spongy layer of the plant collapses, resulting in less reflection of near-infrared light, while the reflection in the visible spectrum remains unchanged. Consequently, the mathematical integration of these two signals can facilitate the distinction between plants and non-plants, as well as between healthy and unhealthy plants.

4. HARDWARE AND SOFTWARE TECHNOLOGIES

The technology employed in drones is highly sophisticated and modern, as it must compensate for the lack of a pilot, hence facilitating the operation of unmanned aerial vehicles and their autonomous functions. They primarily rely on sensors, microcontrollers, communication networks, and artificial intelligence. A drone is an automated system divided into two components. The machine's hardware control unit is referred to as the Autopilot System, utilized for managing flights and numerous attributes. The autopilot unit comprises waypoint navigation with altitude and airspeed control, a fully integrated inertial measurement unit (IMU) including gyroscope and accelerometer, a GPS system, and a barometric pressure sensor. All these micro-electromechanical systems (MEMS) are incorporated into the board (refer to Fig. 3(a)). All sensors possess autonomous fail-safe protocols; in the event of a malfunction such as altitude, position, or communication modem failure the aircraft will redirect to the original takeoff point. The flight path and recorded data will be preserved in the autopilot chipset and can be readily downloaded to the ground station computer for post-flight analysis.

Figure 5. (a.) Autopilot unit, GPS, radio link

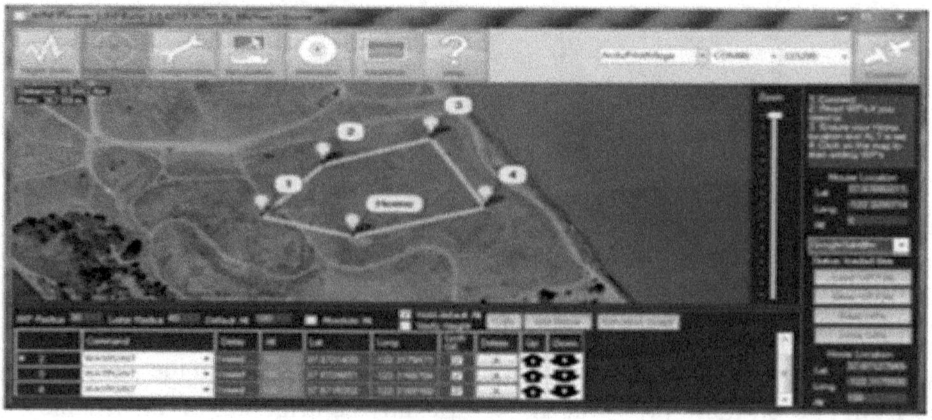

Figure 6. (b.) GCS software for flight planning

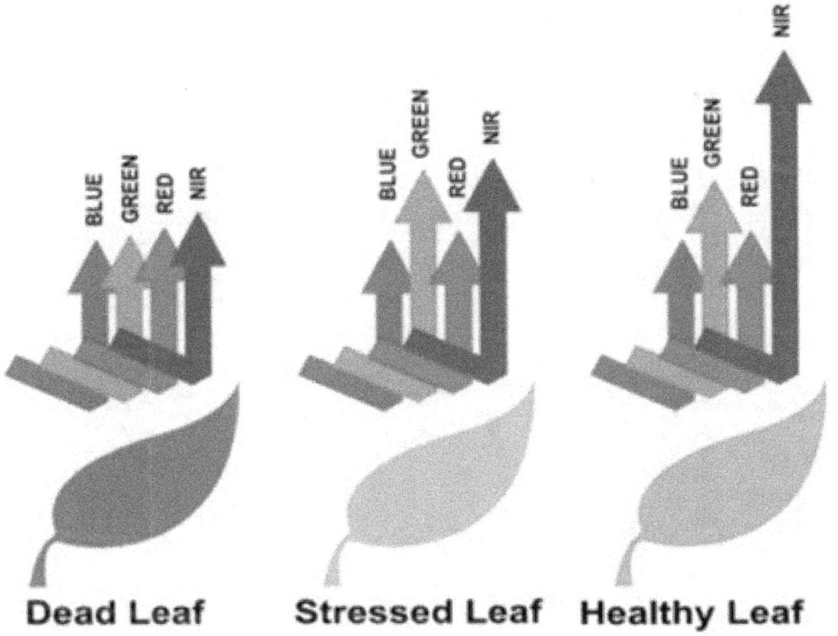

The Ground Control Station Software is updating the aircraft's real-time data, as illustrated in Fig. 4.2 (b). The GCS serves as an interface between the aircraft and the computer. It facilitates flight programming; the pre-flight technology employed in drones is highly sophisticated and advanced, compensating for the lack of a pilot and so enabling the operation of unmanned aerial vehicles and their autonomous behavior. They primarily rely on sensors, microcontrollers, communication networks, and artificial intelligence. A drone is an automated system divided into two components. The machine's hardware control unit is referred to as the Autopilot System, utilized for regulating flight and numerous attributes. The autopilot unit comprises waypoint navigation with altitude and airspeed control, a fully integrated inertial measurement unit (IMU) including gyroscope and accelerometer, a GPS system, and a barometric pressure sensor. All sensors possess autonomous fail-safe protocols; in the event of a failure such as altitude, position, or communication modem failure the aircraft will automatically navigate towards the designated takeoff point. The flight path and recorded data will be preserved in the autopilot chipset and can be readily downloaded to the ground station computer for post-flight analysis. A separate modem (downlink) is linked to the Ground Control Station Software, facilitating the real-time data update of the aircraft. The GCS serves as an interface between the aircraft and the computer. It facilitates flight programming, pre-flight simulations, tracking of the flight path, and monitoring of its heading and position on the map. It additionally produces a log file for post-flight examination.

5. PRECISION AGRICULTURE USING UAS

Figure 7. (a.) NIR analysis of leaf area index

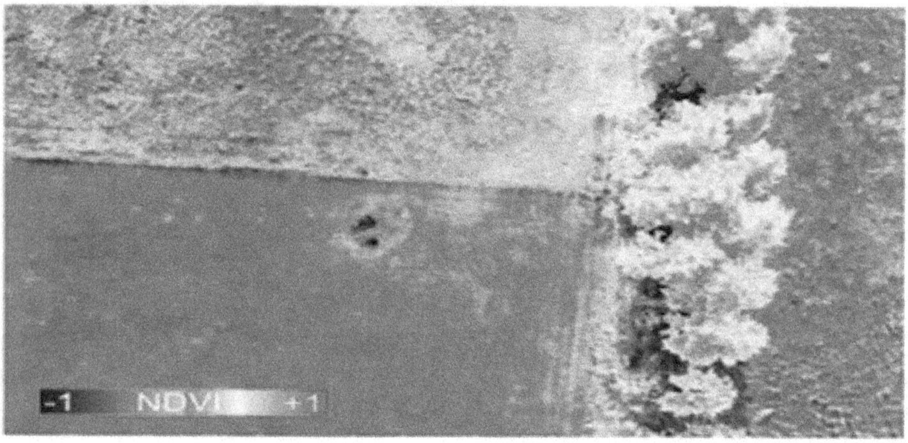

Figure 8. (b.) NDVI image of a playground

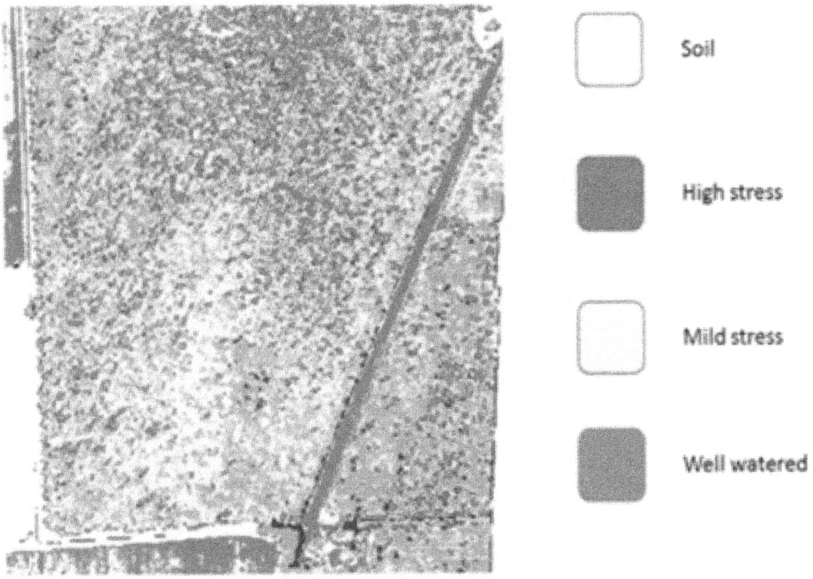

Vegetation indices in remote sensing are prevalent, with certain indices utilizing the RGB spectral bands. The indices employed are the Green Red Ratio Vegetation Index (GRRVI) and the Normalized Green Red Difference Index (NGRDI), which measure reflectance in the green and red portions of the spectrum. The Leaf Area Index (LAI) quantifies plant canopies. Unilateral green leaf area per unit of ground surface area.

(LAI = leaf area/ground area)

NDVI is the ratio of reflectance in the near-infrared and red segments of the electromagnetic spectrum, as seen in Fig. 5(b).

Figure 9. (c.) NDVI image showing stressed and well-watered crops

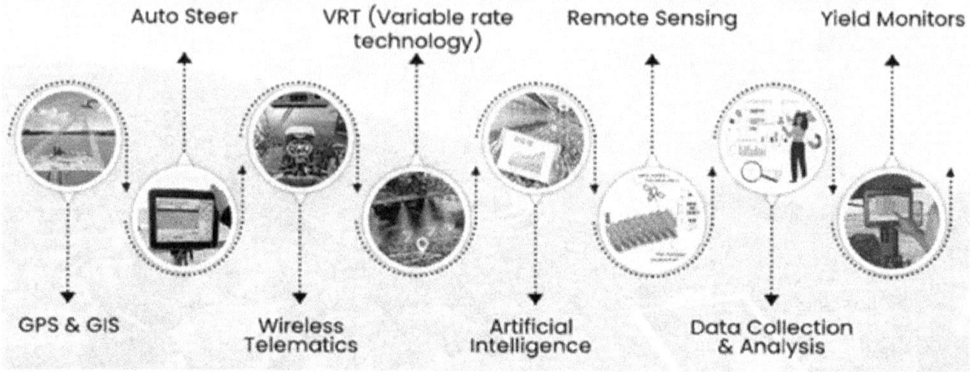

6. INTERNET OF THINGS (IOT) INTEGRATION

Smart farming integrates Artificial Intelligence (AI) and the Internet of Things (IoT) into cyber-physical systems for comprehensive farm management. AI applications cover various areas such as soil management, crop health monitoring, disease detection, and weed control. A significant AI application involves mobile expert systems, which enable farmers to use smartphones for tasks like disease diagnosis, species identification, and soil health analysis through mobile applications. Additionally, AI facilitates real-time analysis of satellite imagery to monitor farming progress. These innovations provide a scientific basis for precision agriculture, enhancing its effectiveness in optimizing agricultural outputs, See in Fig. 6.1 major components of precision farming.

Figure 10. Major components of precision farming

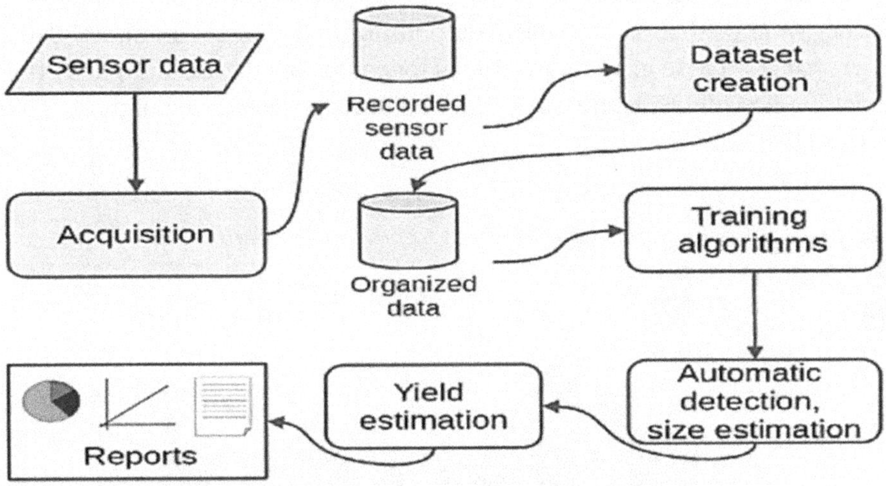

The Internet of Things (IoT) constitutes a network of interconnected devices and technologies that are essential in precision agriculture and smart farming. The IoT architecture in agriculture combines agricultural sensors with information and communication technology (ICT) and unmanned aerial vehicles (UAVs), enabling critical data collection for precision farming (Tomaszewski et al., 2022). The increasing prevalence of IoT and mobile data is essential in the framework of the fourth industrial revolution.

Progress in communication technologies and wireless networks, including 5G, LoRaWAN, NB-IoT, Sigfox, ZigBee, and Wi-Fi, has substantially expanded the utilization of IoT across various sectors. These technologies provide real-time remote management, extensive phenotyping, and enhancements in coverage, bandwidth, connection density, and end-to-end latency. In agriculture, IoT technology, combined with cloud computing, facilitates smart farming applications such as livestock monitoring, intelligent greenhouses, fishery management, and meteorological tracking (Tomaszewski et al., 2022).

Communication for agricultural machinery necessitates characteristics such as minimal power consumption, economic viability, and rapid response times. LoRa distinguishes itself among these technologies because of its large range, low implementation costs, ease of networking, and capacity to support multiple nodes concurrently. Artificial Intelligence is essential to the operation of Robotics and Autonomous Systems (RAS). Its interaction with IoT enables the perpetual creation of data streams. Data mining techniques are essential for transforming agricultural data into useful insights for decision-making (Say et al., 2018). AI examines varied environmental data and previous agricultural records housed in extensive data repositories, revealing concealed patterns essential for pest identification, disease detection, yield prediction, and fertilizer optimization inside agricultural decision support systems. These insights are crucial for enhancing the accuracy and efficacy of agricultural methods, facilitating informed decisions grounded in thorough data analysis in Fig. 6.2.

Figure 11. Data collection and extraction to be applied in precision farming.

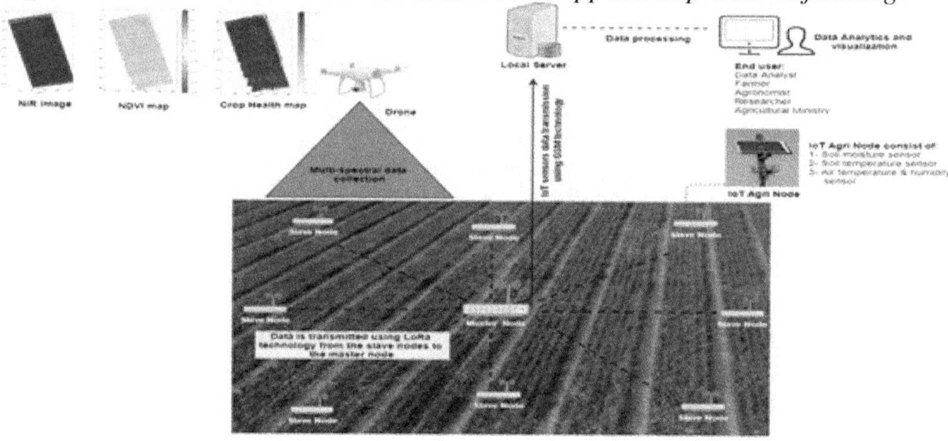

A comprehensive method is employed for monitoring crop health. The primary components of the system include IoT agricultural nodes, a communication channel for data transmission, a drone equipped with a multispectral camera, a local server for data archiving, and a web portal for data display. Figure 3 illustrates the high-level system architecture.

Figure 12. Proposed system architecture.

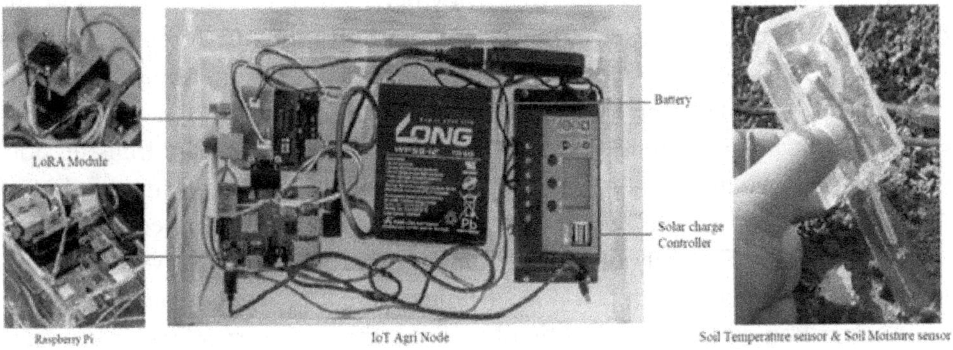

7. DEVELOPMENT OF IOT AGRI NODES

The creation of a standard IoT node chiefly entails the meticulous selection of sensors, communication modules, and the power supply. The specifications of these components pertinent to the present investigation are defined below.

7.1 IoT Sensors

The suggested system utilized three sensors: an air temperature and humidity sensor, a soil temperature sensor, and a soil moisture sensor. The DHT11 digital sensor was employed to record air temperature and humidity, the DS18B20 digital sensor was chosen for soil temperature measurement, and the Capacitive Soil Moisture sensor was utilized to monitor soil moisture levels. Soil moisture sensors are primarily classified into two categories: volumetric and tensiometric. Volumetric sensors quantify the soil's water level, while tensiometric sensors assess the soil's water potential.

7.2 Communication Module

The communication module's objective is to ensure the reliable transmission of data from one endpoint to another, significantly influencing the system's performance characterization. For research purposes, the 'LoRA' (Long Range) communication module was deemed more appropriate due to its extensive transmission range. Furthermore, agricultural land provides an unobstructed line of sight, enhancing the efficacy of LoRa technology in this context. The LoRa modules function on two frequency types: 433MHz and 868MHz. The 433MHz frequency module was utilized for the application, as this frequency band is license-free; conversely, the 868MHz ISM band is not free in Pakistan. The LoRa modules employ patented modulation algorithms that transmit data via Chirps and can communicate with the microcontroller utilizing the Serial Peripheral communicate (SPI) protocol.

The IoT agricultural nodes were set in a star topology for data transmission, comprising 8 slave nodes and 1 master node. The slave nodes transmitted data to the master node via LoRa technology, while the master node relayed data to the web portal using GSM technology. Star topology offers cost-effectiveness; nevertheless, it also presents a single point of failure. A solitary master node is tasked with transmitting data gathered from all nodes to the local server. To mitigate this constraint, the SD card was positioned on the master node to save data for 24 hours in the event of a communication breakdown at the master node.

7.3 Power Module

In IoT-based systems, the uninterrupted supply of power is the principal concern for maintaining operational functionality for real-time monitoring. The primary energy source for powering the planned agricultural system is solar energy. To do this, the slave nodes are equipped with a 10W solar panel and a 4Ah battery to ensure extended battery life during rainfall and overcast conditions. The master node is outfitted with a 40W solar panel and a 5Ah battery to ensure sufficient power supply. Figure 7.3 illustrates an IoT agricultural node equipped with a LoRa and power module.

Figure 13. IoT agri node with its components.

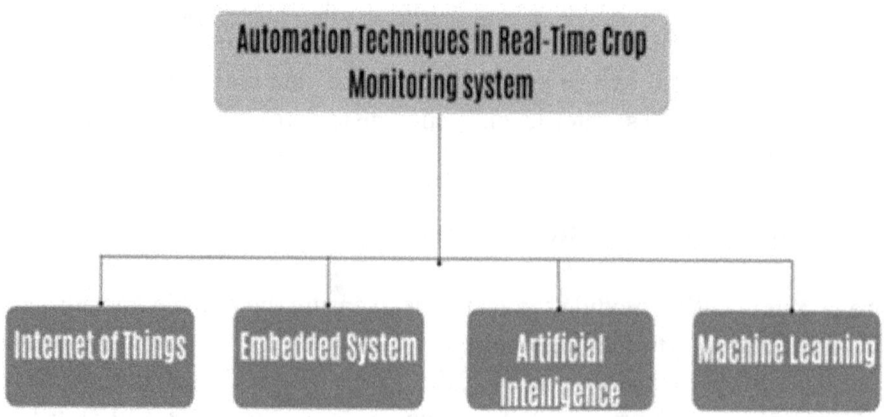

8. DATA ANALYSIS AND AI TOOLS

8.1 Machine Learning for Predictive Analytics

Advanced data mining results in predictive data analytics. The terms data mining and data extraction appear similar; nonetheless, a significant distinction exists between them. The primary goal in data extraction is acquiring data from a single source and loading it into an integrated database. Consequently, one can 'extract' data from a source or legacy system and transfer it into a conventional database or data warehouse. Conversely, the process of extracting subtle or concealed prognostic information from extensive databases or data warehouses is termed data mining. Data mining is the identification of patterns within extensive data repositories from databases. Data mining employs computer methodologies derived from statistics and pattern identification utilizing historical data. The nature of data mining is defined by patterns in data. Mishra and Silakari (2012).

8.2 Predictive Analytics Techniques

Predictive models assess hazards and possibilities through the analysis of historical and transactional data. They delineate the relationships among components, assessing the hazards or potentials linked to a certain set of conditions. Predictive analytics employs three fundamental techniques: data profiling and transformations, sequential pattern analysis, and time series tracking. Data profiling and transformation encompasses data formatting, attribute modification, and dependency analysis. Sequential Pattern Analysis discerns relationships among data rows, whereas Time Series Tracking monitors corporate business KPIs or industrial operations. Advanced predictive analytic methodologies encompass Classification-Regression, Association Analysis, and Time Series Forecasting. These strategies utilize data properties to categorize items into established classifications or forecast the value of a numerical variable.

8.3 Predictive Models

Creating a highly creative and artistic model necessitates exceptional proficiency in predictive analytics, rendering it a highly skilled endeavor acknowledged by specialists. To construct a predictive model, we require several fundamental steps. The procedure is as follows:

a) Project
b) Exploration

c) Data Preparation
d) Model Building
e) Deployment
f) Model Management

Most experts assert that the data preparation phase in the development of predictive models is the most time-intensive aspect of the process. Mishra and Silakari (2012).

8.4 Modeling Process

The modeling process comprises several stages, some of which are detailed below:
Objective: To delineate the aim of the project
Acquire the data: Collecting datasets from several sources pertaining to the project.

1. Examine the data, sanitize it, and prepare it by delineating variables, attributes, tokens, and other terminology employed in the project, offering comprehensive explanations and contextual applications.
2. The procedure entails dividing data into training, validation, and test subsets, simplifying by minimizing variables and terms by the creation of redundant variables.
3. In supervised learning, data is divided into training and validation sets to construct a model. This method is employed in classification and prediction tasks. The data is subsequently utilized across various models, each with multiple configurations. The validation data offers an impartial assessment of the model's efficacy; nonetheless, it becomes integrated into the model upon selecting the optimal performing variant.
4. The data mining task entails constructing a model by partitioning data into training and validation sets through various methodologies.
5. The algorithm identifies fitted and predicted values in training and validation datasets, with fitted values denoting the records where the model has been applied.
6. At this step, multiple prediction algorithms are evaluated and refined for accuracy in performance. The optimal model, characterized by minimal validation error and simplicity, is selected for predicting new data.
7. The selected model must be implemented with new data following its selection.

Prediction error can be quantified by multiple methodologies:

a) Average error
b) Total sum of squared errors

c) RMS error (Root mean squared error)

9. APPLICATIONS FOR AI FOR SUSTAINABLE FARMING

A. Precision Agriculture

Precision agriculture is an essential aspect of smart farming, facilitating the accurate and efficient application of inputs such as seeds, fertilizers, and water. It converts conventional agricultural methods into data-centric systems, minimizing waste and enhancing productivity. Variable rate seeding optimizes planting density for certain field zones, reducing crop competition and enhancing germination (Šarauskis et al., 2022). Fertilizer optimization reduces excessive application and mitigates environmental damage. Targeted irrigation, utilizing sophisticated technologies and soil moisture sensors, delivers precise watering according to plant requirements. In the United States, a maize farmer employing variable rate technology experienced a 15% yield gain while decreasing fertilizer application by 20% (Saavoss et al., 2021). Precision agriculture enhances economic efficiency by decreasing input costs and augmenting output, with research indicating a yield gain of up to 20% and a reduction in input expenses of 15%. The implementation of a solar-powered aquaponic system (Vo et al., 2021) enhances both operational efficiency and sustainability in agriculture.

B. Crop Monitoring in Real Time

IoT-enabled devices, drones, and satellite images facilitate real-time crop monitoring to detect early indicators of illnesses, pests, ripening status, and nutritional deficits. These technologies employ multispectral imaging, which collects data across various wavelengths, to uncover disparities in plant health that are imperceptible to the naked eye. IoT sensors deliver instantaneous data on growing circumstances, whilst AI diagnostics assess irregularities, forecast illness occurrences, and recommend remedial measures. In India, tea farms utilize drones with multispectral cameras to monitor pest infestations, resulting in a 25% reduction in chemical usage (Mahmud et al., 2023). The incorporation of economical tools for small-scale farmers may improve accessibility and increase the adoption of these technologies (Padhiary et al., 2023). Figure 9.1 depicts the amalgamation of IoT devices, drones, and satellite imagery for efficient real-time agricultural surveillance.

Figure 14. Technology in real time crop monitoring system

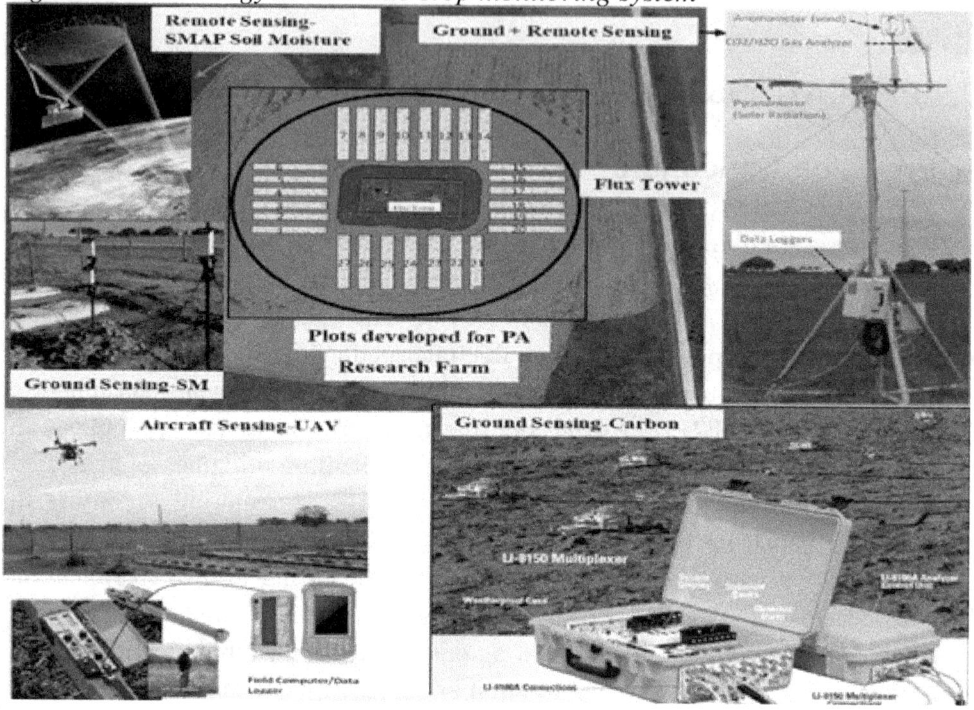

C. Advanced Management of Irrigation

Intelligent irrigation is essential for effective water utilization in arid areas. It employs IoT sensors to supply water directly to the root zone, so it reduces evaporation and runoff. Weather-integrated systems utilize soil moisture data and meteorological forecasts to enhance irrigation scheduling. Automated sprinklers utilize real-time sensor data to modify irrigation schedules. Farmers in Israel have used advanced drip irrigation systems, resulting in a 40% reduction in water usage and enhanced agricultural production. These approaches preserve water, minimize resource depletion, and establish optimal development conditions for crops. In summary, sophisticated irrigation management is crucial for promoting water-efficient agriculture methods.

D. Pest and Disease Management

Intelligent pest and disease management use sophisticated detection, forecasting, and control technology to mitigate yield losses and decrease chemical pesticide application. Drone-based remote sensing acquires high-resolution imagery, detecting initial indicators of disease or pest presence (Abd El-Ghany et al., 2020). IoT traps monitor pest numbers and transmit notifications upon reaching action levels. Drones utilize precision spraying to apply pesticides exclusively to impacted regions, so conserving chemicals. Biological controls incorporate natural agents into intelligent systems for sustainable pest management. In Pakistan, AI-driven systems forecast fungal infections, decreasing losses by 30% (Ali et al., 2024). Cost analyses of conventional and intelligent pest management techniques can illustrate the financial advantages of these technologies.

E. Crop Forecasting and Managing

Produce Farmers can optimize harvest time for superior storage and market strategies through precise yield estimation. Yields could be precisely estimated in the past when real-time data was integrated into intelligent systems. Machine learning systems predict agricultural yield by employing historical data, meteorological factors, and soil health indicators. The application of several vegetation indices, particularly the Normalized Difference Vegetation Index, enables remote spatial yield prediction. On-farm sensors assess stem diameter and leaf area index to determine yield. In Australia, wheat producers employing yield prediction systems have enhanced harvest schedules and diminished post-harvest losses by 15%. Benefits include enhanced readiness for the demand for machinery and manpower. Minimize post-harvest losses through optimal transportation and storage, enhanced supply forecasts, and refined market strategies.

F. Robotics and Autonomous Machinery

The automation of agricultural activities is transforming farming practices. Robotics and autonomous technology are automating chores like harvesting fruits and vegetables, weeding, and tilling, thereby diminishing labor reliance and enhancing efficiency. California strawberry farms have experienced a 20% enhancement in picking efficiency, resulting in time savings and decreased labor costs. These devices enhance precision and diminish the necessity for manual effort. Examples include the Harvey Robot, which automates the harvesting of fruits and vegetables with minimum human intervention, and the weeding robot, which employs artificial intelligence to differentiate between crops and weeds, subsequently eliminating them automatically or applying pesticides (Gupta and Gupta, 2024). The implementation of automation techniques in food processing machinery enhances productivity and processing quality.

G. Climate-Aware Solutions

Climate-conscious solutions are revolutionizing agriculture by mitigating risks and enhancing resilience. This encompasses drought-resistant agriculture, the application of biogenic nanoparticles in organic farming, which identifies crops that flourish with scarce water resources, and frost mitigation systems that issue alerts for the pre-application of anti-frost agents or protective coverings for crops. These technologies enhance productivity despite adverse weather conditions and reduce financial losses resulting from climate-related crop failures. Climate-resilient practices supply suitable cultivars and evaluate the potential alterations in crop performance under changing conditions. Drought-resilient agriculture selects crops that flourish with scarce water resources while optimizing water utilization (Li et al., 2019). Intelligent weather stations offer frost alerts, allowing farmers to apply anti-frost chemicals or protect crops in advance.

10. CHALLENGES AND CONSIDERATIONS

Innovative agricultural technologies, capable of enhancing farming practices, face numerous obstacles. Significant installation expenses, the necessity for contemporary technologies like IoT, AI, and robotics, and the complexity of managing the extensive data generated by these systems provide considerable hurdles (Shafique et al., 2020). Connectivity issues, particularly in rural areas with inadequate internet access, hinder the real-time functioning of IoT devices and cloud-based applications (Padhiary et al., 2024b). Integrating intelligent systems with legacy equipment poses significant challenges, as most antiquated agricultural machinery is incompatible with contemporary technologies. The deficiency of technical skills and training among farmers results in a steep learning curve and hinders adoption. Agriculturists are apprehensive regarding the ownership and possible misuse of sensitive data, which exacerbates their uncertainty. Confronting these challenges is essential to realizing the complete potential of smart agricultural technologies for sustainable and efficient farming.

11. THE FUTURE OF AI IN SUSTAINABLE FARMING

This section illustrates the vast potential of smart systems to fundamentally transform agriculture, given the rapid advancement of technology. The future of agriculture depends on the enhanced integration of technologies such as AI, machine learning, IoT, and data analytics, which are expected to enhance crop output, soil vitality, and sustainability. Numerous prospects exist:

1) Augmented AI and ML functionalities: AI and ML are anticipated to assume a more significant role in automating decision-making processes and enhancing predictive accuracy (Jarrahi, 2018). These advancements will allow farmers to predict weather fluctuations, pest infestations, and crop health concerns with enhanced precision, facilitating proactive and efficient responses.
2) Data integration and advanced analytics: The amalgamation of various data sources, such as meteorological forecasts, soil sensors, and crop health metrics, will yield a comprehensive perspective of agricultural operations. Cloud computing and advanced analytics will facilitate more accurate and thorough farm management strategies (Johnraja et al., 2024).
3) Sustainability and regenerative agriculture: Intelligent systems will facilitate regenerative agricultural practices by assessing soil health, optimizing resource utilization, and minimizing environmental impact. Technologies such as AI and IoT will facilitate the implementation of sophisticated pest management strategies and enhance soil conservation, thereby contributing to sustainable agricultural systems (Rabha et al., 2024).
4) Enhanced accessibility and affordability: As technology becomes increasingly economical and attainable, even small-scale farmers will gain advantages from smart farming solutions (Mizik, 2021). Initiatives to close the digital divide and offer training will enhance the democratization of these systems' adoption.
5) Climate-resilient agriculture: Intelligent systems will bolster resilience to climate change by offering tools to alleviate risks such as droughts, frost, and unpredictable weather patterns (Argyroudis et al., 2022). Agriculturalists will have the capacity to choose climate-suitable crops and implement adaptive farming techniques to ensure productivity.
6) Smart farming systems are positioned to transform the agriculture of tomorrow by accomplishing the dual goals of greater productivity and environmental sustainability (Javaid et al., 2022). Applications for 3D printing in the design of smart farming machinery (Padhiary and Roy, 2024), alongside the development of smart sensors and innovative food products (Padhiary et al., 2024a), will illuminate the future of agricultural practices. These technologies will not only address the increasing global food demand but also alleviate pressure on natural resources. By stimulating innovation, collaboration, and responsible practices, smart systems will generate a more resilient, efficient, and sustainable agriculture business.

REFERENCES

Abd El-Ghany, N. M., Abd El-Aziz, S. E., & Marei, S. S. (2020). A review: Application of remote sensing as a promising strategy for insect pests and diseases management. *Environmental Science and Pollution Research International, 27*(27), 33503–33515. DOI: 10.1007/s11356-020-09517-2 PMID: 32564316

Ali, F., Rehman, A., Hameed, A., Sarfraz, S., Rajput, N. A., & Atiq, M. (2024). Climate Change Impact on Plant Pathogen Emergence: Artificial Intelligence (AI) Approach BT - Plant Quarantine Challenges under Climate Change Anxiety. In Abd-Elsalam, K. A., & Abdel-Momen, S. M. (Eds.), *Springer Nature Switzerland* (pp. 281–303).

Angers, D. A., & Eriksen-Hamel, N. S. (2008). Full-inversion tillage and organic carbon distribution in soil profiles: A meta-analysis. *Soil Science Society of America Journal, 72*(5), 1370–1374. DOI: 10.2136/sssaj2007.0342

Argyroudis, S. A., Mitoulis, S. A., Chatzi, E., Baker, J. W., Brilakis, I., Gkoumas, K., Vousdoukas, M., Hynes, W., Carluccio, S., Keou, O., Frangopol, D. M., & Linkov, I. (2022). Digital technologies can enhance climate resilience of critical infrastructure. *Climate Risk Management, 35*, 100387. DOI: 10.1016/j.crm.2021.100387

Babaeian, E., Sidike, P., Newcomb, M. S., Maimaitijiang, M., White, S. A., Demieville, J., Ward, R. W., Sadeghi, M., LeBauer, D. S., Jones, S. B., Sagan, V., & Tuller, M. (2019). A new optical remote sensing technique for high-resolution mapping of soil moisture. *Frontiers in Big Data, 2*, 37. DOI: 10.3389/fdata.2019.00037 PMID: 33693360

Bhatia, A., Chug, A., & Singh, A. P. (2020). Hybrid SVM-LR classifier for powdery mildew disease prediction in tomato plant. 2020 7th International conference on signal processing and integrated networks (SPIN). IEEE. pp.218–223.

Carleo, G., Cirac, I., Cranmer, K., Daudet, L., Schuld, M., Tishby, N., Vogt-Maranto, L., & Zdeborová, L. (2019). Machine learning and the physical sciences. *Reviews of Modern Physics, 91*(4), 45002. DOI: 10.1103/RevModPhys.91.045002

Chang, C. Y., Zhou, R., Kira, O., Marri, S., Skovira, J., Gu, L., & Sun, Y. (2020). An Unmanned Aerial System (UAS) for concurrent measurements of solar-induced chlorophyll fluorescence and hyperspectral reflectance toward improving crop monitoring. *Agricultural and Forest Meteorology, 294*, 108145. DOI: 10.1016/j.agrformet.2020.108145

Chostner, B. (2017). See & spray: The next generation of weed control. *Resour. Mag., 24*, 4–5.

Deutsch, C.A., J.J. Tewksbury, M. Tigchelaar, D.S. Battisti, S.C. Merrill, R.B. Huey and R.L. Naylor. 2018. Increase in crop losses to insect pests in a warming climate. Science (80-.). 361:916–919.

Dhaka, V. S., & Vikas, L. (2014). Wheat yield prediction using artificial neural network and crop prediction techniques (a survey). *IJRASET*, *2*, 330–341.

Duarte-Carvajalino, J. M., Alzate, D. F., Ramirez, A. A., Santa-Sepulveda, J. D., Fajardo-Rojas, A. E., & Soto-Suárez, M. (2018). Evaluating late blight severity in potato crops using unmanned aerial vehicles and machine learning algorithms. *Remote Sensing (Basel)*, *10*(10), 1513. DOI: 10.3390/rs10101513

Fenu, G., & Malloci, F. M. (2021). Forecasting plant and crop disease: An explorative study on current algorithms. *Big Data and Cognitive Computing*, *5*(1), 2. DOI: 10.3390/bdcc5010002

Fernández-Quintanilla, C., Peña, J. M., Andújar, D., Dorado, J., Ribeiro, A., & López-Granados, F. (2018). Is the current state of the art of weed monitoring suitable for site-specific weed management in arable crops? *Weed Research*, *58*(4), 259–272. DOI: 10.1111/wre.12307

Fróna, D., Szenderák, J., & Harangi-Rákos, M. (2019). The challenge of feeding the world. *Sustainability (Basel)*, *11*(20), 5816. DOI: 10.3390/su11205816

Gupta, N., & Gupta, P. K. (2024). Robotics and Artificial Intelligence (AI) in Agriculture with Major Emphasis on Food Crops BT - Digital Agriculture: A Solution for Sustainable Food and Nutritional Security. In Priyadarshan, P. M., Jain, S. M., Penna, S., & Al-Khayri, J. M. (Eds.), *Springer International Publishing* (pp. 577–605).

Gutierrez, A., Ansuategi, A., Susperregi, L., Tubío, C., Rankić, I., & Lenža, L. (2019). A benchmarking of learning strategies for pest detection and identification on tomato plants for autonomous scouting robots using internal databases. *Journal of Sensors*, *2019*, 5219471. DOI: 10.1155/2019/5219471

Hochreiter, S., & Schmidhuber, J. (1997). Long Short-Term Memory. *Neural Computation*, *9*(8), 1735–1780. DOI: 10.1162/neco.1997.9.8.1735 PMID: 9377276

Jarrahi, M. H. (2018). Artificial intelligence and the future of work: Human-AI symbiosis in organizational decision making. *Business Horizons*, *61*(4), 577–586. DOI: 10.1016/j.bushor.2018.03.007

Javaid, M., Haleem, A., Singh, R. P., & Suman, R. (2022). Enhancing smart farming through the applications of Agriculture 4.0 technologies. *Int. J. Intell. Networks*, *3*, 150–164. DOI: 10.1016/j.ijin.2022.09.004

Johnraja, J. I., Leelipushpam, P. G. J., Shirley, C. P., & Princess, P. J. B. (2024). Impact of Cloud Computing on the Future of Smart Farming BT - Intelligent Robots and Drones for Precision Agriculture. In Balasubramanian, S., Natarajan, G., & Chelliah, P. R. (Eds.), *Springer Nature Switzerland* (pp. 391–420).

Kamilaris, A., Kartakoullis, A., & Prenafeta-Boldú, F. X. (2017). A review on the practice of big data analysis in agriculture. *Computers and Electronics in Agriculture*, *143*, 23–37. DOI: 10.1016/j.compag.2017.09.037

Kartikeyan, P., & Shrivastava, G. (2021). Review on emerging trends in detection of plant diseases using image processing with machine learning. *International Journal of Computer Applications*, *975*(11), 39–48. DOI: 10.5120/ijca2021920990

Karunathilake, E., Le, A. T., Heo, S., Chung, Y. S., & Mansoor, S. (2023). The path to smart farming: Innovations and opportunities in precision agriculture. *Agriculture*, *13*(8), 1593. DOI: 10.3390/agriculture13081593

Khan, S. A., Mulvaney, R. L., Ellsworth, T. R., & Boast, C. W. (2007). The myth of nitrogen fertilization for soil carbon sequestration. *Journal of Environmental Quality*, *36*(6), 1821–1832. DOI: 10.2134/jeq2007.0099 PMID: 17965385

Khanal, S., Fulton, J., & Shearer, S. (2017). An overview of current and potential applications of thermal remote sensing in precision agriculture. *Computers and Electronics in Agriculture*, *139*, 22–32. DOI: 10.1016/j.compag.2017.05.001

Li, K., Huang, G., & Wang, S. (2019). Market-based stochastic optimization of water resources systems for improving drought resilience and economic efficiency in arid regions. *Journal of Cleaner Production*, *233*, 522–537. DOI: 10.1016/j.jclepro.2019.05.379

Mahmud, M.S., A. Zahid and A.K. Das. 2023. Sensing and Automation Technologies for Ornamental Nursery Crop Production: Current Status and Future Prospects.

Mishra, N., & Silakari, S. (2012). Predictive analytics: A survey, trends, applications, oppurtunities & challenges. *International Journal of Computer Science and Information Technologies*, *3*, 4434–4438.

Mizik, T. 2021. Climate-Smart Agriculture on Small-Scale Farms: A Systematic Literature Review.

Mohanty, S. P., Hughes, D. P., & Salathé, M. (2016). Using deep learning for image-based plant disease detection. *Frontiers in Plant Science*, *7*, 1419. DOI: 10.3389/fpls.2016.01419 PMID: 27713752

Nagasubramanian, K., Jones, S., Singh, A. K., Sarkar, S., Singh, A., & Ganapathysubramanian, B. (2019). Plant disease identification using explainable 3D deep learning on hyperspectral images. *Plant Methods*, *15*(1), 1–10. DOI: 10.1186/s13007-019-0479-8 PMID: 31452674

Nascimento, A., De Souza, V., Graciela, M., Faria, I., Edna, C., Philippsen, G. S., Catuzo, G., Silva, C., Ratti, G., Inumaro, S., Gonçalves, J. E., Gazim, Z. C., Aparecida, F., Augusto, F., Seixas, V., Claudia, E., & Lovato, W. n.d. *Pr ep ot pe er re v Pr ep t n ot pe er v* ed. 1–30.

Nigussie, Z., Tsunekawa, A., Haregeweyn, N., Adgo, E., Nohmi, M., Tsubo, M., Aklog, D., Meshesha, D. T., & Abele, S. (2017). Factors influencing small-scale farmers' adoption of sustainable land management technologies in north-western Ethiopia. *Land Use Policy*, *67*, 57–64. DOI: 10.1016/j.landusepol.2017.05.024

Oliver, M.A., T.F.A. Bishop and B.P. Marchant. 2013. Precision agriculture for sustainability and environmental protection.

Padhiary, M., Barbhuiya, J. A., Roy, D., & Roy, P. (2024a). 3D printing applications in smart farming and food processing. *Smart Agricultural Technology*, *9*, 100553. DOI: 10.1016/j.atech.2024.100553

Padhiary, M., & Roy, P. (2024). Advancements in Precision Agriculture: Exploring the Role of 3D Printing in Designing All-Terrain Vehicles for Farming Applications. *International Journal of Scientific Research*, *13*, 861–868.

Padhiary, M., Saha, D., Rani, N., Akhtar Barbhuiya, J., Sethi, L. N., & Tech, M. (2023). Efficient Precision Agriculture with Python-based Raspberry Pi Image Processing for Real-Time Plant Target Identification. IJRAR23C1798. *International Journal of Research and Analytical Reviews*, •••, 539.

Padhiary, M., Sethi, L. N., & Kumar, A. (2024b). Enhancing Hill Farming Efficiency Using Unmanned Agricultural Vehicles: A Comprehensive Review. *Transactions of the Indian National Academy of Engineering : an International Journal of Engineering and Technology*, *9*(2), 253–268. DOI: 10.1007/s41403-024-00458-7

Padhiary, M., Tikute, S. V., Saha, D., Barbhuiya, J. A., & Sethi, L. N. (2024c). Development of an IOT-Based Semi-Autonomous Vehicle Sprayer. *Agricultural Research*. Advance online publication. DOI: 10.1007/s40003-024-00760-4

Palazzi, V., Bonafoni, S., Alimenti, F., Mezzanotte, P., & Roselli, L. (2019). Feeding the world with microwaves: How remote and wireless sensing can help precision agriculture. *IEEE Microwave Magazine*, *20*(12), 72–86. DOI: 10.1109/MMM.2019.2941618

Plant, R., Pettygrove, G., & Reinert, W. (2000). Precision agriculture can increase profits and limit environmental impacts. *California Agriculture*, *54*(4), 66–71. DOI: 10.3733/ca.v054n04p66

Rabha, G., Kumar, K., Kumar, D., & Kumar, D. (2024). *A Comprehensive Review of Integrating AI and IoT in Farm Machinery : Advancements*. Applications, and Sustainability.

Rajpurkar, P. (2017). CheXNet: Radiologist-Level Pneumonia Detection on Chest X-Rays with Deep Learning. ArXiv abs/1711 5225.

Ray, D. K., Mueller, N. D., West, P. C., & Foley, J. A. (2013). Yield trends are insufficient to double global crop production by 2050. *PLoS One*, *8*(6), e66428. DOI: 10.1371/journal.pone.0066428 PMID: 23840465

Roser, M. and L. Rodés-Guirao. (2013). Future population growth. Our world data.

Saavoss, M., T. Capehart, W. McBride and A. Effland. (2021). Trends in Production Practices and Costs of the U.S. Corn Sector. 1–35.

Šarauskis, E., M. Kazlauskas, V. Naujokienė, I. Bručienė, D. Steponavičius, K. Romaneckas and A. Jasinskas. (2022). Variable Rate Seeding in Precision Agriculture: Recent Advances and Future Perspectives.

Savla, A., Israni, N., Dhawan, P., Mandholia, A., Bhadada, H., & Bhardwaj, S. (2015). Survey of classification algorithms for formulating yield prediction accuracy in precision agriculture. *2015 International Conference on Innovations in Information, Embedded and Communication Systems (ICIIECS)*. IEEE. pp.1–7. DOI: 10.1109/ICIIECS.2015.7193120

Say, S. M., Keskin, M., Sehri, M., & Sekerli, Y. E. (2018). Adoption of precision agriculture technologies in developed and developing countries. *The Online Journal of Science and Technology*, *8*, 7–15.

Shafique, K., B.A. Khawaja, F. Sabir, S. Qazi and M. Mustaqim. (2020). Internet of Things (IoT) for Next-Generation Smart Systems: A Review of Current Challenges, Future Trends and Prospects for Emerging 5G-IoT.

Soni, A., D. Dharmacharya, A. Pal, V.K. Srivastava, R.N. Shaw and A. Ghosh. (2021). Design of a machine learning-based self-driving car. Mach. Learn. Robot. Appl. 139–151.

Sundaresan, J., Santosh, K. M., Déri, A., Roggema, R., & Singh, R. (2014). *Geospatial Technologies and Climate Change*. Springer. DOI: 10.1007/978-3-319-01689-4

Suthaharan, S. and S. Suthaharan. (2016). Support vector machine. Mach. Learn. Model. algorithms big data Classif. Think. with examples Eff. Learn. 207–235.

Teke, M., Deveci, H. S., Haliloğlu, O., Gürbüz, S. Z., & Sakarya, U. (2013). A short survey of hyperspectral remote sensing applications in agriculture. 2013 6th international conference on recent advances in space technologies (RAST). IEEE. pp.171–176.

Tomaszewski, L., Kołakowski, R., & Zagórda, M. (2022). Application of mobile networks (5G and beyond) in precision agriculture. *IFIP International Conference on Artificial Intelligence Applications and Innovations*. Springer. pp.71–86. DOI: 10.1007/978-3-031-08341-9_7

Vo, T.T., H. Ko, J.-H. Huh and N. Park. (2021). Overview of Solar Energy for Aquaculture: The Potential and Future Trends.

Waldrop, M. P., Zak, D. R., Sinsabaugh, R. L., Gallo, M., & Lauber, C. (2004). Nitrogen deposition modifies soil carbon storage through changes in microbial enzymatic activity. *Ecological Applications*, *14*(4), 1172–1177. DOI: 10.1890/03-5120

Wang, L., Lee, C.-Y., Tu, Z., & Lazebnik, S. 2015. Training deeper convolutional networks with deep supervision. arXiv Prepr. arXiv1505.02496.

Xiao, Q., Li, W., Kai, Y., Chen, P., Zhang, J., & Wang, B. (2019). Occurrence prediction of pests and diseases in cotton on the basis of weather factors by long short term memory network. *BMC Bioinformatics*, *20*(S25), 1–15. DOI: 10.1186/s12859-019-3262-y PMID: 31874611

Yang, J., Guo, X., Li, Y., Marinello, F., Ercisli, S., & Zhang, Z. (2022). A survey of few-shot learning in smart agriculture: Developments, applications, and challenges. *Plant Methods*, *18*(1), 28. DOI: 10.1186/s13007-022-00866-2 PMID: 35248105

Yu, Y., Si, X., Hu, C., & Zhang, J. (2019). A Review of Recurrent Neural Networks: LSTM Cells and Network Architectures. *Neural Computation*, *31*(7), 1235–1270. DOI: 10.1162/neco_a_01199 PMID: 31113301

Chapter 16
AI-Powered Solutions for Sustainable Water Conservation

Muhammad Umar Farooq
Yunnan Agricultural University, China

Ghulam Murtaza
https://orcid.org/0009-0000-6920-8684
University of Agriculture, Faisalabad, Pakistan

Muhammad Huzaifa Mahmood
https://orcid.org/0000-0001-7303-1500
University of Turin, Italy

Shahid Iqbal Rai
https://orcid.org/0009-0006-1868-8407
The University of Lahore, Pakistan

Hassan Tanveer
https://orcid.org/0009-0003-1786-637X
Depaul University, Chicago, USA

Ejaz Ul Haq
Quaid i Azam University, Islamabad, Pakistan

Hamdan Muhammad
Aitchison College, Lahore, Pakistan

Rehan Mehmood Mehmood Sabir
https://orcid.org/0009-0007-4711-8304
University of Agriculture, Faisalabad, Pakistan

ABSTRACT

Artificial Intelligence (AI) has emerged as a disruptive influence in water management, providing instruments to optimize water utilization, improving conservation initiatives, and foster sustainability. Applications for AI in tackling water-related issues. In water management, AI enhances early leak detection, predictive maintenance, and demand forecasting, hence promoting efficient water distribution, cost reduction, and infrastructure robustness. AI-driven smart irrigation and precision agriculture technologies enhance water efficiency, augment crop yields, and minimize waste in sustainable agriculture. In water quality monitoring, AI facilitates real-time pollutant detection and enhances treatment procedures, thereby protecting public

DOI: 10.4018/979-8-3373-0680-3.ch016

health and reducing chemical usage. Moreover, AI augments public awareness by delivering tailored water conservation techniques and educational resources, facilitating behavioral modification and encouraging sustainable practices. Future breakthroughs in machine learning and data analytics possess significant promise to transform water conservation.

1. INTRODUCTION

Water, essential for life on Earth, is increasingly acknowledged worldwide as a significant economic and social resource (Mishra, 2023). Water comprises 71% of the Earth's surface, although only 2.5% is freshwater, with a scant 1% being readily accessible (Anastasi, 2024). The oceans constitute 97.5% of the Earth's total water. Fresh and atmospheric water, essential for terrestrial life, represents a diminutive and delicate component within a very large system, comprising less than 1% of our total water resources (excluding the 1.7% currently encased in ice). Society's comprehension and management of water-related risks have significantly advanced during the last two centuries. This must now broaden to encompass all scales, from local to global.

Essential freshwater resources necessary for life are threatened by population growth, industrialization, climate change, and their involvement in energy production (Singh, 2024). Worldwide, approximately 40% of the population faces water scarcity, with more than 700 million individuals lacking access to potable water (Greve et al., 2018). Approximately 50% of anthropogenic effluent is released straight into rivers or seas untreated, resulting in significant environmental and health repercussions. The world population is expected to rise by two billion (2×10^9) by 2030 (Ganivet, 2020), making it imperative to address humanity's water needs while safeguarding the essential functions of freshwater ecosystems, which is one of the foremost challenges of the 21st century. A notable scarcity of fresh water in relation to human demands is currently apparent in numerous regions globally (Rosa et al., 2020a). Water shortage denotes a spatial and temporal disparity between the availability of freshwater and its demand. Alongside biophysical issues, economic and institutional contexts are critical components that influence water shortage (Vallino et al., 2020).

Water scarcity includes both physical and economic dimensions. Physical water shortage arises when access to water is restricted. It impacts both blue water (i.e., water from aquatic sources or aquifers) and green water (i.e., soil moisture derived from precipitation). Economic water scarcity arises when water resources are physically accessible, yet institutional and economic deficiencies restrict access to these resources (Rosa et al., 2020a).

Agriculture is the greatest consumer of freshwater globally and depends on consistent and robust freshwater supply for global food security (D'Odorico et al., 2020). These promises require heightened focus on global agricultural water constraints concerning humanity's capacity to satisfy future food demand. Agricultural green water scarcity (GWS) arises when rainfall is insufficient to satisfy crop water requirements (CWR), potentially resulting in water-stressed crop development. In these instances, more blue water is supplied by irrigation to guarantee sufficient crop development and avert water stress. Agricultural blue water scarcity arises when blue water supplies are insufficient to satisfy irrigation demands. In some areas, the shortage of agricultural blue water results in unsustainable irrigation methods that deplete blue water reserves and/or disrupt aquatic ecosystems. Agricultural economic water shortage is characterized by insufficient irrigation resulting from restricted institutional and economic capability rather than hydrological limitations (Dalin et al., 2017).

Agricultural blue water scarcity has become a central issue in the discourse on water scarcity, increasingly regarded as a global socio-environmental menace to food and energy security. Disturbingly, fifty percent of irrigated agricultural lands are experiencing blue water scarcity or are subjected to unsustainable irrigation methods. Recent focus has been directed towards agricultural economic water shortage, indicating that up to one-fourth of worldwide rain-fed croplands are amenable to sustainable irrigation growth; however, insufficient institutional and economic capacity is obstructing irrigation implementation. Water security might be perceived as an acceptable water-related risk to society (Mekonnen and Hoekstra, 2020).

This dilemma transcends individual concerns and pertains to the broader context. Agriculture, industries, and ecosystems are experiencing stress, jeopardizing the foundation of sustainable growth. Experts caution that by 2030, global water demand may exceed supply by 40%. Consequently, identifying more intelligent and effective methods for water management and conservation has emerged as an essential need for all. This scarcity is a critical environmental concern, highlighting the necessity for appropriate water management approaches (Giwa and Dindi, 2017).

The necessity for sustainable water utilization has reached unprecedented urgency. Due to climate change, population increase, and urbanization exerting significant strain on water supplies, the pursuit of innovative solutions has become essential. In the face of contemporary water issues, conventional conservation techniques frequently inadequately support environmental sustainability. The demand for sophisticated, data-driven solutions has intensified significantly. In this context, artificial intelligence (AI) is demonstrating its transformative potential by providing innovative methods for more efficient water management and conservation. Through the real-time analysis of extensive data sets, AI can enhance water distribution, mitigate waste, and accurately forecast future water requirements. Integrating AI into water management facilitates the development of more intelligent and efficient systems that advantage both humanity and the ecosystem. AI can identify leaks in water systems, develop optimized irrigation plans to conserve water in agriculture, and predict floods to safeguard populations. These applications demonstrate the remarkable capacity of AI to tackle one of humanity's most pressing challenges: guaranteeing the sustainable utilization of our invaluable water resources.

Artificial Intelligence is a transformational entity adept at analyzing vast amounts of data, making intelligent decisions, and optimizing processes with unmatched efficiency. The use of AI in water conservation transcends traditional approaches, offering a dynamic and adaptive answer to the complex challenges encountered by global water management systems. The applications of AI range from intelligent water distribution to precise irrigation, showcasing diversity and potential. In the following parts, we will examine the various facets of AI in water conservation, including its applications, advantages, problems, and its critical role in fostering a sustainable future for this vital resource. Researchers are employing sophisticated technology like AI to revolutionize worldwide water resource management, promoting sustainable water supplies. AI, frequently seen as the apex of technology, presents the notion of 'digital water' via the ongoing and adaptive analysis of vast information.

This technology-oriented strategy enables water management and political entities to create efficient water systems, successfully tackling issues associated with water variability and sustainability. Essential responsibilities in water management include identifying new water reservoirs and guaranteeing the sustainable stewardship of current stocks. AI-driven algorithms and programs are being investigated for the creation of water plants that provide real-time data and aid in the advancement of future models. Software-based systems utilising sensors and neural networks dynamically orchestrate water operations, incorporating predictive models, robotic sensors, and blockchain technology for financial transactions. In a notable experiment, researchers from the Key Laboratory of Eco-hydrology of the Inland River Basin employed AI to forecast groundwater reserves, demonstrating the potential of hybrid models to improve global water systems. In a pivotal experiment, researchers from the leading Eco-hydrology research institution in the Inland River Basin employed AI to forecast groundwater reserves, demonstrating the efficacy of hybrid models in improving global water systems. Global collaboration among scientists and analysts is essential for the sustainable use of this unique resource. Utilising AI in water resource management has shown effectiveness in optimizing water distribution and enhancing water conservation systems (Mahardhika and Putriani, 2023). Figure 1.1 illustrates the function of AI in water management to improve economic distribution and conservation methods.

Figure 1. AI's role in water management

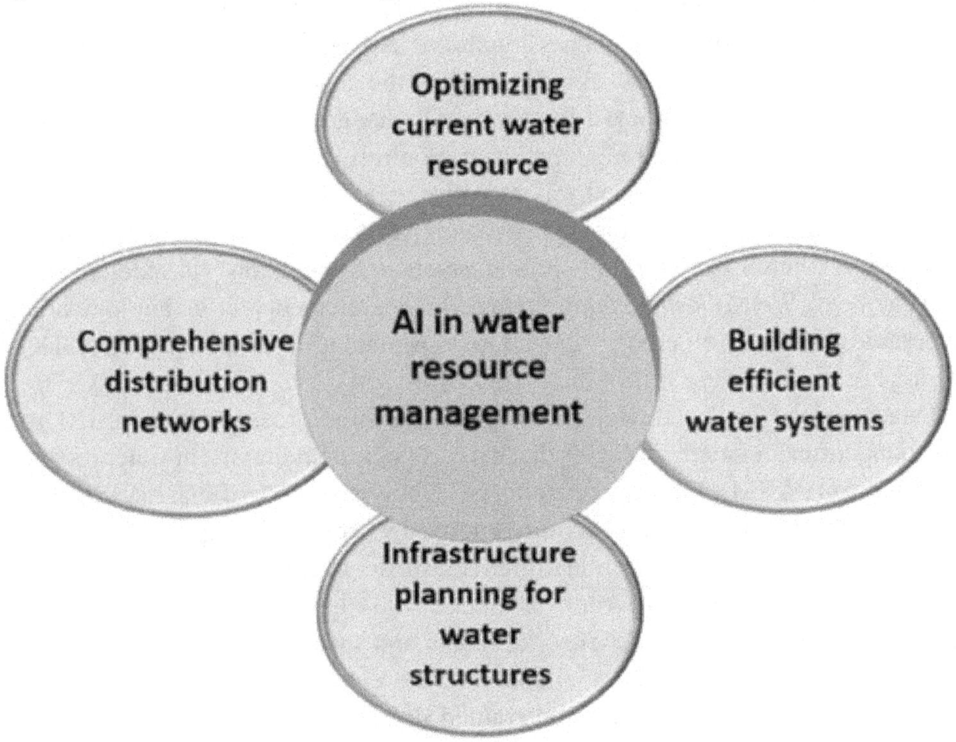

2. AI APPLICATIONS IN SMART WATER MANAGEMENT

Effective management of sustainable water resources depends on high-quality data regarding water resources, the adoption of water-saving, environmentally friendly, and hybrid technologies, particularly in industrial and agricultural sectors, as well as awareness initiatives aimed at minimizing household water consumption and promoting sustainable dietary practices.

For effective water resource management, the initial phase must involve the meticulous examination of hydrologically collected data from historical climatology records (Zabihi et al., 2023). Subsequently, during the prediction phase, the results of the hydrological events are assessed after the documentation of data processing through clustering for any computer models. Ultimately, in the controlling phase, it is imperative to prioritize suitable methods to mitigate risks and guarantee sustainability (Alizadeh and Yazdi, 2023).

Traditional approaches involve simulating hydrological systems through intricate physical models that adhere to mathematical formulae, while optimizing parameters by establishing appropriate boundary conditions. Nonetheless, progress in physically based modelling systems has been impeded by the challenges in obtaining modelling parameters. Furthermore, a framework of changes is necessary in areas such as data-driven decision-making, enhanced monitoring, forecasting and managing hydrological events, customized water management, and task automation in water resource management. Artificial intelligence (AI) possesses the capacity to fundamentally revolutionize hydrological and practical applications of water resource management (WRM), establishing them as pivotal elements for technological and economic progress in attaining water-related sustainable development goals (SDGs) (Kumar et al., 2024).

Machine learning (ML), deep learning (DL), and reinforcement learning (RL) are the foundational AI methodologies that drive diverse applications in water resource management (WRM). Machine Learning is a subdivision of Artificial Intelligence in which machines independently acquire knowledge from data (experience) through statistical methodologies. Deep learning is a subset of machine learning in which machines acquire knowledge from vast datasets through artificial neural networks, facilitating sophisticated functions like image and speech recognition. Reinforcement Learning (RL) is a subdomain of Machine Learning (ML) that is especially advantageous in situations with constrained datasets. In reinforcement learning, agents acquire decision-making skills through trial-and-error interactions with an environment, gaining rewards for favorable choices and penalties for unfavorable ones, ultimately striving to optimize cumulative rewards over time. Water Resource Management (WRM) depends significantly on varied data (Tetzlaff et al., 2017), and disciplines such as hydrology, meteorology, climatology, and geology can be consolidated into a unified platform with the application of AI methodologies. Figure 2.1 illustrates the application of AI subfields in Water Resource Management (WRM). Remote sensing, satellite images, the Internet of Things (IoT), and drones may be employed during the data collection phase. Upon successful data collection, structured data may be analyzed using standard machine learning algorithms; conversely, unstructured and complex data necessitates the use of deep learning techniques. In scenarios with insufficient data, reinforcement learning emerges as the favored methodology. The RL agent acquires knowledge dynamically by interacting with its environment, gathering real-time data, and receiving feedback in the form of rewards or penalties contingent upon its actions. Machine Learning, Deep Learning, and Reinforcement Learning approaches can be employed in Water Resource Management to generate predictions, formulate strategies, and implement actions for risk mitigation and sustainability assurance (Solomatine et al., 2008).

Figure 2. AI subfields applied in water resource management

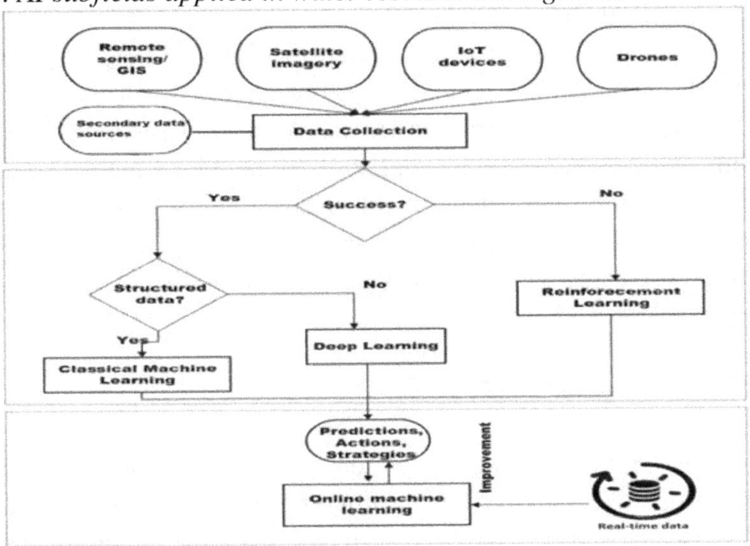

Online Machine Learning (OML) methodologies are essential in Water Resource Management (WRM), facilitating the ongoing enhancement of forecasts, actions, and plans through the dynamic acquisition of data. OML entails the real-time updating of models or systems as new data emerges, promoting adaptability and enhancement. This methodology enables systems to acquire knowledge and adapt according to the most recent data, so improving their efficacy over time.

3. LEAK DETECTION AND PREVENTION

Water management is essential for the effective use and preservation of water resources, especially with rising water scarcity and environmental issues. A primary difficulty in water management is identifying and preventing leaks in water distribution systems, which can result in substantial water loss, infrastructure damage, and environmental risks (Hamamed et al., 2023). Conventional leak detection techniques typically depend on manual inspections or expensive, labor-intensive devices, which may lack scalability and efficiency for extensive water distribution systems. Recent breakthroughs in Computational Fluid Dynamics (CFD) and Deep Learning (DL) have provided significant opportunities to enhance the accuracy and efficiency of leak detection in water distribution systems (B. et al., 2022). Researchers have aimed to create automated and intelligent systems for real-time leak detection and precise localization by simulating water flow in pipes and utilising machine learning algorithms to analyze the resultant data (Ravichandran et al., 2021).

The efficacy of this hybrid method in detecting leaks in pipes composed of diverse materials and sizes, under various operational conditions. The project entails simulating water flow in pipes constructed from copper and steel, varying in diameter, with strategically positioned apertures along their lengths (Romero-Ben et al., 2023). Automating the detection and localization of leaks can potentially reduce water losses, lower infrastructure maintenance expenses, and lessen the environmental consequences of water leaking (Soldevila et al., 2022).

4. PREDICTIVE MAINTENANCE

Water resources, crucial for human existence and production, are currently endangered by detrimental contaminants resulting from anthropogenic activities and natural phenomena (Ray et al., 2023). A significant volume of wastewater is generated daily, predominantly containing hazardous chemicals, and is discharged into the environment without treatment or reuse. Untreated sewage typically contains a high concentration of diverse nutrients, organic matter, suspended particles, organic micropollutants, and both pathogenic and nonpathogenic microorganisms (Chahal et al., 2016). Wastewater can be categorized into six types: municipal, residential, industrial, medical, agricultural, and nuclear, depending on the source. Municipal and household wastewater are the most prevalent, with research on wastewater treatment mostly focused on these two categories of sewage (Jasim, 2020). Wastewater treatment (WWT) eliminates impurities from sewage by a combination of physical, chemical, and biological processes, resulting in clean water that is properly discharged back into the environment. WWT is a crucial method for the reuse of water resources and sustainable development, vital for safeguarding public health and the environment, and has been extensively used in industrial and agricultural sectors. There is an urgent need for effective and novel methods to enhance efficiency, lower costs, and diminish energy usage in wastewater treatment (Malviya and Jaspal, 2021).

Artificial intelligence (AI) facilitates classification and regression analysis of vast data sets generated continuously, thereby invigorating industries and significantly advancing various sectors of society. The WWT process primarily involves water quality assessment, laboratory-scale investigation, and process design. AI models are increasingly prevalent in wastewater-related domains, particularly in recent years (see Figure 4.1), and have been utilized for the prediction and optimization of the wastewater treatment process. Artificial intelligence models have been effectively utilized in wastewater treatment (WWT) process design, water quality monitoring (Wang et al., 2022), optimization of WWT process parameters (Bhagat et al., 2023), and prediction of WWT process performance. These studies have shown that an AI model, as a potent instrument, has attained significant success in the applications of WWT-related domains. Nonetheless, most review articles on AI applications in wastewater treatment mostly concentrate on specific techniques or process design elements, including adsorption processes, membrane bioreactors, membrane processes, and wastewater treatment plants.

Figure 3. Role of AI models in wastewater-related fields

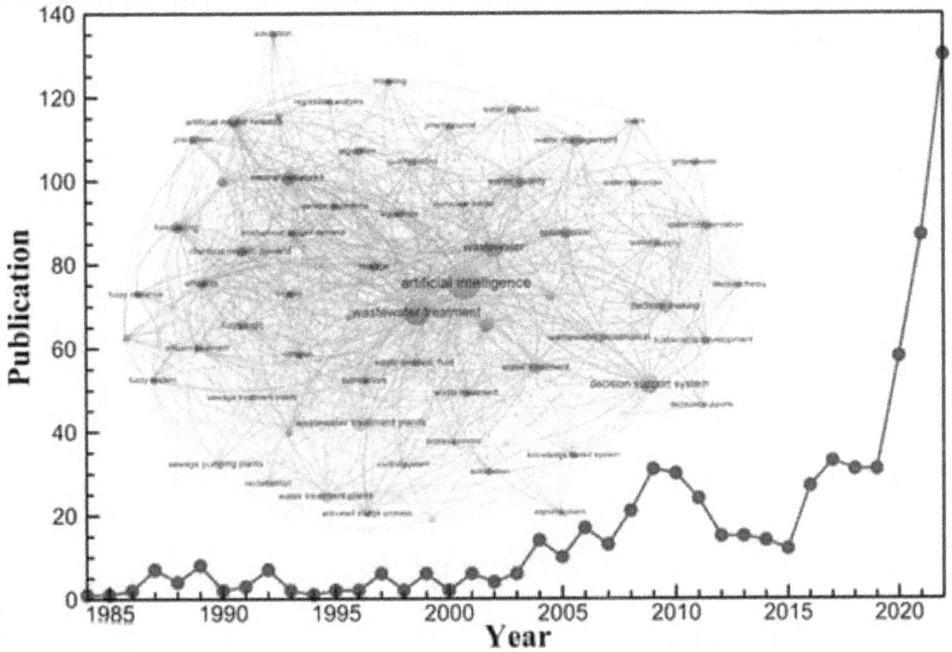

5. DEMAND FORECASTING AND OPTIMIZATION

The swift population growth necessitates AI-driven predictions of sectoral water requirements, which are essential for the planning, allocation, and effective management of water resources. The analysis of AI methodologies in urban water demand forecasting emphasized the necessity of integrating meteorological variables and demographic data into AI models (Muhammad et al., 2019). Demand forecasting may be categorized as either short-term or long-term, based upon the prevailing circumstances. Short-term demand forecasting assists water providers in planning for water consumption, optimizing supply, and minimizing waste on an hourly or daily basis. AI-predicted water consumption can subsequently manage water pressure, enhance pump operation scheduling, and minimize leakages through effective system maintenance (Antunes et al., 2018). Long-term water demand forecasting utilizing AI can incorporate anthropogenic factors and climate change model data for macro-level reservoir volume assessments. Optimal operating plans for reservoir release utilizing AI are implemented through support vector machines, a promising long-term water management strategy (Jalili et al., 2023). The primary benefit of AI in this application is its ability to include intricate patterns in water usage, including sudden weather fluctuations and shifts in consumer behavior. Figure 5.1 shows the methodology adopted.

Figure 4. DL for more efficient water demand forecasting

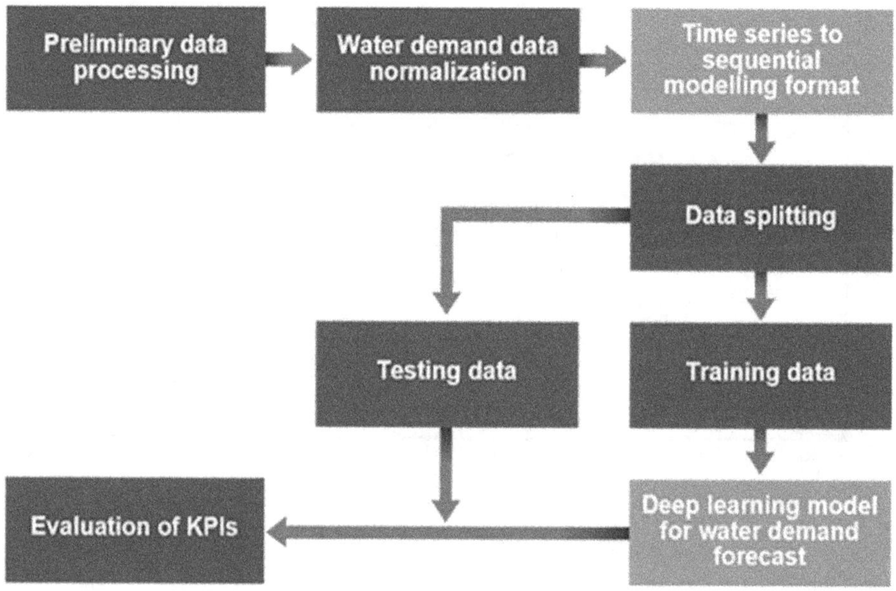

The fundamental steps in the process include:

(i) Data normalization
(ii) Construction of the sequence model/input data format
(iii) Specification of the deep learning model architecture
(iv) Model training, testing, and subsequent evaluation

Step iv is frequently executed concurrently with step iii.

The input size for the deep learning model varied from 2 to 14 data points. The equation below illustrates the input-output data format employed for the deep learning model, as outlined in step (ii) of the technique.

$$Qt = f\,(Qt\text{-}1, Qt\text{-}2, Qt\text{-}3, \ldots, Qt\text{-}n)$$

In this equation, n represents the input window size in hours, Q signifies the flow (demand) in million liters per day (MLD), and t indicates the timestep.

6. AI FOR SMART IRRIGATION SYSTEMS

In smart agriculture, the efficient use of resources is essential for irrigation. According to World Bank figures, around 70% of freshwater will be required for agriculture in the future. Agricultural water optimization and efficient water utilization are critical factors to consider in study employing various approaches. The smart system is an effective idea for resource utilization across various fields, particularly in agriculture, where it significantly contributes to irrigation systems, decision-making, disease prediction, control systems, and energy efficiency, among others. Utilize a sophisticated technology to optimize water usage and enhance production.

Technologically enhanced agricultural and smart irrigation systems reduce freshwater consumption. The intelligent system utilizing IoT and machine learning principles facilitates centralized storage, processing, and analysis of data from many ubiquitous sensors, along with the decision-making process (Gubbi et al., 2013). The method of artificially dispersing resources or water in agriculture is termed an irrigation system. This irrigation technique is mostly utilized in regions with low rainfall to sustain soil moisture and temperature. Another significant application of artificial irrigation systems is to inhibit the proliferation of undesirable plant growth or weeds in agricultural fields. The irrigation system comprises two categories: the conventional irrigation system, which utilizes watering buckets, cans, and canals, and the modern irrigation system, which enhances the conventional approach through the integration of advanced technologies.

The primary techniques of traditional irrigation include basin, furrow, strip, and basin irrigation, and they do not constitute applied knowledge. The costs associated with the traditional irrigation method are exceedingly high. The contemporary irrigation system has three types: sprinkler irrigation, pot irrigation, and drip irrigation (Janani and Jebakumar, 2019).

The weather-based evapotranspiration controller is categorized as signal-based controllers, historical controllers, and on-site weather measurement controllers for irrigation systems, among others. The soil moisture sensor controller is categorized into suspended cycle irrigation systems and water-on-demand irrigation systems. These represent the primary classifications based on the utilization of components and attributes. The diverse control and sensing mechanisms of contemporary smart irrigation technology are illustrated in figure 6.1.

Figure 5. Various controllers and sensors in smart irrigation system

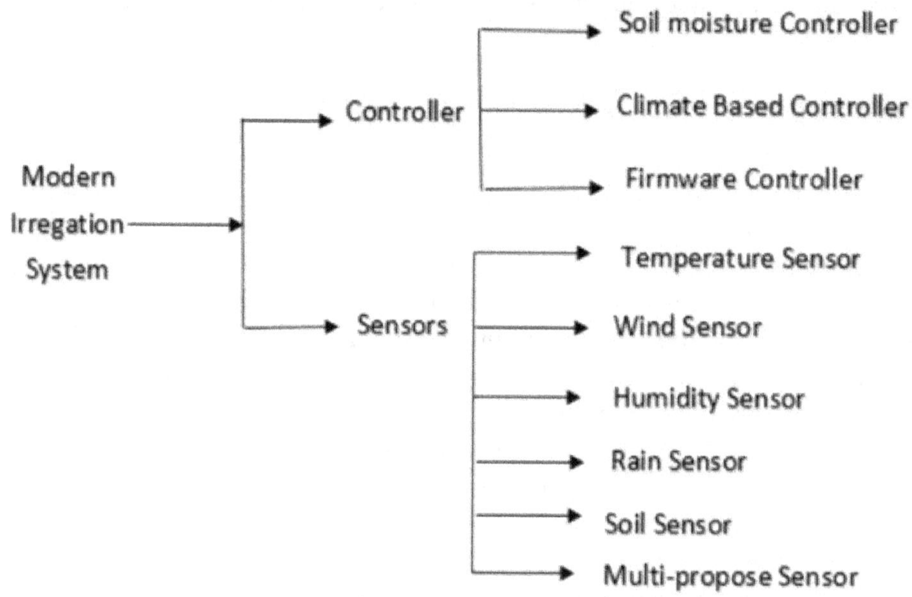

The primary objective of the contemporary smart irrigation system is to minimize water consumption and costs while enhancing yields. The automated smart irrigation system employed various contemporary technologies, including artificial intelligence and machine learning (Liakos et al., 2018), big data, the Internet of Things, wireless sensor networks, and cloud computing. The primary elements of smart irrigation systems are computational algorithms, sensors, automatic controllers, satellite data, flow meters, valves, wireless sensor networks, and batteries. The contemporary irrigation system is depicted in figure 6.2, comprising many stages:

i. Real-time data acquisition
ii. Implementation of methodologies
iii. Classification and forecasting
iv. Regulation and decision-making.

Figure 6. Modern irrigation system

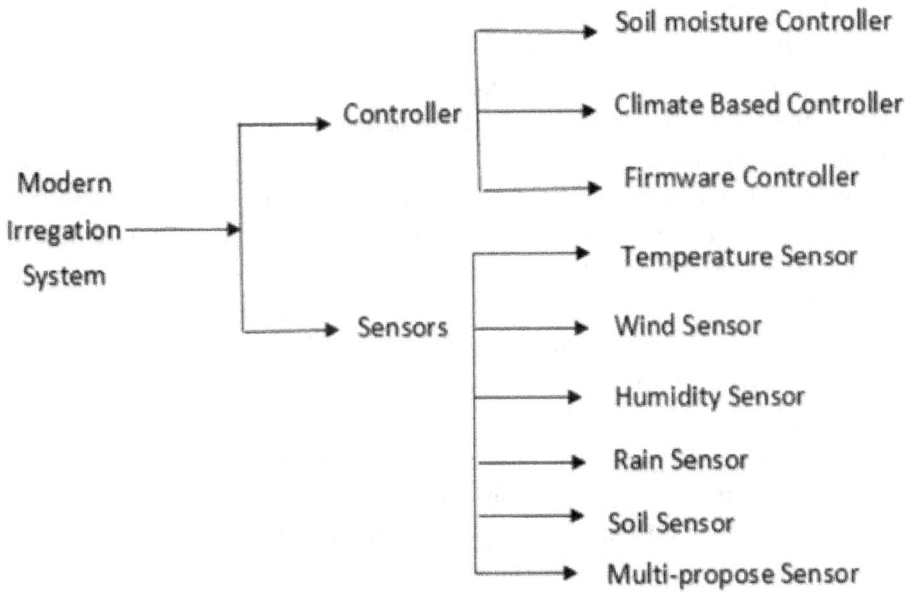

7. AI FOR PRECISION AGRICULTURE

Precision agriculture, driven by the incorporation of advanced Artificial Intelligence (AI) technology, is at the vanguard of a transformative period in contemporary farming (Sharma et al., 2023). As the world population expands and environmental issues escalate, the necessity for sustainable agricultural practices has become increasingly critical. The integration of AI with precision agriculture offers a viable solution to these challenges by optimizing resource use, improving crop management, and promoting a more sustainable and efficient agricultural ecosystem, including water resource management (Karunathilake et al., 2023). The utilization of AI in precision agriculture centers on employing sophisticated computational methods, machine learning algorithms, computer vision, and sensor technologies to enable data-driven decision-making. This chapter seeks to thoroughly examine the diverse role of AI in transforming traditional agricultural methods, highlighting the essential technologies and their combined influence on attaining water sustainability goals. The integration of AI in precision agriculture fundamentally transforms traditional, homogenous farming practices into a more tailored and adaptive methodology (Misra and Ghosh, 2024). This transformation is facilitated by the real-time data collecting capabilities of AI-driven technology, including drones, satellites, and terrestrial sensors, which provide farmers with comprehensive insights into crop health, soil conditions, and environmental factors. Utilizing this extensive knowledge, farmers may make informed choices about irrigation, fertilization, and pest management, thus decreasing waste, optimizing resource distribution, and diminishing agriculture's environmental impact (Patel et al., 2023). The diverse uses of AI in precision agriculture include not only data analysis but also autonomous systems and robots. Intelligent machines utilizing AI algorithms are transforming agricultural practices, encompassing planting, harvesting, and crop maintenance. Artificial intelligence and machine learning in agriculture are revolutionizing farming systems; these autonomous technologies improve operational efficiency and mitigate labor shortages, fostering a more sustainable and economically viable future for agriculture. With the rise of data-centric precision agriculture, ethical considerations, data security, and equal access to technology have become key concerns (Wilgenbusch et al., 2022). This analysis examines the ethical ramifications of AI in agriculture, highlighting the necessity of responsible data management and addressing potential inequalities in technology adoption. Additionally, it examines the obstacles related to the digital divide in rural regions, highlighting the necessity for inclusive methods that guarantee all farmers can reap the benefits of developments in precision agriculture (Robinson et al., 2020). This seeks to deliver a comprehensive analysis of AI in precision agriculture, presenting insights into the technologies influencing sustainable farming methods. This analysis clarifies the present status of AI integration, the challenges encountered, and future possibilities, thereby enriching the ongoing discussion on utilizing advanced technologies to address the increasing demands of global food production while maintaining environmental stewardship and long-term agricultural sustainability.

Satellite technology offers an aerial perspective of agricultural terrains, delivering critical knowledge into crop conditions, growth trends, and overall health (Khan et al., 2023). High-resolution satellite photography allows farmers to easily monitor extensive land areas, pinpointing regions that may need particular attention, such as pest infestations or nutrient deficits. The ongoing developments in satellite technology have improved temporal and geographical resolution, establishing it as an essential instrument for precision agriculture (Fotso Kamga et al., 2021). Unmanned aerial vehicles, or drones, have become valuable instruments for precision agriculture. Drones, outfitted with cameras and sensors, can obtain high-resolution images and gather data with remarkable accuracy. Drones allow farmers to observe crops with greater geographical precision, providing comprehensive data on plant health, growth discrepancies, and possible problems. The nimbleness and availability of drones render them especially effective for prompt and precise interventions (Rejeb et al.,

2021). Implementing ground-based sensors in the field yields immediate, localized data on several agricultural factors. These sensors can quantify soil moisture, nutritional composition, temperature, and other essential parameters affecting crop vitality. The data obtained from these sensors enable accurate decision-making, permitting farmers to customize irrigation and fertilization methods according to the distinct requirements of various sections within the same field (Mutyalamma et al., 2020). The incorporation of AI facilitates real-time data collection from many sources, such as satellites, drones, and terrestrial sensors. This ongoing flow of data facilitates real-time monitoring of crop conditions, allowing farmers to react swiftly to arising problems. Real-time data captures adaptive and responsive agricultural methods, enhancing efficiency and sustainability.

AI algorithms evaluate data from various monitoring sources to determine the overall health of crops (Pimenov et al., 2023). By recognizing patterns linked to healthy and stressed plants, these algorithms can spot early indicators of illnesses, nutrient deficits, or pest infestations. This proactive strategy enables farmers to execute timely actions, reducing the effects of possible hazards and enhancing agricultural production. Timely identification facilitates focused interventions, decreasing reliance on broad-spectrum treatments and lessening the ecological consequences of pest management strategies. These technologies provide farmers with unparalleled insights, enabling informed decision-making and promoting the sustainable and efficient management of agricultural resources (Liang and Shah, 2023).

The integration of Machine Learning (ML) with Decision Support Systems (DSS) in precision agriculture has become a formidable force, providing farmers with data-driven insights and predictive analytics. This synergy promotes informed decision-making, improves resource management, and supports the overall sustainability of agricultural operations. Machine learning algorithms excel at analyzing extensive datasets, identifying significant patterns, and producing predictions. In decision support systems, this functionality allows farmers to make data-informed decisions utilizing historical data, present circumstances, and predictive analytics (Beriya and Saroja, 2019). Machine Learning is essential for enhancing the precision of weather forecasting in decision support systems. Machine learning algorithms examine historical weather data, satellite images, and real-time meteorological information to deliver more accurate and localized weather forecasts (Salcedo-Sanz et al., 2020). Precise weather predictions allow farmers to enhance irrigation planning, prepare for unfavorable weather occurrences, and reduce the effects of climatic fluctuations on agricultural output. Machine learning algorithms can reveal patterns and trends in historical data that may be obscured by conventional methods (Sarker, 2021). Ensuring the dependability of predictive models, rectifying data quality concerns, and delivering user-friendly interfaces are key factors. The ethical ramifications of depending on algorithmic decision-making in agriculture require thorough scrutiny.

The integration of Machine Learning and Decision Support Systems represents a notable progression in precision agriculture (Shorten et al., 2021). As technology advances, the integration of machine learning and decision support systems will likely be pivotal in determining the future of agriculture.

Efficient resource management is fundamental to sustainable agriculture. The incorporation of Artificial Intelligence (AI) technologies in precision agriculture has transformed the methods by which farmers maximize resource utilization, including water, fertilizers, and pesticides (Obi et al., 2024). This thorough examination investigates the role of AI in resource management, promoting a prudent and eco-friendly approach to agricultural activities. AI-driven smart irrigation solutions signify a transformative advancement in agricultural water management. AI-driven systems facilitate the accurate regulation of nutrient levels in the soil. By persistently monitoring and evaluating data pertaining to soil health, crop requirements, and nutrient composition, these systems deliver real-time insights into the nutritional necessities of plants. This detailed method guarantees that crops have the necessary nutrients at the appropriate time and in precise amounts, fostering optimal growth and reducing waste. The integration of AI in resource management enhances environmental sustainability in the agricultural sector (Okunade et al., 2023). AI mitigates environmental contamination, soil deterioration, and water body eutrophication by optimizing water and fertilizer usage through precise applications. The integration of AI in resource management signifies a pivotal transition towards precision agriculture (Chowdhury et al., 2023).

Automation and robots are essential elements of contemporary agriculture, transforming conventional farming methods and enhancing efficiency, production, and sustainability. These vehicles utilize AI algorithms to sow seeds at ideal depths and intervals, facilitating consistent crop development. During harvesting, sophisticated sensors and robotic arms facilitate selective and timely picking, minimizing waste and enhancing total yield efficiency (Rajendran et al., 2024). Autonomous vehicles utilizing robotic technology and AI algorithms accurately identify and target weeds or pests. This focused strategy diminishes the application of herbicides and insecticides, lessening environmental repercussions while safeguarding crop health. Robotics, utilizing computer vision and machine learning, executing autonomous weeding by differentiating between crops and weeds. This not only diminishes the necessity for pesticides but also mitigates manpower shortages, rendering weed management more sustainable and economically viable. Robotic arms fitted with cameras and sensors do meticulous pruning and thinning of crops. This degree of automation guarantees uniform and ideal spacing among plants, fostering healthier growth and enabling effective harvesting. Automation and robots mitigate issues related to labor shortages in agriculture. The utilization of autonomous machines for repetitive activities enables human labor to focus on more skilled and intricate facets of farming, hence enhancing overall operational efficiency. Although the initial expenditure on automation technology may be considerable, long-term financial sustainability is demonstrated by diminished labor costs, enhanced productivity, and superior produce quality. The comprehensive cost-effectiveness enhances the sustainability of contemporary agricultural techniques. The use of IoT technology facilitates uninterrupted connectivity among diverse robotic systems and agricultural machinery (Vermesan et al., 2020). This integrated network facilitates real-time data interchange, promoting adaptive decision-making and improving the overall efficiency of agricultural operations. Agriculturalists can remotely oversee and regulate robotic devices, making modifications depending on real-time data and evolving conditions. This degree of control guarantees that agricultural activities can be precisely adjusted for optimal results, even remotely (Dong et al., 2021). Automation and robots have initiated a novel epoch of accuracy and efficacy in agricultural practices. The amalgamation of AI, robotics, and communication technologies not only resolves conventional difficulties but also enhances the sustainability and economic feasibility of agriculture. As technology progresses, automation is set to play a pivotal role in transforming the future of agriculture worldwide.

8. AI IN WATER QUALITY MONITORING AND PROTECTION

Monitoring water quality is crucial for ensuring environmental sustainability and public health, since it guides efforts to reduce pollution and protect aquatic ecosystems. The UN Sustainable Development Goal 6 seeks to provide the accessibility and sustainable administration of water and sanitation for everyone. It emphasizes the provision of safe and inexpensive drinking water, access to adequate sanitation facilities, and the promotion of healthy hygiene practices. The objective underscores the need for safeguarding and rehabilitating aquatic ecosystems to preserve water quality and promotes the efficient and sustainable management of water resources. Water management is intricate and encompasses the monitoring of water quality and quantity characteristics, biodiversity, and aquatic life, the identification of pollution sources and pollutant removal, sanitation, flood protection, and resource allocation, among other factors.

The primary contributors to water contamination include effluents from metropolitan areas, leaks, and runoff from agricultural and industrial operations. Water monitoring encompasses numerous hydrological, physical, chemical, and biological factors, with some assessed on-site and others analyzed through laboratory sampling. Most nations implement monitoring plans that delineate sampling sites, parameters to be measured, and sampling frequency, which are linked to corresponding efforts and expenditures for labor, reagents, equipment, and other resources.

Every country or region possesses distinct quality standards that establish threshold values for metrics and classification systems to assess the condition of a water body and its suitability for various applications. In the European Union, this sector is governed by the Water Framework Directive (WFD), and all Member States send monitoring data to the European Environment Agency. The categorization of water bodies adheres to the One Out – All Out concept, indicating that the status is determined by the ranking of the most detrimental parameter. A water body cannot attain 'good' classification if any parameter is classified as 'bad'. Other nations have established classification systems utilizing water quality indexes (WQIs), which are dimensionless figures that consolidate the values of various chosen indicators.

There are different WQIs, but the process of calculating the WQI usually includes the following steps:

- Selection of relevant water quality parameters
- Assigning a weight to each parameter
- Calculation of sub-indexes (e.g. based on limit values for a certain quality class)
- Aggregation of sub-indexes into the WQI

According to the value of the WQI, the water is then categorized into quality classes, depending on the calculation method and water uses.

The WQI approach is arduous and possesses various limitations that may result in erroneous outcomes; yet, it has served as a valuable tool for evaluating overall water quality and its long-term trends for many years. Turbidity, a measure of water clarity, is a crucial element in Water Quality Index (WQI) computations as it directly indicates the quantity of suspended particles and sediments in water. Conventional turbidity monitoring techniques, including single-angle optical devices, sometimes encounter difficulties in delivering precise results in intricate and fluctuating situations. These approaches are especially vulnerable to sensor drift, environmental noise, and processing delays, which constrain their efficacy for real-time or large-scale monitoring applications (Ateia et al., 2024).

The current advancement of AI tools for evaluating water quality has the potential to yield substantial enhancements in this field. This may diminish monitoring efforts and expenses while enhancing the precision of the Water Quality Index (WQI) prediction and Water Quality Classification (WQC) in several manners, some of which are outlined below:

- Comprehensive analysis of monitoring data, succeeded by the computation of the Water Quality Index (WQI), may reveal which parameters exert the most significant influence on the WQI value, thereby facilitating a reduction in the number of monitored parameters or their frequency

- Intricate modeling of WQI values may enable the forecasting of WQI
- Artificial intelligence algorithms can categorize datasets into quality classes based on raw monitoring data without necessitating the calculation of WQIs

Furthermore, the integration of AI and remote sensing could potentially supplant conventional monitoring techniques with satellite data and real-time on-site sensor data (IoT – Internet of Things), thereby diminishing the costs and labor associated with water quality monitoring.

9. PUBLIC AWARENESS AND BEHAVIOR CHANGE WITH AI

Global water demand is increasing, particularly in highly populated regions, due to population expansion and urbanization. The spatial concentration of water demand in metropolitan regions affects the size of demand, peak intensity, distribution among usage sectors, and both indoor and outdoor consumption. This, combined with climate change and alterations in land use, is exacerbating the pressure on limited water supplies, resulting in both operational and environmental problems with water delivery. To address this changing landscape, conventional supply and management strategies must be modified to satisfy future demand while averting the unsustainable exploitation of resources. Considering geographical constraints and rising marginal costs that frequently hinder capacity expansion via infrastructural measures, demand-side management strategies are essential to augment supply-side interventions in ensuring a reliable water supply, as well as mitigating shortages and overall utility expenses (Beal et al., 2016). The efficacy of demand-side management interventions has been evidenced in numerous studies, particularly over the past twenty years, propelled by the implementation of various water conservation initiatives globally, especially in regions experiencing extended droughts (e.g., California) or periods of low recharge (e.g., Australia). Among the extensive array of demand-side management interventions available such as technological, financial, legislative, maintenance, and educational strategies numerous recent studies (Sønderlund et al., 2014) have focused on investigating the impact of behavioral change programs designed to encourage water conservation practices and enhance user awareness regarding water usage. A comprehensive examination of behavioral interventions has been feasible, particularly since the late 1990s, due to the advent of smart water meters, which facilitated the monitoring of water consumption data with exceptional spatial (household) and temporal (ranging from several minutes to a few seconds) sampling resolution (Cardell-Oliver, 2013). The success of effective water demand management relies on the cumulative impact of activities done by various individuals. This indicates that behavioral change initiatives based on tailored demand-side management strategies might influence the specific factors driving user demand and encourage environmentally sustainable practices.

Behavioral interventions and instructional tools about water issues, with gamified strategies and persuasive technology for water management, seek to enhance users' understanding and encourage positive behaviors towards water conservation through gaming mechanisms. Computer and social sciences provide an extensive array of gamified instruments to enhance environmental awareness across several sectors (Negruşa et al., 2015), including the water sector. A recent review of serious games in environmental management indicates that water management games represent a significant segment of this broader category, comprising over 20% of the 25 games analyzed, which were developed between 1994 and 2013. These games serve as valuable instruments for educational purposes (game-based learning), as they typically enhance motivation and promote engagement, and for aiding decision-making processes, particularly when multiple stakeholders with divergent interests are involved, or for validating behavioral models, such as agent-based models. Consequently, a diverse array of users, including students, learners, professionals, local stakeholders, and decision-makers, may derive benefits from utilizing such technologies. However, contemporary literature frequently lacks comprehensive descriptions of certain gamified strategies addressing environmental and water issues, along with quantitative metrics that facilitate an equitable evaluation of the efficacy of these tools for educational and awareness objectives, or a comparative analysis of various games based on standardized criteria (Madani et al., 2017). Our objective is to evaluate and critically assess the most pertinent research and industry gamified strategies for water management, emphasizing water awareness and behavioral modification. Furthermore, we provide the first findings about user engagement and water saving outcomes derived from the practical implementation of a gamified strategy.

10. TECHNICAL CONCEPTS AND METHODOLOGIES

Water utilities require support technologies to align water demand with supply. To achieve this objective, it is essential to comprehend the elements that influence demand by analyzing high-frequency mining water consumption data. The advancement of smart metering technologies and Internet of Things (IoT) protocols facilitate real-time data collection, while cloud computing and big data infrastructures enable utilities to store and process this data. Additionally, machine learning and pattern analysis allow for the extraction of valuable insights, including leakage identification and end-use disaggregation. Water utilities address significant concerns, including supplying consumers with educational resources on efficient water usage and developing effective campaigns to modify their behavioral practices. Gamification and persuasive interfaces present an effective approach to facilitate behavioral change, such as by offering consumers visualization tools to enhance their awareness of resource consumption and incorporating incentives and gamified activities designed to steer their behavior towards more sustainable practices. Ultimately, water utilities must navigate rapidly evolving policies, severe weather impacts, and various factors affecting their revenues; thus, they necessitate simulation tools to predict the consequences of diverse pricing strategies and analyze consumer responses. In this context, machine learning techniques, in conjunction with game theory, can be employed to address the issue. In this section, we will present fundamental concepts about smart metering technologies, big data analytics for end-use disaggregation and profiling, as well as persuasive technologies like visualization interfaces and gamification.

11. SMART METERING TECHNOLOGIES

Smart meters are electronic devices installed at consumers' locations to gather comprehensive household usage data, which is then transmitted to utilities for billing and monitoring purposes. Moreover, smart meters facilitate bidirectional communication, so allowing for remote control and monitoring of the grid's overall operation, as well as assisting in the detection of malfunctions and leaks. Smart metering systems utilize a Local Area Network (LAN) to convey individual consumption data to a local data collector. This transmission may occur at various frequencies (e.g., daily or hourly), depending on the meter's capabilities and the specific application of the collected data. The collector acquires the data and may perform some preparation. Data is subsequently delivered across a Wide Area Network (WAN) to the utility for processing and utilization by business applications. From the customer's perspective, a primary advantage of smart metering is the facilitation of real-time or near real-time consumption monitoring (Britton et al., 2013), thereby enhancing comprehension and awareness of energy usage, perhaps resulting in more effective consumption practices. Moreover, smart metering empowers utility companies to enhance customer service through on-demand readings, decreased meter reading service expenses, expedited response times to customer inquiries, improved billing accuracy, and diminished non-revenue water losses due to enhanced leak detection. Ultimately, by analyzing the extensive volume of aggregated consumption data using statistical methodologies, organizations gain a robust instrument for strategizing and predicting the creation and allocation of resources based on real-time information. Notwithstanding the advantages of smart metering systems, challenges associated with their deployment and public perception have impeded their acceptance. A primary problem is the privacy and security of consumption data. Despite the implementation of secure design methodologies across all components of the metering system encompassing access control, data validation, error checking, and data encryption the extensive utilization of utility backhaul communications and public wireless networks has heightened vulnerability to potential security breaches. Utilities are worried not only with unwanted access to consumer data but also with potential unauthorized data manipulation aimed at reducing bills (Castelletti et al., 2017). Another concern is the accuracy of the meters: despite the high precision of smart devices, individual meters remain susceptible to hardware and software malfunctions. Consequently, utilities must establish monitoring techniques and regulate operational procedures to detect anomalies and errors, rectifying them before the execution of the billing process. The primary challenge is the duration and expense associated with establishing the smart metering infrastructure: the implementation phase may span from months to years and necessitates effective planning strategies to mitigate costs, technical difficulties, and logistical concerns. Companies incur additional concealed expenses, such as managing the infrastructure for big data storage, acquiring or developing analytical tools, and adapting existing systems to facilitate integration with the new infrastructure (Castelletti et al., 2017).

12. BIG DATA ANALYTICS FOR END USE DISAGGREGATION AND USER PROFILING

In recent years, the declining cost of storage media and the expansion in network capacity have rendered data storage and transfer comparatively simple and economical. This has compelled numerous firms across various sectors to amass substantial data volumes and derive valuable insights to enhance their operations through "big data" mining and analytical methodologies. The phrase "big data" denotes vast quantities of diverse unstructured data generated at high bit rates across several application scenarios (Cuzzocrea et al., 2011), such as social networks, sensor data streams, and scientific and enterprise applications. Big data analytics encompass a collection of knowledge discovery and data mining approaches employed to reveal structures, trends, and patterns within large datasets. These strategies are employed across various domains, including medicine, scientific research, computer performance enhancement, financial market forecasting, insurance, consumer marketing profiling, and resource optimization. In water management, big data and analytics have gained increasing significance since the introduction of smart metering technologies in the late 1990s, which provided new residential water consumption data at exceptionally high spatial (household) and temporal (from several minutes to a few seconds) resolution. This data may include comprehensive insights into individual consumers' water consumption habits and demand patterns, which can be utilized for the creation and assessment of various consumer-specific water demand management techniques. The necessity for automated tools to extract pertinent information for demand management from extensive smart metered datasets has driven the advancement of data analytics techniques and mathematical models to precisely delineate sub-daily water consumption patterns and end-use water consumption profiles. A recent study by comprehensively reviews multiple studies on high-resolution residential water demand modeling and management, emphasizing that data analysis and modeling are crucial for elucidating water users' behavior, particularly for (i) automatic water end-use characterization and (ii) user modeling. Water end-use characterization entails the disaggregation of total home water use into various end-use categories (e.g., showering, toilet flushing, irrigation, etc.). There is a necessity for automatic and scalable algorithms capable of conducting water end-use characterization, as direct monitoring of end uses is impractical due to the substantial expenses associated with installing multiple sensors and the limited acceptability among water consumers due to its invasive nature. End-use disaggregation remains an unresolved research challenge, with numerous techniques described in the literature (Zoha et al., 2012). Most of them consist of two primary phases:

- **Observation phase:** the actual consumption of appliances must be measured (intrusively) over a specified duration to establish reference consumption patterns for individual end uses, sometimes referred to as "signatures." This information is utilized to establish the "consumption base" or "baseline" and to calibrate the end use disaggregation algorithms.
- **Classification phase:** During this phase, machine learning approaches, often grounded in optimization and pattern recognition methods, are employed to decompose the aggregate consumption data into individual end-use contributions following algorithm training. The predominant algorithms employed in disaggregation problems include decision trees, classifiers, and clustering algorithms (Cominola et al., 2015).

User modeling employs data mining techniques to identify valuable characteristics in water consumption data, facilitating the modeling of diverse water demand profiles at the household level, potentially influenced by natural and socio-psychographic factors, as well as users' reactions to various demand management interventions. posits that user modeling encompasses two categories: descriptive and predictive models. The objective of descriptive models is to characterize patterns and trends in historical water demand time series, while predictive models aim to forecast future demands based on natural factors (e.g., temperature and seasonality) and sociopsychographic determinants (i.e., variables that define user demographics, values, attitudes, interests, or lifestyles, such as age, household size, income level, and conservation attitudes). Data-driven methodologies, multivariate and correlation analyses, data dimensionality reduction techniques (e.g., Principal Component Analysis), with feature extraction and clustering approaches, are frequently employed for the objectives. Consequently, as has traditionally occurred for various marketing objectives, the advancement of big data analytics techniques in the water sector is becoming essential for comprehending and characterizing the behavior of diverse consumers, ultimately facilitating the execution of tailored demand management strategies and personalized feedback to support water conservation and management initiatives.

Persuasive technologies aim to modify user attitudes or behaviors without compulsion or deception and are utilized in visualization interfaces, video games, mobile devices, and social media. Examples of persuasive technology can be observed in online retail platforms, healthcare systems, educational environments, resource management, and collaborative tools. Persuasive applications seek to embed themselves in user habits, functioning through a cycle of triggers stimuli that capture the user's attention actions, which are the user's responses to these triggers, and variable rewards, which recognize the user's actions in a manner that surprises and incites a desire for repetition. Employing persuasive technologies offers several advantages compared to conventional media or human persuaders. The paramount factor is ubiquity: because to the spread of mobile and wearable devices, smart meters, and home gadgets interconnected by IoT technologies, individuals can be targeted anywhere and at any time. The range of modalities is a significant advantage: computers and electronic devices can convey messages through numerous formats, including text, audio, video, graphs, and animations, which enhances persuasive impact. Ultimately, computers have the capacity to store and retrieve vast quantities of information, allowing them to provide new data, statistics, and references in a manner unparalleled by other media. Visualization Interfaces Visualization has emerged as a prominent subject in computer science, defined as the utilization of computer-assisted, interactive visual representations of abstract data to enhance cognitive processes. The visualization of information pertains to the design and aesthetic elements of computer interfaces that display intricate abstract data clearly and understandably, facilitating user analysis through visual aids. Data visualization utilizes technical disciplines such as information retrieval, data mining, computer graphics, and human-computer interaction. Another notable aspect of information visualization is its persuasive potential on users; research (Pandey et al., 2014) indicates that visual data representations can sway personal opinions on a subject by facilitating data interpretation, thereby providing feedback that challenges initial beliefs. While ethical concerns regarding the misuse of visualization for deception or manipulation are still being debated, the persuasive power of visualization can be effectively employed to shape users' behaviors. The notion of gamification has gained significant traction in both academic research and commercial implementations. Despite sharing numerous characteristics with gamification, serious games and gamified applications are fundamentally distinct: serious games are fully functional games designed for specific purposes, such as subject education or skill training, whereas gamified applications enhance existing systems, services, tasks, or processes by incorporating game-based mechanisms that are not essential for their operation or execution. The primary objectives of gamification are to engage users and improve their experience, with the intent of augmenting user participation in operational or social interactions or enhancing user productivity. Gamification can be employed to define objectives for users across many settings (e.g., skill acquisition): forcing users to fulfill a predetermined criterion is recognized as an effective motivational tactic (Hamari et al., 2014). To build a gamified application, it is essential to identify the target audience and comprehend their motives for gameplay. Bartle introduced a player taxonomy grounded in user incentives and interactions with the game and fellow players, comprising four categories:

- **Achievers:** they are competitive and relish overcoming formidable hurdles, whether imposed by the game or self-generated. The more arduous the aim, the greater the sense of reward they typically experience. Most children appreciate witnessing advancement through points, level completion, or a comparable indication of progression.
- **Explorers**: they enjoy investigating the world, encompassing both its geography and the intricate nuances of the game mechanics. They endeavor to master all mechanics, shortcuts, techniques, and glitches within the game, often playing it multiple times to achieve this proficiency.
- **Socializers:** they frequently prioritize interpersonal relationships with other players over engaging in the game itself. The game serves solely as a medium for social interaction, whether within the game or externally, allowing users to engage with friends.
- **Killers:** They are motivated by competition with others and favor human opponents over computer-controlled artificial intelligence.

Figure 7. Bartle's player taxonomy

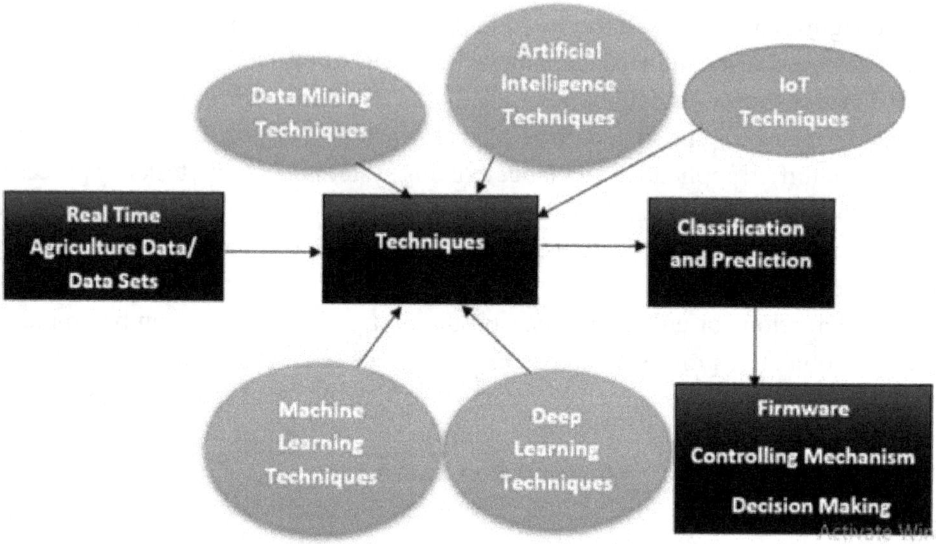

The gamification designer might consider the intended player type to devise game mechanisms that effectively engage them. When targeting an explorer, the designer must consider the scale of the planet and the variety of geographies and interactions. Conversely, if the objective is the cohort of socializers, emphasis should be placed on the game's personalized aspects and the systems of communication and collaboration. Four essential things are required to build a game. The initial aspect is the game mechanism, which delineates the processes and regulations of the game: the mechanics articulate the objective, the methods to attain it, and the consequences of player actions; selecting the mechanic determines the gameplay experience. Numerous strategies are commonly employed to gamify routine chores (Schell, 2008).

Achievements and badges are generally conferred when a player attains a specific level or quantity of points inside the game. They may necessitate the user to execute tasks repeatedly, attain proficiency in a skill, or investigate the environment.

- **Progress Bar:** it signifies the user's proximity to achieving the specified objective. Upon the player's completion of a task or activity, the progress bar is filled to signify advancement towards a goal. The movement of the progress bar is frequently dictated by the task's intensity or by the allocation of points. The progress bar is frequently integrated with experience points, whereby the accumulation of experience points dictates the advancement of the progress bar.

- **Experience Points:** Experience points serve as a metric for experience a player has accumulated within a game or environment. Points may be earned by accomplishing tasks, discovering locations and attributes, or engaging in analogous activities. Experience points are typically associated with a leveling system, wherein attaining specific thresholds unlocks new levels, rewards, or features. Experience points are frequently integrated with leaderboards.
- **Leaderboards:** are lists of participants ranked according to their accumulated points, activities performed, or other specified metrics. Each user possesses a score established by predetermined rules prior to the commencement of a tournament; this score is juxtaposed with that of other participants, and players are ranked according to their scores. Leaderboards can be entirely dynamic or state-dependent, meaning that a new ranking is recalibrated after a specified duration.
- **Contests**: they are classified as either dual-style head-to-head competitions or unrestricted free-for-all contests with no participant capacity limitations. A duel-like event may have a knockout format where participants strive to achieve the greatest score within a specified timeframe or a comparable objective. In a free-for-all competition, the victor is the individual who achieves a designated objective most effectively. The objective that players endeavor to attain is established by a specific set of regulations determined prior to the commencement of the event.
- Real-world Rewards: In conjunction with leaderboards and contests, tangible incentives may be conferred upon the top players of the game. The incentives are typically associated with the application's subject matter.
- **Social networking**: offers users a shared platform to disseminate their accomplishments and current activities, since individuals often derive a sense of gratification when their posts garner responses. It is feasible to connect the game to established social networks such as Facebook or Twitter, or to establish specialized forums akin to those on the Steam platform. The second part is the narrative, encompassing the sequence of events that transpire throughout the game as well as those that transpired before its commencement. The narrative may be linear or subject to alteration based on user activities, so it is intrinsically connected to the game mechanics that facilitate the storytelling. The third part is aesthetics, encompassing the visual design, music, and sound that accompany the storytelling and mechanics. Aesthetic components establish the most immediate connection with user experience, significantly affecting engagement and immersion.

The fourth factor that unifies them is the enabling technology: "it is fundamentally the medium in which the aesthetics manifest, the mechanics transpire, and the narrative unfolds." Schell, 2008. Selecting a certain technology facilitates specific functions while potentially complicating others; hence, making the appropriate choice is a nuanced issue.

Figure 8. Bartle's player taxonomy

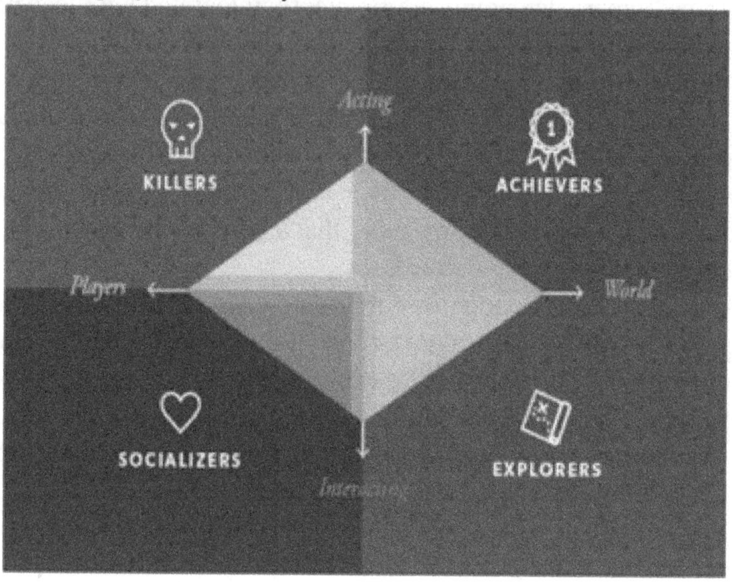

13. CHALLENGES AND CONSIDERATIONS

- Data security and privacy concerns in AI-based water management systems.
- Ensuring equitable access to AI-powered water solutions for all communities.
- The need for human expertise alongside AI for effective water management.

14. THE FUTURE OF AI IN WATER CONSERVATION

The integration of AI in water resource management has seen substantial transformations in recent decades, driven by advancements in AI technology and the growing intricacy of water-related challenges. Since the late twentieth century, AI has been employed in this domain since deterministic or statistical hydrological models were inadequate for managing the intricate and dynamic water systems. These traditional models require additional calibrations and exhibit limited flexibility in response to rapidly changing environmental circumstances resulting from climate change and human activities (Krysanova et al., 2018).

The application of AI in water resource management commenced in the early 1980s and continued until the early 1990s, a period that also witnessed the advent of machine learning. Artificial Neural Networks (ANNs) were among the initial AI techniques employed in hydrology, demonstrating efficacy in estimating complex nonlinear relationships between climatic inputs and hydrological outputs, such as stream flow and groundwater levels. Preliminary studies revealed that artificial neural networks (ANNs) demonstrated enhanced predictive accuracy over traditional models in river flow analysis, particularly in scenarios constrained by limited basins and historical usage data, enabling the network to discern complex, non-linear interrelationships among the variables. These achievements facilitated the comprehensive implementation of AI in water resources management.

The ongoing advancement of AI technology in the 2000s introduced increasingly advanced techniques such as Support Vector Machines (SVMs) and fuzzy logic systems into the domain. These strategies provided enhancements over conventional approaches regarding the management of uncertainties intrinsic to hydrological processes, such as variations in precipitation patterns, soil moisture levels, and alterations in land use that influence water needs. certain AI models began to offer realistic answers to traditional methods where data constraints posed certain challenges in certain domains (Rajaee et al., 2019). The superiority of AI over traditional statistical methods in modeling complex systems marked a pivotal moment in water resources management, enhancing the accuracy of predictions and decision-making.

During the 2010s, deep learning methodologies emerged as a pivotal advancement in artificial intelligence for water resources management. Machine learning is a subset of artificial intelligence that specifically employs multilayer neural networks for the pattern recognition of extensive datasets. LSTM, a member of the RNN category, was specifically developed for time series forecasting. LSTMs have been more effective in hydrology, where they have been utilized for forecasting streamflow, rainfall, and floods at various temporal resolutions (Khand and Senay, 2024). These deep learning algorithms have significantly improved the precision and dependability of hydrological forecasts, particularly in areas with scarce historical data, where conventional models frequently encounter difficulties. The efficacy of these models has resulted in their extensive utilization in diverse hydrological research, underscoring the transformational capacity of AI in this domain.

Besides hydrological modeling, artificial intelligence has profoundly influenced water management strategies, especially in flood prediction and risk mitigation. Machine learning algorithms, including Random Forest (RF) and Gradient Boosting

Machines (GBMs), have been employed to analyze extensive datasets from remote sensing, meteorological stations, and IoT sensors to predict flood risks with exceptional precision. AI-driven models have been utilized in flood-prone areas, including the Yangtze River Basin in China and the Ganges River Basin in India, offering critical insights for disaster preparedness and mitigation methods (Chu et al., 2022).

In the last twenty years, AI methodologies have been developed as solutions to tackle the complexity, dynamism, non-linearity, and non-stationarity inherent in hydrological processes (Apaydin et al., 2021). The increase in extreme natural catastrophes resulting from climate change and global warming has presented substantial challenges to sustainable hydrology and water resource management. As a result, there has been a significant rise in the investigation of AI models to analyze and forecast hydrological variability amid escalating hydro-geo-meteorological uncertainty. Artificial intelligence methods present viable alternatives or enhancements to traditional physically based or statistical hydrological modeling methodologies. Utilizing data from various sources, such as micro sensing, imaging, in situ, and remote sensing devices, AI methodologies are now facilitating the development of dependable and resilient hydrological models at higher spatiotemporal resolutions, essential for tackling complex nonlinear hydro-meteorological processes (Singh et al., 2021). Consequently, investigating novel AI models is crucial for the improved allocation, regulation, and conservation of water resources, thereby enhancing their sustainable management.

15. CONCLUSION

AI-powered solutions have emerged as transformative tools in the pursuit of sustainable water conservation. By integrating artificial intelligence into smart water management, we can significantly enhance efficiency, reduce waste, and optimize resource allocation. This chapter has explored various AI applications, from leak detection and predictive maintenance to demand forecasting, smart irrigation, and precision agriculture, all of which contribute to more effective water conservation strategies.

Furthermore, AI-driven water quality monitoring and public awareness initiatives are fostering informed decision-making and behavioral changes, ensuring that both individuals and industries adopt more responsible water usage practices. The incorporation of smart metering technologies and big data analytics allows for detailed user profiling and end-use disaggregation, enabling a more precise understanding of water consumption patterns.

Despite the significant advantages, challenges such as data privacy concerns, high implementation costs, and the need for robust infrastructure must be addressed to maximize AI's impact. Overcoming these hurdles requires collaboration among governments, industries, and research institutions to develop scalable, cost-effective, and ethical AI solutions.

Looking ahead, AI's role in water conservation will continue to expand, driven by advancements in machine learning, IoT, and real-time analytics. Future innovations will further enhance predictive capabilities, automation, and adaptive water management systems, ensuring a more sustainable and resilient approach to global water challenges. By harnessing AI's full potential, we can pave the way for smarter, more efficient water conservation practices that safeguard this vital resource for future generations.

REFERENCES

Alizadeh, Z., & Yazdi, J. (2023). Calibration of hydrological models for ungauged catchments by automatic clustering using a differential evolution algorithm: The Gorganrood river basin case study. *Journal of Hydroinformatics*, *25*(3), 645–662. DOI: 10.2166/hydro.2023.081

Anastasi, C. (2024). *Desalination and Water Security*. Taylor & Francis.

Antunes, A., Andrade-Campos, A., Sardinha-Lourenço, A., & Oliveira, M. S. (2018). Short-term water demand forecasting using machine learning techniques. *Journal of Hydroinformatics*, *20*(6), 1343–1366. DOI: 10.2166/hydro.2018.163

Apaydin, H., Taghi Sattari, M., Falsafian, K., & Prasad, R. (2021). Artificial intelligence modelling integrated with Singular Spectral analysis and Seasonal-Trend decomposition using Loess approaches for streamflow predictions. *Journal of Hydrology (Amsterdam)*, *600*, 126506. DOI: 10.1016/j.jhydrol.2021.126506

Ateia, M., Wei, H., & Andreescu, S. (2024). Sensors for Emerging Water Contaminants: Overcoming Roadblocks to Innovation. *Environmental Science & Technology*, *58*(6), 2636–2651. DOI: 10.1021/acs.est.3c09889 PMID: 38302436

Steffelbauer, D. B., Deuerlein, J., Gilbert, D., Abraham, E., & Piller, O. (2022). Pressure-leak duality for leak detection and localization in water distribution systems. *Journal of Water Resources Planning and Management*, *148*(3), 04021106.

Beal, C. D., Gurung, T. R., & Stewart, R. A. (2016). Demand-side management for supply-side efficiency: Modeling tailored strategies for reducing peak residential water demand. *Sustainable Production and Consumption*, *6*, 1–11. DOI: 10.1016/j.spc.2015.11.005

Beriya, A., & Saroja, V. N. (2019). *Data-Driven Decision Making for Smart Agriculture*. ICT India Working Paper.

Britton, T. C., Stewart, R. A., & O'Halloran, K. R. (2013). Smart metering: Enabler for rapid and effective post meter leakage identification and water loss management. *Journal of Cleaner Production*, *54*, 166–176. DOI: 10.1016/j.jclepro.2013.05.018

Cardell-Oliver, R. (2013). Water use signature patterns for analyzing household consumption using medium resolution meter data. *Water Resources Research*, *49*(12), 8589–8599. DOI: 10.1002/2013WR014458

Castelletti, A., Cominola, A., Facchini, A., Giuliani, M., Fraternali, P., ORTIZ HERRERA, S., ... & Rottondi, C. (2017). Gamified approaches for water management systems: an overview. Smart Water Grids: A Cyber-Physical Systems Approach, 169-192.

Chahal, C., van den Akker, B., Young, F., Franco, C., Blackbeard, J., & Monis, P. (2016). Pathogen and Particle Associations in Wastewater: Significance and Implications for Treatment and Disinfection Processes. *Advances in Applied Microbiology*, *97*, 63–119. DOI: 10.1016/bs.aambs.2016.08.001 PMID: 27926432

Chowdhury, S., Dey, P., Joel-Edgar, S., Bhattacharya, S., Rodriguez-Espindola, O., Abadie, A., & Truong, L. (2023). Unlocking the value of artificial intelligence in human resource management through AI capability framework. *Human Resource Management Review*, *33*(1), 100899. DOI: 10.1016/j.hrmr.2022.100899

Chu, K. S., Oh, C. H., Choi, J. R., & Kim, B. S. (2022). Estimation of threshold rainfall in ungauged areas using machine learning. *Water (Basel)*, *14*(6), 859.

Cominola, A., Giuliani, M., Piga, D., Castelletti, A., & Rizzoli, A. E. (2015). Benefits and challenges of using smart meters for advancing residential water demand modeling and management: A review. *Environmental Modelling & Software*, *72*, 198–214. DOI: 10.1016/j.envsoft.2015.07.012

Cuzzocrea, A., Song, I.-Y., & Davis, K. C. 2011. Analytics over large-scale multidimensional data: the big data revolution! *Proceedings of the ACM 14th international workshop on Data Warehousing and OLAP*. pp.101–104. DOI: 10.1145/2064676.2064695

D'Odorico, P., Chiarelli, D. D., Rosa, L., Bini, A., Zilberman, D., & Rulli, M. C. (2020). The global value of water in agriculture. *Proceedings of the National Academy of Sciences of the United States of America*, *117*(36), 21985–21993. DOI: 10.1073/pnas.2005835117 PMID: 32839335

Dalin, C., Wada, Y., Kastner, T., & Puma, M. J. (2017). Groundwater depletion embedded in international food trade. *Nature*, *543*(7647), 700–704. DOI: 10.1038/nature21403 PMID: 28358074

Dong, H., Zhang, J., & Zhao, X. (2021). Intelligent wind farm control via deep reinforcement learning and high-fidelity simulations. *Applied Energy*, *292*, 116928. DOI: 10.1016/j.apenergy.2021.116928

Fotso Kamga, G. A., Bitjoka, L., Akram, T., Mengue Mbom, A., Rameez Naqvi, S., & Bouroubi, Y. (2021). Advancements in satellite image classification: Methodologies, techniques, approaches and applications. *International Journal of Remote Sensing*, *42*(20), 7662–7722. DOI: 10.1080/01431161.2021.1954261

Ganivet, E. (2020). Growth in human population and consumption both need to be addressed to reach an ecologically sustainable future. *Environment, Development and Sustainability*, *22*(6), 4979–4998. DOI: 10.1007/s10668-019-00446-w

Giwa, A., & Dindi, A. (2017). An investigation of the feasibility of proposed solutions for water sustainability and security in water-stressed environment. *Journal of Cleaner Production*, *165*, 721–733. DOI: 10.1016/j.jclepro.2017.07.120

Greve, P., Kahil, T., Mochizuki, J., Schinko, T., Satoh, Y., Burek, P., Fischer, G., Tramberend, S., Burtscher, R., Langan, S., & Wada, Y. (2018). Global assessment of water challenges under uncertainty in water scarcity projections. *Nature Sustainability*, *1*(9), 486–494. DOI: 10.1038/s41893-018-0134-9

Gubbi, J., Buyya, R., Marusic, S., & Palaniswami, M. (2013). Internet of Things (IoT): A vision, architectural elements, and future directions. *Future Generation Computer Systems*, *29*(7), 1645–1660. DOI: 10.1016/j.future.2013.01.010

Hamari, J., Koivisto, J., & Sarsa, H. (2014). Does gamification work?--a literature review of empirical studies on gamification. *2014 47th Hawaii international conference on system sciences*. Ieee. pp.3025–3034.

Jalili, A. A., Najarchi, M., Shabanlou, S., & Jafarinia, R. (2023). Multi-objective Optimization of water resources in real time based on integration of NSGA-II and support vector machines. *Environmental Science and Pollution Research International*, *30*(6), 16464–16475. DOI: 10.1007/s11356-022-22723-4 PMID: 36190637

Janani, M., & Jebakumar, R. (2019). A study on smart irrigation using machine learning. *Cell Cell. Life Sci. J.*, *4*, 1–8.

Jasim, N. A. (2020). The design for wastewater treatment plant (WWTP) with GPS X modelling. *Cogent Engineering*, *7*(1), 1723782. DOI: 10.1080/23311916.2020.1723782

Karunathilake, E., Le, A. T., Heo, S., Chung, Y. S., & Mansoor, S. (2023). The path to smart farming: Innovations and opportunities in precision agriculture. *Agriculture*, *13*(8), 1593. DOI: 10.3390/agriculture13081593

Khan, A., Hassan, M., & Shahriyar, A. K. (2023). Optimizing onion crop management: A smart agriculture framework with IoT sensors and cloud technology. *Appl. Res. Artif. Intell. Cloud Comput.*, *6*, 49–67.

Khand, K., & Senay, G. B. (2024). Evaluation of streamflow predictions from LSTM models in water- and energy-limited regions in the United States. *Machine Learning with Applications, 16*, 100551. DOI: 10.1016/j.mlwa.2024.100551

Krysanova, V., Donnelly, C., Gelfan, A., Gerten, D., Arheimer, B., Hattermann, F., & Kundzewicz, Z. W. (2018). How the performance of hydrological models relates to credibility of projections under climate change. *Hydrological Sciences Journal, 63*(5), 696–720. DOI: 10.1080/02626667.2018.1446214

Kumar, V., Unal, S., Bhagat, S. K., & Tiyasha, T. (2024). A data-driven approach to river discharge forecasting in the Himalayan region: Insights from Aglar and Paligaad rivers. *Results in Engineering, 22*, 102044. DOI: 10.1016/j.rineng.2024.102044

Liakos, K. G., Busato, P., Moshou, D., Pearson, S., & Bochtis, D. (2018). Machine learning in agriculture: A review. *Sensors (Basel), 18*(8), 2674. DOI: 10.3390/s18082674 PMID: 30110960

Liang, C., & Shah, T. (2023). IoT in agriculture: The future of precision monitoring and data-driven farming. *Eig. Rev. Sci. Technol., 7*, 85–104.

Madani, K., Pierce, T. W., & Mirchi, A. (2017). Serious games on environmental management. *Sustainable Cities and Society, 29*, 1–11. DOI: 10.1016/j.scs.2016.11.007

Mahardhika, S. P., & Putriani, O. (2023). Deployment and use of Artificial Intelligence (AI) in water resources and water management. *IOP Conference Series. Earth and Environmental Science, 1195*(1), 1195. DOI: 10.1088/1755-1315/1195/1/012056

Malviya, A., & Jaspal, D. (2021). Artificial intelligence as an upcoming technology in wastewater treatment: A comprehensive review. *Environmental Technology Reviews, 10*(1), 177–187. DOI: 10.1080/21622515.2021.1913242

Mekonnen, M. M., & Hoekstra, A. Y. (2020). Blue water footprint linked to national consumption and international trade is unsustainable. *Nature Food, 1*(12), 792–800. DOI: 10.1038/s43016-020-00198-1 PMID: 37128061

Mishra, R. K. (2023). Fresh water availability and its global challenge. *Br. J. Multidiscip. Adv. Stud., 4*(3), 1–78. DOI: 10.37745/bjmas.2022.0208

Misra, S., & Ghosh, A. (2024). *Agriculture paradigm shift: a journey from traditional to modern agriculture. Biodiversity and Bioeconomy*. Elsevier.

Muhammad, A. U., Li, X., & Feng, J. (2019). Artificial Intelligence Approaches for Urban Water Demand Forecasting: A Review BT - Machine Learning and Intelligent Communications. In Zhai, X. B., Chen, B., & Zhu, K. (Eds.), *Springer International Publishing* (pp. 595–622).

Mutyalamma, A. V., Yoshitha, G., Dakshyani, A., & Padmavathi, B. V. (2020). Smart agriculture to measure humidity temperature moisture Ph. and nutrient values of the soil using IoT. *International Journal of Engineering and Advanced Technology, 9*(5), 9. DOI: 10.35940/ijeat.D8940.069520

Negruşa, A. L., Toader, V., Sofică, A., Tutunea, M. F., & Rus, R. V. (2015). Exploring gamification techniques and applications for sustainable tourism. *Sustainability (Basel), 7*(8), 11160–11189. DOI: 10.3390/su70811160

Obi, O. C., Akagha, O. V., Dawodu, S. O., Anyanwu, A. C., Onwusinkwue, S., & Ahmad, I. A. I. (2024). Comprehensive review on cybersecurity: Modern threats and advanced defense strategies. *Comput. Sci. IT Res. J., 5*(2), 293–310. DOI: 10.51594/csitrj.v5i2.758

Okunade, B. A., Adediran, F. E., Maduka, C. P., & Adegoke, A. A.Beatrice Adedayo OkunadeFoluke Eyitayo AdediranChinedu Paschal MadukaAdebukola Adejumoke Adegoke. (2023). Community-based mental health interventions in Africa: A review and its implications for US healthcare practices. *Int. Med. Sci. Res. J., 3*(3), 68–91. DOI: 10.51594/imsrj.v3i3.621

Pandey, A. V., Manivannan, A., Nov, O., Satterthwaite, M., & Bertini, E. (2014). The persuasive power of data visualization. *IEEE Transactions on Visualization and Computer Graphics, 20*(12), 2211–2220. DOI: 10.1109/TVCG.2014.2346419 PMID: 26356935

Patel, A., A. Mahore, R.D. Nalawade, A. Upadhyay and V. Choudhary. (2023). Advancements in Precision Agriculture: Harnessing the Power of Artificial Intelligence and Drones in Indian Agriculture. *World Environ. Day* 43.

Pimenov, D. Y., Bustillo, A., Wojciechowski, S., Sharma, V. S., Gupta, M. K., & Kuntoğlu, M. (2023). Artificial intelligence systems for tool condition monitoring in machining: Analysis and critical review. *Journal of Intelligent Manufacturing, 34*(5), 2079–2121. DOI: 10.1007/s10845-022-01923-2

Rajaee, T., Ebrahimi, H., & Nourani, V. (2019). A review of the artificial intelligence methods in groundwater level modeling. *Journal of Hydrology (Amsterdam), 572*, 336–351. DOI: 10.1016/j.jhydrol.2018.12.037

Rajendran, V., Debnath, B., Mghames, S., Mandil, W., Parsa, S., Parsons, S., & Ghalamzan-E., A. (2024). Towards autonomous selective harvesting: A review of robot perception, robot design, motion planning and control. *Journal of Field Robotics, 41*(7), 2247–2279. DOI: 10.1002/rob.22230

Ravichandran, T., Gavahi, K., Ponnambalam, K., Burtea, V., & Mousavi, J. S. (2021). Ensemble-based machine learning approach for improved leak detection in water mains. *Journal of Hydroinformatics*, *23*(2), 307–323. DOI: 10.2166/hydro.2021.093

Ray, S. S., Verma, R. K., Singh, A., Ganesapillai, M., & Kwon, Y.-N. (2023). A holistic review on how artificial intelligence has redefined water treatment and seawater desalination processes. *Desalination*, *546*, 116221. DOI: 10.1016/j.desal.2022.116221

Rejeb, A., Rejeb, K., Simske, S., & Treiblmaier, H. (2021). Humanitarian drones: A review and research agenda. *Internet of Things : Engineering Cyber Physical Human Systems*, *16*, 100434. DOI: 10.1016/j.iot.2021.100434

Robinson, L., Schulz, J., Dodel, M., Correa, T., Villanueva-Mansilla, E., Leal, S., Magallanes-Blanco, C., Rodriguez-Medina, L., Dunn, H. S., Levine, L., McMahon, R., & Khilnani, A. (2020). Digital inclusion across the Americas and Caribbean. *Social Inclusion (Lisboa)*, *8*(2), 244–259. DOI: 10.17645/si.v8i2.2632

Romero-Ben, L., Alves, D., Blesa, J., Cembrano, G., Puig, V., & Duviella, E. (2023). Leak detection and localization in water distribution networks: Review and perspective. *Annual Reviews in Control*, *55*, 392–419. DOI: 10.1016/j.arcontrol.2023.03.012

Rosa, L., Chiarelli, D. D., Rulli, M. C., Dell'Angelo, J., & D'Odorico, P. (2020a). Global agricultural economic water scarcity. *Science Advances*, *6*(18), eaaz6031. DOI: 10.1126/sciadv.aaz6031 PMID: 32494678

Salcedo-Sanz, S., Ghamisi, P., Piles, M., Werner, M., Cuadra, L., Moreno-Martínez, A., Izquierdo-Verdiguier, E., Muñoz-Marí, J., Mosavi, A., & Camps-Valls, G. (2020). Machine learning information fusion in Earth observation: A comprehensive review of methods, applications and data sources. *Information Fusion*, *63*, 256–272. DOI: 10.1016/j.inffus.2020.07.004

Sarker, I. H. (2021). Machine learning: Algorithms, real-world applications and research directions. *SN Computer Science*, *2*(3), 160. DOI: 10.1007/s42979-021-00592-x PMID: 33778771

Schell, J. (2008). *The Art of Game Design: A book of lenses*. CRC press. DOI: 10.1201/9780080919171

Sharma, A., Sharma, A., Tselykh, A., Bozhenyuk, A., Choudhury, T., Alomar, M. A., & Sánchez-Chero, M. (2023). Artificial intelligence and internet of things oriented sustainable precision farming: Towards modern agriculture. *Open Life Sciences*, *18*(1), 20220713. DOI: 10.1515/biol-2022-0713 PMID: 37854322

Shorten, C., Khoshgoftaar, T. M., & Furht, B. (2021). Deep Learning applications for COVID-19. *Journal of Big Data*, *8*(1), 1–54. DOI: 10.1186/s40537-020-00392-9 PMID: 33457181

Singh, B., Sihag, P., Parsaie, A., & Angelaki, A. (2021). Comparative analysis of artificial intelligence techniques for the prediction of infiltration process. *Geol. Ecol. Landscapes*, *5*(2), 109–118. DOI: 10.1080/24749508.2020.1833641

Singh, V. (2024). *Water Resources. Textbook of Environment and Ecology*. Springer. DOI: 10.1007/978-981-99-8846-4

Soldevila, A., Boracchi, G., Roveri, M., Tornil-Sin, S., & Puig, V. (2022). Leak detection and localization in water distribution networks by combining expert knowledge and data-driven models. *Neural Computing & Applications*, *34*(6), 4759–4779. DOI: 10.1007/s00521-021-06666-4

Solomatine, D., L.M. See and R.J. Abrahart. (2008). Data-driven modelling: concepts, approaches and experiences. *Pract. hydroinformatics Comput. Intell. Technol. Dev. water Appl.* 17–30.

Sønderlund, A. L., Smith, J. R., Hutton, C., & Kapelan, Z. (2014). Using smart meters for household water consumption feedback: Knowns and unknowns. *Procedia Engineering*, *89*, 990–997. DOI: 10.1016/j.proeng.2014.11.216

Tetzlaff, D., Carey, S. K., McNamara, J. P., Laudon, H., & Soulsby, C. (2017). The essential value of long-term experimental data for hydrology and water management. *Water Resources Research*, *53*(4), 2598–2604. DOI: 10.1002/2017WR020838

Vallino, E., Ridolfi, L., & Laio, F. (2020). Measuring economic water scarcity in agriculture: A cross-country empirical investigation. *Environmental Science & Policy*, *114*, 73–85. DOI: 10.1016/j.envsci.2020.07.017

Vermesan, O., Bahr, R., Ottella, M., Serrano, M., Karlsen, T., Wahlstrøm, T., Sand, H. E., Ashwathnarayan, M., & Gamba, M. T. (2020). Internet of robotic things intelligent connectivity and platforms. *Frontiers in Robotics and AI*, *7*, 509753. DOI: 10.3389/frobt.2020.00104 PMID: 33501271

Wang, G., Su, W., Hu, B., Al-Huqail, A., Majdi, H. S., Algethami, J. S., Jiang, Y., & Ali, H. E. (2022). Assessment in carbon-based layered double hydroxides for water and wastewater: Application of artificial intelligence and recent progress. *Chemosphere*, *308*, 136303. DOI: 10.1016/j.chemosphere.2022.136303 PMID: 36084828

Wilgenbusch, J. C., Pardey, P. G., Hospodarsky, N., & Lynch, B. J. (2022). Addressing new data privacy realities affecting agricultural research and development: A tiered-risk, standards-based approach. *Agronomy Journal*, *114*(5), 2653–2668. DOI: 10.1002/agj2.20968

Zabihi, O., Siamaki, M., Gheibi, M., Akrami, M., & Hajiaghaei-Keshteli, M. (2023). A smart sustainable system for flood damage management with the application of artificial intelligence and multi-criteria decision-making computations. *International Journal of Disaster Risk Reduction*, *84*, 103470. DOI: 10.1016/j.ijdrr.2022.103470

Zoha, A., Gluhak, A., Imran, M. A., & Rajasegarar, S. (2012). Non-intrusive load monitoring approaches for disaggregated energy sensing: A survey. *Sensors (Basel)*, *12*(12), 16838–16866. DOI: 10.3390/s121216838 PMID: 23223081

About the Contributors

Jabulani Garwi is a distinguished academic researcher, diplomat, and development practitioner with a focus on sustainable development and African development studies. Currently affiliated with the University of the Free State, South Africa, he possesses a diverse interdisciplinary background. Dr Garwi holds a Bachelor of Arts Honors degree in Economic History from the University of Zimbabwe, a Master of Arts degree in Development Studies from Midlands State University, and a Master of Science in Climate Change and Sustainable Development from Bindura University of Science Education. Furthermore, he has acquired postgraduate qualifications in Sustainable Development and Sustainable Project Management, Monitoring, and Evaluation. His academic journey culminated in the attainment of a Doctor of Philosophy degree in Sustainable Agriculture from the University of the Free State. Dr Garwi's research expertise spans a wide array of critical domains, including sustainable development, agri-tourism, rural development, livelihoods, poverty dynamics, agrarian systems, environmental management, project management, marketing, food security, and the intricate relationship between climate change and agriculture. He has contributed significantly to academia through his publications in peer-reviewed journals, book chapters, and books, which are indexed in reputable databases such as Scopus and other highly regarded platforms

Reason Masengu holds the position of Senior Lecturer of Marketing in the Department of Business Management at Middle East College, located in Muscat, Oman. In this role, he imparts knowledge in courses such as Global Marketing, Marketing Management, Digital Marketing, and other subjects pertaining to the business management domain. He undertakes various responsibilities, including student counseling, developing curricula, ensuring quality standards, and facilitating teaching and learning activities. Reason holds the distinction of being a Fellow of the Chartered Institute of Marketing, a Senior Member of the Marketers Association of

Zimbabwe, and a Professional Member of the Australian Marketing Institute. As an ardent researcher in the fields of marketing, logistics, and supply management, he has acquired international exposure in administering research grants.

***

Mohammed Saleh Al Ansari has been working as an associate professor in the College of Engineering, Department of Chemical Engineering at the University of Bahrain. He earned his bachelor's degree in chemical engineering from King Saud University in Riyadh in 1983. In 1987, he received his MSc degree in desalination engineering, Desalination Technology, Glasgow. In 1998, he received his Ph.D. in corrosion and protection science and engineering. He has over 35 years of experience in teaching, research, and industry

Muhammad Soban Arif is a dedicated scholar and advocate specializing in the geopolitics of the Asia-Pacific region. Holding a Bachelor's degree in International Relations, he has authored scholarly articles in HEC-recognized journals, analyzing regional security, power dynamics, and socio-political trends. Combining academic rigor with practical experience, Soban interned at AWAM, an NGO advocating for marginalized communities, where he contributed to human rights initiatives and grassroots empowerment. Passionate about lifelong learning, he has enhanced his expertise through certifications in geopolitical analysis and global governance via Coursera. Committed to bridging research and policy, Soban strives to advance equitable solutions for global challenges while amplifying underrepresented voices in international discourse.

Ankur Bhardwaj (Ph.D.) is Assistant professor in the Department of Biotechnology, Shri Vaishnav Institute of Sciences, Shri Vaishnav Vidyapeeth Vishwavidyalaya, Indore, Madhya Pradesh, INDIA. He prominently has worked in the field of Plant Biotechnology, Molecular Biology, Immuno-informatics and Microbiology. He has published more than 10 research papers in peer reviewed Journals and 12 book chapters. From 2017 to 2024, he received 03 young scientist award and best thesis award in 2021. Dr. Bhardwaj has successfully guided more than 10 master's students in their final year project. Dr. Bhardwaj is a lifetime member of the Scholars Academic and Scientific Society and American Chemical Society and have published more than 05 conference abstracts. Recently Dr. Bhardwaj received Prof. M.L. Khorana IJPS Best Paper Award in the area of Pharmaceutical & Medicinal Synthetic Chemistry and was a keynote speaker in National webinar series entitled "Current Trends in Infectious disease, Diagnosis and Therapeutics".

Deepak Kumar Chauhan is a teacher and scholar in Constitutional Law and Governance, and Environmental Law. He holds a master's degree in law from Kurukshetra University Kurukshetra and completed his LL.D. from CCS University Meerut in 2010.He is the founding faculty member at Department of Law, School of Legal Studies, Central University of Punjab. He has published one book (edited) and a good number of research papers of international repute. He has a couple of research projects with the Ministry of Consumers Affairs (Department of Food and Civil Supply), Ministry of Social Justice (NISD), to his credit as well. He is an author of e- content modules on Environmental Law. Under his supervision many Ph.D., and LLM scholars have completed their research degrees. Besides academics he is also involved in the administrative work of the university, like NSS Programme coordinator, (Former) Head of the Department, (Former) Public Information officer. Prof. Deepak is on the Board of Management, IQAC, School Board, editorial board of some journals and has been an invited speaker at conferences, seminars, expert lectures in reputed institutions, academic Programmes.

Manjunatha Reddy (Ph.D., M.Sc., M.Phil.) is a Faculty member in the Department of Botany, Bangalore University. He completed his M.Sc. in Botany in 2004, M.Phil. in Biotechnology in 2007, and Ph.D. in Botany in 2016 from Bangalore University, India. His research primarily focuses on plant sciences and in vitro regeneration & influence of AM Fungal species on micropropagated Withania somnifera. He began his academic career as a Lecturer at AMC American PU & UG College, Bangalore in 2004 and later joined Bangalore University as a Guest Faculty in 2007. He also served as Assistant Professor at SRN Adarsh College, Bangalore, and held the position of Principal at both Anjanadri PU College, Devanahalli (2014–2017) and RVS Group of Institutions, Bangalore (2017–2024). With over 20 years of teaching and research experience, He has worked in plant biotechnology, agricultural biotechnology, and Botany. He has published several research papers in peer-reviewed journals and has contributed to national and international conferences. He has also guided numerous students in their academic and research projects. He is a member of the Botanical Society of America and several other professional societies. His contributions to education have been recognized with the Star Teacher Award in 2012, and he continues to foster an interactive learning environment.

Anwesha Ghosh is an Assistant Professor of Law at Amity Law School, Amity University Kolkata. A gold medalist in both LL.B. and LL.M., she holds a UGC NET-JRF qualification and has submitted her Ph.D. thesis on Legislative Governance Towards Gender Diversity in Indian Corporate Boards at the Central University of Punjab, India. Her research emphasizes the role of women directors in fostering inclusive corporate governance. Ms. Ghosh's academic journey includes accolades like the Endowment Medal for excellence in Corporate and Commercial Laws and significant contributions to projects on HIV/AIDS and the criminal justice system in India. She has authored An Introspection of Corporate Criminal Liability in India and published in Scopus-indexed and UGC Care journals.

Oindrilla Ghosh is a Research Scholar hailing from the School of Law at Christ (Deemed to be University) Bangalore, India, focuses her research on health of tribal communities. Her work includes 9 publications on Environmental Disasters that are indexed in SCOPUS and Web of Sciences conference proceedings. She completed her early education at Loreto Day School, Bowbazar and went on to achieve distinction in her undergraduate (B.A/LL.B) and postgraduate (LL.M) studies at Calcutta University. Having served as an Assistant Professor of Law at Kingston Law College,Barasat, Kolkata, Ghosh has actively participated in numerous National Conferences, International Conferences, UN Global Summits, webinars, and seminars organized by prestigious academic institutions. Her expertise lies in health perspectives, environmental studies, human behavior, and related areas. Additionally, she has conducted various workshops, lectures in educational institutions, and is engaged with NGO's to help the needy

Surendra Prakash Gupta is working as Associate Professor in the Department of Biotechnology, SVIS, Shri Vaishnav Vidhyapeeth Viswavidhayalaya, Indore. He has 16 years of academic experience under various capacities in University. He has worked in the area of Drug-DNA interactions, Bioenergy, and Computational Biology. He has qualified GATE in 2006 in Life Science. He has published 05 research papers in peer reviewed Journals, 11 book chapters, 05 conference papers and more than 35 conference abstracts. He has supervised B.Tech and M.Sc Biotechnology students successfully. He is also a member of several reputed professional bodies like, IAENG, Education and Development Research and ASR etc. and Lifetime Member of IARA, and ISTD, Indore Chapter.

H. R. Raveesha is working as a Professor in the Department of Botany, Bangalore University, Bengaluru, where he has been actively involved in teaching and research for over two decades. He completed his M.Sc. in Botany from Bangalore University in 1997 and earned his Ph.D. in Plant Tissue Culture in 2003. He has 20 years of academic experience and has established himself as a Plant Tissue Culture and Phytochemistry. Prof. Raveesha began his academic career as a Guest Lecturer in 2001 and steadily progressed to Assistant Professor, Associate Professor, and ultimately Professor in 2018. He has also held several key administrative roles, such as the Coordinator for the M.Sc. Molecular Biology Course and Bangalore University NSS program Coordinator. Currently, he is the Chairman of the Department of

Botany. His leadership and dedication have significantly contributed to the growth and development of the Department. He has successfully led and participated in numerous research projects, including major government-funded initiatives on endangered species and enhancing anticancer drugs through plant tissue culture. He has published over 55 research papers in peer-reviewed journals and contributed to several book chapters. He has guided eight M.Phil. students and awarded eight Ph. D. students. He currently supervising five Ph.D. candidates.

Binod Kumar is a graduate student known for his academic excellence, with his total 5 scholarly works indexed in Scopus database on Neuroscience in Organizational Behavior, Neuromarketing and AI Applications on Wildlife Conservation. Hailing from Ranchi, Jharkhand in India, Binod is currently pursuing a Master of Laws (LL.M) at the prestigious Radha Govind University. His passion for technology and innovation along with climate changes is evident through his high academic performance and numerous projects. Binod has a keen interest in artificial intelligence, machine learning, neuropsychology, climate changes and disaster management, fields in which he has undertaken several research projects. He is particularly fascinated by the potential of AI to solve real-world problems and is dedicated to learning and applying new techniques in these areas. He volunteers at local NGOs, teaching underprivileged children basic computer skills. With a strong foundation in law and a commitment to making a positive impact, Binod Kumar is poised to become a future leader in the legal industry.

Muhammad Huzaifa Mahmood has graduated from the University of Bologna, Italy with a master's degree in Precise and Sustainable Agriculture with a final score of 110/110 cum Lode. Currently, he is working as a Research Assistant at the University of Turin, where he focused on evaluating precision agriculture technologies, including agronomic management, field trials, and performance data analysis for agricultural systems like spray drift and integrated pest management (IPM). He has been involved in optimizing Fixed-point Spraying Systems in vineyards to enhance efficiency and water sustainability. He did his Bachelor's in Agronomy from the University of Agriculture in Faisalabad, Pakistan, where he developed his interest in agronomy, crop production, climate change, and ecosystem management.

Rampelli Manojkumar is an accomplished academic and researcher specializing in power systems and energy management. He holds a Ph.D. in Power Engineering from the Indian Institute of Technology Guwahati (IIT Guwahati), where he graduated with a CGPA of 9.25 in 2023, focusing on optimal energy management and voltage control using battery storage and smart converters.

Prior to this, he earned an M.Tech in Power and Energy Systems from the National Institute of Technology Karnataka, Surathkal (2015), and a B.E. in Electrical and Electronics Engineering from Vasavi College of Engineering, Hyderabad (2013). With over four years of teaching experience, Dr. Manojkumar currently serves as an Assistant Professor at BVRIT Hyderabad College of Engineering for Women, where he teaches courses such as Renewable Energy Sources, Power Quality, and Energy Audit Conservation. His research expertise spans energy storage systems, optimal power flow, smart grids, and voltage control, with a strong emphasis on integrating power electronics and data-driven solutions for sustainable energy systems. Dr. Manojkumar's research contributions are underscored by around 10 high-impact journal publications (including Q1 journals like IEEE Transactions on Power Systems, Sustainable Cities and Society, and IEEE Transactions on Industrial Informatics), 14 conference papers, and seven patents. His work has garnered 241 citations, an h-index of 10, and recognition such as the Best Paper Award at ICDSAI 2023. He has actively contributed to funded projects, including collaborations with IIT Guwahati, Swinburne University, and Nanyang Technological University, addressing challenges in microgrids and smart transformer systems. A recipient of prestigious scholarships from the Government of India and multiple NPTEL medals, he also serves as an editor for International Journal of Ambient Energy and reviews for leading SCI/Scopus journals. His technical proficiency in MATLAB, Python, and optimization algorithms, combined with his dedication to advancing power systems through AI and IoT applications, positions him as a dynamic contributor to the field of sustainable energy technologies

Rumbidzai Matindike (PhD) is a dedicated researcher who is driven by a strong desire to address new and demanding issues in both scholarly and practical settings. She is a PhD student at the University of Johannesburg, Faculty of Science, Department Geography Environmental Management Energy Studies. Passion for sustainability drives her research focus on issues impacting the relationship between humans and the environment.

Ananya Pandey is a Research Scholar in School of Law at Christ University, Bangalore. She has her research interests in the field of Constitutional Law, Environment Law, Human Rights, Food Security, Intellectual Property Law and Jurisprudence.

Sachin Singh Parihar, a UGC JRF recipient (2020) and Ph.D. Scholar at the Central University of Punjab's Department of Law is an accomplished legal scholar with a focus on renewable energy frameworks and sustainability dimensions. He earned his LL.M. in Criminal Law from Jamia Millia Islamia University, New

Delhi, India. Currently pursuing his Ph.D. at the Central University of Punjab under the guidance of Prof. (Dr.) Deepak Kumar Chauhan, Mr. Parihar's research is centered on the "Implementation of Renewable Energy Framework with Reference to Sustainability Dimensions: A Comparative Analysis of Laws in India, China, and Germany." His commitment to environmental and legal studies is evident in this ambitious and globally relevant project. Mr. Parihar has actively contributed to the academic discourse, having participated in numerous national and international conferences. Furthermore, his scholarly works have found a place in various reputed peer-reviewed journals, reflecting the depth and impact of his research contributions. With a strong foundation in criminal law and a keen interest in the intersection of law and sustainable energy, Mr. Parihar continues to be a noteworthy presence in the field of legal research and environmental law.

Naresh Prajapati is a Doctoral Fellow in Law at the Central University of Punjab. Since July 2024, he has served as the Lead Editor for the Journal of International Women Studies' Special Issue (Scopus Index) and as the Senior Editor for the Africa University Law Review at Africa University, Zimbabwe, since June 2024. He holds a BA (Hons.) in Geography (Gold Medalist), LL.B. from Banaras Hindu University, LL.M. from the Central University of Punjab, and two PG Diplomas in Forensic Science and Human Resource Management. His academic interests include constitutional law, international law, legal theory, Hindu and Muslim law, environmental law, climate change law, and human rights. Naresh has participated in numerous International Conferences.

Manjeri Subin Sunder Raj, a fifth-generation lawyer, enrolled in 2007 and is extremely passionate about Rights of Nature, Environmental Rights and Human Rights. He is involved in Public Interest Litigations and several initiatives involving coordination and conduction of environmental protection campaigns, animal and human rights issues, rural extension programmes and legal aid clinics, in India. He is the first and only Indian to be a part of the Expert Member Group of the Harmony with Nature Network, United Nations. Dr. Subin was a Linnaeus-Palme Fellow at the Royal Institute of Technology (KTH - Kungliga Tekniska högskolan), Stockholm, Sweden, in 2014 and 2015 and delivered lectures on Water Law. He was one among the 15 selected from around the world, and invited to present his doctoral work at the First Postgraduate Colloquium on the Frontiers of International Environmental Law held at the Faculty of Law, University of Oslo, in September 2017.

V. Uma received the M.Tech, and PhD degrees in computer science from Pondicherry University in 2007 and 2014 respectively. She was awarded the Pondicherry University gold medal for M.Tech. degree in Distributed Computing

Systems. She has more than 10 years of teaching experience at PG level. Her research interest includes Data mining, knowledge representation and reasoning (spatial and temporal knowledge),Machine Learning and sentiment analysis. She has authored and co-authored more than 20 peer-reviewed journal papers, which includes publications in Elsevier, IGI, Springer and Inderscience. She has also authored 8 chapters in various books published by IGI Global. She has also authored books on temporal knowledge representation and reasoning, Python Programming and Data Analysis and Programming in C++- A Hands on approach. She received the Best Paper Award in International Conference on Digital Factory in the year 2008 and also in IEEE sponsored International Conference on Smart Structures and Systems in 2019.

V. Subha is a research scholar in the field of Artificial Intelligence at Pondicherry University, India. With a decade of experience as an Assistant Professor in Computer Science, she specializes in Machine Learning, Deep Learning, Image Processing, Video Processing, Computer Vision, and Big Data. She holds a B.E. in Electrical and Electronics (2008) and an M.E. in Computer Science (2010) from Annamalai University. Her research focuses on developing AI-driven solutions for real-world challenges.

Index

A

B

C

D

E

F

G

H

I

L

M

N

P

R

S

T

U

W

Y

Z

www.ingramcontent.com/pod-product-compliance
Lightning Source LLC
LaVergne TN
LVHW080614300926
849162LV00005B/1148